MW00844761

# Chemical Thermodynamics: Principles and Applications

# Chemical Thermodynamics: Principles and Applications

J. Bevan Ott and Juliana Boerio-Goates

*Department of Chemistry and Biochemistry*
*Brigham Young University*
*Provo, Utah*
*USA*

ELSEVIER
ACADEMIC
PRESS

AMSTERDAM • BOSTON • HEIDELBERG • LONDON • NEW YORK • OXFORD
PARIS • SAN DIEGO • SAN FRANCISCO • SINGAPORE • SYDNEY • TOKYO

Elsevier Academic Press
525 B Street, Suite1900, San Diego, California 92101-4495, USA
http://www.elsevier.com

Elsevier Academic Press
84 Theobald's Road, London WC1X 8RR, UK
http://www.elsevier.com

British Library Cataloguing in Publication Data
A catalogue record for this book is available from the British Library

Library of Congress Catalog Number: 99-68232

ISBN 13: 978-0-12-530990-5

Transferred to Digital Printing 2007

# Contents

# Preface to the Two-Volume Series *Chemical Thermodynamics: Principles and Applications* and *Chemical Thermodynamics: Advanced Applications*

We recently completed the construction of a new chemistry building at Brigham Young University. The building is located just below a major geological fault line that runs parallel to the magnificent Wasatch Mountains. As a result, special care was taken to establish a firm foundation for the building to ensure that it would withstand a major earthquake. Massive blocks of concrete, extensively reinforced with metal bars, were poured deep in the earth, and the entire building was built upon these blocks. Resting on this foundation are the many classrooms, offices, and laboratories, with their wide variety of specialized functions. Each of the principal areas of chemistry is housed on a separate floor or wing in the building.

Thermodynamics is, in many ways, much like this modern science building. At the base of the science is a strong foundation. This foundation, which consists of the three laws, has withstood the probing and scrutiny of scientists for over a hundred and fifty years. It is still firm and secure and can be relied upon to support the many applications of the science. Relatively straightforward mathematical relationships based upon these laws tie together a myriad of applications in all branches of science and engineering. In this series, we will focus on chemical applications, but even with this limitation, the list is extensive.

Both our new chemistry building and the science of thermodynamics are functional, but beautiful. The building is a very modern combination of glass, steel, concrete, and brick, set on the edge of a hill, where it projects an image of strength, stability, and beauty. The aesthetic beauty of thermodynamics results from the rigor of the discipline. Thermodynamics is one of the pre-eminent examples of an exact science. The simple mathematical relationships that are obtained from the laws enable one to derive a very large body of mathematical equations that can be used to describe and predict the outcome of the many processes of interest to chemists. One rests assured that if the laws are true, then the equations describing the applications are valid also.

Einstein recognized the fundamental significance of thermodynamics. He said,

> *A theory is the more impressive the greater the simplicity of its premises, the more different are the kinds of things it relates, and the more extended is its range of applicability. Therefore, the deep impression which classical thermodynamics made upon me. It is the only physical theory of universal content which I am convinced, that within the framework of applicability of its basic concepts, will never be overthrown.*[a]

A tension is always present in writing a thermodynamics book, between writing a textbook that the beginning serious student can easily follow, and writing a reference book that the established investigator on the cutting edge of the discipline can find useful. We do not think that the two goals are mutually exclusive and have tried very hard to address both audiences. The division into two volumes represents an attempt to organize material into two levels of sophistication and detail. To continue the metaphor of the chemistry building, we build the exterior and framework of the discipline in the first volume. In the second volume, we furnish the various floors of the "building" with applications of thermodynamic principles to a diverse set of specialized but broadly-defined problems involving chemical processes.

The first volume entitled *Chemical Thermodynamics: Principles and Applications* is appropriate for use as a textbook for an advanced undergraduate level or a beginning graduate level course in chemical thermodynamics. In the ten chapters of this volume, we develop the fundamental thermodynamic relationships for pure-component and variable-composition systems and apply them to a variety of chemical problems.

One does not learn thermodynamics without working problems and we have included an ample supply of exercises and problems at the end of each chapter. The exercises are usually straightforward calculations involving important equations. They are intended to move the reader into an active engagement with the equations so as to more fully grasp their significance. The problems often

[a] Taken from Albert Einstein, Autobiographical Notes, page 33 in The Library of Living Philosophers, Vol. VII; *Albert Einstein: Philosopher-Scientist*, edited by P. A. Schilpp, Evanston, Illinois, 1949.

involve more steps, and possibly data analysis and interpretation of the resulting calculations. Computer manipulation of the data for fitting and graphical representation is encouraged for these. Also, the chapters contain worked out examples within the body of the text. They illustrate problem-solving techniques in thermodynamics, as well as furthering the development of the topic at hand and expanding the discussion, and should be considered as an integral part of the presentation.

The intended audience of the second volume entitled *Chemical Thermodynamics: Advanced Applications* is the advanced student or research scientist. We have used it, independently of the first volume, as the text for an advanced topics graduate level course in chemical thermodynamics. It can also serve as an introduction to thermodynamic studies involving more specialized disciplines, including geology, chemical separations, and biochemistry, for the research scientist in or outside of those disciplines. We hope it will be especially helpful for non-thermodynamicists who might be unfamiliar with the power and utility of thermodynamics in diverse applications. Given the more advanced nature of the material covered here, problems are only provided at the end of the chapters in this volume. Taken together, the two volumes make an excellent reference source for chemical thermodynamics.

Even a thermodynamics book that contains much aqueous chemistry can be dry to read. We have tried to adopt a somewhat informal style of writing that will carry (rather than drag) the reader along through the derivations and reasoning processes. In the first volume, we have kept the beginner in mind by filling in the gaps in derivations to the point that they are easy to follow; as we move along, more and more is left to the reader to fill in the intermediate steps. It is a difficult line to tread — to give enough detail to be informative, but not so much that the discussion becomes repetitive. We hope we have succeeded in providing the proper balance. In order not to interrupt the flow of the dialogue, we have relegated some of the details and reminders to footnotes at the bottom of the pages.

As much as possible, we have used "real" examples in our discussions. We present many examples of contemporary scientific phenomena in which analysis along thermodynamic lines offers a unique and valuable perspective. Examples include laser cooling, properties of high temperature superconductors, theories of continuous phase transitions, theories of electrolyte solutions, and (fluid + fluid) phase equilibria. However, we have also chosen to feature some descriptions of the very old experiments that helped lay the foundation of the science. For example, Linhart's classic 1912 work on the determination of cell $E^\circ$ values is described, along with Haber's ammonia synthesis, and Giauque's 1930 study of the third law applied to glycerine. These are the result of exceptionally high quality investigations by investigators who worked under difficult circumstances. It is humbling to see the quality of the work accomplished by these pioneers and reminds us that the field of thermodynamics has been built on the shoulders of giants.

A complete set of references to all sources of data are included, so that the reader can go to the original source if more detail is needed. We have also tried to include references to more advanced and specialized texts, monographs, reviews, and other compilations that the reader who is looking for more detail, can go to for supplementary reading.

We have generally used SI units throughout both volumes, and, as much as possible, have followed the recommendations of the IUPAC publication *Quantities, Units, and Symbols in Physical Chemistry*. An exception is the use of the bar in describing the standard state pressure for the gas. In our estimation the simplicity gained by using the bar more than compensates for the small compromise of SI units that this substitution entails. As we do this, we are careful to remind the reader continually of what we have done so that confusion can be avoided. It seems to us that IUPAC has set the precedent for such compromises of convenience by retaining the definition of the normal boiling point and normal freezing point as the temperature where the pressure is one atm. We have followed this convention also. In Chapter 10, we have used $\omega = \text{cm}^{-1}$ for energy in statistical thermodynamics calculations. Again, the simplicity introduced by this choice overshadows the advantages of going to SI units. Besides, we think it will be a long time before our spectroscopy friends stop using $\text{cm}^{-1}$ as the unit for expressing energy. Since we are invading their discipline in this chapter, we feel inclined to go along.

With few exceptions, we have used SI notation throughout. One exception is in the use of $\gamma_R$ (instead of $f$) for the activity coefficient with a Raoult's law standard state. It seems to us that using $f$ would cause serious problems by confusing the activity coefficient with fugacity. We do not use the symbol $K^\circ$ (or $K^\theta$) for what IUPAC describes as the "standard equilibrium constant". Such a choice of symbol and name seems confusing and redundant to us. Instead we use the symbol $K$ for what we refer to as the "thermodynamic equilibrium constant", a choice that is acknowledged by IUPAC as acceptable. We have also chosen to keep the "free" in "free energy" while recognizing that many readers have grown up with free energy and would be confused by "Helmholtz energy" or "Gibbs energy".

Our science building at Brigham Young University is not complete. We are still adding equipment and modifying laboratories to accommodate the latest of experiments. In the same way, these two volumes do not represent a completed study of chemical thermodynamics. This is especially true in Chapters 15 and 16 where we have chosen to use the "case study" approach in which we introduce selected examples where we apply thermodynamics to the study of processes of an industrial, geological, and biological nature. It is impossible to cover these broad fields in one book. The examples that we have chosen, some of which are of historical interest while others represent very recent applications, should allow the reader to see how the discipline applies in these areas and be able to extrapolate to other related problems.

Our hope is that the foundation has been built strong enough and the rooms completed to the point that new additions and changes can be easily accommodated and supported. It has been our experience that each time we have taught thermodynamics, we have found a new corridor to follow, leading to a new room to explore. This is one of the things that excites us most about studying thermodynamics. The science is old, but the applications (and implications) are endless.

The collaboration with many scientists over the years has had a major influence on the structure and content of this book. We are especially indebted to J. Rex Goates, who collaborated closely with one of the authors (JBO) for over thirty years, and has a close personal relationship with the other author (JBG). Two giants in the field of thermodynamics, W. F. Giauque and E. F. Westrum, Jr., served as our major professors in graduate school. Their passion for the discipline has been transmitted to us and we have tried in turn to pass it on to our students. One of us (JBG) also acknowledges Patrick A. G. O'Hare who introduced her to thermodynamics as a challenging research area and has served as a mentor and friend for more than twenty years.

We express appreciation to Brigham Young University for providing ongoing support for the thermodynamics related research that has served as the foundation for this project. It is significant that this has happened in an age when it is not fashionable to support research that involves making measurements with a calorimeter, densimeter, or thermometer. We especially appreciate the commitment and support of the university that has led directly to the creation of the two volumes that have resulted from this project. We recognize the help of Samuel Kennedy in composing many of the 137 figures, of Danelle Walker for help in preparing the manuscript, including composing over 1500 equations, and especially to Rebecca Wilford for continual support throughout the project. Finally, we recognize our spouses, RaNae Ott and Steven Goates, for their ongoing support and encouragement throughout what has become a long term project.

J. Bevan Ott and Juliana Boerio-Goates

# Preface to the First Volume *Chemical Thermodynamics: Principles and Applications*

Our attempt in writing this book has been to develop rigorously the thermodynamic foundation and then use this base to describe chemical processes. Because the book is intended primarily for chemists, we use practical examples in the body of the text as a forum for motivating a theoretical development and/or we follow abstract derivations of equations with data from real chemical systems to illustrate their use. We have intentionally come down on the side of providing enough information to carry the beginning student along in these derivations, since one of our primary goals is to provide a textbook that students enjoy reading and can learn from.

In Chapter 1, we describe the fundamental thermodynamic variables: pressure ($p$), volume ($V$), temperature ($T$), internal energy ($U$), entropy ($S$), and moles ($n$). From these fundamental variables we then define the derived variables enthalpy ($H$), Helmholtz free energy ($A$) and Gibbs free energy ($G$). Also included in this chapter is a review of the verbal and mathematical language that we will rely upon for discussions and descriptions in subsequent chapters.

Chapter 2 presents the First and Second Laws of Thermodynamics and their implications. Both the Carnot and Carathéodory approaches to the second law are given. While the first of these two approaches is simpler and more traditional, the second gives a greater generality to the Second Law. We have attempted to present the Carathéodory treatment in sufficient detail to follow its logical development, but we have sacrificed some rigor by relying on the mathematicians for the theorems that lead to the mathematical statements. The reader may choose to ignore this treatment without loss of continuity, but we find it to be one of the more aesthetically satisfying aspects of thermodynamics. With the development of the laws, examples are given using these laws to calculate the internal energy and entropy changes for selected processes.

Chapter 3 starts with the laws, derives the Gibbs equations, and from them, develops the fundamental differential thermodynamic relationships. In some ways, this chapter can be thought of as the core of the book, since the extensions and applications in all the chapters that follow begin with these relationships. Examples are included in this chapter to demonstrate the usefulness and nature of these relationships.

Chapter 4 presents the Third Law, demonstrates its usefulness in generating absolute entropies, and describes its implications and limitations in real systems. Chapter 5 develops the concept of the chemical potential and its importance as a criterion for equilibrium. Partial molar properties are defined and described, and their relationship through the Gibbs–Duhem equation is presented.

In Chapter 6, fugacity and activity are defined and described and related to the chemical potential. The concept of the standard state is introduced and thoroughly explored. In our view, a more aesthetically satisfying concept does not occur in all of science than that of the standard state. Unfortunately, the concept is often poorly understood by non-thermodynamicists and treated by them with suspicion and mistrust. One of the firm goals in writing this book has been to lay a foundation and describe the application of the standard state in such a way that all can understand it and appreciate its significance and usefulness.

Chapters 7 to 9 apply the thermodynamic relationships to mixtures, to phase equilibria, and to chemical equilibrium. In Chapter 7, both nonelectrolyte and electrolyte solutions are described, including the properties of ideal mixtures. The Debye–Hückel theory is developed and applied to the electrolyte solutions. Thermal properties and osmotic pressure are also described. In Chapter 8, the principles of phase equilibria of pure substances and of mixtures are presented. The phase rule, Clapeyron equation, and phase diagrams are used extensively in the description of representative systems. Chapter 9 uses thermodynamics to describe chemical equilibrium. The equilibrium constant and its relationship to pressure, temperature, and activity is developed, as are the basic equations that apply to electrochemical cells. Examples are given that demonstrate the use of thermodynamics in predicting equilibrium conditions and cell voltages.

Chapter 10, the last chapter in this volume, presents the principles and applications of statistical thermodynamics. This chapter, which relates the macroscopic thermodynamic variables to molecular properties, serves as a capstone to the discussion of thermodynamics presented in this volume. It is a most satisfying exercise to calculate the thermodynamic properties of relatively simple gaseous systems where the calculation is often more accurate than the experimental measurement. Useful results can also be obtained for simple atomic solids from the Debye theory. While computer calculations are rapidly approaching the level of sophistication necessary to perform computations of many thermodynamic quantities on a variety of systems, these calculations are not yet, with the exception of those for the ideal gas and the atomic solid, able to generate thermodynamic quantities at the level of accuracy obtainable from experiment. For this reason, we have chosen not to expand the discussion to cover molecular dynamics or molecular simulations calculations of thermodynamic properties in this chapter.

This volume also contains four appendices. The appendices give the mathematical foundation for the thermodynamic derivations (Appendix 1), describe the ITS-90 temperature scale (Appendix 2), describe equations of state for gases (Appendix 3), and summarize the relationships and data needed for calculating thermodynamic properties from statistical mechanics (Appendix 4). We believe that they will prove useful to students and practicing scientists alike.

The thermodynamics treatment followed in this volume strongly reflects our backgrounds as experimental research chemists who have used chemical thermodynamics as a base from which to study phase stabilities and thermodynamic properties of nonelectrolytic mixtures; and phase properties and chemical reactivities in metals, minerals, and biological systems. As much as possible, we have attempted to use actual examples in our presentation. In some instances they are not as pretty as generic examples, but "real-life" is often not pretty. However, understanding it and its complexities is beautiful, and thermodynamics provides a powerful probe for helping with this understanding.

# Chapter 1

# Introduction

## 1.1 Thermodynamics — A Preeminent Example of an Exact Science

In 1830, Auguste Comte, a French philosopher who is generally recognized as the founder of sociology and positivism, wrote the following:[1]

> *Every attempt to employ mathematical methods in the study of chemical questions must be considered profoundly irrational and contrary to the spirit of chemistry ... If mathematical analysis should ever hold a prominent place in chemistry — an aberration which is happily almost impossible — it would occasion a rapid and widespread degeneration of that science.*

If Comte had lived long enough to see the development of thermodynamics and its applications, he might have retracted these words. However, he died well before the work of Black, Rumford, Hess, Carnot, Joule, Clausius, Kelvin, Helmholtz, and Nernst that established different aspects of the sciences, followed by the contributions of Gibbs, Lewis, and Guggenheim that unified the science into a coherent whole.[a]

---

[a] J. Willard Gibbs was the first to build a unified body of thermodynamic theorems from the principles developed by others. His work was first published in the *Transactions of the Connecticut Academy of Sciences* in 1906 and later as Volume 1 of *The Scientific Papers of J. Willard Gibbs* published in 1961 by Dover Publications. G. N. Lewis and Merle Randall were the first to apply these principles specifically to chemical processes in their book, *Thermodynamics and the Free Energy of Chemical Substances*, published in 1923. E. A. Guggenheim published his book *Modern Thermodynamics by the Methods of Willard Gibbs* in 1933, in which he independently developed the same principles. Lewis, Randall and Guggenheim must be considered as the founders of modern chemical thermodynamics because of the major contributions of these two books in unifying the applications of thermodynamics to chemistry.

The importance of mathematics to the study of thermodynamics was well understood by Guggenheim, however. We quote him:[2]

> *Thermodynamics, like classical Mechanics and classical Electromagnetism, is an exact mathematical science. Each such science may be based on a small finite number of premises or laws from which all the remaining laws of the science are deductible by purely logical reasoning.*

Thermodynamics starts with two basic laws stated with elegant simplicity by Clausius.[3]

- *Die Energie der Welt ist konstant.*[aa]
- *Die Entropie der Welt strebt einem Maximum zu.*[aa]

These statements are "laws of experience". That is, no one has been able to find exceptions to them (although many have tried). If one assumes that these two laws are valid, then four fundamental equations, referred to as the **Four Fundamental Equations of Gibbs**, can be obtained. From these four, more than 50,000,000 equations relating the thermodynamic properties of the system can be derived using relatively simple mathematics. The derivations are rigorous. Thus, if the two laws are true, then the four equations are correct, and hence, the 50,000,000 equations are valid. These are the conditions Guggenheim was referring to that qualify a discipline as an exact science. By starting with a very few basic laws or postulates, a large body of rigorous mathematical relationships can be derived.

Most of the 50,000,000 equations have little use. However, a significant number are invaluable in describing and predicting the properties of chemical systems in terms of thermodynamic variables. They serve as the basis for deriving equations that apply under experimental conditions, some of which may be difficult to achieve in the laboratory. Their applications will form the focus of several chapters.

But just what are the thermodynamic variables that we use to describe a system? And what is a system? What are Energie (energy) and Entropie (entropy) as described by Clausius? We will soon describe the thermodynamic variables of interest. But first we need to be conversant in the language of thermodynamics.

---

[aa] Translation: The Energy of the Universe is constant.
The Entropy of the Universe is increasing to a maximum.

## 1.2 The Language of Thermodynamics

### 1.2a The Thermodynamic System

A **system** is the region in space that is the subject of the thermodynamic study. It can be as large or small, or as simple or complex, as we want it to be, but it must be carefully and consistently defined. Sometimes the system has definite and precise physical boundaries, such as a gas enclosed in a cylinder so that it can be compressed or expanded by a piston. However, it may be also something as diffuse as the gaseous atmosphere surrounding the earth.

Systems may be classified as **homogeneous** or **heterogeneous**. A **homogeneous system** is one whose properties are either the same throughout the system, or vary continuously from point to point with no discontinuities. Thus, a gas-filled balloon and an ice crystal are examples of homogeneous systems because their properties are the same throughout. A vacuum flask filled with crushed ice also qualifies as homogeneous, because, from a thermodynamic point of view, it is immaterial if we have one piece of ice or several pieces, as long as all have the same physical properties.[b] An aqueous solution of sodium chloride in a very long vertical tube is considered to be homogeneous even though concentration gradients due to the influence of the gravitational field will be present. Smoothly varying changes such as a concentration gradient are governed by simple thermodynamic relationships, and are homogeneous.

A **heterogeneous** system consists of two or more homogeneous systems separated by physical boundaries or surfaces of discontinuity. Thus, a mixture of liquid water and ice form a heterogeneous system, as does the atmosphere with clouds, fog, or smoke. The homogeneous regions are called **phases**. Thermodynamics defines the conditions under which different phases can exist together in a state of equilibrium. Again, from a thermodynamic point of view, it is immaterial whether one or more pieces of each phase are present. Thus, a heterogeneous mixture of liquid water and ice consists of two phases whether we have one or several pieces of ice in contact with the liquid water.[c]

Phases are often described in terms of the physical state — solid, liquid, or gas, and the number of components. The classification of the physical state is usually unambiguous. Thus, at **ambient** (near room temperature and pressure)

---

[b] This remains true unless the crystals become very small. If the surface to volume ratio becomes large, particles of different sizes may have different properties, and surface effects may become important.

[c] The boundaries between phases are actually not surfaces, but instead are thin regions in which properties change rapidly from those of one phase to those of another. These surface effects can be neglected unless the subdivision is very small so that the surface area becomes large.

conditions, metallic iron and sodium chloride are unquestionably solids, while mercury is a liquid and helium is a gas. Sometimes the distinction is not so clear. For example, is glass at room temperature a solid, or a liquid of very high viscosity? Carbon dioxide at ambient temperature and high pressure is liquid; heating to a temperature not too far above ambient converts the liquid to gas, but no obvious phase boundary accompanies the change of phase. The term **fluid** is often used to describe the liquid or gaseous state under conditions where it is difficult to distinguish between the two.[d]

A distinction between a solid and liquid is often made in terms of the presence of a **crystalline** or **noncrystalline** state. Crystals have definite lines of cleavage and an orderly geometric structure. Thus, diamond is crystalline and solid, while glass is not. The hardness of the substance does not determine the physical state. Soft crystals such as sodium metal, naphthalene, and ice are solid while supercooled glycerine or supercooled quartz are not crystalline and are better considered to be supercooled liquids. Intermediate between the solid and liquid are **liquid crystals**, which have orderly structures in one or two dimensions,[4] but not all three. These demonstrate that science is never as simple as we try to make it through our classification schemes. We will see that thermodynamics handles such exceptions with ease.

### 1.2b Isolated, Closed, and Adiabatic Systems: Surroundings and the Universe

Associated with a system are its **surroundings**. Thermodynamic systems are often classified as to whether mass or energy flow in to or out of the system and hence, out of or into the surroundings. An **isolated system** is one in which neither energy nor mass can flow in or out. In a **closed system**, mass cannot flow in or out but energy can be added or removed. This is in contrast to an **open system** for which both mass and energy can flow in or out. Often we work with heterogeneous closed systems that have **open phases**. That is, mass cannot flow in or out of the system, but can flow from one homogeneous phase within the system to another.

Energy is usually added or removed from a system as heat or work. An **adiabatic process** is one in which energy, in the form of heat,[e] is not allowed to

---

[d] A substance is often referred to as a fluid (instead of a gas or liquid) at pressure and temperature conditions near the critical point. We will describe the critical (and near-critical) region in detail in later chapters. For now, we define the critical temperature as the temperature above which it is not possible to change a gas into a liquid through compression, and the critical pressure as the lowest pressure for which a liquid will change to a gas without a phase boundary while heating past the critical temperature. A **supercritical fluid** is a fluid at a temperature above the critical temperature. At high pressures, supercritical fluids have properties that approach those of liquids, while at low pressures, they behave more gas-like.

[e] In a later chapter we will mathematically define and describe the quantities heat and work.

flow in or out of the system.[f] Different conditions apply to limit the flow of energy in or out of a system as work. For example, we shall see that in a constant volume process, pressure–volume work cannot flow in or out of the system.[g]

The combination of the system plus the surroundings makes up the **universe**. Strictly speaking, the universe is unbounded, but in practice, the universe needs only to be large enough so that, for all intents and purposes, the thermodynamic process under study with all its exchanges of mass and energy are contained. Thus, the universe would be large when considering the flow of energy from the sun to the earth, but could be considered as being small when considering the metabolism of a microorganism.

### 1.2c Components and Mixtures

A homogeneous mixture of two or more components, whether solid, liquid, or gaseous, is called a **solution**. Solutions have variable composition while pure substances do not. That is, the relative amounts of the various components in a solution can vary. Thus, air, salt water, and sixteen carat gold are each solutions. The gemstone, ruby, is also a solution since it consists of the mineral corundum ($Al_2O_3$) with some of the aluminum replaced by chromium to give the crystal its characteristic color. Since the amount of chromium present can be varied, ruby is a solution.

A **component** in a mixture is a substance of fixed composition that can be mixed with other components to form a solution. For thermodynamic purposes, the choice of components is often arbitrary, but the number is not. Thus, aqueous sulfuric acid solutions consist of two components, usually designated as $H_2SO_4$ and $H_2O$. But $SO_3$ and $H_2O$ could also be considered as the components since $SO_3$, $H_2SO_4$, and $H_2O$ are related through the equation

$$SO_3 + H_2O = H_2SO_4.$$

The fact that aqueous sulfuric acid consists of $H_3O^+$, $HSO_4^-$, and $SO_4^{2-}$, as well as $H_2SO_4$ (perhaps) and $H_2O$ does not allow one to consider this solution to be a mixture of four (or five) components. Species that are in equilibrium and, hence,

---

[f] It is not possible to exactly reproduce the conditions described above in the laboratory and approximations must be made. For example, a liquid in a closed silvered Dewar (vacuum) flask approximates an isolated system, but not exactly so because heat will slowly be exchanged between the system and the surroundings if the two are not at the same temperature. Adiabatic conditions come closest to being achieved when attempts are made to keep the system and surroundings at the same temperature, although a good insulating material helps also.

[g] Constant volume processes are referred to as **isochoric**. Thus, heat does not flow in an adiabatic process and work does not flow in an isochoric process.

related by an equilibrium reaction cannot be considered as independent components. As another example, gaseous HF consists of many different hydrogen bonded molecular species, but it is a single component, since all of the species are related through equilibrium expressions such as

$$2HF(g) = H_2F_2(g)$$
$$H_2F_2(g) + HF(g) = H_3F_3(g)$$

The number of components in a system can change with experimental conditions, and one must exercise care in defining the system. For example, a mixture of $H_2$, $O_2$, and $H_2O(g)$ at low temperature is a three-component mixture.[h] However, heating the mixture to a high temperature causes the three species to be in rapid equilibrium through the reaction

$$2H_2(g) + O_2(g) = 2H_2O(g)$$

and the number of components reduces to two. Heating to very high temperature results in the presence of significant amounts of H, O, and OH through reactions of the type

$$H_2 = 2H$$
$$O_2 = 2O$$
$$H_2O = H + OH$$

but these reactions do not cause the formation of additional components, since all are related through the equilibrium reactions.

When working with solutions, one often talks about the various components as being the **solvent** or a **solute**. The solvent is the component in which the other components (solutes) are considered to be dissolved. The solvent is usually the component present in largest amounts, but this is not always so. Thus, water may be considered to be the solvent in concentrated sulfuric acid solution, although very little water is present. Water is usually considered to be the solvent in solutions in which it is present. Such solutions are designated as **aqueous**. Sometimes no one component plays a role that would justify calling it a solvent, in which case, the solvent and solute designation lose

[h] Thermodynamics does not tell us how long it takes for a process to occur. Thus, thermodynamics predicts that diamond is unstable and should convert to graphite. In a similar manner, thermodynamics predicts that $O_2(g)$ and $H_2(g)$ should react at room temperature to form $H_2O$. But the process is very slow, and $H_2$, $O_2$, and $H_2O$ can be treated as separate components. A spark or heating to high temperature causes the reaction to proceed to equilibrium, and under these conditions, three components become two.

their significance, and the different substances in the mixture are designated simply as components.

### 1.2d Chemical Processes

The thermodynamic properties of a chemical substance are dependent upon its state and, therefore, it is important to indicate conditions when writing chemical reactions. For example, in the burning of methane to form carbon dioxide and water, it is important to specify whether each reactant and product are solid, liquid, or gaseous since different changes in the thermodynamic property will occur depending upon the state of each substance. Thus, different volume and energy changes occur in the reactions

$$\mathrm{CH_4(g) + 2O_2(g) = CO_2(g) + 2H_2O(l)}$$

and

$$\mathrm{CH_4(g) + 2O_2(g) = CO_2(g) + 2H_2O(g)}.$$

Note that the state is put in parentheses after the formula for the specific reactant or product. The notation (aq) (for "aqueous") is used to denote reactants or products in solution with water as the solvent. Thus the reaction for the dissolving of solid NaCl into water is written as

$$\mathrm{NaCl(s) = NaCl(aq)}$$

An equivalent reaction is

$$\mathrm{NaCl(s) = Na^+(aq) + Cl^-(aq)}$$

since NaCl(aq) is completely dissociated in aqueous solution. Later we will understand better why these two reactions are equivalent. We will have cause to write a generalized chemical reaction as

$$\nu_1 \mathrm{A_1} + \nu_2 \mathrm{A_2} + \cdots = \nu_m \mathrm{A_m} + \nu_{m+1} \mathrm{A_{m+1}} + \cdots \tag{1.1}$$

In reaction (1.1), $\nu_1$ moles of reactant $\mathrm{A_1}$, $\nu_2$ moles of reactant $\mathrm{A_2}$, ... form $\nu_m$ moles of product $\mathrm{A_m}$, $\nu_{m+1}$ moles of product $\mathrm{A_{m+1}}$, ... A shorthand notation for reaction (1.1) is

$$\sum \nu_i \mathrm{A_i} = 0. \tag{1.2}$$

In reaction (1.2) the coefficients $\nu_i$ are positive for the products and negative for the reactants.

## 1.3 Thermodynamic Variables

In most applications, thermodynamics is concerned with five fundamental properties of matter: volume ($V$), pressure ($p$), temperature ($T$), internal energy ($U$) and entropy ($S$). In addition, three derived properties that are combinations of the fundamental properties are commonly encountered. The derived properties are enthalpy ($H$), Helmholtz free energy ($A$) and Gibbs free energy ($G$).[i]

Before describing these thermodynamic variables, we must talk about their properties. The variables are classified as **intensive** or **extensive**. Extensive variables depend upon the amount while intensive variables do not. Density is an example of an intensive variable. The density of an ice crystal in an iceberg is the same as the density of the entire iceberg. Volume, on the other hand, is an extensive variable. The volume of the ocean is very different from the volume of a drop of sea water. When we talk about an extensive thermodynamic variable $Z$ we must be careful to specify the amount. This is usually done in terms of the molar property $Z_m$, defined as

$$Z_m = \frac{Z}{n} \tag{1.3}$$

where $n$ is the number of moles. Thus $V_m$ is the molar volume and $S_m$ is the molar entropy.[j] Note that molar properties are intensive, since they do not depend upon amount, once the definition of the mole is established.

The second thing to note about the thermodynamic variables is that, since they are properties of the system, they are **state functions**. A state function $Z$ is one in which $\Delta Z = Z_2 - Z_1$; that is, a change in $Z$ going from state (1) to state (2), is independent of the path. If we add together all of the changes $\Delta Z_i$ in going from state (1) to state (2), the sum must be the same no matter how many steps are involved and what path we take. Mathematically, the condition of being a state function is expressed by the relationship

$$\oint \mathrm{d}Z = 0. \tag{1.4}$$

---

[i] Other variables come into play for specific kinds of problems. For example, surface area $A_s$ and surface tension $\gamma$ are used when surface effects are considered, and electrical potential $E$ and quantity of electrical charge $Q$ are included when electrical work is involved.

[j] An alternative is to consider the value of the thermodynamic property per unit mass. Such quantities are called **specific** properties. Thus the specific volume is the volume per unit mass. It is the reciprocal of the density and is an intensive property.

Equation (1.4) states that if we add together all of the infinitesimal changes d$Z$ over a closed path, the sum is equal to zero. This is a necessary condition for a state function.

In equation (1.4), the infinitesimal change d$Z$ is an exact differential. Later, we will describe the mathematical test and condition to determine if a differential is exact.

The following are the variables we will use as we apply thermodynamics to chemical systems.

### 1.3a Number of Moles (*n*)

A mole contains Avogadro's number ($N_A = 6.0221317 \times 10^{23}$) of units. Thus we can talk about a mole of potassium containing $N_A$, K atoms, or a mole of potassium chloride, which contains $N_A$, KCl units.[k]

For pure substances, $n$ is usually held constant. We will usually be working with molar quantities so that $n = 1$. The number of moles $n$ will become a variable when we work with solutions. Then, the number of moles will be used to express the effect of concentration (usually mole fraction, molality, or molarity) on the other thermodynamic properties.

### 1.3b Volume (*V*)

The volume $V$ is the space occupied by the system. It is usually expressed in cubic meters ($m^3$) or cubic decimeters ($dm^3$). A $dm^3$ is the same volume as a liter (L), but $dm^3$ is preferred to the liter because it is a part of the SI (Système International d'Unités) system of units.

Volume is an extensive property. Usually, we will be working with $V_m$, the molar volume. In solution, we will work with the partial molar volume $\overline{V}_i$, which is the contribution per mole of component i in the mixture to the total volume. We will give the mathematical definition of partial molar quantities later when we describe how to measure them and use them. Volume is a property of the state of the system, and hence is a state function.[l] That is

$$\oint \mathrm{d}V = 0.$$

If one adds together all of the volume changes in going from state 1 to state 2, the sum $\Delta V = V_2 - V_1$, is the same no matter what changes $\Delta V_1$, $\Delta V_2$, $\Delta V_3$, ... are encountered.

---

[k] KCl in the solid state consists of $K^+$ ions and $Cl^-$ ions. A mole of KCl contains $N_A K^+$ ions and $N_A Cl^-$ ions, but no KCl molecules.

[l] In Appendix 1 we will describe in more detail the properties of state functions and the resulting exact differentials.

## 1.3c Pressure (*p*)

Pressure is force per unit area. If force is expressed in Newtons,[m] and area in square meters then the pressure is Newtons · $m^{-2}$ (or Newtons/$m^2$) or Pascals. The Pascal (Pa) is the SI unit of pressure. It is the one we will use almost exclusively. The Pa is small and often the kiloPascal (kPa) or megaPascal (MPa) are used. Other units are the bar (1 bar = 0.1 MPa) and the atmosphere and Torr (1 atm = 101,325 Pa = 760 Torr).

Pressure is a potential, and as such is an intensive quantity. We will see later that it is a driving force that transfers energy in the form of work into or out of a system.

## 1.3d Temperature (*T*)

Like pressure, temperature is an intensive variable. In a qualitative sense it may be thought of as the potential that drives the flow of heat. This can be seen by referring to Figure 1.1. If two systems are in thermal contact, one at temperature $T_1$ and the other at temperature $T_2$, then heat will be exchanged between the two systems so that $q_1$ flows from system 1 to system 2 and $q_2$ flows from system 2 to system 1. If $T_2 > T_1$, then the rate of flow of heat from system 2 to system 1 will be greater than the rate of flow of heat from system 1 to system 2. The net effect will be that system 1 will increase in temperature and system 2 will decrease in temperature. With time, the difference between the two heat flow rates decreases until it becomes zero. When this occurs, $T_2 = T_1$ and the two systems are said to be in **thermal equilibrium**; the flow of heat from system 1 to 2 balances the flow of heat from system 2 to 1.

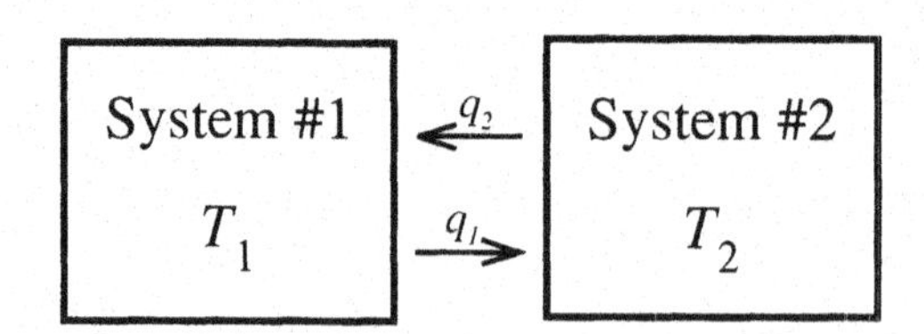


**Figure 1.1** Heat is exchanged between two systems. If $T_1 < T_2$ then $q_1 < q_2$ and $T_1$ will increase with time while $T_2$ decreases. If $T_1 = T_2$, then $q_1 = q_2$ and no net flow of heat occurs. This is a state of thermal equilibrium.

---

[m] The Newton (N) is the SI unit of force. In terms of units of mass, length, and time, 1N = 1 kg · m · $sec^{-2}$.

**The Zeroth Law of Thermodynamics:** An extension of the principle of thermal equilibrium is known as the **zeroth law of thermodynamics**, which states that two systems in thermal equilibrium with a third system are in equilibrium with each other. In other words, if $T_1$, $T_2$, and $T_3$ are the temperatures of three systems, with $T_1 = T_3$ and $T_2 = T_3$, then $T_1 = T_2$. This statement, which seems almost trivial, serves as the basis of all temperature measurement. Thermometers, which are used to measure temperature, measure their own temperature. We are justified in saying that the temperature $T_3$ of a thermometer is the same as the temperature $T_1$ of a system if the thermometer and system are in thermal equilibrium.

**Temperature Scales:** A quantitative description of temperature requires the definition of a temperature scale. The two most commonly encountered in thermodynamics are the **absolute or ideal gas (°A) scale** and the **thermodynamic or Kelvin (K) scale.**[n]

**The Thermodynamic or Kelvin Temperature Scale**: Description of the Kelvin temperature scale must wait for the laws of thermodynamics. We will see that the Kelvin temperature is linearly related to the absolute or ideal gas temperature, even though the basic premises leading to the scales are very different, so that

$$T(\text{Kelvin}) = cT(\text{Absolute}). \tag{1.5}$$

The constant $c$ is arbitrary and can be set equal to one so that

$$T(\text{Kelvin}) = T(\text{Absolute}). \tag{1.6}$$

**The Absolute Temperature Scale:** The absolute temperature scale is based on the $(p, V, T)$ relationships for an ideal gas as given by equation (1.7)

$$pV = nRT \tag{1.7}$$

where $T$ is the absolute temperature, $p$, $V$, and $n$ are the pressure, volume, and number of moles as described, earlier and $R$ is the gas constant.[o] An ideal gas thermometer compares the temperatures $T_1$ and $T_2$ along two different

[n] Note that we do not write the Kelvin temperature as degrees Kelvin or °K. Temperature on this scale is expressed as Kelvin and temperature changes as Kelvins, but not °K.

[o] We will more thoroughly define the properties of the ideal gas later.

isotherms (constant temperature). From equation (1.7)

$$T_2 = T_1 \frac{(pV)_2}{(pV)_1} \tag{1.8}$$

where $(pV)_2$ is the $pV$ product of the ideal gas at $T_2$ and $(pV)_1$ is the $pV$ product at $T_1$. The size of the temperature increment (or degree) is obtained by dividing the temperature interval between the absolute zero of temperature (where the volume of the ideal gas becomes zero) and the temperature of the triple point of water into exactly 273.16 units. With this definition, the temperature $T$ can be obtained by comparing $(pV)_T$, the $(pV)$ product at temperature $T$, with $(pV)_{t.p.}$, the $(pV)$ product at the triple point through the relationship

$$T = 273.16 \frac{(pV)_T}{(pV)_{t.p.}}. \tag{1.9}$$

It is not possible to construct an ideal gas thermometer. Instead, a real gas thermometer must be used under conditions where the real gas behaves as an ideal gas. This is done by extrapolating the $pV$ product to zero pressure (where all gases behave ideally), and equation (1.9) becomes

$$T = 273.16 \lim_{p \to 0} \frac{(pV)_T}{(pV)_{t.p.}}. \tag{1.10}$$

Gas thermometers that employ equation (1.10) can be constructed to measure either pressure while holding the volume constant (the most common procedure) or volume while holding the pressure constant. The $(pV)$ product can be extrapolated to zero $p$, but this is an involved procedure. More often, an equation of state or experimental gas imperfection data are used to correct to ideal behavior. Helium is the usual choice of gas for a gas thermometer, since gas imperfection is small, although other gases such as hydrogen have also been used. In any event, measurement of absolute temperature with a gas thermometer is a difficult procedure. Instead, temperatures are usually referred to a secondary scale known as the International Temperature Scale or ITS-90.

**The International Temperature Scale — ITS-90**: For ITS-90, temperatures of a series of fixed points are measured as accurately as possible with a gas thermometer. A complete description of ITS-90 is given in the literature[5-7] and in Appendix 2, where Table A2.1 summarizes these fixed points in terms of $T_{90}$, the ITS-90 temperature.[p]

These fixed points are used to calibrate a different kind of thermometer that is easier to use than a gas thermometer. Over the temperature range from 13.8033 to 1234.93 °A (or K), which is the temperature interval most commonly encountered, the thermometer used for ITS-90 is a platinum resistance thermometer. In this thermometer, the resistance of a specially wound coil of platinum wire is measured and related to temperature. More specifically, temperatures are expressed in terms of $W(T_{90})$, the ratio of the resistance $R(T_{90})$ of the thermometer at temperature $T_{90}$ to the resistance at the triple point of water $R$ (273.16 K), as given in equation (1.11)

$$W(T_{90}) = \frac{R(T_{90})}{R\,(273.16\ \text{K})}. \tag{1.11}$$

Temperatures $T_{90}$ for a specific thermometer are calculated from the equation

$$W(T_{90}) = W_r(T_{90}) + \Delta W(T_{90}) \tag{1.12}$$

where $W(T_{90})$ is the resistance ratio of the thermometer measured at the temperature $T_{90}$, $W_r(T_{90})$ is the resistance ratio calculated from a reference function, and $\Delta W(T_{90})$ is a difference function obtained from the calibration of the thermometer using the fixed points given in Table A2.1. The reference function $W_r(T_{90})$ is a complex power series expansion of $T_{90}$. The mathematical form of this function in the temperature ranges from 13.8033 to 273.16 K, and from 273.16 to 1234.93 K, along with values for the fitting coefficients, are given in Appendix 2.

The form of the difference function and the fixed points used for calibration depend upon the temperature region. For example, in the temperature range from 24.5561 to 273.16 K,

$$\Delta W(T_{90}) = a_2[W(T_{90}) - 1] + b_2[W(T_{90}) - 1]^2 + \sum_{i=1}^{3} c_i[\ln W(T_{90})]^i. \tag{1.13}$$

---

[p] $T_{90}$ is usually expressed in terms of the thermodynamic temperature in Kelvins. As we indicated earlier, this scale is the same as the absolute scale and is the one most commonly used. Thus, temperatures in thermodynamics are usually referred to in Kelvins rather than in degrees absolute, even though the measurements are ultimately referred back to the gas thermometer.

The fixed points used to determine the constants $a_2$, $b_2$, and $c_1$ to $c_3$ are the triple points of $H_2$, Ne, $O_2$, Ar, and Hg. In the temperature range from 273.15 to 692.677 K,

$$\Delta W(T_{90}) = a_8[W(T_{90}) - 1] + b_8[W(T_{90}) - 1]^2, \tag{1.14}$$

with $a_8$ and $b_8$ determined from the freezing points of Sn and Zn. The functional form of $\Delta W(T_{90})$ and the fixed points used for calibration for all of the temperature ranges are summarized in Appendix 2.

ITS-90 is the latest in a series of international temperature scales. Each represents an improvement over an earlier scale as more reliable values for the fixed points are established. It has been the practice to update the ITS temperature scale approximately every twenty years. Earlier scales include (ITS-27), (ITS-48), (IPTS-68), and a provisional scale in 1976.

### 1.3e Internal Energy (*U*)

A system may contain energy in many different forms. The system as a whole may be moving, in which case it possesses kinetic energy associated with the motion. It may contain gravitational, electric, or magnetic potential energy by virtue of its position relative to the Earth's surface or its position inside an electric or magnetic field. It is not possible to assign a value for the total energy of a system on an absolute scale. Rather, differences from some reference point are the best one can do.

In thermodynamics, the energy of interest is generally the **internal energy**, $U$, the energy contained within the system, rather than the kinetic or potential energy of the system as a whole. Consider, for example, a balloon attached to a table that is moving with the Earth's rotation. Thermodynamic studies focus on the energy associated with the gas molecules inside the balloon, and ignore the kinetic energy associated with movement of the balloon as a whole.

While thermodynamics does not describe the nature of this internal energy, it is helpful to consider the insights gained from kinetic molecular theory. According to this theory, the internal energy can be partitioned into kinetic and potential energy terms associated with various motions and positions of the nuclei of the atoms or molecules that make up the gas, and with energies associated with their electrons.

In an atomic gas, internal energy is present as translational kinetic energy arising from the motions of the atoms within their container, and as electronic energy. In a molecular gas, molecules possess translational kinetic energy arising from the motion of the molecular unit, but also have rotational and vibrational kinetic energy associated with the internal motions of the nuclei. Potential energy is also associated with the vibrational motion. (In the classical picture, the vibrating molecule constantly exchanges kinetic and potential energy.) Molecular gases also possess electronic energy.

Atoms and molecules are said to possess degrees of freedom associated with the different kinds of motion they undergo. The number of degrees of freedom arise from the number of coordinates needed to describe a particular kind of motion. In a collection of $\eta$ independent atoms, $3\eta$ coordinates are needed to describe the translational motion of all of the atoms: three Cartesian coordinates $(x, y, z)$ for each atom. When those $\eta$ atoms are brought together in a molecule, $3\eta$ coordinates still must be specified to completely describe the motion of the molecule, but the nature of the coordinates are changed. Only three Cartesian coordinates and hence, three degrees of freedom are associated with the translational motion of the molecule as a whole. Angular coordinates specify the rotational motion of the molecule; two are needed for a linear molecule, and three for a nonlinear one. Therefore, linear molecules are said to possess two rotational degrees of freedom, while nonlinear ones have three. The remaining numbers of coordinates (and degrees of freedom) are associated with vibrational motions. Thus a linear molecule has $3\eta - 5$ vibrational degrees of freedom while a nonlinear molecule has $3\eta - 6$.

Let us use NO(g) and $NO_2$(g) to consider in some detail the various kinds of internal energies. For NO(g), $\eta = 2$, so six degrees of freedom are present. Three are translational, two are rotational, and one vibrational. The single vibrational degree of freedom is associated with stretching of the NO bond. Unlike most diatomic molecules, the ground electronic state of NO contains an unpaired electron. This unpaired electron gives rise to the existence of an excited electronic state at a very low energy relative to the ground state ($121.1\ cm^{-1}$).[q] Thus, when energy is added to molecules of NO gas, it must be partitioned not only among the various degrees of freedom, but also among electronic states as well. In the classical picture, energy divides equally among the various degrees of freedom. However, as we will see in Chapter 10, statistical mechanics predicts that the partitioning is unequal, with the relative amounts depending upon temperature and the differences in energy within the translational, rotational, vibrational and electronic energy levels.

The $NO_2$ molecule is nonlinear. It has nine degrees of freedom: three translational, three rotational, and $3\eta - 6 = 3$ vibrational. The complex vibrational motion of this molecule can be resolved into three fundamental

[q] Spectroscopists use the non-SI unit $cm^{-1}$ to express energy. We will describe it more fully in Chapter 10. With the exception of energy on the molecular level, we will use the SI unit Joule to express energy.

vibrations:

Vibrational Modes of $NO_2$

Symmetrical Stretch    Asymmetrical Stretch    Bending

The nearest excited electronic state of $NO_2$ is more than 10,000 $cm^{-1}$ above the ground electronic state. Because of the large energy difference, energy partitioning into the excited electronic states will not take place until the temperature is very very high.

Solids and liquids also have internal energy. In the case of solids, translational motion is usually very limited and rotational motion is only present in special circumstances; the common form of internal energy is usually vibrational. In liquids, all three forms of energy are usually present, although in some instances, some forms of motion may be restricted.

The attractive and repulsive forces between molecules contribute to the energy of the system as potential energy. The zero of potential energy is usually defined for an infinite separation between the molecules. At the intermolecular distances found in solids and liquids, attractive forces predominate, and the potential energy is considerably less than zero. In gases, on the other hand, the molecules are generally separated by large distances,[r] and the potential energy is nearly zero. An alternate definition of an ideal gas other than $pV = nRT$ is one in which the attractions and repulsions between molecules in the gas are zero, and hence, the potential energy is zero. Mathematically this definition is expressed as

$$\left(\frac{\partial U}{\partial V}\right)_T = 0. \tag{1.15}$$

Equation (1.15) states that the internal energy of the ideal gas does not change with volume. This is true when the attractive and repulsive forces between the molecules in the gas are zero.

[r] The molecular separation in gases is large only if the pressure is small. At high pressures, attractive and repulsive forces become important in gases.

Temperature must be held constant in equation (1.15), since changing the temperature changes the energy. The internal kinetic and potential energy of the molecules in a system is often referred to as the thermal energy. Kinetic-molecular theory predicts that motion will stop at the absolute zero of temperature[s] where the thermal energy will be zero.

In Chapter 10 we will use statistical mechanical methods to calculate the thermodynamic properties of a system from a knowledge of the properties of the molecules that make up the system.[t] In classical thermodynamics we are interested in knowing $(U - U_0)$, the internal energy above absolute zero (where the energy is $U_0$), or $\Delta U$, the change in energy resulting from a change in state. The important things to note are that (1) internal energy is an extensive variable and (2) internal energy is a property of the state of the system so that it is a state function. That is

$$\oint \mathrm{d}U = 0.$$

As with other extensive variables, we will usually work with the molar function $U_\mathrm{m}$ defined as

$$U_\mathrm{m} = \frac{U}{n}. \tag{1.16}$$

Calculations involving $U$ or $U_\mathrm{m}$ must wait until the laws of thermodynamics are formulated.

### 1.3f Entropy (*S*)

Entropy $S$ like internal energy, volume, pressure, and temperature is a fundamental property of a system. As such, it is a function of the state of the system and a state function so that

$$\oint \mathrm{d}S = 0.$$

Entropy is an extensive property and $S_\mathrm{m}$ the molar entropy is often used.

---

[s] Quantum mechanics predicts that a zero point vibrational energy is present in a solid at the absolute zero of energy. This energy is usually not included as a part of the thermal energy.

[t] In Chapter 10, we will make quantitative calculations of $U - U_0$ and the other thermodynamic properties for a gas, based on the molecular parameters of the molecules such as mass, bond angles, bond lengths, fundamental vibrational frequencies, and electronic energy levels and degeneracies.

The nature of this fundamental property $S$ can be understood by looking on a molecular level, although once again this is beyond the scope of classical thermodynamics. Figure 1.2 is a photograph of Ludwig Boltzmann's tomb in the Zentral Friedhof in Vienna, Austria. The equation written across the top

$$S = k \log W \tag{1.17}$$

helps us to understand the nature of entropy[u] from a molecular point of view.

In equation (1.17), $S$ is entropy, $k$ is a constant known as the Boltzmann constant, and $W$ is the thermodynamic probability.[v] In Chapter 10 we will see how to calculate $W$. For now, it is sufficient to know that it is equal to the number of arrangements or microstates that a molecule can be in for a particular macrostate. Macrostates with many microstates are those of high probability. Hence, the name thermodynamic probability for $W$. But macrostates with many microstates are states of high disorder. Thus, on a molecular basis, $W$, and hence $S$, is a measure of the disorder in the system. We will wait for the second law of thermodynamics to make quantitative calculations of $\Delta S$, the change in $S$, at which time we will verify the relationship between entropy and disorder. For example, we will show that

- Expanding a gas increases the disorder. Hence, entropy increases with the expansion.
- Heating a solid increases the amplitude of vibrations, and hence, the disorder, and entropy increases.
- Melting a solid increases the disorder with a corresponding increase in entropy.
- Mixing two gases at constant $T$ and $p$ increases the disorder, and hence, the entropy.

Unlike $U$ for which there is no absolute base for energy, there is a state of complete order in which $W$ is equal to one and therefore, $S$ is equal to zero. Thus, absolute values of $S$ can be determined. The procedure for doing so is the subject of the Third Law of Thermodynamics described in a later chapter.

In addition to the fundamental variables $p$, $V$, $T$, $U$, and $S$ that we have described so far, three other thermodynamic variables are commonly encountered: enthalpy; Helmholtz free energy; and Gibbs free energy. They are extensive variables that do not represent fundamental properties of the

---

[u] Equation (1.17) is the basis for calculating $S$ from molecular properties using statistical mechanical techniques. The procedure is described in Chapter 10.

[v] Boltzmann used the mathematical operation log in equation (1.17) to represent the natural logarithm. We will use ln instead.

**Figure 1.2** $S = k \log W$ written across Ludwig Boltzmann's tomb in Vienna, Austria.

system in the sense that volume, energy, and entropy do. They are known as derived functions since they are obtained as combinations of fundamental variables.

## 1.3g Enthalpy (*H*)

By definition

$$H = U + pV. \tag{1.18}$$

Students often ask, "What is enthalpy?" The answer is simple. Enthalpy is a mathematical function defined in terms of fundamental thermodynamic properties as $H = U + pV$. This combination occurs frequently in thermodynamic equations and it is convenient to write it as a single symbol. We will show later that it does have the useful property that in a constant pressure process in which only pressure–volume work is involved, the change in enthalpy $\Delta H$ is equal to the heat $q$ that flows in or out of a system during a thermodynamic process. This equality is convenient since it provides a way to calculate $q$. Heat flow is not a state function and is often not easy to calculate. In the next chapter, we will make calculations that demonstrate this path dependence. On the other hand, since $H$ is a function of extensive state variables it must also be an extensive state variable, and $\oint \mathrm{d}H = 0$. As a result, $\Delta H$ is the same regardless of the path or series of steps followed in getting from the initial to final state and

$$\Delta H = \sum \Delta H_{\mathrm{i}}$$

where $\sum \Delta H_{\mathrm{i}}$ is the sum of $\Delta H$ for all the steps used in getting from the initial to final state.

As with other state functions, the molar enthalpy defined by

$$H_{\mathrm{m}} = \frac{H}{n} \tag{1.19}$$

is often used. As a result of the relationship to $U$ given by equation (1.18), it is not possible to obtain absolute values of $H$. Only $\Delta H$ or $(H - H_0)$ values can be obtained, where $H_0$ is the enthalpy at the absolute zero of temperature.

### 1.3h Helmholtz Free Energy (*A*)

The Helmholtz free energy $A$ is the second of the three derived thermodynamic properties. It is defined as

$$A = U - TS. \tag{1.20}$$

It is an extensive property and state function so that

$$\oint \mathrm{d}A = 0. \tag{1.21}$$

Only differences in $A$ can be calculated because of its relationship to $U$ in equation (1.20). As with the other extensive properties, the molar (intensive) property is often used as given by

$$A_{\mathrm{m}} = \frac{A}{n}. \tag{1.22}$$

In earlier days, $A$ was called the **work function** because it equals the work performed on or by a system in a reversible process conducted at constant temperature. In the next chapter we will quantitatively define work, describe the reversible process and prove this equality. The name "free energy" for $A$ results from this equality. That is, $\Delta A$ is the energy "free" or available to do work. Work is not a state function and depends upon the path and hence, is often not easy to calculate. Under the conditions of reversibility and constant temperature, however, calculation of $\Delta A$ provides a useful procedure for calculating $w$.

### 1.3i Gibbs Free Energy (*G*)

The Gibbs free energy is defined as

$$G = U + pV - TS \tag{1.23}$$

or

$$G = H - TS. \tag{1.24}$$

The combination of fundamental variables in equation (1.23) that leads to the variable we call $G$ turns out to be very useful. We will see later that $\Delta G$ for a reversible constant temperature and pressure process is equal to any work other than pressure–volume work that occurs in the process. When only pressure–volume work occurs in a reversible process at constant temperature and pressure, $\Delta G = 0$. Thus $\Delta G$ provides a criterion for determining if a process is reversible.

Again, since $G$ is a combination of extensive state functions

$$\oint \mathrm{d}G = 0 \tag{1.25}$$

and

$$G_{\mathrm{m}} = \frac{G}{n}. \tag{1.26}$$

The intensive function $G_{\mathrm{m}}$ for a pure substance is known as the **chemical potential**. We will see that it is the potential that drives the flow of mass during a chemical process or a phase change.

We have now described the thermodynamic variables $n$, $p$, $T$, $V$, $U$, $S$, $H$, $A$, and $G$ and expressed expectations for their usefulness. We will be ready to start deriving relationships between these variables after we review briefly the mathematical operations we will employ.

## 1.4 The Mathematics Of Thermodynamics

One of the pleasant aspects of the study of thermodynamics is to find that the mathematical operations leading to the derivation and manipulation of the equations relating the thermodynamic variables we have just described are relatively simple. In most instances basic operations from the calculus are all that are required. Appendix 1 reviews these relationships.

Of special interest are the properties of the **exact differential**. We have seen that our thermodynamic variables are state functions. That is, for a thermodynamic variable $Z$

$$\oint \mathrm{d}Z = 0.$$

If one causes an infinitismal change $\mathrm{d}Z$ to occur, the quantity $\mathrm{d}Z$ is an exact differential (whenever $Z$ is a state function). We will now describe the test that determines if a differential is exact and summarize the relationship between exact differentials.

### 1.4a The Pfaffian Differential and the Test for Exactness

A mathematical expression of the form

$$\delta Z(X, Y) = M(X, Y)\,\mathrm{d}X + N(X, Y)\,\mathrm{d}Y \tag{1.27}$$

is a two-dimensional Pfaffian differential expression.[w] There are two possible behaviors for such an expression: (a), $\delta Z$ is an exact differential, in which case we replace $\delta Z$ by $\mathrm{d}Z$, with the differential $\mathrm{d}Z$ implying exactness, or (b), $\delta Z$ is inexact, but integrable. That is, an integrating factor $\lambda(X, Y)$ exists such that $\lambda\delta Z = \mathrm{d}W$, where $\mathrm{d}W$ is exact.[x]

---

[w] An extended discussion of two- and three-dimensional Pfaffian differential expressions and other mathematical operations commonly encountered in thermodynamics is given in Appendix 1: Mathematics for Thermodynamics.

[x] The inexact, but integrable, Pfaffian will become important in the next chapter as we calculate work $w$ and heat $q$, which form inexact differentials $\delta w$ and $\delta q$, and define the thermodynamic (Kelvin) temperature.

The differential $\delta Z$ (or $dZ$) can be tested for exactness by applying the **Maxwell Relation** given by equation (1.28):

$$\left(\frac{\partial M}{\partial Y}\right)_X = \left(\frac{\partial N}{\partial X}\right)_Y \tag{1.28}$$

If the coefficients $M$ and $N$ are related through equation (1.28), then $dZ$ is exact.

**Example 1.1:** Test the following differential expressions to determine which is an exact differential

(a) $dz = x^2\,dx + y^2\,dy$
(b) $dz = x^2y\,dx + xy^2\,dy$
(c) $dz = xy^2\,dx + x^2y\,dy$

**Solution:**
For (a),

$$\left(\frac{\partial x^2}{\partial y}\right)_x = 0 = \left(\frac{\partial y^2}{\partial x}\right)_y$$

and $dz$ is exact.

For (b),

$$\left(\frac{\partial x^2 y}{\partial y}\right)_x = x^2 \qquad \text{and} \qquad \left(\frac{\partial xy^2}{\partial x}\right)_y = y^2$$

and $dz$ is not exact.

For (c),

$$\left(\frac{\partial xy^2}{\partial y}\right)_x = 2xy = \left(\frac{\partial x^2 y}{\partial x}\right)_y$$

and $dz$ is exact.

## 1.4b Relationships Between Exact Differentials

As we have seen earlier, the thermodynamic variables $p$, $V$, $T$, $U$, $S$, $H$, $A$, and $G$ (that we will represent in the following discussion as $W$, $X$, $Y$, and $Z$) are state functions. If one holds the number of moles and hence composition constant, the thermodynamic variables are related through two-dimensional Pfaffian equations. The differential for these functions in the Pfaff expression is an exact differential, since state functions form exact differentials. Thus, the relationships that we now give (and derive where necessary) apply to our thermodynamic variables.

We start by noting that any dependent thermodynamic variable $Z$ is completely specified by two — and only two — independent variables $X$ and $Y$ (if $n$ held constant). As an example, the molar volume of the ideal gas depends upon the pressure and temperature. Setting $p$ and $T$ fixes the value of $V_m$ through the equation

$$V_m = \frac{RT}{p}.$$

This is shown graphically by the ($V_m$, $p$, $T$) surface given in Figure 1.3. The three variables $p$, $V_m$, and $T$ exist together only on this surface. Cutting through the surface with isothermal (constant $T$) planes generates hyperbolic curves

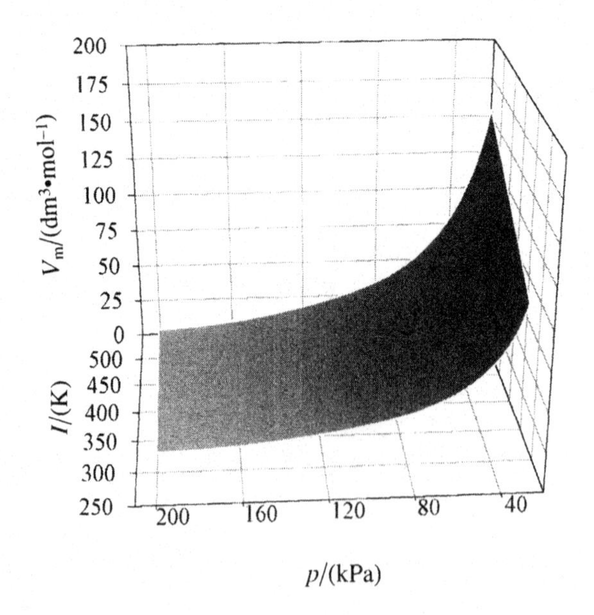


**Figure 1.3** The volume of ideal gas as a function of pressure and temperature is restricted to the surface shown in the figure. Thus, specifying $p$ and $T$ fixes $V_m$.

relating $p$ and $V_m$, while cutting through with isobaric (constant pressure) planes generates straight lines with intercepts at the origin that relate $V_m$ and $T$.

Similar types of relationships can be found between the other thermodynamic variables. In general, specifying two variables fixes the state of the third.[y] Thus specifying $V_m$ and $T$ fixes the value of $S_m$, specifying $H_m$ and $G_m$ fixes $U_m$, and so on. As another example, Figure 1.4 shows the ($S_m$, $p$, $T$) surface for an ideal monatomic gas.[z] The entropy, $S_m$ is restricted to values of $p$ and $T$ on the surface.

When the state of a variable $Z$ that forms an exact differential is fixed by specifying two others, $X$ and $Y$, one can write

$$Z = f(X, Y) \tag{1.29}$$

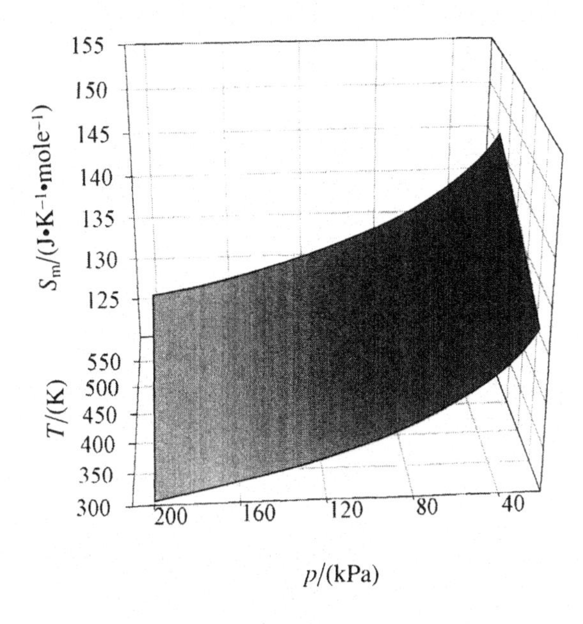


**Figure 1.4** The entropy of ideal He gas as a function of pressure and temperature is restricted to the surface shown in the figure. Thus, specifying $p$ and $T$ fixes $S_m$.

[y] Occasional exceptions can be found to the statement that the state of one thermodynamic variable is completely fixed by specifying two others. An example occurs in liquid water at a temperature near the maximum density (277 K). Specifying the temperature and pressure fixes the molar volume, but the reverse is not true. Specifying the molar volume and the pressure does not completely establish the temperature, since there are two temperatures, one above and one below the temperature of minimum density, that have the same volume at a given pressure.

[z] In Chapter 10, we will derive the relationship between $S_m$, $p$, and $T$ that is used to generate the surface in Figure 1.4.

in which case[aa]

$$dZ = \left(\frac{\partial Z}{\partial X}\right)_Y dX + \left(\frac{\partial Z}{\partial Y}\right)_X dY. \tag{1.30}$$

Other relationships between the partial derivatives of the variables $W$, $X$, $Y$, and $Z$ that form exact differentials are as follows:

$$\left(\frac{\partial Z}{\partial X}\right)_Y = \left(\frac{\partial Z}{\partial W}\right)_Y \left(\frac{\partial W}{\partial X}\right)_Y \tag{1.31}$$

$$\left(\frac{\partial Z}{\partial X}\right)_Y = \frac{\left(\frac{\partial Z}{\partial W}\right)_Y}{\left(\frac{\partial X}{\partial W}\right)_Y} \tag{1.32}$$

$$\left(\frac{\partial Z}{\partial X}\right)_Y = \frac{1}{\left(\frac{\partial X}{\partial Z}\right)_Y} \tag{1.33}$$

$$\left(\frac{\partial}{\partial X}\left(\frac{\partial Z}{\partial Y}\right)_X\right)_Y = \left(\frac{\partial}{\partial Y}\left(\frac{\partial Z}{\partial X}\right)_Y\right)_X. \tag{1.34}$$

Equation (1.34) states that the order of differentiation is immaterial for the exact differential. The Maxwell relation follows directly from this property,

[aa] The Pfaffian expressions given by equations (1.27) and (1.30) are equivalent equations for the exact differential, with

$$M = \left(\frac{\partial Z}{\partial X}\right)_Y \text{ and } N = \left(\frac{\partial Z}{\partial Y}\right)_X.$$

since

$$\left(\frac{\partial N}{\partial X}\right)_Y = \left(\frac{\partial}{\partial X}\left(\frac{\partial Z}{\partial Y}\right)_X\right)_Y = \left(\frac{\partial}{\partial Y}\left(\frac{\partial Z}{\partial X}\right)_Y\right)_X = \left(\frac{\partial M}{\partial Y}\right)_X.$$

**Example 1.2:** Show that the Pfaffian

$$\mathrm{d}V_{\mathrm{m}} = -\frac{RT}{p^2}\,\mathrm{d}p + \frac{R}{p}\,\mathrm{d}T$$

can be written for the ideal gas, and apply the Maxwell relation to show that $\mathrm{d}V_{\mathrm{m}}$ is an exact differential.

**Solution:** We start with

$$V_{\mathrm{m}} = f(p, T)$$

so that we can write

$$\mathrm{d}V_{\mathrm{m}} = \left(\frac{\partial V_{\mathrm{m}}}{\partial p}\right)_T \mathrm{d}p + \left(\frac{\partial V_{\mathrm{m}}}{\partial T}\right)_p \mathrm{d}T.$$

With

$$V_{\mathrm{m}} = \frac{RT}{p}$$

we get

$$\left(\frac{\partial V_{\mathrm{m}}}{\partial p}\right)_T = -\frac{RT}{p^2}$$

and

$$\left(\frac{\partial V_{\mathrm{m}}}{\partial T}\right)_p = \frac{R}{p}.$$

Substitution of these two derivatives into the equation for $\mathrm{d}V_\mathrm{m}$ gives

$$\mathrm{d}V_\mathrm{m} = -\frac{RT}{p^2}\,\mathrm{d}p + \frac{R}{p}\,\mathrm{d}T,$$

which is the differential relationship we wanted to derive. Applying the Maxwell relation gives

$$\left(\frac{\partial\left(-\dfrac{RT}{p^2}\right)}{\partial T}\right)_p = -\frac{R}{p^2}, \quad \left(\frac{\partial\left(\dfrac{R}{p}\right)}{\partial p}\right)_T = -\frac{R}{p^2}.$$

The two derivatives give the same result, proving that $\mathrm{d}V_\mathrm{m}$ is exact.

## 1.5 Derivation of Thermodynamic Equations Using the Properties of the Exact Differential

Equation (1.30) is the starting point for deriving a number of useful relationships.[bb] For example, equation (1.30) can be divided by $\mathrm{d}Y$ and restricted to the condition of constant $Z$ to give

$$\left(\frac{\mathrm{d}Z}{\mathrm{d}Y}\right)_Z = \left(\frac{\partial Z}{\partial X}\right)_Y \left(\frac{\mathrm{d}X}{\mathrm{d}Y}\right)_Z + \left(\frac{\partial Z}{\partial Y}\right)_X \left(\frac{\mathrm{d}Y}{\mathrm{d}Y}\right)_Z. \tag{1.35}$$

But

$$\left(\frac{\mathrm{d}Z}{\mathrm{d}Y}\right)_Z = 0, \quad \left(\frac{\mathrm{d}Y}{\mathrm{d}Y}\right)_Z = 1, \quad \left(\frac{\mathrm{d}X}{\mathrm{d}Y}\right)_Z = \left(\frac{\partial X}{\partial Y}\right)_Z,$$

in which case equation (1.35) becomes

$$\left(\frac{\partial Z}{\partial Y}\right)_Y \left(\frac{\partial X}{\partial Y}\right)_Z + \left(\frac{\partial Z}{\partial Y}\right)_X = 0. \tag{1.36}$$

---

[bb] When the number of moles becomes a variable, equation (1.30) must be expanded to include terms of the type $(\partial Z/\partial n_i)_{T,p,n_j \neq n_i}\,\mathrm{d}n_i$. We will use these relationships in later chapters when we consider the thermodynamics of mixtures.

Rearranging gives

$$\left(\frac{\partial X}{\partial Y}\right)_Z = -\frac{\left(\frac{\partial Z}{\partial Y}\right)_X}{\left(\frac{\partial Z}{\partial X}\right)_Y}. \tag{1.37}$$

Rearranging again and making use of equation (1.33) gives

$$\left(\frac{\partial Z}{\partial X}\right)_Y \left(\frac{\partial Y}{\partial Z}\right)_X \left(\frac{\partial X}{\partial Y}\right)_Z = -1. \tag{1.38}$$

## 1.5a Examples of the Application of Exact Differential Relationships

By using relationships for an exact differential, equations that relate thermodynamic variables in useful ways can be derived. The following are examples.

**Example 1.3:** The isothermal[cc] compressibility $\kappa$, coefficient of cubic expansion $\alpha$, and pressure coefficient $\beta$ are defined as follows:[dd]

$$\kappa = -\frac{1}{V}\left(\frac{\partial V}{\partial p}\right)_T \tag{1.39}$$

$$\alpha = \frac{1}{V}\left(\frac{\partial V}{\partial T}\right)_p \tag{1.40}$$

$$\beta = \left(\frac{\partial p}{\partial T}\right)_V. \tag{1.41}$$

---

[cc] $\kappa$, the compressibility at constant temperature is often designated as $\kappa_T$ to distinguish it from the adiabatic compressibility (compressibility at constant entropy) designated as $\kappa_S$. We use $\kappa_S$ so infrequently that we will designate $\kappa_T$ simply as $\kappa$.

[dd] The definitions of compressibility and coefficient of expansion given by equations (1.39) and (1.40) apply rigorously only for gases, liquids including liquid solutions, and isotropic solids, where the effect of pressure or temperature on volume would be the same in each of the three perpendicular directions. For anisotropic solids, this is not true, and more sophisticated relationships are required to describe the effects of pressure and temperature on volume. An approximate $\kappa$ or $\alpha$ is often used for these solids that represents an average effect in the three directions. The bulk modulus is sometimes used in place of the compressibility. It is defined as the reciprocal of the average compressibility.

Show that

$$\beta = \frac{\alpha}{\kappa}. \tag{1.42}$$

**Solution:**

$$V = f(p, T)$$

$$\mathrm{d}V = \left(\frac{\partial V}{\partial T}\right)_p \mathrm{d}T + \left(\frac{\partial V}{\partial p}\right)_T \mathrm{d}p.$$

Dividing by $\mathrm{d}T$ and specifying constant $V$ gives

$$\left(\frac{\mathrm{d}V}{\mathrm{d}T}\right)_V = \left(\frac{\partial V}{\partial T}\right)_p \left(\frac{\mathrm{d}T}{\mathrm{d}T}\right)_V + \left(\frac{\partial V}{\partial p}\right)_T \left(\frac{\mathrm{d}p}{\mathrm{d}T}\right)_V = 0.$$

Solving for

$$\left(\frac{\mathrm{d}p}{\mathrm{d}T}\right)_V$$

and writing it as

$$\left(\frac{\partial p}{\partial T}\right)_V$$

gives

$$\left(\frac{\partial p}{\partial T}\right)_V = \frac{\left(\frac{\partial V}{\partial T}\right)_p}{-\left(\frac{\partial V}{\partial p}\right)_T}. \tag{1.43}$$

Multiplying and dividing the right side of equation (1.43) by $1/V$ gives

$$\left(\frac{\partial p}{\partial T}\right)_V = \frac{\frac{1}{V}\left(\frac{\partial V}{\partial T}\right)_p}{-\frac{1}{V}\left(\frac{\partial V}{\partial p}\right)_T}, \tag{1.44}$$

which, by the previous definitions becomes

$$\beta = \frac{\alpha}{\kappa}.$$

A final example relates the temperature derivatives of $U$ and $H$. We will find this relationship to be useful later when we relate heat capacities at constant $p$ and at constant $V$.

**Example 1.4:** Start with equation (1.18) relating $H$ and $U$ and the properties of the exact differential and prove that

$$\left(\frac{\partial H}{\partial T}\right)_p = \left(\frac{\partial U}{\partial T}\right)_V + \left[p + \left(\frac{\partial U}{\partial V}\right)_T\right]\left(\frac{\partial V}{\partial T}\right)_p. \tag{1.45}$$

**Solution:** From equation (1.18)

$$H = U + pV.$$

Differentiating with respect to $T$ at constant $p$ gives

$$\left(\frac{\partial H}{\partial T}\right)_p = \left(\frac{\partial U}{\partial T}\right)_p + p\left(\frac{\partial V}{\partial T}\right)_p. \tag{1.46}$$

But

$$U = f(V, T)$$

so that

$$dU = \left(\frac{\partial U}{\partial V}\right)_T dV + \left(\frac{\partial U}{\partial T}\right)_V dT. \tag{1.47}$$

Dividing equation (1.47) by $dT$ and specifying constant $p$ gives

$$\left(\frac{\partial U}{\partial T}\right)_p = \left(\frac{\partial U}{\partial V}\right)_T \left(\frac{\partial V}{\partial T}\right)_p + \left(\frac{\partial U}{\partial T}\right)_V \left(\frac{\partial T}{\partial T}\right)_p. \tag{1.48}$$

Substituting equation (1.48) into equation (1.46) with

$$\left(\frac{\partial T}{\partial T}\right)_p = 1$$

gives

$$\left(\frac{\partial H}{\partial T}\right)_p = \left(\frac{\partial U}{\partial V}\right)_T \left(\frac{\partial V}{\partial T}\right)_p + \left(\frac{\partial U}{\partial T}\right)_V + p\left(\frac{\partial V}{\partial T}\right)_p.$$

Collecting terms results in the equation we are trying to derive

$$\left(\frac{\partial H}{\partial T}\right)_p = \left(\frac{\partial U}{\partial T}\right)_V + \left[p + \left(\frac{\partial U}{\partial V}\right)_T\right] \left(\frac{\partial V}{\partial T}\right)_p. \tag{1.45}$$

## 1.6 Calculation of Changes in the Thermodynamic Variable

We have seen that a thermodynamic variable, $Z$, can be completely specified by two other thermodynamic variables, $X$ and $Y$, in which case, we can write

$$Z = f(X, Y)$$

so that

$$dZ = \left(\frac{\partial Z}{\partial X}\right)_Y dX + \left(\frac{\partial Z}{\partial Y}\right)_X dY. \tag{1.30}$$

Equation (1.30) can be integrated to calculate $\Delta Z$, the change in $Z$ for a thermodynamic process. Thus

$$\Delta Z = \int dZ = \int \left(\frac{\partial Z}{\partial X}\right)_Y dX + \int \left(\frac{\partial Z}{\partial Y}\right)_X dY. \tag{1.49}$$

Any of the thermodynamic variables can be chosen to perform the integration in equation (1.49). However, the most useful choices are usually to let $X$ and $Y$ be either ($p$ and $T$), or ($V$ and $T$), in which case, equation (1.49) becomes

$$\Delta Z = \int \left(\frac{\partial Z}{\partial p}\right)_T dp + \int \left(\frac{\partial Z}{\partial T}\right)_p dT. \tag{1.50}$$

or

$$\Delta Z = \int \left(\frac{\partial Z}{\partial V}\right)_T dV + \int \left(\frac{\partial Z}{\partial T}\right)_V dT. \tag{1.51}$$

To integrate equations (1.50) and (1.51), we must know how $Z$ changes with $p$, $V$, and $T$. The laws of thermodynamics that we will discuss next in Chapter 2 will allow us to derive the necessary relationships; we will apply the resulting equations in Chapter 3.

## 1.7 Use Of Units

In this book, we will express our thermodynamic quantities in SI units as much as possible. Thus, length will be expressed in meters (m), mass in kilograms (kg), time in seconds (s), temperature in Kelvins (K), electric current in amperes (A), amount in moles (mol), and luminous intensity in candella (cd). Related units are cubic meters ($m^3$) for volume, Pascals (Pa) for pressure, Joules (J) for energy, and Newtons (N) for force. The gas constant $R$ in SI units has the value of $8.314510\ J \cdot K^{-1} \cdot mol^{-1}$, and this is the value we will use almost exclusively in our calculations.

When making calculations, we will generally convert all data to their values in SI units, since calculations involving SI units give answers in SI units. For example, when calculating the volume of an ideal gas from the equation

$$V = \frac{nRT}{p},$$

expressing $n$ in mol, $R$ as $8.314\ \text{J}\cdot\text{K}^{-1}\cdot\text{mol}^{-1}$, $T$ in K, and $p$ in Pa gives $V$ in $\text{m}^3$. Unless there are unusual circumstances, we will not attempt to carry and cancel the units in the calculation when we use these SI units. An exception is that we will sometimes include in the calculation, quantities expressed in terms of decimal multiples or submultiples of SI units, in which case we will show the conversion factor with units. For example, if we include in the calculation an enthalpy change expressed in kJ, or a pressure in MPa, we would show the factors $(10^3\ \text{J}\cdot\text{kJ}^{-1})$, or $(10^6\ \text{Pa}\cdot\text{MPa}^{-1})$, in the calculation.[ee]

Exceptions to the use of SI units are found in Chapter 10 where we work with molecules instead of moles, and units such as $\text{cm}^{-1}$ for energy are common. We will also find the bar unit for pressure to be very useful as we define standard state conditions, but a pressure of one atmosphere (atm) is still the condition that defines the **normal boiling point** and the **normal freezing point** of a liquid.

## Exercises

E1.1 Use the Maxwell's relation to determine if $\mathrm{d}z$ is an exact differential in the following examples:

(a) $\mathrm{d}z = (2x + y^2)\,\mathrm{d}x + (x^2 + 2y)\,\mathrm{d}y$
(b) $\mathrm{d}z = (x^2 + 2y)\,\mathrm{d}x + (y^2 + 2x)\,\mathrm{d}y$
(c) $\mathrm{d}z = xy\mathrm{d}x + xy\mathrm{d}y$
(d) $\mathrm{d}z = (x^2 + y^2)(x\mathrm{d}x + y\mathrm{d}y)$
(e) $\mathrm{d}z = (x + y)\,\mathrm{d}x + (x + y)\,\mathrm{d}y$

E1.2 Given the equation of state for a gas

$$pV_{\mathrm{m}} = RT + \frac{bp}{T^2}$$

[ee] Due to a fortuitous cancellation of units, volume can be expressed in $\text{cm}^3$ and pressure in MPa to give the same results as volume in $\text{m}^3$ with pressure in Pa, without invoking any conversion factors. Thus $pV$ in $\text{Pa}\cdot\text{m}^3$ gives Joules as does $pV$ in $\text{MPa}\cdot\text{cm}^3$.

where $R$ and $b$ are constants. Show that

(a) $$\left(\frac{\partial V_{\mathrm{m}}}{\partial T}\right)_p = \frac{R}{p} - \frac{2b}{T^3}$$

(b) $$\left(\frac{\partial V_{\mathrm{m}}}{\partial p}\right)_T = -\frac{RT}{p^2}$$

(c) $\mathrm{d}V_{\mathrm{m}}$ is an exact differential

E1.3 The volume $V$ of a right circular cylinder is given by

$$V = \pi r^2 h$$

where $r$ is the radius and $h$ is the height.
Show that:

(a) $\mathrm{d}V = \pi r^2\, \mathrm{d}h + 2\pi\, rh\, \mathrm{d}r$
(b) $\mathrm{d}V$ is an exact differential

## Problems

P1.1 Use the properties of the exact differential and the defining equations for the derived thermodynamic variables as needed to prove the following relationships:

(a) $$\left(\frac{\partial U}{\partial V}\right)_p \left(\frac{\partial V}{\partial p}\right)_T = -\left(\frac{\partial U}{\partial T}\right)_p \left(\frac{\partial T}{\partial p}\right)_V$$

(b) $$\left(\frac{\partial H}{\partial T}\right)_p = \left(\frac{\partial U}{\partial T}\right)_V + \left[V - \left(\frac{\partial H}{\partial p}\right)_T\right] \left(\frac{\partial p}{\partial T}\right)_V$$

(c) $$\left(\frac{\partial U}{\partial T}\right)_V = \left(\frac{\partial H}{\partial T}\right)_p - \left[\left(\frac{\partial H}{\partial T}\right)_p \left(\frac{\partial T}{\partial p}\right)_H + V\right] \left(\frac{\partial p}{\partial T}\right)_V$$

P1.2 The compressibility, $\kappa$, and the coefficient of expansion, $\alpha$, are defined by the partial derivatives:

$$\kappa = -\frac{1}{V}\left(\frac{\partial V}{\partial p}\right)_T$$

$$\alpha = \frac{1}{V}\left(\frac{\partial V}{\partial T}\right)_p$$

Show that

$$\left(\frac{\partial \alpha}{\partial p}\right)_T + \left(\frac{\partial \kappa}{\partial T}\right)_p = 0$$

## References


1. A. Comte, *Philosophie Positive*, Third Edition, Volume 3, p. 29, 1869.
2. E. A. Guggenheim, *Modern Thermodynamics by the Method of Willard Gibbs*, Methuen & Co., Ltd., London, 1933, p. 1.
3. R. J. E. Clausius, *Pogg. Ann.*, **125**, 400 (1865). Quoted in *The Scientific Papers of J. Willard Gibbs, Vol. 1. Thermodynamics*, Dover, New York, 1961, p. 55.
4. General reviews of the structure and properties of liquid crystals can be found in the following: G. H. Brown, J. W. Doane, and V. D. Neff, "A Review of the Structure and Physical Properties of Liquid Crystals," CRC Press, Cleveland, Ohio, 1971; P. J. Collings and M. Hind, "Introduction to Liquid Crystals. Nature's Delicate Phase of Matter," Taylor and Francis, Inc., Bristol, Pennsylvania, 1997; P. J. Collins, "Liquid Crystals. Nature's Delicate Phase of Matter," Princeton University Press, Princeton, New Jersey, 1990. A thermodynamic description of the phase properties of liquid crystals can be found in S. Kumar, editor, "Liquid Crystals in the Nineties and Beyond," World Scientific, Riven Edge, New Jersey, 1995.
5. J. B. Ott and J. Rex Goates, "Temperature Measurement with Applications to Phase Equilibria Studies," in *Physical Methods of Chemistry, Vol. VI, Determination of Thermodynamic Properties*; B. W. Rossiter and R. C. Baetzold, eds, John Wiley & Sons, New York, 1992, pp. 463–471.
6. H. Preston-Thomas, "The International Temperature Scale of 1990 (ITS-90)," *Metrologia*, **27**, 3–10, 1990.
7. M. L. McGlashan, "The International Temperature Scale of 1990 (ITS-90)," *J. Chem. Thermodyn.*, **22**, 653–663, 1990.

# Chapter 2

# The First and Second Laws of Thermodynamics

In Chapter 1 we described the fundamental thermodynamic properties' internal energy $U$ and entropy $S$. They are the subjects of the First and Second Laws of Thermodynamics. These laws not only provide the mathematical relationships we need to calculate changes in $U, S, H, A$, and $G$, but also allow us to predict spontaneity and the point of equilibrium in a chemical process. The mathematical relationships provided by the laws are numerous, and we want to move ahead now to develop these equations.[1]

## 2.1 The First Law of Thermodynamics

We earlier referred to an 1865 statement of the First Law given by Clausius,

*Die Energie der Welt ist Konstant.*

That is, the energy of the universe is constant. The First Law is often referred to as the **Law of Conservation of Energy**; energy cannot be created or destroyed.[a] Mathematically, the First Law can be written as

$$\sum \Delta U = 0 \text{ (Universe)}. \tag{2.1}$$

Different processes involving energy changes can occur in the universe, but their sum must be zero. That is,

$$\Delta U_1 + \Delta U_2 + \Delta U_3 + \cdots = 0 \text{ (Universe)} \tag{2.2}$$

where the $\Delta U_i$'s are the changes in internal energy for the various processes.

---

[a] With the advent of atomic energy in the twentieth century, the Law of Conservation of Energy needed to be modified to include mass as a form of stored energy, with the equivalence given by the equation $E = mc^2$.

The First Law predicts that energy added to or removed from a system must be accounted for by a change in the internal energy $\Delta U$. Energy is added to or removed from a system as work, $w$ or heat, $q$, in which case

$$\Delta U = q + w. \tag{2.3}$$

For an infinitesimal process, equation (2.3) becomes

$$\mathrm{d}U = \delta q + \delta w \tag{2.4}$$

where $\delta q$ and $\delta w$ are infinitesimal amounts of heat or work. The symbols $\delta q$ and $\delta w$ are used to represent the infinitesimal amount (rather than $\mathrm{d}q$ and $\mathrm{d}w$) to emphasize that $q$ and $w$ are not state functions, and hence, the infinitesimal quantities are not exact differentials.

Heat and work represent ways that energy can be transferred from one system to another, and they are dependent upon the path followed in making the transfer. It is interesting to note that although $q$ and $w$ change with the path, their sum, in going from an initial state (1) to a final state (2) by two different paths, produces the same $\Delta U$, since $U$ is a state function and $\Delta U$ is independent of the path. Thus, as a path is modified to change $q$, $w$ changes accordingly so that the sum stays constant.[b] We now want to calculate $w$ and $q$ for different processes so that we can calculate $\Delta U$.

## 2.1a Work

A fundamental definition of work comes from classical mechanics; all other expressions can be derived from it. The relationship is

$$\delta w = \vec{f} \cdot \mathrm{d}s \tag{2.5}$$

where $\delta w$ is the work resulting from an infinitesimal displacement $\mathrm{d}\vec{s}$ in a system operating against a force $\vec{f}$.

The force $\vec{f}$ and displacement $\vec{s}$ are vector quantities, and equation (2.5) indicates that the vector dot product of the two gives $\delta w$, a scalar quantity. The result of this operation is equation (2.6)

$$\delta w = f \cos \theta \, \mathrm{d}s, \tag{2.6}$$

where $\theta$ is the angle between the vectors. In equation (2.6), $f$ and $\mathrm{d}s$ are the scalar magnitudes of $\vec{f}$ and $\mathrm{d}\vec{s}$. We will only consider applications where $\theta$ will be

---

[b] We will see later when we discuss the second law that a relationship exists between $q$ and $S$ that converts the inexact differential $\delta q$ into the exact differential $\mathrm{d}S$.

0 so that $\cos\theta = 1$ and

$$\delta w = f\,\mathrm{d}s. \tag{2.7}$$

Equation (2.7) can be applied to processes involving many forms of work. Our present interest is in calculating pressure–volume work, that is, the work done in changing the volume of a system by an amount $\mathrm{d}V$ against an external pressure $p_{\mathrm{ext}}$. The process is shown in Figure 2.1. A fluid of volume $V$ is confined in a cylinder. The fluid exerts a pressure $p$ on a piston; this pressure is balanced by an external pressure $p_{\mathrm{ext}}$ so that the piston does not move. The external pressure is then decreased by an amount $\mathrm{d}p$.[c] This causes the piston to move by an amount $\mathrm{d}x$ until the internal pressure decreases by $\mathrm{d}p$ and equilibrium is again established.

The work accomplished by this process is given by equation (2.8),

$$\delta w = f_{\mathrm{ext}}\,\mathrm{d}x \tag{2.8}$$

where $f_{\mathrm{ext}}$ is the external force opposing the displacement $\mathrm{d}x$. Since pressure is force per unit area, we can write

$$f_{\mathrm{ext}} = p_{\mathrm{ext}}A \tag{2.9}$$

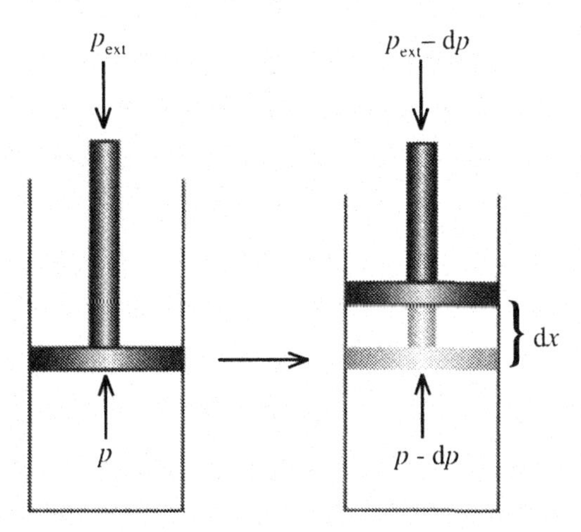


**Figure 2.1** Decreasing the external pressure by $\mathrm{d}p$ causes the piston to move a distance $\mathrm{d}x$. This increase in volume reduces the internal pressure by $\mathrm{d}p$ and re-establishes mechanical equilibrium. The process produces pressure–volume work.

[c] If $\mathrm{d}p$ is negative, the pressure increases and a compression occurs instead of an expansion. This changes the arithmetic sign of $\delta w$.

where $A$ is the cross-sectional area of the piston in Figure 2.1. Combining equations (2.8) and (2.9) and recognizing that $A\mathrm{d}x = \mathrm{d}V$, the volume change, gives

$$\delta w = -p_{\mathrm{ext}}\, \mathrm{d}V. \tag{2.10}$$

The negative sign is included because it is convenient to know whether work is done on or by the system. The convention we follow in writing equations (2.3) and (2.4) is to give a positive sign to work done **on** the system. That is, work added **to** the system increases the internal energy, and must be positive. At the same time, work done **by** the system decreases the internal energy and must be negative. The negative sign in equation (2.10) gives $\delta w$ the correct arithmetic sign. For example, in an expansion, work is done by the system and so must be negative by our convention. But $\mathrm{d}V$, and hence $p_{\mathrm{ext}}\mathrm{d}V$, is positive, so that we need a negative sign in equation (2.10). In the compression, $\mathrm{d}V$ is negative, but this is the condition where we want $\delta w$ to be positive. Again this requires the negative sign in equation (2.10).

## 2.1b Calculation of Work

Equation (2.10) can be integrated to calculate $w$.

$$\int \delta w = w$$

or

$$w = -\int_{V_1}^{V_2} p_{\mathrm{ext}}\, \mathrm{d}V. \tag{2.11}$$

The integral in equation (2.11) is a line integral.[d] To integrate equation (2.11), one must know the relationship between $p_{\mathrm{ext}}$ and $V$. That is, one must specify the path for the process leading to the volume change. Figure 2.2 is a graph of $p_{\mathrm{ext}}$ against $V$ for a particular process. The area under the curve is the negative of the work associated with the process. The magnitude of this area, and therefore the work, depends upon how $p_{\mathrm{ext}}$ changes with $V$. For this reason, we must always specify the path to calculate $w$. We will now calculate $w$ for several special processes where $p_{\mathrm{ext}}$ is known for every point of the path. While these

[d] See the discussion in Appendix 1 to review the properties of line integrals.

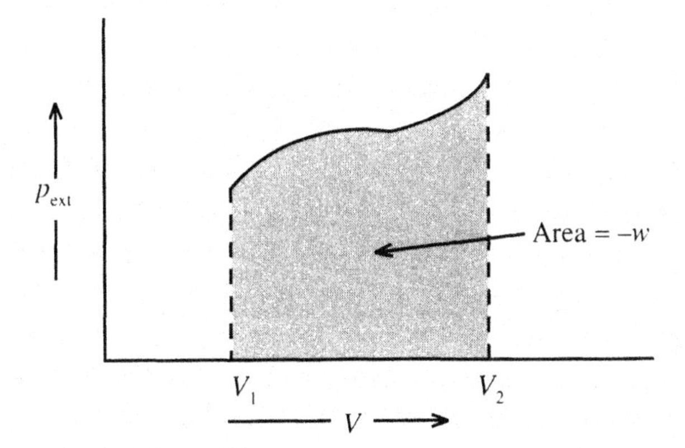


**Figure 2.2** Graph of $p_{ext}$ against $V$. The area under the curve is the magnitude of the work accomplished in the expansion from $V_1$ to $V_2$ against an external pressure $p_{ext}$.

processes may seem to be artificial or limiting in their applicability, we will see that they are very useful.

**The Isobaric Process:** Figure 2.3 shows the relationship between $p_{ext}$ and $V$ during an isobaric (constant pressure) process. In this expansion, $p_{ext}$ is constant and usually equal to $p$, the pressure of the fluid.[e] When this is true, equation (2.11) becomes

$$w = -\int_{V_1}^{V_2} p \, dV$$

$$= -p\Delta V \tag{2.12}$$

where

$$\Delta V = V_2 - V_1.$$

[e] Usually the isobaric expansion is accomplished by adding heat to a system. When $\delta q$ is added, the pressure $p$ of the system increases by d$p$. This pressure is now larger than $p_{ext}$ and the piston moves from $V_1$ to $(V_1 + dV)$, which decreases $p$ by d$p$ so that the balance between $p$ and $p_{ext}$ is re-established. This process is repeated until $V_2$ is obtained. In the overall process, $p$ never differs from $p_{ext}$ by more than an infinitesimal d$p$, and $p$ can be substituted for $p_{ext}$ in equation (2.11) to give equation (2.12). A consequence of the process is that a total quantity of heat $q$ is added and the temperature increases. The heat capacity, which we will describe in the next section, must be known in order to calculate the temperature change.

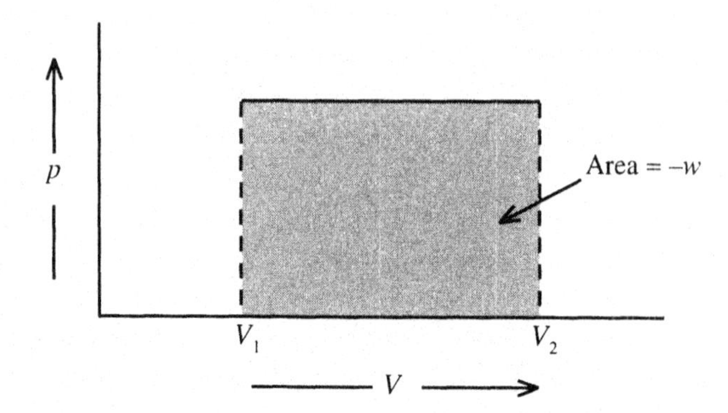


**Figure 2.3** Work of expansion for an isobaric process. In the expansion, $p_{ext}$ is constant and equal to $p$, the pressure of the fluid, unless a mechanical constraint prevents the two pressures from being equal.

**The Isochoric Process:** Figure 2.4 shows the relationship between $p_{ext}$ and $V$ for an isochoric (constant volume) process going from state $a$ to state $b$. In this process $\mathrm{d}V = 0$ and $w = 0$.[f]

**The Isothermal Process:** In an isothermal (constant temperature) expansion, heat is added to balance the work removed, so that the temperature of the system does not change. The amount of work can be calculated from the line

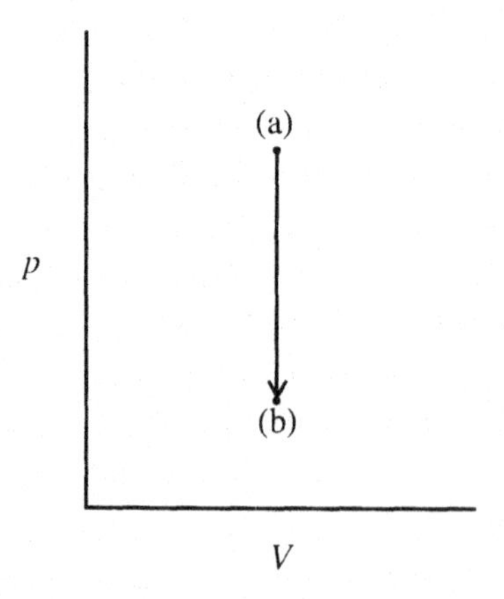


**Figure 2.4** In the change from state (a) to state (b), $\Delta V = 0$ and $w = 0$.

[f] In an isochoric process, heat is added to change the temperature without changing the volume. A calculation of the temperature change requires a knowledge of the heat capacity at constant volume. We will describe this calculation in the next section.

integral given in equation (2.11). This calculation depends upon the relationship between $p_{\text{ext}}$ and $V$. Figure 2.5 compares four examples in which we start at the same initial pressure $p_{\text{ext},1}$ and volume $V_1$ and final pressure $p_{\text{ext},2}$ and volume $V_2$, but follow different paths to get from the initial to final state.

In Figure 2.5(i), the pressure is reduced in one step from $p_{\text{ext},1}$ to $p_{\text{ext},2}$ and the piston in Figure 2.1 moves from $V_1$ to $V_2$. The work is the area (A) shown and can be calculated from

$$\begin{aligned} w &= -\int_{V_1}^{V_2} p_{\text{ext},2}\, \mathrm{d}V \\ &= -p_{\text{ext},2}\Delta V. \end{aligned}$$

Process (i) can be accomplished by representing $p_{\text{ext}}$ by two large weights placed on top of the piston in Figure 2.1, of sufficient mass to give a pressure $p_{\text{ext},1}$.

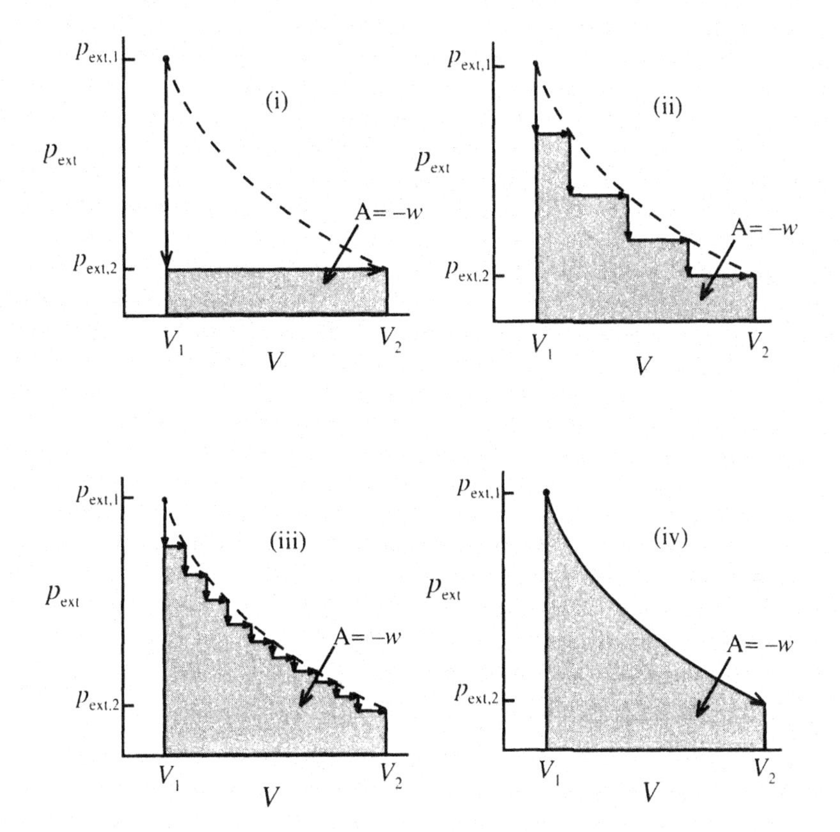


**Figure 2.5** Comparison of the work accomplished following four different paths during an isothermal expansion.

Removing one of the weights changes the pressure to $p_{\text{ext},2}$ and the expansion occurs at a pressure equal to $p_{\text{ext},2}$.

Note that in process (i), removing the weight causes an expansion that is essentially out of control. The pressure of the fluid during the expansion is always larger than $p_{\text{ext}}$ until $V_2$ is obtained, and the process cannot be altered or stopped during the expansion. The process proceeds until $p$ again equals $p_{\text{ext}}$, in this case $p_{\text{ext},2}$.

The external pressure in process (ii) can be represented by four weights stacked on top of the piston in Figure 2.1 with a total mass sufficient to give a pressure $p_{\text{ext},1}$. Removing one of the weights reduces the pressure to an intermediate value and an expansion occurs, again out of control, until a volume is obtained in which $p_{\text{ext}}$ and $p$ are again equal. This process is repeated, leaving a weight on top of the piston equivalent to $p_{\text{ext},2}$. Expansion, again out of control, occurs each time successive weights are removed until $V_2$, is reached, at which point equilibrium is established between $p_{\text{ext}}$ and $p$. The work in this process, which is the area shown under the curve could be easily calculated if the intermediate pressure and volume were known, along with $p_{\text{ext},2}$ and $V_2$.

In process (iii), the expansion is accomplished by removing a large number of small weights. This could be done by stacking a pile of sand on the piston and removing the sand grains one at a time. The work for this process, which is the area shown in the figure, could be calculated if the mass of the grains of sand were known.

If the grains of sand are small, each step does not represent a very large departure from equilibrium between $p$ and $p_{\text{ext}}$. This process is an example of a **quasi-static process**; that is, one in which the process is never far from equilibrium during the expansion.

**The Reversible Process:** Process (iv) is a representation of a hypothetical reversible expansion in which $p$ and $p_{\text{ext}}$ never differ by more than an infinitesimal d$p$. To carry out the process, $p_{\text{ext}}$ is decreased by an amount d$p$. An expansion then occurs so that $V_1$ goes to $(V_1 + \mathrm{d}V)$. This causes the internal pressure to decrease by an amount d$p$ and equilibrium is re-established. This process is repeated an infinite number of times until all of the d$V$ changes add up to $\Delta V$ and $V_2 = V_1 + \Delta V$ is reached.[g] In each step, $p$ never differs from $p_{\text{ext}}$ by more than an infinitesimal amount. The process is under control and can be reversed at any time by increasing the pressure by an infinitesimal amount. Hence, the name "reversible" is applied to this process.

[g] An infinite number of infinitesimal steps requires an infinite time to accomplish, and therefore, such a reversible process is not possible except in a hypothetical sense.

Since in the reversible process, $p$ and $p_{\text{ext}}$ never differ by more than an infinitesimal amount, $p$ can replace $p_{\text{ext}}$ in equation (2.11) to calculate $w$.

$$\delta w = -p\,\mathrm{d}V \text{ (reversible process)} \tag{2.13}$$

$$w = -\int_{V_1}^{V_2} p\,\mathrm{d}V \text{ (reversible process)} \tag{2.14}$$

If a relationship is known between the pressure and volume of the fluid, the work can be calculated. For example, if the fluid is the ideal gas, then $pV = nRT$ and equation (2.14) for the isothermal reversible expansion of ideal gas becomes

$$w = -\int_{V_1}^{V_2} \frac{nRT}{V}\mathrm{d}V. \tag{2.15}$$

Integration gives

$$w = -nRT\ln\frac{V_2}{V_1}, \tag{2.16}$$

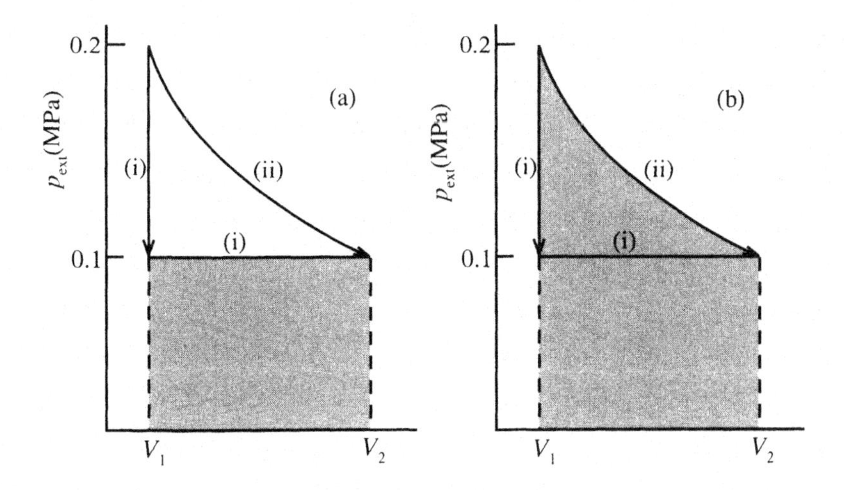


**Figure 2.6** Comparison of the work obtained from the two isothermal expansions at 300 K of ideal gas following path (i) in (a) and path (ii) in (b). In each instance, the initial pressure is 0.200 MPa and the final pressure is 0.100 MPa.

which for one mole becomes

$$w = -RT\ln\frac{V_{m,2}}{V_{m,1}}.$$

**Example 2.1:** Compare the work obtained from the two isothermal expansions at 300 K of one mole of ideal gas following the paths shown in Figure 2.6.

**Solution:** For process (i), $p_{ext}$ is constant, and

$$w_i = -\int p_{ext}\,dV_m = -p_{ext}(V_{m,2} - V_{m,1}).$$

The volumes $V_{m,1}$ and $V_{m,2}$ can be calculated from their positions along the isotherm. For the ideal gas shown as curve (ii) in Figure 2.6,

$$V_{m,1} = \frac{RT}{p_1}.$$

$$V_{m,2} = \frac{RT}{p_2}.$$

$$w_i = -p_2\left(\frac{RT}{p_2} - \frac{RT}{p_1}\right) = -RT\left(1 - \frac{p_2}{p_1}\right).$$

$$= -RT\left(1 - \frac{0.100\text{ MPa}}{0.200\text{ MPa}}\right) = -RT(1 - 0.500).$$

$$= -(8.314)(300)(0.500).$$

$$= -1247\text{ J}\cdot\text{mol}^{-1}.$$

For process (ii), equation (2.16) applied to one mole gives

$$w_{ii} = -RT\ln\frac{V_{m,2}}{V_{m,1}}.$$

$$\frac{V_{m,2}}{V_{m,1}} = \frac{RT/p_2}{RT/p_1} = \frac{p_1}{p_2}.$$

$$w_{ii} = -RT\ln\frac{p_1}{p_2}.$$

$$= -(8.314)(300)\ln\frac{0.200\ \text{MPa}}{0.100\ \text{MPa}}.$$

$$= -1729\ \text{J}\cdot\text{mol}^{-1}.$$

Note that the reversible expansion results in an amount of work that is larger in magnitude than for the nonreversible process. A comparison of the areas in Figure 2.5 verifies that the area under the reversible process is larger in magnitude, and in fact would be the largest possible for an expansion. Because of the negative sign in equation (2.13) we can say that the reversible expansion results in an amount of work that is a maximum in magnitude, but a minimum in value (largest negative number).

For a compression, the process is opposite from that of an expansion. Figure 2.7 compares the work added to a system during isothermal compressions. In Figure 2.7(a), two weights are added to change the pressure from $p_{\text{ext},1}$ to $p_{\text{ext},2}$ in two steps while the gas compresses from $V_1$ to $V_2$. The shaded area (A) is the work that must be added. In Figure 2.7(b), the compression is reversible. Work is positive since $\mathrm{d}V$ is negative in equation (2.13). Note that in the reversible isothermal compression, the work added is a minimum in magnitude and in value. Thus, the reversible isothermal work is a minimum for both a compression and an expansion, with $w$ positive for the compression and negative for the expansion.

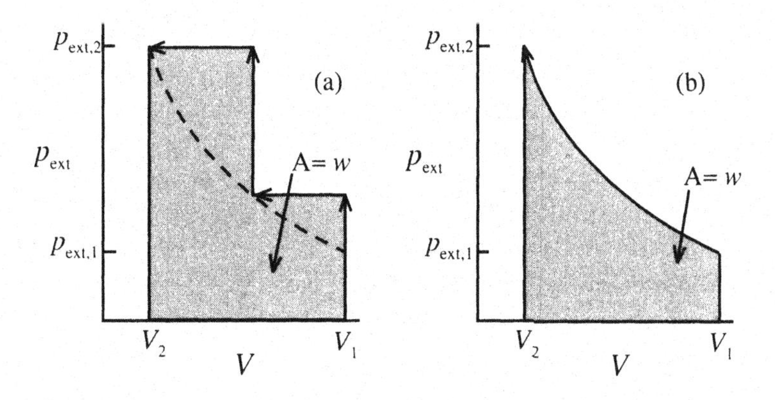


**Figure 2.7** Comparison of isothermal work of compression. Graph (b) represents the reversible process for which the work required for the compression is a minimum.

In an isothermal process, heat must be added during an expansion and removed during a compression to keep the temperature constant. We will describe this more fully as we now calculate the heat added or removed in isobaric, isochoric, and isothermal processes.

## 2.1c Calculation of Heat

Heat, like work, is energy in transit and is not a function of the state of a system. Heat and work are interconvertible. A steam engine is an example of a machine designed to convert heat into work.[h] The turning of a paddle wheel in a tank of water to produce heat from friction represents the reverse process, the conversion of work into heat.

**Heat Capacity:** When the flow of heat in to or out of a system causes a temperature change, the heat is calculated from the heat capacity $C$ defined as

$$C = \frac{\delta q}{\mathrm{d}T} \tag{2.17}$$

where $\delta q$ is the flow of heat resulting in a temperature change $\mathrm{d}T$. Rearranging and integrating gives

$$\int \delta q = q$$

$$q = \int_{T_1}^{T_2} C\,\mathrm{d}T. \tag{2.18}$$

Equation (2.18) is another example of a line integral, demonstrating that $\delta q$ is not an exact differential. To calculate $q$, one must know the heat capacity as a function of temperature. If one graphs $C$ against $T$ as shown in Figure 2.8, the area under the curve is $q$. The dependence of $C$ upon $T$ is determined by the path followed. The calculation of $q$ thus requires that we specify the path. Heat is often calculated for an isobaric or an isochoric process in which the heat capacity is represented as $C_p$ or $C_V$, respectively. If molar quantities are involved, the heat capacities are $C_{p,\mathrm{m}}$ or $C_{V,\mathrm{m}}$. Isobaric heat capacities are more

[h] The laws of thermodynamics govern the conversion of heat into work and work into heat. We will see that a consequence of the Second Law is that heat cannot be completely converted into work.

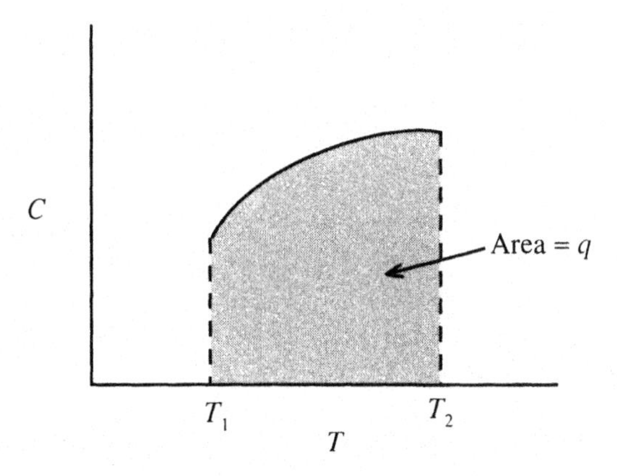


**Figure 2.8** The area under the curve between $T_1$, and $T_2$ obtained by graphing heat capacity $C$ against $T$ gives the heat added in raising the temperature from $T_1$ to $T_2$.

commonly encountered since most chemical processes are carried out at constant pressure.[i]

To integrate the line integral in equation (2.18), heat capacity is often expressed as a power series expansion in $T$ of the type

$$C_{p,\mathrm{m}} = a + bT + \frac{c}{T^2} + dT^2 + \cdots \tag{2.19}$$

The number of terms required in equation (2.19) depends upon the substance and the temperature interval. For small temperature differences, heat capacity changes with temperature are small and $C_p$ can be represented reasonably well by the constant term $a$. For larger temperature changes $a$ and $b$ are used in equation (2.19), and over very large temperature ranges, higher power terms are included. Tables of the coefficients for equation (2.19) for a number of substances are summarized in the literature,[2] with an abbreviated list[3] summarized in Table 2.1.

**Example 2.2:** In the temperature range from 298 to 2500 K and at a pressure of 0.1 MPa, the heat capacity of $CO_2$(g) in $\mathrm{J \cdot K^{-1} \cdot mol^{-1}}$, with coefficients obtained from Table 2.1, is given by

$$C_{p,\mathrm{m}} = 44.14 + 9.04 \times 10^{-3}\ T - \frac{8.54 \times 10^5}{T^2}.$$

[i] Thermodynamics provides a way to calculate the difference between $C_p$ and $C_V$. The procedure will be described later in this chapter and more generally in Chapter 3.

**Table 2.1** Coefficients at $p$ = 0.1 MPa for the heat capacity equation $C^{\circ}_{p,\mathrm{m}}/(\mathrm{J \cdot K^{-1} \cdot mol^{-1}}) = a + 10^{-3}b \cdot T + 10^{5}c \cdot T^{-2} + 10^{-6}d \cdot T^{2}$

| Substance | State | Temperature range (K) | $a$ | $b$ | $c$ | $d$* |
|---|---|---|---|---|---|---|
| Ag | solid | 298–1234 | 21.30 | 8.54 | 1.51 | |
| Ag | liquid | 1234–1600 | 30.54 | | | |
| Au | solid | 298–1336 | 23.68 | 5.113 | 0.142 | |
| Au | liquid | 1336–3081 | 29.3 | | | |
| $Br_2$ | liquid | 273–334 | 71.5 | | | |
| $Br_2$ | gas | 298–2000 | 37.359 | 0.464 | −1.293 | |
| Br | gas | 298–2000 | 19.874 | 1.490 | 0.423 | |
| C | graphite | 298–1100 | 0.109 | 38.940 | −1.481 | −17.385 |
| C | graphite | 1100–4073 | 24.439 | 0.435 | −31.627 | |
| C | diamond | 298–1200 | 9.12 | 13.22 | −6.19 | |
| $CH_4$ | gas | 298–2000 | 12.447 | 76.689 | 1.448 | −18.004 |
| CO | gas | 298–2500 | 28.41 | 4.10 | −0.46 | |
| $CO_2$ | gas | 298–2500 | 44.14 | 9.04 | −8.54 | |
| $Cl_2$ | gas | 298–3000 | 36.90 | 0.25 | −2.85 | |
| Cl | gas | 298–3000 | 23.033 | −0.749 | −0.695 | |
| $F_2$ | gas | 298–2000 | 34.69 | 1.84 | −3.35 | |
| F | gas | 298–2000 | 21.686 | −0.444 | 1.159 | |
| $H_2$ | gas | 298–3000 | 27.28 | 3.26 | 0.50 | |
| H | gas | 298–6000 | 20.786 | | | |
| HF | gas | 298–2000 | 26.90 | 3.43 | 1.09 | |
| HCl | gas | 298–2000 | 26.53 | 4.60 | 1.09 | |
| HBr | gas | 298–1600 | 26.15 | 5.86 | 1.09 | |
| HI | gas | 298–2000 | 26.32 | 5.94 | 0.92 | |
| $H_2O$ | gas | 298–2500 | 30.00 | 10.71 | 0.33 | |
| Hg | liquid | 298–630 | 30.38 | −11.46 | | 10.155 |
| Hg | gas | 298–3000 | 20.786 | | | |
| $I_2$ | solid | 298–387 | −50.647 | 246.906 | 27.974 | |
| $I_2$ | liquid | 387–458 | 80.672 | | | |
| $I_2$ | gas | 298–2000 | 37.405 | 0.569 | −0.619 | |
| I | gas | 290–2000 | 20.393 | 0.402 | 0.280 | |
| K | liquid | 336–1037 | 37.179 | −19.12 | | 12.318 |
| $N_2$ | gas | 298–2500 | 27.87 | 4.27 | | |
| N | gas | 298–1800 | 20.786 | | | |
| $NH_3$ | gas | 298–1800 | 29.75 | 25.10 | −1.55 | |
| $N_2O$* | gas | 298–3000 | 58.099 | 1.385 | −54.116 | 10.920 |
| NO | gas | 298–3000 | 27.681 | 7.439 | −0.151 | −1.431 |
| $NO_2$ | gas | 298–1500 | 35.685 | 22.907 | −4.703 | −6.335 |
| $N_2O_4^*$ | gas | 298–3000 | 128.323 | 1.598 | −128.633 | 24.782 |

(*continued*)

**Table 2.1** *Continued*

| Substance | State | Temperature range (K) | $a$ | $b$ | $c$ | $d$* |
|---|---|---|---|---|---|---|
| Na | solid | 298–371 | 14.790 | 44.229 | | |
| Na | liquid | 371–1156 | 37.468 | −19.154 | | 10.636 |
| $O_2$ | gas | 298–3000 | 29.96 | 4.18 | −1.67 | |
| O | gas | 298–2000 | 20.874 | −0.050 | 0.975 | |
| $O_3$ | gas | 298–2000 | 44.346 | 15.594 | −8.611 | −4.347 |
| P | white (solid) | 298–317 | 19.12 | 15.82 | | |
| P | white (liquid) | 317–870 | 26.326 | | | |
| P | gas | 298–2000 | 20.669 | 0.172 | | |
| $P_2$ | gas | 298–2000 | 36.296 | 0.799 | −4.159 | |
| S | rhombic | 298–369 | 14.98 | 26.11 | | |
| S | monoclinic | 369–388 | 14.90 | 29.12 | | |
| S | liquid | 388–718 | 449.734 | −959.973 | −208.844 | 607.140 |
| S | gas | 298–2000 | 21.916 | −0.460 | 1.862 | |
| $SO_2$ | gas | 298–1800 | 43.43 | 10.63 | −5.94 | |
| $SO_3$ | gas | 298–2000 | 57.145 | 27.347 | −12.912 | −7.728 |
| Si | solid | 298–1685 | 22.824 | 3.858 | −3.540 | |

* For $N_2O$(g), and $N_2O_4$(g), the last term of the heat capacity equation is $10^{-8}d \cdot T^{-3}$ (instead of $10^{-6}d \cdot T^2$).

Calculate the amount of heat necessary to isobarically raise the temperature of one mole of $CO_2$(g) from 500 K to 2000 K at $p = 0.1$ MPa.

**Solution:** A graph of $C_{p,\mathrm{m}}$ against $T$ is shown in Figure 2.9. The area under the curve is $q$. It can be obtained by integrating equation (2.18) with $C_{p,\mathrm{m}}$ given by equation (2.19)

$$q = \int_{T_1}^{T_2} \left( a + bT + \frac{c}{T^2} \right) \mathrm{d}T$$

$$q = a(T_2 - T_1) + \frac{b}{2}(T_2^2 - T_1^2) - c\left( \frac{1}{T_2} - \frac{1}{T_1} \right).$$

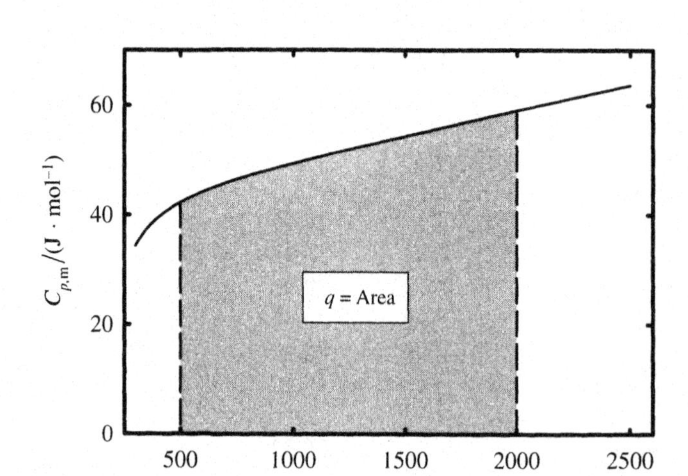


**Figure 2.9** Graph of $C_{p,m}$ against $T$ for $CO_2$(g) using equation (2.19).

With $a = 44.14$, $b = 9.04 \times 10^{-3}$, $c = -8.54 \times 10^5$ and $T_1 = 500$ K, $T_2 = 2000$ K, we get[j]

$$q = (44.14)(2000 - 500) + \frac{9.04 \times 10^{-3}}{2}(2000^2 - 500^2)$$

$$+ 8.54 \times 10^5 \left(\frac{1}{2000} - \frac{1}{500}\right).$$

$$q = 8.188 \times 10^4 \text{ J} \cdot \text{mol}^{-1}.$$

**Relationships between *U, H, q, $C_p$*, and *$C_V$*:**

Equations (2.10) and (2.4) can be combined to give

$$dU = \delta q - p_{\text{ext}}\, dV \tag{2.20}$$

At constant volume, $dV = 0$ and

$$dU = \delta q \text{ (constant volume).} \tag{2.21}$$

---

[j] Note that when $C_p$ is expressed in SI units, the units for $a$ would be $J \cdot K^{-1} \cdot mol^{-1}$, $b$ would have the units $J \cdot K^{-2} \cdot mol^{-1}$, and the units on $c$ would be $J \cdot K \cdot mol^{-1}$. Following the convention decided upon in Chapter 1, we will not express the units for this (and similar expressions) during the calculation, and will only give the units for the final answer.

A similar relationship at constant pressure relating $\mathrm{d}H$ and $\delta q$ can be derived. By definition {see equation (1.18)}

$$H = U + pV.$$

Differentiating gives

$$\mathrm{d}H = \mathrm{d}U + p\,\mathrm{d}V + V\,\mathrm{d}p. \tag{2.22}$$

Combining equations (2.20) and (2.22) and equating $p_{\mathrm{ext}}$ to $p$, the usual case for an isobaric process,[k] gives

$$\mathrm{d}H = \delta q + V\,\mathrm{d}p$$

which, at constant pressure becomes

$$\mathrm{d}H = \delta q \text{ (constant pressure)}$$

since $\mathrm{d}p = 0$.
In summary

$$\mathrm{d}U = \delta q \text{ (isochoric process)} \tag{2.23}$$

$$\mathrm{d}H = \delta q \text{ (isobaric process).}^{l} \tag{2.24}$$

These two equations are the starting point for a number of thermodynamic derivations and calculations.

Equations (2.23) and (2.24) can be combined with equation (2.17) to relate $C_V$ and $C_p$ to $U$ and $V$. For an isochoric process

$$C_V = \left(\frac{\delta q}{\mathrm{d}T}\right)_V = \left(\frac{\mathrm{d}U}{\mathrm{d}T}\right)_V \tag{2.25}$$

$$U = f(V, T)$$

---

[k] As discussed earlier, usually in the isobaric process, $p_{\mathrm{ext}}$ and $p$ never differ from one another by more than an infinitesimal amount and $p_{\mathrm{ext}}$ can be equated to $p$ in the combination of equations (2.20) and (2.22). Processes can be designed in which mechanical constraints make this not so, in which case, equation (2.24) is not rigorously true. In such processes, the difference between $p_{\mathrm{ext}}\,\mathrm{d}V$ abd $p\,\mathrm{d}V$ is usually small, and equation (2.24) is still a good approximation.

[l] Equation (2.24) is the basis for the name **heat content** for $H$. This name was in common use a number of years ago, but is now replaced by enthalpy. Heat content implies that a substance contains heat, which it cannot since heat is energy flowing from one system to another.

$$dU = \left(\frac{\partial U}{\partial V}\right)_T dV + \left(\frac{\partial U}{\partial T}\right)_V dT.$$

At constant $V$, $dV = 0$ and

$$dU = \left(\frac{\partial U}{\partial T}\right)_V dT$$

or

$$\left(\frac{dU}{dT}\right)_V = \left(\frac{\partial U}{\partial T}\right)_V.$$

Substituting the partial derivative for the total derivative in equation (2.25) gives

$$C_V = \left(\frac{\partial U}{\partial T}\right)_V. \tag{2.26}$$

A comparable derivation using equation (2.23) gives

$$C_p = \left(\frac{\delta q}{dT}\right)_p = \left(\frac{dH}{dT}\right)_p$$

or

$$C_p = \left(\frac{\partial H}{\partial T}\right)_p. \tag{2.27}$$

In Chapter 1 we derived the equation

$$\left(\frac{\partial H}{\partial T}\right)_p = \left(\frac{\partial U}{\partial T}\right)_V + \left[p + \left(\frac{\partial U}{\partial V}\right)_T\right]\left(\frac{\partial V}{\partial T}\right)_p. \tag{1.45}$$

Substituting equations (2.26) and (2.27) into equation (1.45) gives

$$C_p = C_V + \left[p + \left(\frac{\partial U}{\partial V}\right)_T\right]\left(\frac{\partial V}{\partial T}\right)_p. \tag{2.28}$$

For one mole, equation (2.28) becomes

$$C_{p,\mathrm{m}} = C_{V,\mathrm{m}} + \left[p + \left(\frac{\partial U_\mathrm{m}}{\partial V_\mathrm{m}}\right)_T\right]\left(\frac{\partial V_\mathrm{m}}{\partial T}\right)_p. \tag{2.29}$$

Equation (2.29) relates $C_{p,\mathrm{m}}$ and $C_{V,\mathrm{m}}$ and can be used to calculate the difference for a specific substance. For example, for the ideal gas

$$V_\mathrm{m} = \frac{RT}{P}$$

so that

$$\left(\frac{\partial V_\mathrm{m}}{\partial T}\right)_p = \frac{R}{p}. \tag{2.30}$$

Also, for the ideal gas

$$\left(\frac{\partial U_\mathrm{m}}{\partial V_\mathrm{m}}\right)_T = 0. \tag{2.31}$$

Substituting equations (2.30) and (2.31), into equation (2.29) gives

$$C_{p,\mathrm{m}} = C_{V,\mathrm{m}} + [p + 0]\frac{R}{p}$$

or

$$C_{p,\mathrm{m}} = C_{V,\mathrm{m}} + R \qquad \text{(ideal gas)} \tag{2.32}$$

Thus, for the ideal gas the molar heat capacity at constant pressure is greater than the molar heat capacity at constant volume by the gas constant $R$. In Chapter 3 we will derive a more general relationship between $C_{p,\mathrm{m}}$ and $C_{V,\mathrm{m}}$ that applies to all gases, liquids, and solids.

### 2.1d Calculation of *q* for Other Processes

We have seen how to calculate $q$ for the isochoric and isobaric processes. We indicated in Chapter 1 that $q = 0$ for an adiabatic process (by definition). For an isothermal process, the calculation of $q$ requires the application of other thermodynamic equations. For example, $q$ can be obtained from equation (2.3) if $\Delta U$ and $w$ can be calculated. The result is

$$q = \Delta U - w. \tag{2.33}$$

**Example 2.3:** Calculate $q$ for the isothermal reversible expansion of the ideal gas under the conditions given in Example 2.1.

**Solution:** $\Delta U = 0$ for the isothermal expansion of the ideal gas. Hence, from equation (2.33)

$$q = -w = nRT\ln\frac{V_2}{V_1}.$$

But from Example 2.1

$$w = -1729 \text{ J} \cdot \text{mol}^{-1}.$$

Hence

$$q = 1729 \text{ J} \cdot \text{mol}^{-1}.$$

In summary, in the isothermal expansion of ideal gas, work flowing out of a system is balanced by heat flowing into the system so that $\Delta U = 0$.

## 2.2 The Second Law Of Thermodynamics

In Chapter 1, we gave the 1865 Clausius statement of the Second Law as:

*Die Entropie der Welt strebt einem Maximum zu,*

or, the entropy of the universe increases to a maximum. In fact, this succinct statement represents a culmination in the development of concepts frequently associated with the Second Law. Earlier statements, including one enunciated by Clausius himself, summarized experiences and observations associated with the performance of engines and cyclic processes. They deal specifically with limitations on what can be accomplished, even by idealized engines, in such processes. To begin our discussion of the Second Law, we will review briefly several of these statements, but we will not discuss them in detail, since our interest is in getting to the statement in terms of entropy, and to its implications.

The **Clausius statement** of the Second Law involves cyclic processes and engines. It can be stated as:[m]

> *It is impossible to construct a device which, operating in a cycle, will produce no other effect than the transfer of heat from a cooler to a warmer body.*

Experience shows that work must be done to transfer heat from a cold reservoir to a warm one. Refrigeration systems operate by removing heat from cold objects and expelling it to a warm room, and they require the operation of a compressor that performs work during the process.

The **Kelvin–Planck statement** of the Second Law also focuses on cyclic devices and limitations. It may be stated as:

> *It is impossible to construct a device which, operating in a cycle, will produce no other effect than the extraction of heat from a reservoir and the performance of an equivalent amount of work.*

In all observed processes, only a fraction of the heat extracted from a heat reservoir can be converted into work, with the remainder returned to a lower temperature reservoir.

A third version of the Second Law, called the **Carnot principle** after the French engineer, Sadi Carnot, also deals with engine performance. It states that:

> *No engine can be more efficient than a reversible engine operating between the same temperature limits, and all reversible engines operating between the same temperature limits have the same efficiency.*

The Clausius and Kelvin–Planck statements and the Carnot principle reflect a historical interest in increasing the efficiency of engines. While the

---

[m] Clausius' original statement is as follows. "Die Wärme kann nicht von selbst aus einem kälteren in einem wärmoren Körpen übergehen." It was later popularized in song by Flanders and Swann: "Heat won't pass from a cooler to a hotter/you can try it if you like but you'd far better not-ter".

statements appear to describe very different aspects of engine performance, one can show that they are equivalent. If one of the statements is taken as true, the other statements can be shown to be true. We will leave the rigorous derivations that show this equivalence to other texts.[4]

### 2.2a The Carnot Cycle: A Hypothetical Engine of Fundamental Importance

The study of engines led Carnot to devise a hypothetical engine that is of fundamental significance in formulating the Second Law in terms of entropy. He conceived a hypothetical engine operating around a reversible cycle that consisted of alternating reversible isothermal and adiabatic expansions and compressions. We are not yet prepared to quantitatively describe the cycle, but we can represent it schematically as shown in Figure 2.10a. The four steps are as follows:

(1) An isothermal expansion of the working fluid of the engine is made from an initial volume $V_1$ to a volume $V_2$ at an empirical high temperature[n] that we will designate as $\theta_2$. Pressure–volume work is done by the system.[o] To maintain isothermal conditions during this process, a quantity of heat $q_2$ is absorbed from a high temperature heat reservoir operating at $\theta_2$.
(2) An adiabatic expansion of the fluid is made from $V_2$ to a volume $V_3$. Work is done by the system, and the temperature drops to a lower empirical temperature $\theta_1$, since no heat is added to the system.
(3) An isothermal compression of the fluid is made at $\theta_1$ from $V_3$ to $V_4$. Work is added to the system, and, to maintain isothermal conditions, a quantity of heat $q_1$ is removed from the system and absorbed in a heat reservoir maintained at the temperature $\theta_1$.
(4) An adiabatic compression of the fluid returns the volume to $V_1$. Since work is added to the system and heat is not allowed to escape, the temperature increases to the initial temperature $\theta_2$.

Since the cyclic process returns the system to the initial state, $\Delta U = 0$, and we can write

$$\Delta U = w + q_1 + q_2 = 0,$$

[n] An empirical temperature scale is based on some arbitrary physical property (such as density, electrical resistance, magnetic susceptibility, etc.) that changes in a way that is continuous and single valued. The ITS-90 temperature scale described in Appendix 2 is an empirical scale that is designed to closely approximate the absolute (ideal gas) temperature scale.

[o] See the discussion of isothermal work in Section 2.1b.

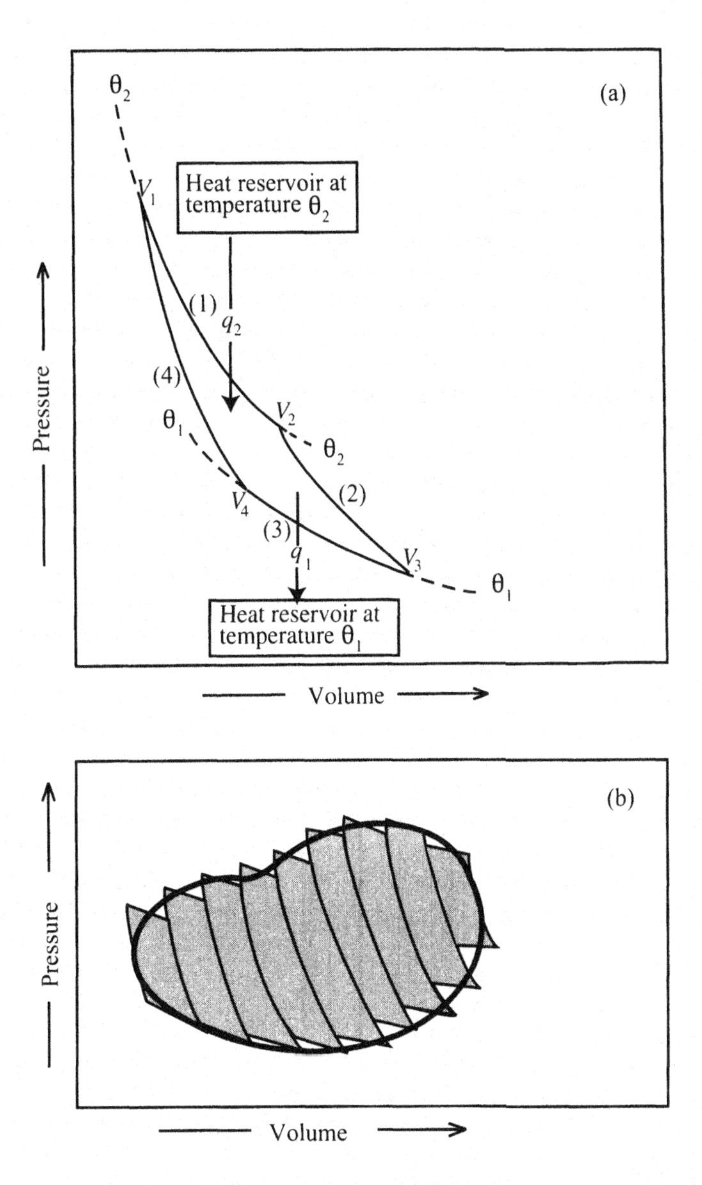


**Figure 2.10** (a) A schematic Carnot cycle in which isotherms at empirical temperatures $\theta_1$ and $\theta_2$ alternate with adiabatics in a reversible closed path. The enclosed area gives the net work produced in the cycle. (b) The area enclosed by a reversible cyclic process can be approximated by the zig-zag closed path of the isothermal and adiabatic lines of many small Carnot cycles.

where $w$ is the net work resulting from the cycle. Solving for $w$ gives

$$w = -(q_2 + q_1).$$

Since $q_2 > 0$ and $q_1 < 0$, and $q_2$ is greater in magnitude than $q_1$, the cyclic engine produces a net amount of (negative) work. From our earlier discussion of pressure–volume work, we know that the area enclosed by the cycle in a $p$–$V$ plot gives the magnitude of this work.[p]

## 2.2b The Kelvin Temperature and Its Role in Calculating an Entropy Change

It was Lord Kelvin who recognized that Carnot's hypothetical engine was of fundamental importance, and used it to define a thermodynamic scale of temperature that has become known as the Kelvin temperature. He set the thermodynamic temperature $T$ of the reservoirs proportional to the amount of heat exchanged at each; that is,

$$T_2 = kq_2 \tag{2.34}$$

and

$$T_1 = -kq_1. \tag{2.35}$$

The minus sign must be included in the second equation, since $q_1 < 0$. Eliminating the proportionality constant $k$ between the two equations gives

$$\frac{q_1}{T_1} + \frac{q_2}{T_2} = \sum \frac{q_{\text{rev}}}{T} = 0.$$

The significance of this relationship is that although $q_1$ and $q_2$ are not state functions, $q/T$ is a state function, since the sum of the $q/T$ terms in the cycle add to zero. This important observation led to the formulation of the entropy function.

Consider any reversible cyclic process that involves the exchange of heat and work. Again, the net area enclosed by the cycle on a $p$–$V$ plot gives the work. This work can be approximated by taking the areas enclosed within a series of Carnot cycles that overlap the area enclosed by the cycle as closely as possible as shown in Figure 2.10b. For each of the Carnot cycles, the sum of the $q/T$ terms

---

[p] We will quantitatively describe the individual steps (alternating reversible isothermal and adiabatic expansions and compressions) of the Carnot cycle in Chapter 3.

is zero, so that for the entire zig-zag cycle, we have

$$\sum \frac{q_{\text{rev}}}{T} = 0. \tag{2.36}$$

This approximation improves as we take more and more, smaller and smaller, Carnot cycles, until in the limit of an infinite number of infinitesimal cycles, the agreement is exact. When this occurs, the sum in equation (2.36) is replaced by an integral over the cycle. That is,

$$\oint \frac{\delta q_{\text{rev}}}{T} = 0. \tag{2.37}$$

The implication of equation (2.37) is that $\delta q_{\text{rev}}/T$ is the differential of a state function. This state function is the entropy $S$, with the differential $\mathrm{d}S$ given by[q]

$$\mathrm{d}S = \frac{\delta q_{\text{rev}}}{T}, \tag{2.38}$$

where $\delta q_{\text{rev}}$ is an infinitesimal quantity of heat transferred reversibly during the cycle, and $T$ is the thermodynamic temperature expressed in Kelvins. We remember that the two symbols for differential changes in a quantity, d and $\delta$ in equation (2.38), distinguish between exact and inexact differentials. Thus, equation (2.38) implies that $T$ serves as an **integrating denominator** to convert the inexact differential element of heat $\delta q_{\text{rev}}$, into the exact differential of the state function $S$.

In the next chapter, we will return to the Carnot cycle, describe it quantitatively for an ideal gas with constant heat capacity as the working fluid in the engine, and show that the thermodynamic temperature defined through equation (2.34) or (2.35) is proportional to the absolute temperature, defined through the ideal gas equation $pV_{\text{m}} = RT$. The proportionality constant between the two scales can be set equal to one, so that temperatures on the two scales are the same. That is, $T(°\text{Absolute}) = T(\text{Kelvin})$.[r]

---

[q] Note that equation (2.38) does not define entropy. Instead it provides a method for the calculation of a change in $S$. We call such a relationship an operational definition. We saw in Chapter 2 that the physical interpretation of entropy is that it is a measure of the disorder in a system. We will see examples of this later in this chapter.

[r] Section 3.3d of Chapter 3 shows the equivalence between the two temperature scales.

### 2.2c The Second Law Expressed in Terms of an Entropy Change

Equation (2.38) relates an entropy change to the flow of an infinitesimal quantity of heat in a reversible process. Earlier in this chapter, we have shown that in the reversible process, the flow of work $\delta w$ is a minimum for the reversible process.[s] Since $w$ and $q$ are related through the first law expression

$$dU = \delta q + \delta w$$

or

$$\Delta U = q + w,$$

and $dU$ or $\Delta U$ is independent of the path, $q$, or $\delta q$ must be a maximum for the reversible process. The conclusion that we reach when we compare a spontaneous or naturally occurring process with a reversible process is that the heat flows are related by

$$\delta q_{\text{nat}} < \delta q_{\text{rev}}.$$

When this relationship is substituted into equation (2.38), we get

$$dS > \frac{\delta q_{\text{nat}}}{T} \tag{2.39}$$

We can generalize by combining equations (2.38) and (2.39) to obtain

$$dS \geqslant \frac{\delta q}{T}. \tag{2.40}$$

In equation (2.40), the equality applies to the reversible process and the inequality to the spontaneous or natural process.

For an adiabatic process, $\delta q = 0$, and equation (2.40) becomes

$$dS \geqslant 0 \qquad \text{(adiabatic process).}$$

---

[s] We recall that for an expansion, the magnitude of the work is a maximum for the reversible process, but since $w < 0$, the reversible work is a minimum when one takes into account the arithmetic sign. For a compression, $w > 0$, and has the minimum positive value for the reversible process. Thus, for both the compression and the expansion, $w$ is a minimum for the reversible process.

Integration gives

$$\Delta S \geqslant 0 \qquad \text{(adiabatic process).}$$

The universe is the pre-eminent example of an adiabatic process. That is, heat cannot flow in or out of the universe. The result is that the above equations apply to the universe and we can write

$$\Delta S \geqslant 0 \qquad \text{(universe).} \tag{2.41}$$

Today, the Second Law, as applied to chemical systems, is firmly associated with the concept of entropy as expressed in the 1865 statement of Clausius and given mathematically by equation (2.41).

In summary, we have seen how the introduction of the idealized Carnot engine leads to the definition of the thermodynamic temperature, an equation for calculating an entropy change from the flow of heat in a reversible process, and to the mathematical formulation of the Second Law in terms of entropy changes. Furthermore, in the next chapter we will apply the Carnot cycle while using an ideal gas as the working fluid to show that the thermodynamic (Kelvin) temperature scale and the ideal gas (Absolute) temperature scale are the same. However, it is possible to obtain equations (2.38) and (2.41) and show the equivalence of the thermodynamic and Absolute temperature scales without relying upon such idealized devices. This approach, which relies upon a fundamental mathematical theorem known as the Carathéodory theorem, is a most intellectually stimulating and satisfying exercise that leads to a much deeper understanding of the Second Law and the significance of entropy. We now present this derivation and encourage the reader to follow it through, and in the process, gain a deeper appreciation for the Second Law of Thermodynamics.

### 2.2d Carathéodory and Pfaffian Differentials

Because equations (2.38) and (2.41) have profound significance for all thermodynamic systems, the reliance on imaginary devices and ideal gases to derive them was considered to be an unsatisfactory state of affairs by many of the early scientists. Constantin Carathéodory, a German mathematician, set about establishing the Second Law without reference to specific thermodynamic systems, real or hypothetical. He was able to do so in a series of arguments, elegant in their logic and general applicability to all thermodynamic systems, and which are worthwhile for us to consider in some detail. We will do so[1] by

[1] We acknowledge S.M. Blinder's treatment in *Advanced Physical Chemistry, A Survey of Modern Physical Principles*, Macmillan, London, 1969, pp. 300–336, as a primary source for the Carathéodory development presented here.

giving indications of the reasoning behind the approach, but will not work through the details of the rigorous mathematics. And, while a strength of the Carathéodory approach is in its generality, we will employ from time to time the specific example of an ideal gas undergoing a reversible adiabatic process to illustrate the line of reasoning Carathéodory used.

Carathéodory recognized that the essential features of equations (2.38) and (2.41) were (a), the existence of an integrating denominator for the differential expression for a reversible heat transfer that transformed the inexact differential into an exact differential, and thereby providing an operational definition for some state function, and (b), the general tendency of this state function to increase. The logic leading to his conclusions can be considered in four parts. The first part of his argument outlines the conditions necessary for the existence of an integrating denominator for a differential equation of three or more variables. The second asserts that these conditions exist for thermodynamic systems undergoing reversible adiabatic processes. The third leads to the identification of the thermodynamic temperature as an appropriate integrating denominator for the differential element of reversible heat in such a system, thus defining operationally a differential element of some state function. The last part leads to the conclusion that, given our conventions for temperature and heat, this state function must increase for spontaneous adiabatic processes, but remain constant for reversible adiabatic ones.

**Pfaffian Differential Expressions With Two Variables** Before we undertake the arguments generalized for three or more variables, we digress to consider some examples involving only two variables. These do not provide the generality we must have to treat thermodynamic systems of three or more variables, but will provide concrete illustrations of the general behavior we will invoke in the development.

In Chapter 1, we wrote the two-dimensional Pfaffian differential $\delta Z$ as

$$\delta Z(X, Y) = M(X, Y)\,\mathrm{d}X + N(X, Y)\,\mathrm{d}Y \qquad (1.27)$$

and saw as an example equation (1.47)

$$\mathrm{d}U = \left(\frac{\partial U}{\partial V}\right)_T \mathrm{d}V + \left(\frac{\partial U}{\partial T}\right)_V \mathrm{d}T \qquad (1.47)$$

in which the differential $\mathrm{d}U$ is related to changes in volume and temperature.

A second Pfaffian differential of interest to us now is the one for the differential quantity of heat, $\delta q_{\text{rev}}$, associated with a **reversible** process.[u] We obtain it by combining equation (1.47) with the first law statement {equation (2.4)} that relates $U$, $w$, and $q$

$$\mathrm{d}U = \delta q + \delta w, \tag{2.4}$$

and the expression {equation (2.13)} that calculates reversible pressure–volume work

$$\delta w = -p\,\mathrm{d}V. \tag{2.13}$$

The result is

$$\delta q_{\text{rev}} = \left[\left(\frac{\partial U}{\partial V}\right)_T + p\right]\mathrm{d}V + \left(\frac{\partial U}{\partial T}\right)_V \mathrm{d}T. \tag{2.42}$$

Equations (1.47) and (2.42) differ in that $\mathrm{d}U$ is an exact differential while $\delta q_{\text{rev}}$ is inexact. We again use the designations d and $\delta$ to distinguish the two types of differentials.

As an example, we can prove that $\delta q_{\text{rev}}$ in equation (2.42) is inexact when applied to the ideal gas. For one mole of an ideal gas, $(\partial U_{\text{m}}/\partial V_{\text{m}})_T = 0$, $p = RT/V_{\text{m}}$ and $(\partial U_{\text{m}}/\partial T)_{V_{\text{m}}} = C_{V,\text{m}}$ is a function of $T$ only[v]

---

[u] In this discussion, we will limit our writing of the Pfaffian differential expression $\delta q$, for the differential element of heat flow in thermodynamic systems, to reversible processes. It is not possible, generally, to write an expression for $\delta q$ for an irreversible process in terms of state variables. The irreversible process may involve passage through conditions that are not true "states" of the system. For example, in an irreversible expansion of a gas, the values of $p$, $V$, and $T$ may not correspond to those dictated by the equation of state of the gas.

[v] To show that $C_V$ depends only upon temperature for the ideal gas, we start with

$$\left(\frac{\partial C_V}{\partial V}\right)_T = \left(\frac{\partial}{\partial V}\left(\frac{\partial U}{\partial T}\right)_V\right)_T = \left(\frac{\partial}{\partial T}\left(\frac{\partial U}{\partial V}\right)_T\right)_V.$$

For the ideal gas,

$$\left(\frac{\partial C_V}{\partial V}\right)_T = 0 \text{ since } \left(\frac{\partial U}{\partial V}\right)_T = 0.$$

so that

$$\delta q_{\text{rev}} = \frac{RT}{V_{\text{m}}}\,\mathrm{d}V_{\text{m}} + \left(\frac{\partial U_{\text{m}}}{\partial T}\right)_{V_{\text{m}}}\mathrm{d}T. \tag{2.43}$$

Applying the condition for exactness gives[w]

$$\left(\frac{\partial\left(\frac{RT}{V_{\text{m}}}\right)}{\partial T}\right)_{V_{\text{m}}} = \frac{R}{V_{\text{m}}}$$

and

$$\left(\frac{\partial}{\partial V_{\text{m}}}\left(\frac{\partial U_{\text{m}}}{\partial T}\right)_{V_{\text{m}}}\right)_T = \left(\frac{\partial}{\partial T}\left(\frac{\partial U_{\text{m}}}{\partial V_{\text{m}}}\right)_T\right)_{V_{\text{m}}} = 0.$$

Because the two derivatives are not equal, the differential expression is inexact.

It can be shown mathematically that a two-dimensional Pfaffian {equation (1.27)} is either exact, or, if it is inexact, an integrating denominator can always be found to convert it into a new, exact, differential. (Such Pfaffians are said to be integrable.) When three or more independent variables are involved, however, a third possibility can occur: the Pfaff differential can be inexact, but possesses no integrating denominator.[x] Carathéodory showed that expressions for $\delta q_{\text{rev}}$ appropriate to thermodynamic systems fall into the class of inexact but integrable differential expressions. That is, an integrating denominator exists that can convert the inexact differential into an exact differential.

---

[w] The condition for exactness for the Pfaffian differential $\mathrm{d}Z = M\,\mathrm{d}X + N\,\mathrm{d}Y$ is the Maxwell relation

$$\left(\frac{\partial M}{\partial Y}\right)_X = \left(\frac{\partial N}{\partial X}\right)_Y$$

[x] The properties of two-dimensional Pfaff differentials are described in Chapter 1. Appendix 1 extends the discussion to include three-dimensional, and by inference, $n$-dimensional Pfaff differentials.

**Pfaffian Differential Expressions with Three or More Variables and the Conditions for the Existence of an Integrating Denominator** We extend the expression for the Pfaffian differentials in three or more variables, by writing it as

$$\delta q = \sum X_i \, \mathrm{d}x_i, \tag{2.44}$$

where the $x_i$'s are the independent variables that define the state of the system, the $X_i$'s are functions of the $x_i$'s, and the Pfaff differential $\delta q$ can be (a), exact; (b), inexact, but possess an integrating denominator that converts the inexact differential to a new exact one; or (c), inexact and possess no integrating denominator.

As we have seen before, exact differentials correspond to the total differential of a state function, while inexact differentials are associated with quantities that are not state functions, but are path-dependent. Carathéodory proved a purely mathematical theorem, with no reference to physical systems, that establishes the condition for the existence of an integrating denominator for differential expressions of the form of equation (2.44). Called the **Carathéodory theorem**, it asserts that an integrating denominator exists for Pfaffian differentials, $\delta q$, when there exist final states specified by $(x_1, \ldots x_n)_f$ that are inaccessible from some initial state $(x_1, \ldots x_n)_{\text{in}}$ by a path for which $\delta q = 0$. Such paths are called **solution curves** of the differential expression.[y] The connection from the purely mathematical realm to thermodynamic systems is established by recognizing that we can express the differential expressions for heat transfer during a reversible thermodynamic process, $\delta q_{\text{rev}}$ as Pfaffian differentials of the form given by equation (2.44). Then, solution curves (for which $\delta q_{\text{rev}} = 0$) correspond to reversible adiabatic processes in which no heat is absorbed or released.

For thermodynamic systems, $\delta q_{\text{rev}}$ will often be completely specified by two of the three variables $p$, $T$, or $V$. When work other than pressure–volume work is involved in the process, additional terms of the form $X_i \, \mathrm{d}x_i$ must be included in equation (2.44). For example, gravitational work introduces the term $mg\mathrm{d}h$, where $m$ is the mass of the system, $g$ is the gravitational acceleration constant, and $\mathrm{d}h$ represents the change in height of the system in the gravitational field. To keep our derivation general, we will include the possibility that other independent variables such as these can be present, and include them in the inexact Pfaffian differential expression for $\delta q_{\text{rev}}$.

---

[y] Solution curves of integrated Pfaffian differential equations are discussed in Appendix 1.

## 2.2e The Carathéodory Principle and Inaccessible States

According to the Carathéodory theorem, the existence of an integrating denominator that creates an exact differential (state function) out of any inexact differential is tied to the existence of points (specified by the values of their $x_i$'s) that cannot be reached from a given point by an adiabatic path (a solution curve). Carathéodory showed that, based upon the earlier statements of the Second Law, such states exist for the flow of heat in a reversible process, so that the theorem becomes applicable to this physical process. This conclusion, which is still another way of stating the Second Law, is known as the **Carathéodory principle**. It can be stated as

> *In the neighborhood*[z] *of every equilibrium state of a thermodynamic system, there exist states unattainable from it by any adiabatic process (reversible or irreversible).*

Like the engine-based statements, Carathéodory's statement invokes limitations. From a given thermodynamic state of the system, there are states that cannot be reached from the initial state by way of any adiabatic process. We will show that this statement is consistent with the Kelvin–Planck statement of the Second Law.

Let us define the system as being characterized by an empirical temperature,[aa] $\theta$ and any other number of variables, $x_1, x_2, \ldots, x_n$. Initially, we will change only two independent variables, $\theta$ and $x_1$. Our generalized example will be set up in such a way as to be consistent with the behavior of gas isotherms. That is, $\theta$ will behave in a way analogous to temperature $T$, and the other variable, $x_1$, will be the analogue of pressure $p$. Then, we will consider the implications when $\theta$ and another variable are allowed to change.

We can represent states of the system (with constant values specified for all the variables except $\theta$ and $x_1$) by a set of isotherms as shown in Figure 2.11a. Two isotherms, $\theta_1$ and $\theta_2$ are shown, with $\theta_2 < \theta_1$. State 1, which is defined by $\theta_1$ and $x_1$, can be connected to states 1′ and 1″ by a series of reversible isothermal processes (horizontal lines in the figure). We remember that heat is absorbed or evolved along a reversible isothermal path, and we will assume that this flow of heat is a continuous function of $x_1$ along the isotherms, with the absorption or liberation depending upon the direction in which $x_1$ is varied. That is, suppose

---

[z] In mathematical terms a neighborhood of a point is a region of arbitrary extent containing the point in its interior.

[aa] We remember from the earlier discussion of the Carnot cycle that an empirical temperature scale is based on some arbitrary physical property (such as density, electrical resistance, magnetic susceptibility, etc.) that changes in a way that is continuous and single valued.

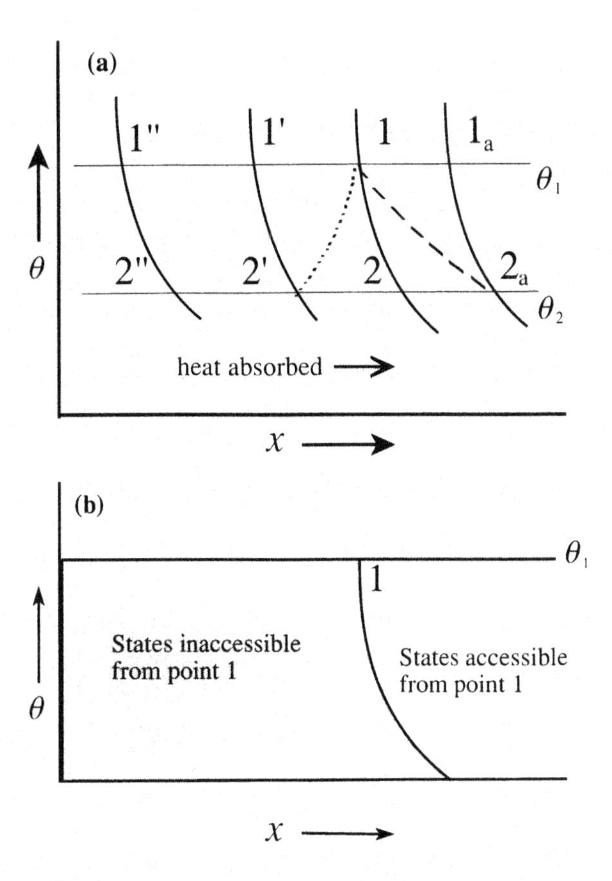


**Figure 2.11** Graph of empirical temperature $\theta$ against $x$, a state variable such as pressure. (a) Reversible adiabatic paths (solid lines: $1'' \rightarrow 2''$; $1' \rightarrow 2'$; and $1 \rightarrow 2$) connect states on two isothermal lines of a general thermodynamic system: An irreversible adiabatic path (dotted line: $1 \rightarrow 2'$) can be shown to be forbidden, while an irreversible adiabatic path (dashed line: $1 \rightarrow 2a$) is allowed. (b) A reversible adiabatic path through state 1 at temperature $\theta_1$ divides the states at lower temperatures into two regions, those that are accessible via irreversible adiabatic paths and those that are not.

heat is liberated in going from 1 to $1'$, then increasing amounts of heat are liberated on going from 1 to states farther and farther removed from 1 such as $1''$. If the direction of the process is switched and the system proceeds from $1''$ to $1'$ or on to 1, heat would be absorbed in increasing amounts.

For each state identified on the $\theta_1$ isothermal as 1, $1'$, $1''$, ..., let us draw paths representing **reversible** adiabatic processes that intersect a second isotherm at $\theta_2$. The intersections of the reversible adiabatic paths from states 1, $1'$ and $1''$ on $\theta_1$ with those on $\theta_2$ are denoted by 2, $2'$ and $2''$, respectively. Along the three paths, 1–2, $1'$–$2'$, and $1''$–$2''$, no heat is absorbed or liberated because the processes that connect these points are defined to be adiabatic.

We wish to show that no points to the left[bb] of 2 on the isotherm $\theta_2$ are accessible from point 1 via **any adiabatic path, reversible or irreversible**. Suppose we assume that some adiabatic path does exist between 1 and 2′. We represent this path as a dotted curve in Figure 2.11a. We then consider the cycle $1 \rightarrow 2' \rightarrow 1' \rightarrow 1$. The net heat associated with this cycle would be that arising from the last step $1' \rightarrow 1$, since the other two steps are defined to be adiabatic. We have defined the direction $1' \rightarrow 1$ to correspond to an absorption of heat, which we will call $q_{1'1}$. From the first law, the net work $w$ done in the cycle, is given by $w = -q_{1'1}$, since $\Delta U$ for the cycle is zero. Thus, for this process, $w$ is negative (and therefore performed by the system), since $q_{1'1}$ is positive, having been absorbed from the reservoir. The net effect of this cycle, then, is to completely convert heat absorbed at a high temperature reservoir into work. This is a phenomenon forbidden by the Kelvin–Planck statement of the Second Law. Hence, points to the left of 2 cannot be reached from point 1 by way of any adiabatic path.

What about points to the right of 2? Can they be reached? Consider an adiabatic path from point 1 to point 2a that is also located on the isothermal $\theta_2$. The cycle of interest is $1 \rightarrow 2a \rightarrow 2 \rightarrow 1$. Again, two of the three steps are adiabatic. In this case, however, heat is evolved during the $2a \rightarrow 2$ step from the conversion of work into heat. The complete conversion of work into heat is a well-known phenomenon and is not forbidden by the laws of thermodynamics. Thus, there are states to the right of 2 on the isotherm $\theta_2$ that are accessible from 1 via an adiabatic path.

Since the only constraint we have placed on $\theta_2$ is that it be less than $\theta_1$, the second isotherm can be as arbitrarily close to the first as we wish. The conclusion that states exist on this second isotherm that cannot be reached from a point on the first isotherm by any adiabatic path is therefore general. Thus, we can argue that there are states located in the plane defined by $\theta$ and $x_1$ that are inaccessible from state 1.

In addition, we have established that there is a sense of direction to the location of the inaccessible states. State 2, the state reached from 1 by a **reversible** adiabatic path, represents the division between the states on the second isotherm that are accessible and inaccessible from state 1. We represent this schematically in Figure 2.11b, where the reversible adiabatic path separates states that are accessible from state 1 from those that are inaccessible. The observation that the reversible path serves as the boundary between the two sets of states will be useful later when we show the direction of allowed processes in terms of the sign of $\Delta S$(universe).

---

[bb] The side of state 2 for which states inaccessible by way of adiabatic paths is determined by our convention established for the direction of heat flow. Reversing that direction would put the inaccessible points on the other side of 2.

Now we can consider the effect of variations of $\theta$ with a second variable, $x_2$. Since we have been general about the nature of the variable, $x_1$, we can expect to obtain similar behavior for the variable $x_2$. We construct isotherms using the same $\theta_1$ and $\theta_2$, but this time in the direction of $x_2$. Our initial point will be the same state 1 as earlier. The value of $x_2$ in state 1 will fix the location of this state on the isotherm in the new direction. A reversible adiabatic path can be constructed that connects state 1 with a state on the second isothermal in the $x_2$ direction. Irreversible states located on one side of this point will be inaccessible from state 1 by adiabatic paths, while states located on the other side of that point will be accessible. Thus, there exist states located on the plane defined by $\theta$ and $x_2$ that are inaccessible from point 1. Similar conclusions can be drawn by considering isotherms localized on the planes formed by $\theta$ and each of the $x_i$.

Thus, we can conclude that, within the neighborhood of every state in this thermodynamic system, there are states that cannot be reached via adiabatic paths. Given the existence of these states, then, the existence of an integrating denominator for the differential element of reversible heat, $\delta q_{\mathrm{rev}}$, is guaranteed from Carathéodory's theorem. Our next task is to identify this integrating denominator.

## 2.2f The Identification of the Absolute (Ideal Gas) Temperature as the Integrating Denominator

We have previously shown that the Pfaff differential $\delta q_{\mathrm{rev}}$ for an ideal gas undergoing only pressure–volume work {equation (2.43)} is an inexact differential. It is easy to show that division of equation (2.43) by the absolute temperature $T$ yields an exact differential expression. The division gives

$$\frac{\delta q_{\mathrm{rev}}}{T} = \frac{R}{V_{\mathrm{m}}}\,\mathrm{d}V_{\mathrm{m}} + \frac{1}{T}\left(\frac{\partial U_{\mathrm{m}}}{\partial T}\right)_{V_{\mathrm{m}}}\mathrm{d}T. \tag{2.45}$$

Application of the condition for exactness shows that both derivatives equal zero, so equation (2.45) must be exact. Thus, we have determined that when an ideal gas is involved, $T$ is an integrating denominator for $\delta q_{\mathrm{rev}}$, and the right hand side of equation (2.45) is the total differential for a state function that we will represent as $\mathrm{d}S$.[cc]

What we must consider now is the generality of the result obtained for the special case of the ideal gas. We define a new thermodynamic system that is the

[cc] We should remember that an integrating denominator for all inexact differential expressions in two dimensions is guaranteed. Hence, its existence for the differential reversible heat element is not unexpected.

composite of two subsystems, each allowed to interact with the other by heat transfer only. We will let each subsystem be a function of three variables[dd], an empirical temperature $\theta$, and two others, $y$ and $z$. States of the first subsystem will be specified by $(\theta_1, y_1, z_1)$ and the second by $(\theta_2, y_2, z_2)$. Since the composite system contains both subsystems, the specification of its state requires all the variables of both subsystems. Thus, the composite system is a function of $(\theta_1, y_1, z_1, \theta_2, y_2, z_2)$. However, we will require that the two subsystems be maintained in thermal equilibrium at a common temperature $\theta$ so that $\theta_1 = \theta_2 = \theta$, and the composite system is a function of $(\theta, y_1, z_1, y_2, z_2)$.

Consider an arbitrary differential reversible process in which we let $\delta q_{\text{rev}}$ be the amount of heat that flows in the composite system. This heat flow must equal the sum of the flow of heat $\delta q_{\text{rev},1}$ and $\delta q_{\text{rev},2}$ in the subsystems that we designate as subsystem 1 and subsystem 2. That is,

$$\delta q_{\text{rev}} = \delta q_{\text{rev},1} + \delta q_{\text{rev},2}. \tag{2.46}$$

Our analysis above has shown that an integrating denominator must exist for the composite system and for the two subsystems. Let us designate these integrating denominators by $\Theta$, $\Theta_1$, and $\Theta_2$ for the composite, subsystem 1 and subsystem 2, respectively. In general, we would expect the integrating denominator to be a function of the variables associated with each entity. Thus,

$$\Theta = \Theta(\theta, y_1, z_1, y_2, z_2),$$

$$\Theta_1 = \Theta_1(\theta, y_1, z_1),$$

and

$$\Theta_2 = \Theta_2(\theta, y_2, z_2).$$

For each entity, we can define a new state function designated by $\Sigma$, $\Sigma_1$, and $\Sigma_2$ that is given by[ee]

$$\mathrm{d}\Sigma = \frac{\delta q_{\text{rev}}}{\Theta}, \tag{2.47}$$

---

[dd] We use three variables to ensure generality of the result, since an integrating denominator is guaranteed to exist for a two-variable system, while the arguments leading to the guarantee of an integrating denominator for $\delta q_{\text{rev}}$ apply equally as well for four or more independent variables as for three.

[ee] We should not be surprised when $\Theta$ turns out to be the temperature $T$ and $\sum$ becomes the entropy $S$.

$$d\Sigma_1 = \frac{\delta q_{\text{rev},1}}{\Theta_1}, \tag{2.48}$$

and

$$d\Sigma_2 = \frac{\delta q_{\text{rev},2}}{\Theta_2}. \tag{2.49}$$

Each of these new functions would be expected to depend on the same variables as its corresponding $\Theta$. Thus

$$\Sigma = \Sigma(\theta, y_1, z_1, y_2, z_2),$$

$$\Sigma_1 = \Sigma_1(\theta, y_1, z_1),$$

and

$$\Sigma_2 = \Sigma_2(\theta, y_2, z_2).$$

Equations (2.47) to (2.49) can be solved for the respective differential heat element, and substitutions can be made into equation (2.46) to yield

$$\Theta(\theta, y_1, z_1, y_2, z_2)\, d\Sigma = \Theta_1(\theta, y_1, z_1)\, d\Sigma_1 + \Theta_2(\theta, y_2, z_2)\, d\Sigma_2. \tag{2.50}$$

We have included the functional dependence for $\Theta$, $\Theta_1$, and $\Theta_2$ in equation (2.50) because it helps us follow what happens next. It turns out to be convenient to make a change in what variables we take as independent in the composite system and either of the subsystems. Since $\Sigma_1$ and $\Sigma_2$ have been identified as state functions, we can invert the order and make them independent variables while one of the independent variables becomes dependent. For example, we could express $y_1$ as a function of $(\theta, \Sigma_1, z_1)$, and $y_2$ as a function of $(\theta, \Sigma_2, z_2)$. Then we can use these relationships to replace the functional dependence in $\Theta$, $\Theta_1$, and $\Theta_2$ and obtain $\Theta = \Theta(\theta, \Sigma_1, z_1, \Sigma_2, z_2)$, $\Theta_1 = \Theta_1(\theta, \Sigma_1, z_1)$ and $\Theta_2 = \Theta_2(\theta, \Sigma_2, z_2)$. We can now divide equation (2.50) by $\Theta$, while incorporating the switch of state variables, to get

$$d\Sigma = \frac{\Theta_1(\theta, \Sigma_1, z_1)}{\Theta(\theta, \Sigma_1, z_1, \Sigma_2, z_2, )}\, d\Sigma_1 + \frac{\Theta_2(\theta, \Sigma_2, z_2)}{\Theta(\theta, \Sigma_1, z_1, \Sigma_2, z_2, )}\, d\Sigma_2. \tag{2.51}$$

The coefficients in front of the $d\Sigma_i$'s are ratios of $\Theta$ functions that individually depend upon different variables. Since $\Sigma$ is a state function, its differential $d\Sigma$

should be a total differential. Thus, from the form of equation (2.51) we can conclude that it is sufficient to regard $\Sigma$ as a function of $\Sigma_1$ and $\Sigma_2$ only. That is, all changes of $d\Sigma$ in the composite system can be considered as arising from the changes $d\Sigma_i$ in the individual subsystems. The arguments leading to this conclusion are as follows.

If equation (2.51) is the total differential for $\Sigma$ as a function of two variables, $\Sigma_1$ and $\Sigma_2$, we can expect that its partial derivatives $(\partial\Sigma/\partial\Sigma_1)_{\Sigma 2}$ and $(\partial\Sigma/\partial\Sigma_2)_{\Sigma 1}$ can be expressed as functions of only those two variables. That is, $\Sigma = \Sigma(\Sigma_1, \Sigma_2)$. Thus, derivatives of $(\partial\Sigma/\partial\Sigma_1)$ and $(\partial\Sigma/\partial\Sigma_2)$, with respect to variables other than $\Sigma_1$ and $\Sigma_2$ should be zero. As we consider the implications of this statement, it is important to note that a change can be made independently in the $z$ variable of one subsystem without affecting that of the other, but a change in $\theta$ will affect both subsystems (since $\theta$ is the same in both subsystems). Therefore, we must consider the implications for $z$ and $\theta$ separately in the analysis that follows.

We first want to consider the implications of having the derivatives $(\partial^2\Sigma/\partial z_i\partial\Sigma_1)$ and $(\partial^2\Sigma/\partial z_i\partial\Sigma_2)$ (with $i = 1$ or 2) equal to zero. Since $\Sigma = \Sigma(\Sigma_1, \Sigma_2)$, we can write

$$d\Sigma = \left(\frac{\partial\Sigma}{\partial\Sigma_1}\right)_{\Sigma_2} d\Sigma_1 + \left(\frac{\partial\Sigma}{\partial\Sigma_2}\right)_{\Sigma_1} d\Sigma_2.$$

Comparing coefficients with those in equation (2.51) gives

$$\frac{\Theta_1}{\Theta} = \left(\frac{\partial\Sigma}{\partial\Sigma_1}\right)_{\Sigma_2}$$

and

$$\frac{\Theta_2}{\Theta} = \left(\frac{\partial\Sigma}{\partial\Sigma_2}\right)_{\Sigma_1}.$$

The conclusion we reach is that these theta ratios depend only upon $\Sigma_1$ and $\Sigma_2$ so that

$$\left(\frac{\partial\left(\dfrac{\Theta_1}{\Theta}\right)}{\partial z_i}\right) = 0 \text{ and } \left(\frac{\partial\left(\dfrac{\Theta_2}{\Theta}\right)}{\partial z_i}\right) = 0, \text{ where } i = 1 \text{ or } 2.$$

Given these relationships, we can show that neither $\Theta$, $\Theta_1$, nor $\Theta_2$ can depend upon the $z$ variables.[ff] Thus, the integrating denominators, $\Theta$, $\Theta_1$, and $\Theta_2$, assumed originally for generality to include a dependence on the $z$ variables, are in fact, independent of them. This leaves a dependency only on $\theta$ and the respective $\Sigma$'s.

The $\Theta_1/\Theta$ and $\Theta_2/\Theta$ ratios also do not depend upon $\theta$, the empirical temperature. However, in the case of this variable, the independence does not arise because both the numerator and denominator are independent of $\theta$, but rather because the dependence cancels out in the ratio. This arises most generally when both numerator and denominator in each ratio are of the following form:

$$\Theta_1(\theta, \Sigma_1) = KT(\theta)F_1(\Sigma_1) \tag{2.54}$$

$$\Theta_2(\theta, \Sigma_2) = KT(\theta)F_2(\Sigma_2) \tag{2.55}$$

$$\Theta(\theta, \Sigma_1, \Sigma_2) = KT(\theta)F(\Sigma_1, \Sigma_2) \tag{2.56}$$

---

[ff] We show this by writing the derivatives of the $\Theta_1/\Theta$ ratio with respect to $z_1$ and $z_2$ and setting them equal to zero. That is

$$\left(\frac{\partial\left(\frac{\Theta_1}{\Theta}\right)}{\partial z_1}\right) = \frac{1}{\Theta}\left(\frac{\partial\Theta_1}{\partial z_1}\right) - \frac{\Theta_1}{\Theta^2}\left(\frac{\partial\Theta}{\partial z_1}\right) = 0 \tag{2.52}$$

and

$$\left(\frac{\partial\left(\frac{\Theta_1}{\Theta}\right)}{\partial z_2}\right) = \frac{1}{\Theta}\left(\frac{\partial\Theta_1}{\partial z_2}\right) - \frac{\Theta_1}{\Theta^2}\left(\frac{\partial\Theta}{\partial z_2}\right) = 0. \tag{2.53}$$

Equivalent equations can be written for derivatives of $\Theta_2/\Theta$.

The first derivative on the right-hand side of equation (2.53) must be zero because subsystems 1 and 2 have been defined to be independent of each other. Therefore, $(\partial\Theta/\partial z_2)$ in the second term of equation (2.53) must also equal zero in order for this equation to be true for all conditions. In a similar manner, starting with the equivalent equations involving the derivative of $\Theta_2/\Theta$, one can show that $(\partial\Theta/\partial z_1) = 0$. If one substitutes this last result into equation (2.52), one gets $(\partial\Theta_1/\partial z_1) = 0$, and the conclusion that $\Theta_1$ is independent of $z_1$. From a similar treatment, one can also show that $(\partial\Theta_2/\partial z_2) = 0$, so that $\Theta_2$ is independent of $z_2$. Thus, since the two ratios $\Theta_1/\Theta$ and $\Theta_2/\Theta$ are independent of $z_i$, neither $\Theta$, $\Theta_1$, nor $\Theta_2$ can depend upon the $z$ variables.

where $K$ is a constant and $T(\theta)$ is some universal function of the empirical temperature scale. One can further simplify equation (2.56) by noting that since the composite system has been defined with the same generality as the two subsystems, and each subsystem function $F_i$ depends only on its own $\Sigma_i$, we can expect that the composite system would depend only upon $\Sigma$. That is, $F(\Sigma_1, \Sigma_2) = F(\Sigma)$.

Equations (2.54) to (2.56) can be substituted into equations (2.47) to (2.49), which after rearrangement, gives

$$\delta q_{\text{rev}} = T(\theta) K F(\Sigma)\, \text{d}\Sigma \tag{2.57}$$

$$\delta q_{\text{rev},1} = T(\theta) K F_1(\Sigma_1)\, \text{d}\Sigma_1 \tag{2.58}$$

$$\delta q_{\text{rev},2} = T(\theta) K F_2(\Sigma_2)\, \text{d}\Sigma_2. \tag{2.59}$$

A comparison of equation (2.38) with equations (2.57) to (2.59) suggests that we describe the entropy, $S$, in terms of its differentials as

$$\text{d}S = K F(\Sigma)\, \text{d}\Sigma \tag{2.60}$$

$$\text{d}S_1 = K F_1(\Sigma_1)\, \text{d}\Sigma_1 \tag{2.61}$$

$$\text{d}S_2 = K F_2(\Sigma_2)\, \text{d}\Sigma_2, \tag{2.62}$$

and equations (2.57) to (2.59) become

$$\delta q_{\text{rev}} = T(\theta)\, \text{d}S \tag{2.63}$$

$$\delta q_{\text{rev},1} = T(\theta)\, \text{d}S_1 \tag{2.64}$$

$$\delta q_{\text{rev},2} = T(\theta)\, \text{d}S_2. \tag{2.65}$$

Equation (2.46) can then be rewritten as

$$T(\theta)\, \text{d}S = T(\theta)\, \text{d}S_1 + T(\theta)\, \text{d}S_2$$

The factor $T(\theta)$ can be eliminated from both sides to yield

$$\text{d}S = \text{d}S_1 + \text{d}S_2 = \text{d}(S_1 + S_2).$$

This equation can be integrated to give

$$S = S_1 + S_2 + \text{a constant of integration}.$$

Since we expect entropy to be extensive and behave like the other extensive thermodynamic properties, the integration constant must be equal to zero so that

$$S = S_1 + S_2. \tag{2.66}$$

Equation (2.66) indicates that the entropy for a multipart system is the sum of the entropies of its constituent parts, a result that is almost intuitively obvious. While it has been derived from a calculation involving only reversible processes, entropy is a state function, so that the property of additivity must be completely general, and it must apply to irreversible processes as well.

The Carathéodory theorem establishes the existence of an integrating denominator for systems in which the Carathéodory principle identifies appropriate conditions — the existence of states inaccessible from one another by way of adiabatic paths. The uniqueness of such an integrating denominator is not established, however. In fact, one can show (but we will not) that an infinite number of such denominators exist, each leading to the existence of a different state function, and that these denominators differ by arbitrary factors of $\Sigma$. Thus, we can make the assignment that $KF_1(\Sigma_1) = KF_2(\Sigma_2) = KF(\Sigma) = 1$. With this assignment, equations (2.57) to (2.59) show that $\Theta$ collapses to $T$, and equations (2.60) to (2.62) show that $S$ is equivalent to $\Sigma$.

By starting with two completely general subsystems existing in thermal equilibrium that together comprise a composite system, we have demonstrated that a single functional form $T(\theta)$ serves as an integrating denominator for both subsystems and the composite system. Since the nature of either subsystem has remained unspecified, we must conclude that the functional form of $T(\theta)$ is also universal for all thermodynamic systems, once an empirical temperature scale has been established. That is, while the particular numerical values of $T$ required to function as an integrating denominator for a particular state of the system are not subject to choice, the functional form of $T(\theta)$ required to yield that value will be determined by the choice of the physical property used to establish $\theta$.

We have already shown that the absolute temperature is an integrating denominator for an ideal gas. Given the universality of $T(\theta)$ that we have just established, we argue that this temperature scale can serve as the thermodynamic temperature scale for all systems, regardless of their microscopic condition. Therefore, we define $T$, the ideal gas temperature scale that we express in degrees absolute, to be equal to $T(\theta)$, the thermodynamic temperature scale that we express in Kelvins. That this temperature scale, defined on the basis of the simplest of systems, should function equally well as an integrating denominator for the most complex of systems is a most remarkable occurrence.

The Carathéodory treatment is grounded in the mathematical behavior of Pfaffian differential expressions {equation (2.44)}, and the observation that a differential element of heat for a reversible process can be written in this form, where the $\mathrm{d}x_i$'s represent state variables. As we have noted earlier, it is not generally possible to write such expressions for the differential element of heat for irreversible processes, and therefore, an integrating denominator does not exist for irreversible processes. The ability to calculate entropy changes by the integration of $\delta q_{\mathrm{rev}}/T$ is strictly limited to reversible processes. A comparable integration using $\delta q_{\mathrm{irr}}/T$ between states connected by an irreversible process would not yield an entropy change for the process. However, because entropy is a state function, the changes associated with irreversible processes can still be calculated if a reversible path, perhaps a purely hypothetical one, can be conceived for which $\delta q_{\mathrm{rev}}/T$ can be evaluated. We will give examples in subsequent sections that demonstrate how this is done.

### 2.2g Entropy Changes for Reversible and Irreversible Paths

We will now consider the last stage in Carathéodory's development of the second law — the establishment of Clausius' statement represented by equation (2.41). To do so, we return to the realm of Pfaffian differential expressions. We have seen that for reversible processes, $\delta q_{\mathrm{rev}}$ is an inexact Pfaffian differential expression, and $\mathrm{d}S$ is an exact one. As an inexact differential expression, equations of the form, $\delta q_{\mathrm{rev}} = 0$, cannot be integrated to yield a general solution for a surface $q_{\mathrm{rev}} = q_{\mathrm{rev}}(x_1, \ldots x_n)$. But it is still possible to identify reversible adiabatic paths for which the equality $\delta q_{\mathrm{rev}} = 0$ is satisfied.[gg] The solid curves in Figure 2.11a and 2.11b are examples of these. In more complicated systems, the paths may be obtained by point-to-point numerical integrations.

Since $\mathrm{d}S$ is an exact differential, equations for $\mathrm{d}S = 0$ can be integrated. The integration yields a family of solution surfaces, $S = S(x_1, \ldots x_n) = \text{constant}$. Each solution surface contains a set of thermodynamic states for which the entropy is constant.[hh]

We distinguish the solution curves from solution surfaces in Figure 2.12. The curved solid path marked $\delta q_{\mathrm{rev}} = 0$ is a solution curve, while the solution surfaces are designated by $S_1$, $S_2$, and $S_3$. Each surface corresponds to a different value for the constant entropy. From equation (2.38)

$$\mathrm{d}S = \frac{\delta q_{\mathrm{rev}}}{T}, \tag{2.38}$$

---

[gg] Exercise E2.16 illustrates how this can be done for the ideal gas.

[hh] In a two-dimensional case, such as those involving the expansion of an ideal gas, these sets form a curve in ($p$ and $T$) or ($V$ and $T$) or ($p$ and $V$) space. In systems with a larger number of independent variables, we obtain mathematical surfaces whose points represent isentropic (constant entropy) thermodynamic states.

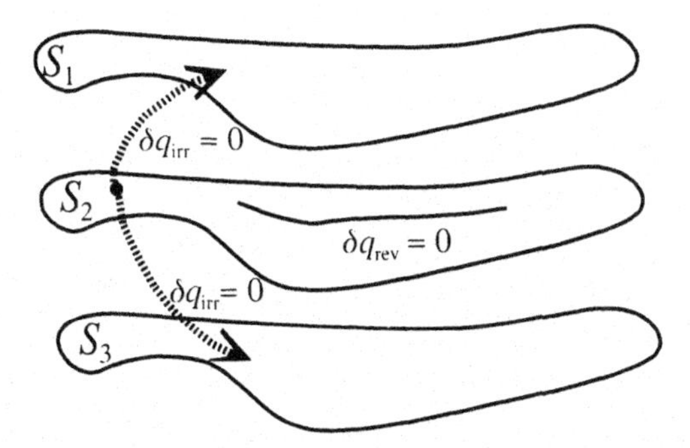


**Figure 2.12** A set of parallel, isentropic surfaces ordered so that $S_1 > S_2 > S_3$. The solid curve marked $\delta q_{rev} = 0$ represents a reversible adiabatic path that connects two states that lie on the entropy surface, $S_2$. The dashed curves marked $\delta q_{irr} = 0$ are irreversible paths that connect states on different entropy surfaces. Only one of these two paths will be allowed; the other will be forbidden.

it is easy to see that $dS = 0$ for a reversible adiabatic process because $\delta q_{rev} = 0$. Two points which can be connected by a reversible adiabatic path must lie on the same entropy surface and a solution curve must lie wholly within a solution surface.

The solution surfaces cannot intersect. If they did, states located at the points of intersection would have multiple values of entropy, and this would violate a fundamental property of state functions. Thus, the surfaces can be expected to be ordered monotonically, either systematically increasing (or decreasing) as one proceeds in a given direction from surface to surface. For our purposes, let us assume $S_1 > S_2 > S_3$ in Figure 2.12.

Presumably all points on the same surface can be connected by some solution curve (reversible adiabatic process). However, states on surface $S_2$, for example, cannot be connected to states on either $S_1$ or $S_3$ by any reversible adiabatic path. Rather, if they can be connected, it must be through irreversible adiabatic paths for which $dS \neq 0$. We represent two such paths in Figure 2.12 by dashed lines.

In the analysis in Section 2.2b, we showed that for any given initial state there are states that are accessible via irreversible adiabatic paths from the initial state, as well as states that are inaccessible from that initial state by way of irreversible adiabatic paths. Figure 2.11b showed that a reversible adiabatic path containing the initial state marked the division between the states that were accessible or inaccessible from that state, with all accessible states lying on one side of the reversible adiabatic path, and all inaccessible states lying on the other side of it.

Given our original assumptions about the continuity of heat flow as one chooses starting points along a particular isotherm, and our understanding that the entropy surfaces are continuously ordered and nonintersecting, we can conclude that the entropy surface on which an initial state resides represents a division between accessible and inaccessible entropy surfaces. States that could, in our earlier analysis be reached by way of irreversible adiabatic processes, must lie on entropy surfaces that are on one side of the surface where the initial state resides, while those that could not be reached by any adiabatic process must lie on the other side. Since our initial state can be on any surface, there must be a constant direction to the allowed pathways. All allowed irreversible adiabatic processes must go to surfaces with either consistently higher or consistently lower entropy. Thus, we see that an irreversible adiabatic process has, as its name implies, a direction to it. The process cannot be reversed. If going from state 1 to state 2 is allowed, then starting at state 2 and trying to go to state 1 would mean traversing entropy surfaces in a direction that is not allowed.

The Carathéodory analysis has shown that a fundamental aspect of the Second Law is that the allowed entropy changes in irreversible adiabatic processes can occur in only one direction. Whether the allowed direction is increasing or decreasing turns out to be inherent in the conventions we adopt for heat and temperature as we will now show.

Since our conclusions reached to this point apply to all thermodynamic systems, we can choose a simple case to evaluate the direction of the inequality. Let us consider two thermal reservoirs maintained at temperatures $T_1$ and $T_2$, respectively, with $T_2 > T_1$. The reservoirs are in thermal contact with each other through a large number of intermediate temperature reservoirs, with the whole system thermally insulated. The intermediate reservoirs are established so that the temperature of each differs from that of its neighbors by increments small enough that heat transfer can be regarded as taking place reversibly from intermediate reservoir to intermediate reservoir. However, because the temperature difference between the outer reservoirs is a measurable one, heat will flow spontaneously and irreversibly from the high-temperature reservoir through the intermediate reservoirs to the low-temperature one. Let us call the amount of heat leaving the high-temperature reservoir $q_2$. It will pass into and out of each of the intermediate reservoirs, so that their net entropy change is zero. Given the convention that we have established for the arithmetic sign of heat flow, we can state that $q_2 < 0$.

Because the overall system is isolated, $q_1$, the heat flowing into the low-temperature reservoir is given by

$$q_1 = -q_2.$$

Since the heat was transferred reversibly out of the high-temperature reservoir and into the low-temperature reservoir, we may use these quantities to calculate the entropy change for each reservoir. The result is

$$\Delta S_1 = \frac{q_1}{T_1}$$

and

$$\Delta S_2 = \frac{q_2}{T_2}.$$

We have previously shown with complete generality that the entropy of a composite system is the sum of the entropies of its components, and so we can write that

$$\Delta S = \frac{q_1}{T_1} + \frac{q_2}{T_2},$$

which can be simplified to

$$\Delta S = \frac{q_1}{T_1} - \frac{q_1}{T_2} = q_1 \left[ \frac{1}{T_1} - \frac{1}{T_2} \right].$$

Since $q_1 > 0$, and $1/T_1 > 1/T_2$ with $T_2 > T_1$, we conclude that $\Delta S$ for this allowed, spontaneous process is greater than zero. Having obtained this result for the specific case, we can extend it to the general case, because our earlier conclusion that there is an allowed direction to spontaneous adiabatic processes applies to all thermodynamic systems.

To summarize

$$dS(\text{reversible adiabatic processes}) = 0$$

and

$$dS(\text{irreversible adiabatic processes}) > 0.$$

When we apply these statements to the universe, the ultimate adiabatically enclosed system, the result is the Clausius statement

$$\Delta S(\text{universe}) \geqslant 0.$$

That is, all processes occurring within the universe must constitute a net process that is adiabatic.

To consider explicitly the possibility of multiple processes, we write

$$\Delta S(\text{universe}) = \sum_i \Delta S_i \geqslant 0 \tag{2.67}$$

where each $\Delta S_i$ represents the entropy change for an individual process. The conclusion we reach from equation (2.67) is that the entropy change associated with an individual process, $\Delta S_i$, may be either positive or negative, so long as the sum of the entropy changes for all the individual processes that occur is greater than zero.

Equation (2.67) provides a way to determine whether a process is reversible, irreversible and allowed, or irreversible and forbidden. One simply adds up all the entropy changes in the universe resulting from the process. When the net entropy change is zero, the process is reversible, and the system is said to be at equilibrium.[ii] If the total entropy is greater than zero, the process will occur. Such a process is often referred to as a **spontaneous** or **natural** process. If the net entropy change is negative, the process is **unnatural** and will not occur. Thus, the Second Law provides an important method for predicting the spontaneity or reversibility of a process. However, it proves to be a difficult one to apply, since one must keep track of all of the entropy changes in the universe. Later we will use the Second Law to show that under the conditions of constant $T$ and $p$, $\Delta G$ for a system can be used to predict if a process is spontaneous or at equilibrium. This criterion is easier to use because it involves making calculations only for the system of interest, and not for all of its external interactions with the rest of the universe.

## 2.2h Calculation of an Entropy Change

To summarize, the Carnot cycle or the Carathéodory principle leads to an integrating denominator that converts the inexact differential $\delta q_{\text{rev}}$ into an exact differential. This integrating denominator can assume an infinite number of forms, one of which is the thermodynamic (Kelvin) temperature $T$ that is equal to the ideal gas (absolute) temperature. The result is

$$\mathrm{d}S = \frac{\delta q_{\text{rev}}}{T}, \qquad (2.38) \text{ or } (2.68)$$

where $\mathrm{d}S$ is the exact differential of the entropy $S$. Equation (2.38) or (2.68) can be used as the starting point for deriving thermodynamic equations for

[ii] Remember that thermodynamics does not tell us anything about how long it takes for a process to occur. Some naturally occurring processes for which $\Delta S(\text{universe}) > 0$ may be very slow.

calculating entropy changes.[jj] We will defer most calculations to the next chapter, but useful relationships can be derived now.

**Calculation of $\Delta S$ for the Reversible Isothermal Expansion of an Ideal Gas:** Integration of equation (2.38) gives

$$\int_{S_1}^{S_2} \mathrm{d}S = \int \frac{\delta q_{\mathrm{rev}}}{T}.$$

Since $T$ is constant

$$\Delta S = \frac{q_{\mathrm{rev}}}{T}.$$

From example 2.3 we saw that for the reversible isothermal expansion of ideal gas

$$q = nRT \ln \frac{V_2}{V_1}.$$

Hence

$$\Delta S = nR \ln \frac{V_2}{V_1}. \tag{2.69}$$

**Calculation of $\Delta S$ for the Reversible Adiabatic Expansion:**

$$\delta q = 0 \tag{2.70}$$

$$\mathrm{d}S = 0$$

$$\Delta S = 0. \tag{2.71}$$

**Calculation of $\Delta S$ for the Isobaric Temperature Change:**[kk]

$$\delta q = C_p \, \mathrm{d}T$$

$$\mathrm{d}S = \frac{\delta q_{\mathrm{rev}}}{T} = \frac{C_p \, \mathrm{d}T}{T}$$

---

[jj] In Chapter 4, we will see that $S$, as well as $\Delta S$, can be calculated.

[kk] In the isobaric and isochoric process, it is not necessary to specify reversibility, since $q$ is equal to the change $\Delta H$ or $\Delta U$, and hence, is independent of the path.

$$\left(\frac{\mathrm{d}S}{\mathrm{d}T}\right)_p = \frac{C_p}{T}$$

$$\left(\frac{\partial S}{\partial T}\right)_p = \frac{C_p}{T} \tag{2.72}$$

$$\Delta S = \int_{T_1}^{T_2} \frac{C_p}{T}\,\mathrm{d}T.$$

**Calculation of $\Delta S$ for the Isochoric Temperature Change:**

$$\delta q = C_V\,\mathrm{d}T$$

$$\mathrm{d}S = \frac{\delta q_{\mathrm{rev}}}{T} = \frac{C_V\,\mathrm{d}T}{T}$$

$$\left(\frac{\mathrm{d}S}{\mathrm{d}T}\right)_V = \frac{C_V}{T}$$

$$\left(\frac{\partial S}{\partial T}\right)_V = \frac{C_V}{T} \tag{2.73}$$

$$\Delta S = \int_{T_1}^{T_2} \frac{C_V}{T}\,\mathrm{d}T.$$

**Calculation of $\Delta S$ for the Reversible (Equilibrium) Phase Change:** The freezing of a liquid at its melting point, a (solid + solid) phase transition at its transition temperature, and the boiling of a liquid or subliming of a solid to a gas at a temperature where the vapor pressure of the liquid or solid is equal to the confining atmospheric pressure, are reversible processes that occur at constant temperature and pressure. The entropy change for these processes is easily calculated. Integrating equation (2.38) at constant $T$ gives

$$\Delta S = \int \frac{\delta q_{\mathrm{rev}}}{T} = \frac{q_{\mathrm{rev}}}{T}.$$

At constant $p$

$$q_{\text{rev}} = \Delta H$$

so that

$$\Delta S = \frac{\Delta H}{T}, \tag{2.74}$$

where $\Delta H$, the change in enthalpy for the change in phase, is $\Delta_{\text{fus}}H$, the enthalpy of fusion, $\Delta_{\text{vap}}H$, the enthalpy of vaporization, $\Delta_{\text{trans}}H$, the enthalpy of transition, or $\Delta_{\text{sub}}H$, the enthalpy of sublimation, depending upon the type of phase change involved. Selected values of $\Delta_{\text{fus}}H$ and $\Delta_{\text{vap}}H$, along with the corresponding equilibrium temperatures, are given in Table 2.2 for a number of substances.

**Calculation of $\Delta S$ for the Mixing of Ideal Gases at Constant $T$ and $p$:** Consider the process shown in Figure 2.13, where $n_A$ moles of ideal gas A are confined in a bulb of volume $V_A$ at a pressure $p$ and temperature $T$. This bulb is separated by a valve or stopcock from bulb B of volume $V_B$ that contains $n_B$ moles of ideal gas B at the same pressure $p$ and temperature $T$. When the stopcock is opened, the gas molecules mix spontaneously and irreversibly, and an increase in entropy $\Delta_{\text{mix}}S$ occurs. The entropy change can be calculated by recognizing that the gas molecules do not interact, since the gases are ideal. $\Delta_{\text{mix}}S$ is then simply the sum of $\Delta S_A$, the entropy change for the expansion of gas A from $V_A$ to $(V_A + V_B)$ and $\Delta S_B$, the entropy change for the expansion of gas B from $V_B$ to $(V_A + V_B)$. That is,

$$\Delta_{\text{mix}}S = \Delta S_A + \Delta S_B.$$

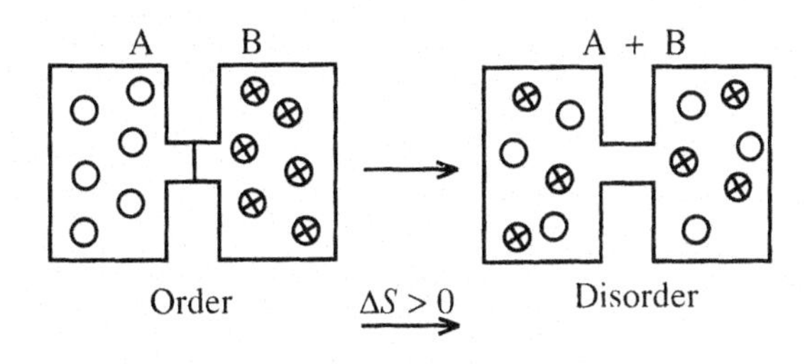


**Figure 2.13** Mixing of ideal gas A with ideal gas B at constant temperature and constant total pressure. The entropy change $\Delta S$ is given by equation (2.78).

**Table 2.2** Enthalpies and temperatures of fusion and vaporization. Normal melting and boiling points and enthalpies of fusion and vaporization are tabulated by type of compound

| Compound[a] | $T_m/(K)$ | $\Delta_{fus}H_m/(J \cdot mol^{-1})$ | $T_b/(K)$ | $\Delta_{vap}H_m/(J \cdot mol^{-1})$ |
|---|---|---|---|---|
| *Elements* | | | | |
| He | – | – | 4.216 | 84 |
| Ne | 24.57 | 335.1 | 27.07 | 1766 |
| Ar | 83.85 | 1176 | 87.29 | 6519 |
| Kr | 115.95 | 1636 | 119.93 | 9029 |
| Xe | 161.3 | 2298 | 165.1 | 12640 |
| Rn | 202 | 2900 | 211 | 16401 |
| $H_2$ | 13.96 | 117 | 20.39 | 902.9 |
| $F_2$ | 55.20 | 1556 | 85.24 | 6318 |
| $Cl_2$ | 172.18 | 6406 | 239.11 | 20410 |
| $Br_2$ | 265.95 | 10544 | 331.4 | 29999 |
| $I_2$ | 386.8 | 15648 | 456 | 41714 |
| $O_2$ | 54.39 | 444.8 | 90.19 | 6820 |
| $S_8$ (equilibrium S) | 392 | 1226 | 717.76 | 10460 |
| $N_2$ | 63.15 | 720 | 77.36 | 5577 |
| $P_4$ (white) | 317.4 | 628 | 553 | 12426 |
| Hg | 234.29 | 2295.3 | 629.88 | 59149 |
| Li | 453.70 | 3024.1 | 1604 | 134683 |
| Na | 370.97 | 2601.6 | 1163 | 89036 |
| K | 336.4 | 2318 | 1039 | 77530 |
| Rb | 312.0 | 2343 | 974 | 69203 |
| Cs | 301.8 | 2134 | 958 | 65898 |
| *Carbon compounds* | | | | |
| $CH_4$ | 90.68 | 941 | 111.67 | 8180 |
| $C_2H_6$ | 89.88 | 2859.3 | 184.52 | 14715 |
| $C_3H_8$ | 85.46 | 3523.8 | 231.08 | 18774 |
| $n\text{-}C_4H_{10}$ | 134.80 | 4661.0 | 272.65 | 22393 |
| $n\text{-}C_5H_{12}$ | 143.43 | 8393.1 | 309.22 | 25773 |
| $n\text{-}C_6H_{14}$ | 177.80 | 13029.0 | 341.89 | 28853 |
| $n\text{-}C_7H_{16}$ | 182.54 | 14033.1 | 371.58 | 31694 |
| $n\text{-}C_8H_{18}$ | 216.36 | 20740.1 | 398.82 | 34367 |
| $C_6H_{12}$ (cyclohexane) | 279.70 | 2665.2 | 353.89 | 30083 |
| $C(CH_3)_4$ | 256.60 | 3256.0 | 282.65 | 22753 |
| $C_2H_4$ | 103.97 | 3350.5 | 169.45 | 13544 |
| $C_2H_2$ | – | – | 189.2 | 21338 |
| $C_6H_6$ | 278.683 | 9832 | 353.25 | 30765 |
| $C_6H_5CH_3$ | 178.16 | 6619 | 383.78 | 33459 |
| $1,2\text{-}C_6H_4(CH_3)_2$ | 247.97 | 13598 | 417.56 | 36823 |

**Table 2.2** *Continued*

| Compound[a] | $T_m/(K)$ | $\Delta_{fus}H_m/(J \cdot mol^{-1})$ | $T_b/(K)$ | $\Delta_{vap}H_m/(J \cdot mol^{-1})$ |
|---|---|---|---|---|
| 1,3–$C_6H_4(CH_3)_2$ | 225.28 | 11565 | 412.25 | 36443 |
| 1,4–$C_6H_4(CH_3)_2$ | 286.41 | 17113 | 411.50 | 36066 |
| $C_{10}H_8$ (naphthalene) | 353.36 | 18054 | 491.11 | 42173 |
| $CH_3OH$ | 175.26 | 3167 | 337.9 | 35271 |
| $C_2H_5OH$ | 158.6 | 5021 | 351.7 | 38576 |
| HCHO | 154.9 | – | 253.9 | 24476 |
| $CH_3CHO$ | 155 | 3222 | 293.3 | 27196 |
| HCOOH | 281.46 | 12678 | 373.7 | 22259 |
| $CH_3COOH$ | 289.77 | 11715 | 391.4 | 24393 |
| $CH_3NH_2$ | 179.70 | 6134 | 266.84 | 25815 |
| $(CH_3)_2NCHO$ | 212.67 | 7895 | 422.71 | 38342 |
| $CCl_4$ | 250.3 | 2510 | 349.9 | 29999 |
| $CHCl_3$ | 209.7 | 9205 | 334.4 | 29372 |
| $COCl_2$ | 145.39 | 5740 | 280.71 | 24401 |
| $CS_2$ | 161.1 | 4393 | 319.41 | 26778 |
| *Oxides* | | | | |
| $SO_2$ | 197.70 | 7401 | 263.14 | 24895 |
| $N_2O$ | 182.34 | 6540 | 184.68 | 16552 |
| NO | 109.55 | 2299 | 121.42 | 13774 |
| $N_2O_4$ | 261.96 | 14652 | 294.31 | 38116 |
| CO | 68.10 | 837.6 | 81.67 | 6042 |
| $CO_2$ | – | – | 194.68[b] | 25234[b] |
| *Hydrides* | | | | |
| HF | 189.39 | 4577 | 292.67 | 6732 |
| HCl | 158.97 | 1991.6 | 188.13 | 16150 |
| HBr | 186.30 | 2406.2 | 206.44 | 17615 |
| HI | 222.37 | 2871.5 | 237.81 | 19765 |
| $H_2O$ | 273.150 | 6009.5 | 373.150 | 40656 |
| $H_2S$ | 187.63 | 2377.3 | 212.85 | 18673 |
| $H_2Se$ | 207.43 | 2515 | 231.9 | 19330 |
| $H_2Te$ | 222 | 4184 | 270.9 | 23221 |
| $NH_3$ | 195.42 | 5653 | 239.76 | 23351 |
| $PH_3$ | 139.41 | 1130 | 185.44 | 14602 |
| $AsH_3$ | 156.9 | 2343 | 210.7 | 17489 |

[a] Values were reproduced from a tabulation by J. R. Goates and J. B. Ott in *Chemical Thermodynamics: An Introduction*, Harcourt Brace, Jovanovich, Inc., New York, 1971, pp. 137–139. References to the original sources for these data are given there.

[b] Normal sublimation point.

The entropy changes $\Delta S_A$ and $\Delta S_B$ can be calculated from equation (2.69), which applies to the isothermal reversible expansion of ideal gas, since $\Delta S$ is independent of the path and the same result is obtained for the expansion during the spontaneous mixing process as during the controlled reversible expansion. Equation (2.69) gives

$$\Delta S_A = n_A R \ln \frac{V_A + V_B}{V_A}$$

$$\Delta S_B = n_B R \ln \frac{V_A + V_B}{V_B}.$$

Hence

$$\Delta_{mix} S = R \left[ n_A \ln \frac{V_A + V_B}{V_A} + n_B \ln \frac{V_A + V_B}{V_B} \right]. \tag{2.75}$$

Using the ideal gas equation, we can write

$$n_A = \frac{pV_A}{RT}$$

$$n_B = \frac{pV_B}{RT}$$

$$n_A + n_B = \frac{p(V_A + V_B)}{RT}.$$

Combining the last three equations gives

$$\frac{n_A}{n_A + n_B} = \frac{V_A}{V_A + V_B} = x_A \tag{2.76}$$

$$\frac{n_B}{n_A + n_B} = \frac{V_B}{V_A + V_B} = x_B \tag{2.77}$$

where $x_A$ and $x_B$ are the mole fractions of A and B in the mixture. Substituting equations (2.76) and (2.77) into equation (2.75) and rearranging gives the equation we want for calculating $\Delta_{mix}S$ for ideal gases.[ll]

$$\Delta_{mix}S = -R[n_A \ln x_A + n_B \ln x_B] \tag{2.78}$$

To find $\Delta_{mix}S_m$, the entropy of mixing for a mole of mixture, we divide equation (2.78) by the total moles, $(n_A + n_B)$, to get

$$\Delta_{mix}S_m = -R[x_A \ln x_A + x_B \ln x_B]. \tag{2.79}$$

### 2.2i Entropy and Disorder

The calculation of $\Delta S$ from equations (2.69) to (2.74), along with equations (2.78) or (2.79), all demonstrate that an increase in entropy causes an increase in disorder. For example:

(a) *For the isothermal expansion of ideal gas*

$$\Delta S = nR \ln \frac{V_2}{V_1}.$$

With $V_2 > V_1$, $\Delta S > 0$.

On a molecular scale, expanding a gas causes the molecules to occupy a larger volume, leading to disorder.

(b) *For the isobaric heating of a solid*

$$\Delta S = \int_{T_1}^{T_2} \frac{C_p}{T} \, dT.$$

If $T_2 > T_1$, then $\Delta S > 0$ since $C_p$ and $T$ are >0. From a molecular point of view, heating a solid increases the amplitudes and energy distributions of the vibrations of the molecules in the solid, resulting in increased disorder.

(c) *For the melting of a solid*

$$\Delta S = \frac{\Delta_{fus}H}{T}.$$

Since $\Delta_{fus}H$ and $T$ are >0, then $\Delta S > 0$. A liquid is less ordered than a solid. Hence, melting a solid leads to increased disorder.

---

[ll] We will see later that this same equation applies to the mixing of liquids or solids when ideal solutions form.

(d) *Mixing two ideal gases*

$$\Delta_{\text{mix}}S = -R[n_A \ln x_A + n_B \ln x_B].$$

Since $x_A$ and $x_B$ are less than 1, $\ln x_A$ and $\ln x_B$ are $<0$ and $\Delta_{\text{mix}}S > 0$.[mm] Figure 2.13 demonstrates that the mixing of gases causes increased disorder on the molecular level.

(e) *Reversible adiabatic expansion*
For this process, $\Delta S = 0$. The increased disorder caused by the expansion of the gas is just balanced by the increased order due to the cooling of the gas.

In Chapter 1 we saw that the Boltzmann equation $S = k \log W$ gives the same qualitative relationship between entropy and disorder and suggested that a fundamental property of entropy is a measure of the disorder in a system. In Chapter 10 we will explore this relationship in more detail on the molecular level, and use the Boltzmann expression to develop quantitative relationships between entropy and disorder.

## 2.3 Implications Of The Laws

As we have seen earlier, the mathematical statement of the First Law of Thermodynamics in terms of the universe can be written as

$$\sum \Delta U = 0 \text{ (universe)}.$$

The law implies the existence of a finite amount of energy in the universe. The energy can be changed from one form into another and transferred from one part of the universe to another, but it can neither be created nor destroyed. The table below summarizes the general types of energy changes that are possible.

| | $\Delta U$ | | |
|---|---|---|---|
| | System + | Surroundings = | Universe |
| (a) | + | − | 0 |
| (b) | − | + | 0 |
| (c) | 0 | 0 | 0 |

[mm] We note from the derivation of equation (2.79) that the increase in entropy is not really due to the mixing of gases so much as it is to the expansion of each gas into a larger volume. The disorder does not result from the presence of a second kind of gas as much as it does from the increase in space available for the molecules. For each of the examples (a) through (d), the increases in entropy translate into an increase in the space (volume) accessible to the molecules.

The vaporization of a liquid is an example of (a). In this process, heat flows from the surroundings into the liquid to give the molecules the increased internal energy they need to exist as a gas. As the liquid vaporizes, energy is added back into the surroundings through the performance of pressure–volume work on the surroundings. The net amount of energy removed from the surroundings — that is, the heat given up minus the work done — must equal the increase in energy content of the system. As a specific example, consider the vaporization of a mole of liquid water at 373.15 K and 0.1 MPa. In the process 40,660 J of heat are absorbed from the surroundings. In the expansion, 3100 J of energy are added back into the surroundings by the work done in the expansion from liquid to gas. The net of these two processes exactly balances the 37,560 J of energy needed in the system to change the liquid into a gas.

The adiabatic expansion of a gas is an example of (b). In the reversible adiabatic expansion of one mole of an ideal monatomic gas, initially at 298.15 K, from a volume of 25 $dm^3$ to a final volume of 50 $dm^3$, 2343 J of energy are added into the surroundings from the work done in the expansion. Since no heat can be exchanged (in an adiabatic process, $q = 0$), the internal energy of the gas must decrease by 2343 J. As a result, the temperature of the gas falls to 188 K.

Finally, (c) is represented by any cyclic process occurring within an isolated system. With the system returning to its initial state, and with no opportunity for transfer of energy between system and surroundings, all energy changes in the process must add to zero.

The mathematical statement of the Second Law in terms of the universe is

$$\sum \Delta S > 0 \text{ (universe)},$$

which states that when all the entropy changes in the universe resulting from some natural process are added together, the sum is greater than zero. The four possible ways in which this can be accomplished are listed below:

| | System | + | $\Delta U$ Surroundings | = | Universe |
|---|---|---|---|---|---|
| (a) | + | | + | | + |
| (b) | − | | + | | + |
| (c) | + | | − | | + |
| (d) | + | | 0 | | + |

Processes represented by (a) are large entropy producers. Burning and decay reactions in which the energy released as heat is not utilized in producing work are of this type. For example, at a temperature of 298.15 K and a pressure of 0.1 MPa, the combustion of a mole of carbon to form carbon dioxide results in a $\Delta S$ in the system of 2.97 $J \cdot K^{-1} \cdot mol^{-1}$. In addition, heat from the reaction flows into the surroundings and increases its entropy by 1320 $J \cdot K^{-1} \cdot mol^{-1}$. The total increase in the universe is 1323 $J \cdot K^{-1} \cdot mol^{-1}$.

The spontaneous freezing of a supercooled liquid is an illustration of (b). When one mole of supercooled liquid water at 263.15 K freezes, the entropy of the water decreases by 20.63 $J \cdot K^{-1} \cdot mol^{-1}$. This decrease in entropy, and the spontaneous ordering of the water molecules, may at first seem to be a violation of the Second Law. However, when one takes into account the accompanying entropy change in the surroundings, the Second Law is still found to be valid. If the heat released from the solidification of one mole of liquid water enters the surroundings at 263.15 K, the entropy of the surroundings will increase by 21.38 $J \cdot K^{-1} \cdot mol^{-1}$. The total change in the universe caused by the solidification process is, therefore, positive:

$$-20.63 + 21.38 = 0.75 \ J \cdot K^{-1} \cdot mol^{-1}.$$

Phase changes are examples of (c). In order to melt a mole of ice in contact with air at 298.15 K, heat must flow into the system from the air. The increase in entropy of the system is 22.00 $J \cdot K^{-1} \cdot mol^{-1}$. The heat leaving the air decreases its entropy by 20.15 $J \cdot K^{-1} \cdot mol^{-1}$. The net change in the universe is once again positive:

$$22.00 - 20.15 = 1.85 \ J \cdot K^{-1} \cdot mol^{-1}.$$

Adiabatic processes are examples of (d). If a mole of ideal gas is allowed to expand adiabatically into an evacuated bulb to twice its initial volume, the entropy of the gas increases by 5.76 $J \cdot K^{-1} \cdot mol^{-1}$. No entropy change occurs in the surroundings, since there is no exchange of heat. Hence, 5.76 $J \cdot K^{-1} \cdot mol^{-1}$ is the net increase in entropy in the universe.

To summarize, in all kinds of naturally occurring processes for which the total energy and entropy changes in the universe can be calculated, energy is conserved and entropy is produced. The laws, formulated on the basis of experience, were proposed at about the same time. The First Law gained almost immediate acceptance, probably because it appeared reasonable and in accord with human intuition as to the permanence of nature. This was not so with the Second Law, however. Because of its far-reaching implication that every natural process permanently alters the universe, the Second Law was slower in receiving general acceptance.

The production of entropy as predicted by the Second Law might be looked upon as an effect working against the best interests of life. We have seen previously that entropy is related directly to disorder. Any naturally occurring process results in some permanent disorder in the universe. Thus, the Second Law predicts an inevitable state of complete disorder in the future.[nn] It is an interesting paradox that as humanity struggles to improve their surroundings by increasing the complexity of the environment, they speed up the decay of the universe.

The build-up of entropy in the universe affects the ease with which energy can be converted into work. Combining the equations

$$\mathrm{d}U = \delta q + \delta w$$

and

$$\delta q = T\,\mathrm{d}S$$

enables one to write

$$\delta w = \mathrm{d}U - T\,\mathrm{d}S,$$

which states that the work obtainable from an amount of energy $\mathrm{d}U$ is limited by the size of the entropy change $\mathrm{d}S$. Clausius looked upon the increase in entropy of the universe as a process of degrading energy, and he envisioned the demise of the universe occurring when a state of maximum entropy is achieved. He referred to this condition as the heat death (Tod Wärme). In this state the same amount of energy would be present in the universe, but none of it would be available for work. In other words, the universe simply "runs down" and eventually everything, animate and inanimate alike, comes to a stop.[oo]

---

[nn] A popular version of the second law: "If you think things are mixed up now, just wait awhile."

[oo] A pessimistic view of this final state is given by T. S. Eliot in "The Hollow Men", Reprinted by permission of Harcourt Brace Jovanovich, Inc., and Faber and Faber Ltd.

> This is the way the world ends
> This is the way the world ends
> This is the way the world ends
> Not with a bang, but a whimper.

A more optimistic view is that of Robert Frost in "Fire and Ice". Reprinted from "The Poetry of Robert Frost", edited by Edward Connery Lathem, with permission from the Estate of Robert Frost, Henry Holt and Co. and Jonathan Cape.

> Some say the world will end in fire,
> Some say in ice.
> From what I've tasted of desire
> I hold with those who favor fire.
> But if I had to perish twice,
> I think I know enough of hate
> To say that for destruction ice
> Is also great
> And would suffice.

## 2.3a The Laws of Thermodynamics and Cyclic Engines

As we have seen, the earliest statements of the Laws of Thermodynamics were not in terms of $U$ and $S$. The formulation of the laws was to some extent a by-product of the early interest in machines and perpetual motion. These statements (that we have referred to in Section 2.2b) generally described the impossibility of accomplishing certain cyclic processes. The First Law denied the possibility of creating energy in a cyclic process that returned both system and surroundings to their initial state (perpetual motion of the first kind). The Second Law denied the possibility of converting heat completely into work in a cyclic process (perpetual motion of the second kind). Perpetual motion machines that would allow a ship to extract heat from the ocean, convert the heat into the work of running a propeller, and finally return the energy to the ocean via heat from the friction of the propeller (none of which violates the First Law) have long intrigued some people (those who are unaware of the Laws of Thermodynamics). All schemes to produce perpetual motion of either variety, (a) cyclic processes that create their own energy or (b) cyclic processes that attempt to exchange heat with only one reservoir, have failed without exception.[pp]

As we have seen, statements of the laws in terms of the conversion of heat into work in a cyclic engine are equivalent to the statements involving energy and entropy. The conversion from one to the other is not always obvious, but the relationships predicting the efficiency of the conversion of heat into work can be derived from the $U$ and $S$ statements.

Consider a process shown in Figure 2.14 in which a quantity of heat $q_2$ flows from a high temperature reservoir $T_2$ into a reversible cyclic engine. Part of the heat is converted into work $w$ while the remainder $q_1$ flows into a low temperature reservoir. The efficiency of this process is defined as[qq] $\eta = -w/q_2$, and it is a function of the reservoir temperatures. The relationship can be derived as follows:

Consider as the universe, the cyclic engine plus the two heat reservoirs for which

$$\sum \Delta U = 0.$$

---

[pp] Popular versions of the laws are the statements:

First Law: "You can't get something for nothing."
Second Law: "You not only can't get something for nothing, you can't even break even."

[qq] The minus sign is included in the definition of $\eta$ since $w$ is negative and we want the efficiency to be positive.

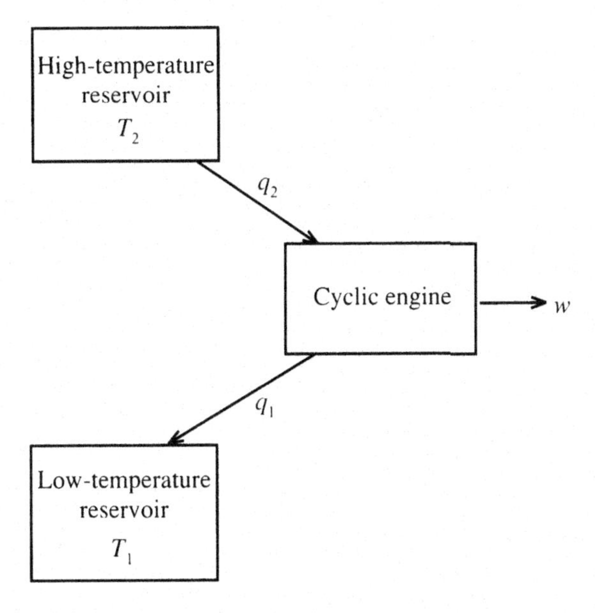


**Figure 2.14** A quantity of heat $q_2$ flows into a cyclic engine from a heat reservoir at a high temperature $T_2$. Part of the heat is converted to work $w$ and the remainder flows into a low-temperature heat reservoir at temperature $T_1$.

Since the process is reversible, for the universe

$$\sum \Delta S = 0.$$[rr]

Since the engine is cyclic, no net change occurs in it so that for the universe

$$\sum \Delta U = q_1 + q_2 + w = 0. \tag{2.80}$$

The reservoirs are at constant temperature in which case

$$\Delta S_2 = \int \frac{\delta q_2}{T_2} = \frac{q_2}{T_2}$$

$$\Delta S_1 = \int \frac{\delta q_1}{T_1} = \frac{q_1}{T_1}.$$

[rr] For the reversible process, $\sum \Delta S = 0$ rather than $\sum \Delta S > 0$.

Adding $\Delta S_1$ and $\Delta S_2$ gives

$$\Delta S = \Delta S_1 + \Delta S_2 = \frac{q_2}{T_2} + \frac{q_1}{T_1} = 0. \tag{2.81}$$

Combining equations (2.80) and (2.81) gives

$$\frac{-w}{q_2} = \eta = \frac{T_2 - T_1}{T_2}. \tag{2.82}$$

Still another statement of the second law is the **Clausius inequality** which states that

$$\oint \frac{\delta q}{T} \leqslant 0 \tag{2.83}$$

for any cyclic process, where the equality applies to the reversible process and the inequality to the natural or spontaneous process.

Equation (2.83) modifies equation (2.82) by adding the inequality so that

$$\eta \leqslant \frac{T_2 - T_1}{T_2}. \tag{2.84}$$

The reversible process (for which the equal sign applies) gives the maximum efficiency for the conversion of heat into work, but even the reversible engine is limited in the extent to which heat can be converted into work.

**Example 2.4:** A typical 500 megawatt coal-fired power plant burns about 200 tons of bituminous coal per hour ($50\ \text{kg} \cdot \text{s}^{-1}$) with a heating value of 12,500 $\text{btu} \cdot \text{lb}^{-1}$ ($28{,}700\ \text{kJ} \cdot \text{kg}^{-1}$). The high-temperature boiler operates at 538 °C while the heat exchanger cools the condensed steam to about 50 °C. Calculate the theoretical (maximum) efficiency of the process and compare with the efficiency that is actually achieved.

**Solution:** The theoretical efficiency is obtained from equation (2.82)

$$\eta = \frac{T_2 - \mathrm{T}_1}{T_2}$$

$$T_2 = 538 + 273 = 811\ \text{K}$$

$$T_1 = 50 + 273 = 323 \text{ K}$$

$$\eta = \frac{811 - 323}{811} = 0.60$$

or 60% efficiency.

The actual efficiency is obtained from $w = 500$ megawatts $= 500 \text{ MJ} \cdot \text{sec}^{-1}$

$$\begin{aligned} q_2 &= (50 \text{ kg} \cdot \text{sec}^{-1})(28{,}700 \text{ kJ} \cdot \text{kg}^{-1}) \\ &= (1.44 \times 10^6 \text{ kJ} \cdot \text{sec}^{-1})(10^{-3} \text{ MJ} \cdot \text{kJ}^{-1}) \\ &= 1.44 \times 10^3 \text{ MJ} \cdot \text{sec}^{-1}. \end{aligned}$$

$$\text{Actual efficiency} = \frac{500 \text{ MJ} \cdot \text{sec}^{-1}}{1.44 \times 10^3 \text{ MJ} \cdot \text{sec}^{-1}} = 0.35$$

or 35% efficient.

This practical problem is of the type that engineers worry about a lot. It demonstrates the difficulty of converting heat into work and the efficiency of actual processes compared with the theoretical efficiency.

## Exercises

E2.1 Find the work in joules performed in each of the following processes:

(a) Two moles of an ideal gas at 298.15 K and initially at a pressure of 0.10 MPa are expanded by suddenly lowering the external pressure to 0.050 MPa and letting the gas expand against this reduced pressure until equilibrium is established. Assume that the final temperature is 298.15 K.

(b) The same gas under the same initial conditions as in part (a) is expanded isothermally and reversibly at 298.15 K until the pressure is 0.050 MPa.

Compare the results of the two cases.

E2.2 Two moles of hydrogen gas are produced at 298.15 K and a pressure of 1 atm (0.101 MPa) by the action of acid on a metal. Calculate the work done by the gas in pushing back the atmosphere. (Assume $H_2$ is an ideal gas.)

E2.3 One mole of $SF_6$ gas is expanded isothermally and reversibly at $T = 300$ K from an initial volume of 30 $\text{dm}^3$ to a final volume of 90 $\text{dm}^3$.

Calculate the work done by the gas in the expansion assuming:
(a) The $SF_6$ (g) obeys the ideal gas equation.
(b) The $SF_6$ (g) obeys the van der Waals equation of state (see Appendix 3). The van der Waals constants for $SF_6$ can be obtained from the critical constants, which are $T_c = 318.7$ K, $p_c = 3.76$ MPa, and $V_c = 198\ \mathrm{cm^3 \cdot mol^{-1}}$. The relationships for a van der Waals gas are $T_c = 8a/27bR$, $p_c = a/27b^2$, and $V_c = 3b$.

E2.4 Helium is a monatomic gas with $C_{V,\mathrm{m}} = \frac{3}{2}R$ and $C_{p,\mathrm{m}} = \frac{5}{2}R$ (ideal gas behavior). Calculate the amount of heat that must be added to one mole of He(g) to change the temperature from 100 K to 1000 K in:
(a) an isochoric process
(b) an isobaric process.

E2.5 Using the constants given in Table 2.1, we can summarize the heat capacity ($\mathrm{J \cdot K \cdot mol^{-1}}$) of elemental silicon by the expressions:

$$\text{solid: } C_{p,\mathrm{m}} = 22.824 + 3.858 \times 10^{-3}\,\mathrm{T} - \frac{3.54 \times 10^5}{T^2} \qquad (298.15\ \mathrm{K} \leqslant T \leqslant 1683\ \mathrm{K})$$

$$\text{liquid: } C_{p,\mathrm{m}} = 27.2 \qquad (1683\ \mathrm{K} \leqslant T \leqslant 3492\ \mathrm{K})$$

The melting temperature of Si is 1683 K where the enthalpy of fusion is $46.4\ \mathrm{kJ \cdot mol^{-1}}$. Calculate the amount of heat that must be added to isobarically raise the temperature of one mole of silicon from 298.15 K to 3000 K.

E2.6 For the compound $CHClF_2$,

$$C_{V,\mathrm{m}} = 43.68 + 0.0962(T - 273.15)$$

where $C_{V,\mathrm{m}}$ is expressed in $\mathrm{J \cdot K^{-1} \cdot mol^{-1}}$ and $T$ in Kelvins. This relationship holds from $T = 298$ K to 413 K. Calculate the change in internal energy $\Delta U_{\mathrm{m}}$ of the gas in going from 300 K to 400 K (assume ideal gas behavior).

E2.7 One mole of an ideal gas with $C_{p,\mathrm{m}} = \frac{5}{2}R$ is expanded isothermally and reversibly at $T = 300$ K from a volume of 10 dm$^3$ to a volume of 20 dm$^3$. The gas is then cooled reversibly at constant pressure until its volume is again 10 dm$^3$. What are the net values of $\Delta U_{\mathrm{m}}$, $q$ and $w$ for the overall process?

E2.8 Two moles of an ideal monatomic gas with $C_{V,\mathrm{m}} = 12.6\ \mathrm{J \cdot K^{-1} \cdot mol^{-1}}$, initially at $p = 0.6$ MPa and $T = 300$ K, are placed in contact with a

large heat reservoir at T = 290 K and expanded against a constant pressure of 0.2 MPa. The temperature of the gas drops to 290 K.
(a) Find $q$ and $w$ for the gas.
(b) Find $\Delta U$ and $\Delta H$ for the gas.

E2.9 One mole of a monatomic ideal gas ($C_{V,\mathrm{m}} = \frac{3}{2} R$, $C_{p,\mathrm{m}} = \frac{5}{2} R$) is changed in state as follows:

| $p_1 = 101.3$ kPa, $T_1 = 273.1$ K | → | $p_2 = 50.6$ kPa, $T_2 = 546.2$ K |
|---|---|---|

The change may be brought about in an infinite variety of ways. Consider for this problem the following two reversible paths (I and II), each consisting of two parts (1 and 2):

(I) (1) An isothermal expansion followed by (2) an isobaric temperature rise.
(II) (1) An isobaric expansion followed by (2) an isochoric temperature decrease.

(a) Determine $p$, $V_\mathrm{m}$, and $T$ for 1 mole of the gas for each of the steps of each of the two paths. Plot the paths on a $p$–$V$ diagram.
(b) Calculate $w$, $q$, and $\Delta U$ for each part [(1) and (2)] of each path [(I) and II)] and for the total paths (I) and (II). Tabulate and compare your results.

E2.10 Two moles of an ideal monatomic gas ($C_{V,\mathrm{m}} = \frac{3}{2} R$) initially at $p = 101$ kPa and $T = 300$ K are put through the following cycle, all stages of which are reversible: (i) an isothermal compression to 202 kPa, (ii) an isobaric temperature increase to 400 K, (iii) a return to the initial state by the path $p = a + bT$, where $a$ and $b$ are constants. Sketch the cycle on a $p$–$T$ plot and evaluate numerically $\Delta U$, $q$, and $w$ for the working substance for each stage of the cycle and for the total process.

E2.11 Calculate $w$ and $\Delta U_\mathrm{m}$ for the fusion of 18.02 g of ice at 273.15 K and 0.101 MPa. At this temperature and pressure, the densities of ice and liquid water are 0.917 $\mathrm{g \cdot cm^{-3}}$ and 1.00 $\mathrm{g \cdot cm^{-3}}$, respectively. For the fusion process, $q = 6009.5\ \mathrm{J \cdot mol^{-1}}$ (see Table 2.2).

E2.12 Calculate $\Delta S_\mathrm{mix}$ for the mixing of 0.25 moles of $D_2$ (deuterium gas) with 0.75 moles of $H_2$ at a total constant pressure of 100 kPa. Assume ideal gas behavior.

E2.13 (a) 18.02 g of liquid water are enclosed under a frictionless weightless piston at a temperature of 373.15 K and a pressure of 0.101 MPa. The pressure above the piston is lowered slightly and the water allowed to vaporize isothermally until all is vaporized. For this process, $q = 40{,}656\ \mathrm{J \cdot mol^{-1}}$. The specific volume of liquid water at 373.15 K is $1.043\ \mathrm{cm^3 \cdot g^{-1}}$, and the specific volume of steam is $1677\ \mathrm{cm^3 \cdot g^{-1}}$ at $T = 373.15$ K and $p = 0.101$ MPa. Calculate the work $w$ attending this vaporization and $\Delta U$ for the process. (b) Find the values of $\Delta U$, $w$ and $q$ for the process where the piston is just removed and the water is allowed to vaporize freely (isothermally) into an evacuated space of such volume that the pressure has built up to 0.101 MPa with all of the water just vaporized.

E2.14 A flask initially containing benzene ($C_6H_6$) at its freezing temperature of 278.6 K is brought into contact with an ice-water bath until 1 mole of benzene has frozen. The heat that must be added to melt the benzene is $9832\ \mathrm{J \cdot mol^{-1}}$ (see Table 2.2). For the process, calculate the decrease in the entropy of the benzene and the net increase in entropy for the combined system (universe).

E2.15 One mole of ideal gas is expanded into an evacuated bulb of such a size that the volume of the gas is doubled.

(a) Calculate $q$, $w$, $\Delta S_{\mathrm{m}}$, and $\Delta U_{\mathrm{m}}$ for this process.

(b) Calculate the entropy change in the universe for this process to show that it is greater than zero for this spontaneous process.

E2.16 For an ideal gas under reversible conditions, a differential element of heat can be expressed in the form

$$\delta q = \frac{V_{\mathrm{m}}\,\mathrm{d}P}{\gamma - 1} + \frac{\gamma}{\gamma - 1} p\,\mathrm{d}V_{\mathrm{m}}$$

where $\gamma = C_{p,\mathrm{m}}/C_{V,\mathrm{m}}$.

(a) Assume $\gamma$ is constant and show that this expression for $\delta q$ is inexact.

(b) By solving the Pfaff differential equation, $\delta q = 0$, obtain the familiar expression for a reversible, adiabatic path:

$$pV_{\mathrm{m}}^{\gamma} = \text{constant}.$$

E2.17 A small hot spring at 373.15 K runs into a large river at 293.15 K at the rate of $100\ \mathrm{dm^3 \cdot min^{-1}}$. Assume the density of $H_2O$ at 373.15 K = $0.95\ \mathrm{g \cdot cm^{-3}}$ and $c_p$ of liquid $H_2O$ is $4.0\ \mathrm{J \cdot g^{-1}}$ and calculate the total entropy change per minute in the universe resulting from this process.

E2.18 A spring is placed in a large thermostat at 300 K and stretched isothermally and reversibly from its equilibrium length $L_0$ to a length of $10\ L_0$. During this reversible stretching 1.00 J of heat is absorbed by the spring. The stretched spring, still in the constant temperature thermostat, is then released without any restraining back-tension and allowed to return to its initial length $L_0$. During this spontaneous process the spring evolves 2.50 J of heat.

(a) What is the entropy change for the stretching of the spring?

(b) What is the entropy change for the collapse of the spring?

(c) What is the entropy change for the universe (spring plus surrounding thermostat) for the total process of stretching plus the return to the initial $L_0$?

(d) How much work was done on the spring in the stretching process?

E2.19 In a heat pump, a quantity of heat $q_1$ is removed from a low temperature reservoir at a temperature $T_1$ and added to a high-temperature reservoir at a temperature $T_2$.

(a) Show that the work required to pump the heat is given by

$$w = q_1 \frac{T_2 - T_1}{T_1}.$$

(b) Assume that the total cost of operation of a heat pump is four times the theoretical power cost for perfect efficiency, whereas the cost of direct electrical heating is just the power cost. If room temperature is 300 K, at what outdoor temperature would the two systems yield the same total cost?

## Problems

P2.1 Given the following specific heat and density information for $H_2O$ (l) at $p = 0.101$ MPa:

| $(T - 273.15)/(\text{K})$ | $c_p/(\text{J} \cdot \text{K}^{-1} \cdot \text{g}^{-1})$ | $\rho/(\text{kg} \cdot \text{m}^{-3})$ |
|---|---|---|
| 0 | 4.2177 | 999.84 |
| 10 | 4.1922 | 999.70 |
| 20 | 4.1819 | 998.20 |
| 30 | 4.1785 | 995.65 |
| 40 | 4.1786 | 992.22 |
| 50 | 4.1807 | 988.04 |
| 60 | 4.1844 | 983.20 |
| 70 | 4.1896 | 977.77 |
| 80 | 4.1964 | 971.80 |
| 90 | 4.2051 | 965.32 |
| 100 | 4.2160 | 958.36 |

Calculate $q$, $w$, $\Delta U_m$, $\Delta H_m$, and $\Delta S_m$, when one mole of $H_2O$ (l) is heated from 273.15 K to 373.15 K at an atmospheric pressure of 0.101 MPa.

P2.2 For the spontaneous process:

$$H_2O\ (l,\ 263.2\ K,\ 0.101\ MPa) = H_2O\ (s,\ 263.2\ K,\ 0.101\ MPa).$$

(a) Choose a reversible path and calculate $\Delta H_m$ and $\Delta S_m$, given that $C_{p,m}$ (s) = 38 $J \cdot K^{-1} \cdot mol^{-1}$, $C_{p,m}$ (l) = 76 $J \cdot K^{-1} \cdot mol^{-1}$, and $\Delta_{fus}H_m$ = 6009.5 $J \cdot mol^{-1}$ for water at 273.2 K.

(b) Assume that the surroundings are at $T$ = 263.2 K and calculate $\Delta S$ (universe) resulting from this process.

P2.3 In the temperature range from 100 K to 300 K, $C_{p,m}$ for $O_2(g)$ is given by

$$C_{p,m}/(J \cdot K^{-1} \cdot mol^{-1}) = 28.966 + 1.40 \times 10^{-3}\ T$$

where $T$ is the Kelvin temperature. The enthalpy of vaporization at the normal boiling point of 90.19 K is 6820 $J \cdot mol^{-1}$ (see Table 2.2).

(a) Use this information to calculate $\Delta S_m$ for the process at $p$ = 101 kPa:

$$O_2(g,\ 298.15\ K) = O_2(l,\ 90.19\ K).$$

(b) Calculate the minimum amount of work required to convert 1 mole of oxygen gas at 298.15 K and 101 kPa to liquid oxygen at 90.19 K and 101 kPa, in a reversible process in which heat is transferred to a heat reservoir at 298.15 K. Note that the entropy decrease of the oxygen was calculated in (a) and that the entropy increase of the heat reservoir must be equal so that the net entropy change will be zero. The First Law may then be used to calculate the work.

## References


1. For interesting discussions of the laws of thermodynamics, see J. R. Goates and J. B. Ott, *Chemical Thermodynamics: An Introduction*, Harcourt Brace, Jovanovich, Inc., New York, 1971; R. Battino and S. E. Wood, *Thermodynamics: An Introduction*, Academic Press, New York, 1968; P. A. Rock, *Chemical Thermodynamics*, University Science Books, Mill Valley, California, 1983; H. A. Bent, *The Second Law: An Introduction to Classical and Statistical Thermodynamics*, Oxford University Press, New York, 1965.
2. See, for example: R. C. Reid, J. M. Prausnitz, and B. E. Poling, "The Properties of Gases and Liquids", Fourth Edition, McGraw-Hill, Inc., New York, 1987.
3. Table taken from I. Barin and O. Knacke, "Thermochemical Properties of Inorganic Substances", Springer-Verlag, Berlin, 1973.

4. See, for example: Jui Sheng Hsieh, *Principles of Thermodynamics*, Scripta Book Company, Washington, 1975; Kenneth Wark, *Thermodynamics*, Fourth Edition, McGraw Hill Book Company, New York, 1983; James Coull and Edward B. Stuart, *Equilibrium Thermodynamics*, John Wiley & Sons, Inc., New York, 1964.

Chapter 3

# Thermodynamic Relationships and Applications

In Chapter 1 we gave the defining equations for enthalpy, Helmholtz free energy, and Gibbs free energy:

$$H = U + pV \tag{3.1}$$

$$A = U - TS \tag{3.2}$$

$$G = U + pV - TS = H - TS. \tag{3.3}$$

In Chapter 2 we used the laws of thermodynamics to write equations that relate internal energy and entropy to heat and work.

$$\mathrm{d}U = \delta q + \delta w \tag{3.4}$$

with

$$\delta q \leqslant T\,\mathrm{d}S \tag{3.5}$$

$$\delta w \geqslant -p\,\mathrm{d}V \tag{3.6}$$

where the inequality applies to the spontaneous process and the equality to the reversible process.

These equations can be used to derive the four fundamental equations of Gibbs and then the 50,000,000 equations alluded to in Chapter 1 that relate $p$, $T$, $V$, $U$, $S$, $H$, $A$, and $G$. We should keep in mind that these equations apply to a reversible process involving pressure–volume work only. This limitation does not restrict their usefulness, however. Since all of the thermodynamic variables are state functions, calculation of $\Delta Z$ ($Z$ is any of these variables) by a reversible path between two states gives the same value as would be obtained for all other paths between those states. When other forms of work are involved, additions can be made to the equations to account for the additional work. The result is that thermodynamic equations can be derived to describe the effect of such things as surface area, electrical field, gravitational field, and magnetic field, on the thermodynamic variables.

## 3.1 The Gibbs Equations

Combining equations (3.4), (3.5), and (3.6), with the equal sign (reversible process) used in equations (3.5) and (3.6), gives

$$dU = T\,dS - p\,dV. \tag{3.7}$$

This is the first Gibbs equation.

Next, $d(pV)$ is added to both sides of equation (3.7) to give

$$dU + d(pV) = T\,dS - p\,dV + d(pV). \tag{3.8}$$

On the left side of the equation, we recognize that $dU + d(pV) = dH$. On the right side, we expand $d(pV)$ to $(p\,dV + V\,dp)$. After cancelling terms we get

$$dH = T\,dS + V\,dp. \tag{3.9}$$

This is the second Gibbs equation.

We now return to equation (3.7) and subtract $d(TS)$ from both sides. After recognizing that $dU - d(TS) = dA$ and expanding $d(TS)$ to $(T\,dS + S\,dT)$ and cancelling terms, we get

$$dA = -S\,dT - p\,dV. \tag{3.10}$$

This is the third Gibbs equation.

Finally, we add $d(pV)$ and subtract $d(TS)$ from both sides of equation (3.7). Substituting $dU + d(pV) - d(TS) = dG$, expanding $d(pV)$ and $d(TS)$, and cancelling terms gives

$$dG = -S\,dT + V\,dp. \tag{3.11}$$

This is the fourth Gibbs equation.

In summary:

$$dU = T\,dS - p\,dV \tag{3.7}$$

$$dH = T\,dS + V\,dp \tag{3.9}$$

$$dA = -S\,dT - p\,dV \tag{3.10}$$

$$dG = -S\,dT + V\,dp. \tag{3.11}$$

The importance of these four equations cannot be overemphasized. They are total differentials for $U$ as $f(S, V)$, $H$ as $f(S, p)$, $A$ as $f(V, T)$, and $G$ as $f(p, T)$. Although they were derived assuming a reversible process, as total differentials they apply to both reversible and irreversible processes. They are the starting points for the derivation of general differential expressions in which we express $U$, $H$, $A$ and $G$ as a function of $p$, $V$, $T$, $C_p$ and $C_V$.[a] These are the relationships that we will now derive.

## 3.2 Partial Differential Relationships

We want to express the partial derivatives

$$\left(\frac{\partial Z}{\partial T}\right)_V, \left(\frac{\partial Z}{\partial T}\right)_p, \left(\frac{\partial Z}{\partial V}\right)_T, \left(\frac{\partial Z}{\partial p}\right)_T, \left(\frac{\partial Z}{\partial V}\right)_p, \text{ and } \left(\frac{\partial Z}{\partial p}\right)_V$$

as functions of $p$, $V$, $T$, $C_V$ and $C_p$, where $Z$ is $U$, $S$, $H$, $A$ and $G$. We already have a start. In Chapter 2 we derived the equations

$$\left(\frac{\partial U}{\partial T}\right)_V = C_V \tag{3.12}$$

$$\left(\frac{\partial H}{\partial T}\right)_p = C_p \tag{3.13}$$

$$\left(\frac{\partial S}{\partial T}\right)_V = \frac{C_V}{T} \tag{3.14}$$

$$\left(\frac{\partial S}{\partial T}\right)_p = \frac{C_p}{T}. \tag{3.15}$$

[a] We will find that in several cases, we cannot eliminate $S$ from the equation. We handle the problem by deriving alternate expressions involving $H$ and $U$ instead of $S$.

The next two equations we derive are obtained as the Maxwell relationships[b] for equations (3.10) and (3.11)

$$\left(\frac{\partial S}{\partial V}\right)_T = \left(\frac{\partial p}{\partial T}\right)_V \tag{3.16}$$

$$\left(\frac{\partial S}{\partial p}\right)_T = -\left(\frac{\partial V}{\partial T}\right)_p. \tag{3.17}$$

Next we derive the equations for

$$\left(\frac{\partial U}{\partial T}\right)_p, \left(\frac{\partial U}{\partial V}\right)_T \quad \text{and} \quad \left(\frac{\partial U}{\partial p}\right)_T.$$

In each case we start with the first Gibbs equation

$$\mathrm{d}U = T\,\mathrm{d}S - p\,\mathrm{d}V. \tag{3.7}$$

To find the first of the three derivatives, we divide by $\mathrm{d}T$ to obtain

$$\frac{\mathrm{d}U}{\mathrm{d}T} = T\frac{\mathrm{d}S}{\mathrm{d}T} - p\frac{\mathrm{d}V}{\mathrm{d}T}. \tag{3.18}$$

Equation (3.18) is true for any condition, including the condition of constant $p$. Specifying constant $p$ gives

$$\left(\frac{\mathrm{d}U}{\mathrm{d}T}\right)_p = T\left(\frac{\mathrm{d}S}{\mathrm{d}T}\right)_p - p\left(\frac{\mathrm{d}V}{\mathrm{d}T}\right)_p. \tag{3.19}$$

We can substitute partial derivatives for the ratio of differentials in equation (3.19). For example,

$$\left(\frac{\mathrm{d}U}{\mathrm{d}T}\right)_p = \left(\frac{\partial U}{\partial T}\right)_p. \tag{3.20}$$

---

[b] The Maxwell relationships can be applied to the Gibbs equations because $\mathrm{d}U$, $\mathrm{d}H$, and $\mathrm{d}G$ are total differentials.

This can be shown as follows:

$$U = f(p, T)$$

so that

$$\mathrm{d}U = \left(\frac{\partial U}{\partial T}\right)_p \mathrm{d}T + \left(\frac{\partial U}{\partial p}\right)_T \mathrm{d}p.$$

If we specify constant $p$, then $\mathrm{d}p = 0$ so that

$$\mathrm{d}U = \left(\frac{\partial U}{\partial T}\right)_p \mathrm{d}T \qquad \text{(constant } p\text{)}.$$

Dividing by $\mathrm{d}T$ while indicating the condition of constant $p$ gives

$$\left(\frac{\mathrm{d}U}{\mathrm{d}T}\right)_p = \left(\frac{\partial U}{\partial T}\right)_p. \tag{3.21}$$

Henceforth, we will make substitutions of the type represented by equation (3.21) without proof.

Substituting equation (3.20) into equation (3.19) gives

$$\left(\frac{\partial U}{\partial T}\right)_p = T\left(\frac{\partial S}{\partial T}\right)_p - p\left(\frac{\partial V}{\partial T}\right)_p. \tag{3.22}$$

Substitution of equation (3.15) into equation (3.22) and simplifying gives the equation that we want

$$\left(\frac{\partial U}{\partial T}\right)_p = C_p - p\left(\frac{\partial V}{\partial T}\right)_p. \tag{3.23}$$

The expression for $(\partial U/\partial V)_T$ is obtained by starting with the same Gibbs equation, dividing by $\mathrm{d}V$ and specifying constant $T$. The result is

$$\left(\frac{\mathrm{d}U}{\mathrm{d}V}\right)_T = T\left(\frac{\mathrm{d}S}{\mathrm{d}V}\right)_T - p.$$

We next substitute the partial derivatives to get

$$\left(\frac{\partial U}{\partial V}\right)_T = T\left(\frac{\partial S}{\partial V}\right)_T - p. \tag{3.24}$$

Substituting equation (3.16) into equation (3.24) gives the final equation.

$$\left(\frac{\partial U}{\partial V}\right)_T = T\left(\frac{\partial p}{\partial T}\right)_V - p. \tag{3.25}$$

The expression for $(\partial U/\partial p)_T$ is also easily obtained. Again we start with the Gibbs equation involving $dU$:

$$dU = T\,dS - p\,dV.$$

Dividing by $dp$, specifying constant $T$, and substituting partial derivatives gives

$$\left(\frac{\partial U}{\partial p}\right)_T = T\left(\frac{\partial S}{\partial p}\right)_T - p\left(\frac{\partial V}{\partial p}\right)_T. \tag{3.26}$$

Substitution of equation (3.17) into equation (3.26) gives the final expression

$$\left(\frac{\partial U}{\partial p}\right)_T = -T\left(\frac{\partial V}{\partial T}\right)_p - p\left(\frac{\partial V}{\partial p}\right)_T. \tag{3.27}$$

We can start with the other Gibbs equations to obtain equivalent relationships involving $H$, $A$, and $G$.

**Example 3.1:** Use the second Gibbs equation to derive equations relating $(\partial H/\partial V)_T$ and $(\partial H/\partial p)_T$ to $p$, $V$, and $T$.

**Solution:** To obtain the expression for $(\partial H/\partial V)_T$ we start with the second Gibbs equation

$$dH = T\,dS + V\,dp.$$

Dividing by $\mathrm{d}V$, specifying constant $T$, and substituting partial derivatives gives

$$\left(\frac{\partial H}{\partial V}\right)_T = T\left(\frac{\partial S}{\partial V}\right)_T + V\left(\frac{\partial p}{\partial V}\right)_T. \tag{3.28}$$

Substitution of equation (3.16) into equation (3.28) gives

$$\left(\frac{\partial H}{\partial V}\right)_T = T\left(\frac{\partial p}{\partial T}\right)_V + V\left(\frac{\partial p}{\partial V}\right)_T. \tag{3.29}$$

The derivative $(\partial H/\partial p)_T$ is obtained next. We start with the second Gibbs equation, divide by $\mathrm{d}p$, specify constant $T$, substitute partial derivatives, and substitute the Maxwell relation given in equation (3.17). The result is

$$\left(\frac{\partial H}{\partial p}\right)_T = V - T\left(\frac{\partial V}{\partial T}\right)_p. \tag{3.30}$$

**Example 3.2:** Derive the relationships that give the effect of $p$ and $V$ on $A$ and $G$ with $T$ held constant.

**Solution:** We start with the third and fourth Gibbs equations. To get $(\partial A/\partial V)_T$ we write

$$\mathrm{d}A = -S\,\mathrm{d}T - p\,\mathrm{d}V.$$

Dividing by $\mathrm{d}V$ and specifying constant $T$ so that $\mathrm{d}T = 0$ gives

$$\left(\frac{\partial A}{\partial V}\right)_T = -p. \tag{3.31}$$

Dividing the Gibbs equation by $\mathrm{d}p$ and specifying constant $T$ gives

$$\left(\frac{\partial A}{\partial p}\right)_T = -p\left(\frac{\partial V}{\partial p}\right)_T. \tag{3.32}$$

Similar relationships can be obtained involving $G$ by starting with the fourth Gibbs equation. The results are

$$\left(\frac{\partial G}{\partial p}\right)_T = V \tag{3.33}$$

and

$$\left(\frac{\partial G}{\partial V}\right)_T = V\left(\frac{\partial p}{\partial V}\right)_T. \tag{3.34}$$

Finally, we want to find equations expressing the effect of temperature on $A$ and $G$ with $p$ or $V$ constant, but these derivatives present problems. For the expressions involving $A$, we start with

$$\mathrm{d}A = -S\,\mathrm{d}T - p\,\mathrm{d}V.$$

Dividing by $\mathrm{d}T$ and specifying constant $V$ gives

$$\left(\frac{\partial A}{\partial T}\right)_V = -S. \tag{3.35}$$

Dividing the Gibbs equation by $\mathrm{d}T$ and specifying constant $p$ gives

$$\left(\frac{\partial A}{\partial T}\right)_p = -S - p\left(\frac{\partial V}{\partial T}\right)_p. \tag{3.36}$$

For $G$, we start with the fourth Gibbs equation and obtain

$$\left(\frac{\partial G}{\partial T}\right)_p = -S \tag{3.37}$$

and

$$\left(\frac{\partial G}{\partial T}\right)_V = -S + V\left(\frac{\partial p}{\partial T}\right)_V. \tag{3.38}$$

### 3.2a The Gibbs–Helmholtz Equation

The temperature derivatives of $G$ and $A$ {equations (3.35) to (3.38)} can be seen to contain $S$. Because we usually do not know $S$ as a function of $p$, $V$, and $T$, these derivatives are of limited value. However, expressions can be derived for the temperature derivatives of $G/T$ and $A/T$ that do not involve $S$, and they can often be used in applications where the original derivative was desired. We derive the equation for $(\partial(G/T)/\partial T)_p$ to illustrate the process.

We start with

$$G = H - TS.$$

Dividing by $T$ and differentiating gives

$$\left(\frac{\partial(G/T)}{\partial T}\right)_p = \left(\frac{\partial(H/T)}{\partial T}\right)_p - \left(\frac{\partial S}{\partial T}\right)_p. \tag{3.39}$$

We can expand $(\partial(H/T)/\partial T)_p$ and substitute $C_p$ for $(\partial H/\partial T)_p$ to obtain

$$\left(\frac{\partial(H/T)}{\partial T}\right)_p = \frac{TC_p - H}{T^2} = \frac{C_p}{T} - \frac{H}{T^2}. \tag{3.40}$$

Substitution of equations (3.40) and (3.15) into equation (3.39) gives

$$\left(\frac{\partial(G/T)}{\partial T}\right)_p = -\frac{H}{T^2}. \tag{3.41}$$

Equation (3.42) is called the **Gibbs–Helmholtz equation**. We will find it to be a very useful relationship. A similar derivation would show that

$$\left(\frac{\partial(A/T)}{\partial T}\right)_V = -\frac{U}{T^2}. \tag{3.42}$$

The derivatives $(\partial Z/\partial p)_V$ and $(\partial Z/\partial V)_p$ are easily obtained by making use of the properties of the exact differential

$$\left(\frac{\partial Z}{\partial p}\right)_V = \left(\frac{\partial Z}{\partial T}\right)_V\left(\frac{\partial T}{\partial p}\right)_V \tag{3.43}$$

and

$$\left(\frac{\partial Z}{\partial V}\right)_p = \left(\frac{\partial Z}{\partial T}\right)_p \left(\frac{\partial T}{\partial V}\right)_p. \tag{3.44}$$

For example:

$$\left(\frac{\partial S}{\partial V}\right)_p = \left(\frac{\partial S}{\partial T}\right)_p \left(\frac{\partial T}{\partial V}\right)_p.$$

Substituting equation (3.15) into the above gives

$$\left(\frac{\partial S}{\partial V}\right)_p = \frac{C_p}{T}\left(\frac{\partial T}{\partial V}\right)_p. \tag{3.45}$$

Similar expressions for the other derivatives can be as easily derived. However, the equations with $p$ and $V$ as the independent variables are not as useful as those involving $p$ and $T$ or $V$ and $T$, and we will not take the time to obtain all of these relationships.

Table 3.1 summarizes the equations we have derived. Any of the equations of interest can be obtained from the table. However, it is almost as easy to remember the equations relating

$$\left(\frac{\partial S}{\partial T}\right)_V, \left(\frac{\partial S}{\partial T}\right)_p, \left(\frac{\partial U}{\partial T}\right)_V, \text{ and } \left(\frac{\partial H}{\partial T}\right)_p$$

to heat capacity (which are very simple relationships) and the four Gibbs equations and derive the other equations as needed.

## 3.2b Observations About the Differential Relationships

The relationships summarized in Table 3.1 can be used to solve many problems, as we will illustrate. But before we do so, it is useful to make three general observations about Table 3.1. First, we note that the relationships for $U$ and $A$ are simpler when $V$ and $T$ are chosen as the independent variables. Whenever possible, we will choose $V$ and $T$ as the independent variables for calculating $\Delta U$ and $\Delta A$. For $H$ and $G$, the relationships are simpler when $p$ and $T$ are the independent variables, and these are the ones we will generally choose to

**Table 3.1** Thermodynamic relationships

| $Z$ | $\left(\frac{\partial Z}{\partial T}\right)_V$ | $\left(\frac{\partial Z}{\partial T}\right)_p$ | $\left(\frac{\partial Z}{\partial V}\right)_T$ | $\left(\frac{\partial Z}{\partial p}\right)_T$ |
|---|---|---|---|---|
| $S$ | $\frac{C_V}{T}$ | $\frac{C_p}{T}$ | $\left(\frac{\partial p}{\partial T}\right)_V$ | $-\left(\frac{\partial V}{\partial T}\right)_p$ |
| $U$ | $C_V$ | $C_p - p\left(\frac{\partial V}{\partial T}\right)_p$ | $T\left(\frac{\partial p}{\partial T}\right)_V - p$ | $-T\left(\frac{\partial V}{\partial T}\right)_p - p\left(\frac{\partial V}{\partial p}\right)_T$ |
| $H$ | $C_V + V\left(\frac{\partial p}{\partial T}\right)_V$ | $C_p$ | $T\left(\frac{\partial p}{\partial T}\right)_V + V\left(\frac{\partial p}{\partial V}\right)_T$ | $V - T\left(\frac{\partial V}{\partial T}\right)_p$ |
| $A$ | $-S^*$ | $-S - p\left(\frac{\partial V}{\partial T}\right)_p$ | $-p$ | $-p\left(\frac{\partial V}{\partial p}\right)_T$ |
| $G$ | $-S + V\left(\frac{\partial p}{\partial T}\right)_V$ | $-S^*$ | $V\left(\frac{\partial p}{\partial V}\right)_T$ | $V$ |

* Alternative expressions are $\left(\frac{\partial(A/T)}{\partial T}\right)_V = -\frac{U}{T^2}$ and $\left(\frac{\partial(G/T)}{\partial T}\right)_p = -\frac{H}{T^2}$

calculate $\Delta H$ and $\Delta G$. For calculations of $\Delta S$, the choice of variables will depend upon the details of the process.

The second observation is that the equations we have derived for extensive variables $Z$ can be applied to the difference $\Delta Z$. For example:

$$\Delta H = H_2 - H_1.$$

Differentiating gives

$$\left(\frac{\partial \Delta H}{\partial T}\right)_p = \left(\frac{\partial H_2}{\partial T}\right)_p - \left(\frac{\partial H_1}{\partial T}\right)_p.$$

But

$$\left(\frac{\partial H_2}{\partial T}\right)_p = C_{p,2}$$

$$\left(\frac{\partial H_1}{\partial T}\right)_p = C_{p,1}.$$

Substitution gives

$$\left(\frac{\partial \Delta H}{\partial T}\right)_p = C_{p,2} - C_{p,1}$$

or

$$\left(\frac{\partial \Delta H}{\partial T}\right)_p = \Delta C_p. \tag{3.46}$$

As a second example:

$$\frac{\Delta G}{T} = \frac{G_2}{T} - \frac{G_1}{T}.$$

Differentiating gives

$$\left(\frac{\partial(\Delta G/T)}{\partial T}\right)_p = \left(\frac{\partial(G_2/T)}{\partial T}\right)_p - \left(\frac{\partial(G_1/T)}{\partial T}\right)_p.$$

Substitution of equation (3.41) gives

$$\left(\frac{\partial(\Delta G/T)}{\partial T}\right)_p = \left(-\frac{H_2}{T^2}\right) - \left(-\frac{H_1}{T^2}\right)$$

or

$$\left(\frac{\partial(\Delta G/T)}{\partial T}\right)_p = -\frac{\Delta H}{T^2}. \tag{3.47}$$

The third observation is that the equations apply as well when the extensive properties are replaced by the corresponding intensive molar properties.[c] For example, if

$$\left(\frac{\partial H}{\partial p}\right)_T = V - T\left(\frac{\partial V}{\partial T}\right)_p, \tag{3.30}$$

then

$$\left(\frac{\partial H_m}{\partial p}\right)_T = V_m - T\left(\frac{\partial V_m}{\partial T}\right)_p. \tag{3.48}$$

The substitutions can be made because the extensive thermodynamic variables in the equations are homogeneous of degree one.[d] Thus, dividing the equation by $n$ converts the extensive variable to the corresponding molar intensive variable. For example, to prove that equation (3.48) follows from equation

[c] In Chapter 5 we will see that the thermodynamic equations also apply to partial molar properties, which we will define at that time, and show that they represent the molar thermodynamic properties of components in solution.

[d] See the discussion of homogeneous functions in Appendix 1.

(3.30) we divide equation (3.30) by $n$ to obtain

$$\frac{1}{n}\left(\frac{\partial H}{\partial p}\right)_T = \frac{V}{n} - \frac{T}{n}\left(\frac{\partial V}{\partial T}\right)_p. \tag{3.49}$$

Since $n$ is constant in the equations we have derived, we can write equation (3.49) as

$$\left(\frac{\partial(H/n)}{\partial p}\right)_T = \frac{V}{n} - T\left(\frac{\partial(V/n)}{\partial T}\right)_p$$

or

$$\left(\frac{\partial H_m}{\partial p}\right)_T = V_m - T\left(\frac{\partial V_m}{\partial T}\right)_p.$$

## 3.3 Applications of the Differential Relationships

The relationships summarized in Table 3.1, expanded to include differences and molar properties, serve as the starting point for many useful thermodynamic calculations. An example is the calculation of $\Delta Z$ for a variety of processes in which $p$, $V$, and $T$ are changed.[e] For any of the extensive variables $Z = S, U, H, A$ or $G$, we can write

$$Z = f(V, T) \quad \text{(usually for } S, U \text{ or } A\text{)}$$

so that

$$\mathrm{d}Z = \left(\frac{\partial Z}{\partial V}\right)_T \mathrm{d}V + \left(\frac{\partial Z}{\partial T}\right)_V \mathrm{d}T. \tag{3.50}$$

Or

$$Z = f(p, T) \quad \text{(usually for } S, H \text{ or } G\text{)}$$

[e] We will see later in this chapter, that $S$ is a useful independent variable in an adiabatic process, while constant $H$ is associated with the Joule–Thomson expansion.

so that

$$dZ = \left(\frac{\partial Z}{\partial p}\right)_T dp + \left(\frac{\partial Z}{\partial T}\right)_p dT. \tag{3.51}$$

Integration gives

$$\Delta Z = \int_{V_1}^{V_2} \left(\frac{\partial Z}{\partial V}\right)_T dV + \int_{T_1}^{T_2} \left(\frac{\partial Z}{\partial T}\right)_V dT \tag{3.52}$$

or

$$\Delta Z = \int_{p_1}^{p_2} \left(\frac{\partial Z}{\partial p}\right)_T dp + \int_{T_1}^{T_2} \left(\frac{\partial Z}{\partial T}\right)_p dT. \tag{3.53}$$

We then substitute our differential expressions from Table 3.1 into equations (3.52) or (3.53) and integrate to get $\Delta Z$.

Usually, we use equations (3.52) or (3.53) under the condition where one variable is changed at a time. This leads to the three commonly encountered processes:

(a) An **isochoric process** in which $dV = 0$. Equation (3.52) becomes

$$\Delta Z = \int_{T_1}^{T_2} \left(\frac{\partial Z}{\partial T}\right)_V dT \quad \text{(isochoric process)}. \tag{3.54}$$

(b) An **isobaric process** in which $dp = 0$, so that equation (3.53) becomes

$$\Delta Z = \int_{T_1}^{T_2} \left(\frac{\partial Z}{\partial T}\right)_p dT \quad \text{(isobaric process)}. \tag{3.55}$$

(c) An **isothermal process** where $dT = 0$, and equations (3.52) and (3.53) become

$$\Delta Z = \int_{V_1}^{V_2} \left(\frac{\partial Z}{\partial V}\right)_T dV \quad \text{(isothermal process)} \tag{3.56}$$

or

$$\Delta Z = \int_{p_1}^{p_2} \left(\frac{\partial Z}{\partial p}\right)_T \mathrm{d}p \quad \text{(isothermal process).} \tag{3.57}$$

## 3.3a Examples of the Application of the Differential Relationships

The following examples demonstrate the usefulness of the differential relationships.

**Example 3.3:** Calculate $\Delta U$, $\Delta S$, $\Delta H$, and $\Delta G$ for the isothermal expansion of one mole of ideal gas at $T = 300$ K, from $p_1 = 0.100$ MPa to $p_2 = 0.200$ MPa.

**Solution:**

$$p = \frac{nRT}{V}.$$

Calculating $\Delta U$:

$$\Delta U = \int_{V_1}^{V_2} \left(\frac{\partial U}{\partial V}\right)_T \mathrm{d}V.$$

From Table 3.1

$$\left(\frac{\partial U}{\partial V}\right)_T = T\left(\frac{\partial p}{\partial T}\right)_V - p.$$

Differentiating the ideal gas equation gives

$$\left(\frac{\partial p}{\partial T}\right)_V = \frac{nR}{V}.$$

Combining gives

$$\left(\frac{\partial U}{\partial V}\right)_T = T\frac{nR}{V} - p$$

$$= \frac{nRT}{V} - \frac{nRT}{V} = 0.$$

$$\Delta U = 0.$$

Calculating $\Delta S$:

$$\Delta S = \int_{V_1}^{V_2} \left(\frac{\partial S}{\partial V}\right)_T \mathrm{d}V$$

$$\left(\frac{\partial S}{\partial V}\right)_T = \left(\frac{\partial p}{\partial T}\right)_V = \frac{nR}{V}$$

$$\Delta S = \int_{V_1}^{V_2} \frac{nR}{V}\,\mathrm{d}V = nR\ln\frac{V_2}{V_1}.$$

At constant $T$,

$$\frac{V_2}{V_1} = \frac{p_1}{p_2}.$$

Substitution gives

$$\Delta S = nR\ln\frac{p_1}{p_2}$$

$$= (1)(8.314)\ln\frac{0.100}{0.200}.$$

$$\Delta S = -5.762\ \mathrm{J{\cdot}K^{-1}}.$$

Calculating $\Delta H$:

$$\Delta H = \int_{p_1}^{p_2} \left(\frac{\partial H}{\partial p}\right)_T \mathrm{d}p$$

$$\left(\frac{\partial H}{\partial p}\right)_T = V - T\left(\frac{\partial V}{\partial T}\right)_p$$

$$V = \frac{nRT}{p}$$

$$\left(\frac{\partial V}{\partial T}\right)_p = \frac{nR}{p}$$

$$\left(\frac{\partial H}{\partial p}\right)_T = V - T\frac{nR}{p} = V - V.$$

$$\Delta H = 0.$$

Calculating $\Delta G$:

$$\Delta G = \int_{p_1}^{p_2} \left(\frac{\partial G}{\partial p}\right)_T \mathrm{d}p$$

$$\left(\frac{\partial G}{\partial p}\right)_T = V = \frac{nRT}{p}$$

$$\Delta G = \int_{p_1}^{p_2} \frac{nRT}{p} \mathrm{d}p$$

$$= nRT \ln \frac{p_2}{p_1}$$

$$= (1)(8.314)(300)\left(\ln \frac{0.200}{0.100}\right)$$

$$\Delta G = 1729 \text{ J}\cdot\text{K}^{-1}.$$

This example provides the proof that we promised earlier that $(\partial U/\partial V)_T = 0$ for the ideal gas. We also obtain the same equation for $\Delta S$ as we derived earlier using $\Delta S = q_{\text{rev}}/T$. It also shows that $\Delta H = 0$ for the isothermal expansion of ideal gas and gives a thermodynamically consistent value[f] for $\Delta G$. That is, in the final state

$$G_2 = H_2 - TS_2.$$

In the initial state

$$G_1 = H_1 - TS_1.$$

Subtracting gives

$$G_2 - G_1 = H_2 - H_1 - T(S_2 - S_1)$$

or

$$\begin{aligned}\Delta G &= \Delta H - T\Delta S \\ &= 0 - (300\ \text{K})(-5.762\ \text{J·K}^{-1}) \\ &= 1729\ \text{J}.\end{aligned}$$

**Example 3.4:** Prove that for the isothermal expansion of one mole of a van der Waals gas from $V_{\text{m},1}$ to $V_{\text{m},2}$,[g]

$$\Delta U_{\text{m}} = a\left(\frac{1}{V_{\text{m},1}} - \frac{1}{V_{\text{m},2}}\right). \tag{3.58}$$

**Solution:** The equation of state for the van der Waals gas is

$$\left(p + \frac{a}{V_{\text{m}}^2}\right)(\text{V}_{\text{m}} - b) = RT \tag{3.59}$$

[f] Thermodynamic calculations are frequently checked to see that they are thermodynamically consistent. That is, the calculation of a thermodynamic quantity using different thermodynamic relationships must give the same answer. We will see a number of examples of this as we calculate thermodynamic results.

[g] Appendix 3 describes various equations of state that apply to fluids, including the van der Waals equation.

where $a$ and $b$ are constants for a particular gas. Solving for $p$ gives

$$p = \frac{RT}{V_m - b} - \frac{a}{V_m^2}. \tag{3.60}$$

The change in internal energy is given by

$$\Delta U_m = \int_{V_1}^{V_2} \left(\frac{\partial U_m}{\partial V_m}\right)_T \mathrm{d}V_m \tag{3.61}$$

with

$$\left(\frac{\partial U_m}{\partial V_m}\right)_T = T\left(\frac{\partial p}{\partial T}\right)_{V_m} - p. \tag{3.62}$$

Differentiation of equation (3.60) gives

$$\left(\frac{\partial p}{\partial T}\right)_{V_m} = \frac{R}{V_m - b}. \tag{3.63}$$

Substitution of equations (3.60) and (3.63) into equation (3.62) gives

$$\left(\frac{\partial U_m}{\partial V_m}\right)_T = (T)\left(\frac{R}{V_m - b}\right) - \left(\frac{RT}{V_m - b} - \frac{a}{V_m^2}\right)$$

$$= \frac{a}{V_m^2}. \tag{3.64}$$

Substitution of equation (3.64) into equation (3.61) gives

$$\Delta U_m = \int_{V_1}^{V_2} \frac{a}{V_m^2} \mathrm{d}V_m.$$

Integration gives

$$\Delta U_{\mathrm{m}} = a\left(\frac{1}{V_{\mathrm{m},1}} - \frac{1}{V_{\mathrm{m},2}}\right). \tag{3.65}$$

**Example 3.5:** Calculate $\Delta H_{\mathrm{m}}$ and $\Delta S_{\mathrm{m}}$ for heating diamond at atmospheric pressure from 300 K to 1000 K. The heat capacity ($\mathrm{J \cdot K^{-1} \cdot mol^{-1}}$) of diamond in this temperature interval can be represented by the equation (see Table 2.1 in Chapter 2)

$$C_{p,\mathrm{m}}/(\mathrm{J \cdot K^{-1} \cdot mol^{-1}}) = 9.12 + 1.32 \times 10^{-2}\mathrm{T} - \frac{6.19 \times 10^5}{\mathrm{T}^2}.$$

**Solution:** For this isobaric process

$$\Delta H_{\mathrm{m}} = \int_{T_1}^{T_2} C_{p,\mathrm{m}}\, \mathrm{d}T$$

$$= \int_{T_1}^{T_2} \left(9.12 + 1.32 \times 10^{-2}T - \frac{6.19 \times 10^5}{T^2}\right) \mathrm{d}T$$

$$= 9.12(T_2 - T_1) + \frac{1.32 \times 10^{-2}}{2}(T_2^2 - T_1^2)$$

$$+ 6.19 \times 10^5 \left(\frac{1}{T_2} - \frac{1}{T_1}\right)$$

$$= 9.12(1000 - 300) + \frac{1.32 \times 10^{-2}}{2}((1000)^2 - (300)^2)$$

$$+ 6.19 \times 10^5 \left(\frac{1}{1000} - \frac{1}{300}\right).$$

$$\Delta H_{\mathrm{m}} = 1.09 \times 10^4\ \mathrm{J \cdot mol^{-1}}.$$

Next, we calculate $\Delta S_m$.

$$\Delta S_m = \int_{T_1}^{T_2} \frac{C_{p,m}}{\mathrm{T}}\, dT$$

$$= \int_{T_1}^{T_2} \left(9.12 + 1.32 \times 10^{-2} T - \frac{6.19 \times 10^5}{T^2}\right) \frac{dT}{T}$$

$$= \int_{T_1}^{T_2} \left(\frac{9.12}{T} + 1.32 \times 10^{-2} - \frac{6.19 \times 10^5}{T^3}\right) dT$$

$$= 9.12 \ln \frac{T_2}{T_1} + 1.32 \times 10^{-2}(T_2 - T_1) + \tfrac{1}{2}(6.19 \times 10^5)\left(\frac{1}{T_2^2} - \frac{1}{T_1^2}\right)$$

$$= 9.12 \ln \frac{1000}{300} + 1.32 \times 10^{-2}(1000 - 300)$$

$$+ \tfrac{1}{2}(6.19 \times 10^5)\left(\frac{1}{(1000)^2} - \frac{1}{(300)^2}\right)$$

$$= 17.09\ \mathrm{J{\cdot}K^{-1}{\cdot}mol^{-1}}.$$

**Example 3.6:** Calculate $\Delta H_m$ and $\Delta S_m$ for the process

| $H_2O$ (l)<br>$T = 263.15$ K<br>$p = 0.100$ MPa | → | $H_2O$ (s)<br>$T = 263.15$ K<br>$p = 0.100$ MPa |
|---|---|---|

Given[1] that at $T = 273.15$ K, $\Delta_{fus}H_m$, the molar enthalpy of fusion of ice, is 6010 $\mathrm{J{\cdot}mol^{-1}}$ (see Table 2.2), the heat capacity $C_{p,m}$ (s) of ice is 37.9 $\mathrm{J{\cdot}K^{-1}{\cdot}mol^{-1}}$ and the heat capacity $C_{p,m}$ (l) of liquid water is 75.9 $\mathrm{J{\cdot}K^{-1}{\cdot}mol^{-1}}$.

**Solution:** For the isobaric process

$$\mathrm{d}\Delta H_{\mathrm{m}} = \left(\frac{\partial \Delta H_{\mathrm{m}}}{\partial T}\right)_p \mathrm{d}T.$$

From Table 3.1

$$\left(\frac{\partial \Delta H_{\mathrm{m}}}{\partial T}\right)_p = \Delta C_{p,\mathrm{m}}.$$

Substitution and integration between $T_1$ and $T_2$ gives

$$\int_{\Delta H_{\mathrm{m},1}}^{\Delta H_{\mathrm{m},2}} \mathrm{d}\Delta H_{\mathrm{m}} = \int_{T_1}^{T_2} \Delta C_{p,\mathrm{m}}\,\mathrm{d}T \tag{3.66}$$

or

$$\Delta H_{\mathrm{m},2} - \Delta H_{\mathrm{m},1} = \int_{T_1}^{T_2} \Delta C_{p,\mathrm{m}}\,\mathrm{d}T$$

where

$$\Delta H_{\mathrm{m},1} = -\Delta_{\mathrm{fus}} H_{\mathrm{m}}.$$

The heat capacity difference

$$\Delta C_{p,\mathrm{m}} = C_{p,\mathrm{m}}(\mathrm{s}) - C_{p,\mathrm{m}}(\mathrm{l})$$

is calculated from

$$\Delta C_{p,\mathrm{m}} = 37.9\ \mathrm{J{\cdot}K^{-1}{\cdot}mol^{-1}} - 75.9\ \mathrm{J{\cdot}K^{-1}{\cdot}mol^{-1}}$$

$$= -38.0\ \mathrm{J{\cdot}K^{-1}{\cdot}mol^{-1}}$$

and is assumed to be constant at that value.
Integration of equation (3.66) with this value for $\Delta C_{p,\mathrm{m}}$ gives

$$\Delta H_{\mathrm{m},2} = \Delta H_{\mathrm{m},1} - 38.0(T_2 - T_1).$$

Substitution of $\Delta H_{\mathrm{m},1} = -6010\ \mathrm{J{\cdot}mol^{-1}}$, $T_1 = 273.15$ K, and $T_2 = 263.15$ K gives

$$\Delta H_{\mathrm{m},2} = -6010 - (38.0)(263.15 - 273.15)$$

$$= -5630\ \mathrm{J{\cdot}mol^{-1}}.$$

Similarly, for $\Delta S_{\mathrm{m}}$

$$\mathrm{d}\Delta S_{\mathrm{m}} = \left(\frac{\partial \Delta S_{\mathrm{m}}}{\partial T}\right)_p \mathrm{d}T$$

$$\left(\frac{\partial \Delta S_{\mathrm{m}}}{\partial T}\right)_p = \frac{\Delta C_{p,\mathrm{m}}}{T}.$$

Integration with constant $\Delta C_{p,\mathrm{m}}$ gives

$$\Delta S_{\mathrm{m},2} = \Delta S_{\mathrm{m},1} + \Delta C_{p,\mathrm{m}} \ln \frac{T_2}{T_1}. \tag{3.67}$$

The change from liquid to solid is a reversible process at $T = 273.15$ K so that

$$\Delta S_{\mathrm{m},1} = \frac{q_{\mathrm{rev}}}{T}.$$

Since $p$ is constant,

$$q_{\mathrm{rev}} = \Delta H_{\mathrm{m},1} = -\Delta_{\mathrm{fus}} H_{\mathrm{m}}$$

and

$$\Delta S_{\mathrm{m},1} = \frac{-6010\ \mathrm{J{\cdot}mol^{-1}}}{273.15\ \mathrm{K}} = -22.00\ \mathrm{J{\cdot}K^{-1}{\cdot}mol^{-1}}.$$

Substitution into equation (3.67) gives

$$\Delta S_{\mathrm{m},2} = -22.00 - 38.0 \ln \frac{263.15}{273.15}$$

$$= -20.58\ \mathrm{J{\cdot}K^{-1}{\cdot}mol^{-1}}.$$

The results obtained in Example 3.6 provide a verification of the Second Law. At $T = 263.15$ K, the change from liquid water to ice is spontaneous. Since the process is isobaric, $q_2 = \Delta H_2$, and hence,

$$\frac{q_2}{T_2} = \frac{\Delta H_{\mathrm{m},2}}{T_2} = \frac{-5630}{263.15} = -21.39\ \mathrm{J{\cdot}K^{-1}{\cdot}mol^{-1}}.$$

Comparing $\Delta S_\mathrm{m}$ and $q/T$, we see that, $\Delta S_\mathrm{m} > q/T$ at this temperature, as expected from the second law for a spontaneous process. Putting it another way, at $T = 263.15$ K, $\Delta S_\mathrm{m} = 20.58\ \mathrm{J{\cdot}K^{-1}{\cdot}mol^{-1}}$ for the system. A quantity of heat $q = 5630$ J flows into the surroundings and

$$\Delta S_\mathrm{surr} = \frac{5630}{263.15} = 21.39\ \mathrm{J{\cdot}K^{-1}{\cdot}mol^{-1}}.$$

Hence,

$$\begin{aligned}\Delta S_\mathrm{universe} &= \Delta S_\mathrm{system} + \Delta S_\mathrm{surr}\\ &= -20.58\ \mathrm{J{\cdot}K^{-1}{\cdot}mol^{-1}} + 21.39\ \mathrm{J{\cdot}K^{-1}{\cdot}mol^{-1}}\\ &= 0.81\ \mathrm{J{\cdot}K^{-1}{\cdot}mol^{-1}},\end{aligned}$$

and $\Delta S_\mathrm{universe} > 0$ for this spontaneous (natural) process.

### 3.3b Difference Between $C_p$ and $C_V$

The differential relationships can be used to derive a general relationship for the difference between $C_p$ and $C_V$.

**Example 3.7:** Prove that

$$C_{p,\mathrm{m}} - C_{V,\mathrm{m}} = \frac{\alpha^2 V_\mathrm{m} T}{\kappa} \tag{3.68}$$

where $\alpha$ is the coefficient of expansion and $\kappa$ is the compressibility,[h] and calculate $(C_{p,\mathrm{m}} - C_{V,\mathrm{m}})$ for liquid mercury at 293.15 K. Given that for

---

[h] Equation (3.68) applies rigorously only for isotropic materials. See Section 1.5a for definitions of $\alpha$ and $\kappa$.

Hg(l) at this temperature, $\alpha = 0.182 \times 10^{-3}\ \mathrm{K}^{-1}$, $\kappa = 0.40 \times 10^{-10}\ \mathrm{Pa}^{-1}$, and $V_\mathrm{m} = 14.76 \times 10^{-6}\ \mathrm{m}^3\cdot\mathrm{mol}^{-1}$.

**Solution:** From equation (2.29)

$$C_{p,\mathrm{m}} = C_{V,\mathrm{m}} + \left[p + \left(\frac{\partial U_\mathrm{m}}{\partial V_\mathrm{m}}\right)_T\right]\left(\frac{\partial V_\mathrm{m}}{\partial T}\right)_p.$$

But

$$\left(\frac{\partial U_\mathrm{m}}{\partial V_\mathrm{m}}\right)_T = T\left(\frac{\partial p}{\partial T}\right)_{V_\mathrm{m}} - p.$$

Substitution of the second equation into the first and cancelling terms gives

$$C_{p,\mathrm{m}} - C_{V,\mathrm{m}} = T\left(\frac{\partial p}{\partial T}\right)_{V_\mathrm{m}}\left(\frac{\partial V_\mathrm{m}}{\partial T}\right)_p.$$

From the properties of an exact differential [equation (1.28)] and the definitions of $\alpha$ and $\kappa$ [equations (1.40) and (1.39)] we get

$$\left(\frac{\partial p}{\partial T}\right)_{V_\mathrm{m}} = -\frac{\left(\dfrac{\partial V_\mathrm{m}}{\partial T}\right)_p}{\left(\dfrac{\partial V_\mathrm{m}}{\partial p}\right)_T} = \frac{\alpha V_\mathrm{m}}{\kappa V_\mathrm{m}} = \frac{\alpha}{\kappa}.$$

Combining gives

$$C_{p,\mathrm{m}} - C_{V,\mathrm{m}} = T\alpha V_\mathrm{m}\frac{\alpha}{\kappa}$$

or

$$C_{p,\mathrm{m}} - C_{V,\mathrm{m}} = \frac{\alpha^2 V_\mathrm{m} T}{\kappa}. \tag{3.68}$$

For Hg(l)

$$C_{p,\mathrm{m}} - C_{V,\mathrm{m}} = \frac{(0.182 \times 10^{-3})^2(14.76 \times 10^{-6})(293.15)}{(0.40 \times 10^{-10})}$$

$$= 3.58\ \mathrm{J{\cdot}K^{-1}{\cdot}mol^{-1}}.$$

For ideal gases, we have shown earlier that

$$C_{p,\mathrm{m}} - C_{V,\mathrm{m}} = R = 8.314\ \mathrm{J{\cdot}K^{-1}{\cdot}mol^{-1}}.$$

Solids and liquids have smaller $\alpha$, $\kappa$, and $V_{\mathrm{m}}$ than do gases, and the difference between $C_{p,\mathrm{m}}$ and $C_{V,\mathrm{m}}$ is smaller than for a gas.

### 3.3c The Reversible Adiabatic Expansion or Compression

In an adiabatic expansion or compression, the system is thermally isolated from the surroundings so that $q = 0$. If the change is reversible, we can derive a general relationship between $p$, $V$, and $T$, that can then be applied to a fluid (such as an ideal gas) by knowing the equation of state relating $p$, $V$, and $T$.

**Example 3.8:** Show that for the reversible adiabatic expansion of ideal gas with constant heat capacity

$$pV_{\mathrm{m}}^{C_p/C_V} = k_{\mathrm{ad}} \tag{3.69}$$

where $k_{\mathrm{ad}}$ is a constant.

**Solution:** In a reversible adiabatic expansion, $\delta q_{\mathrm{rev}} = T\,\mathrm{d}S = 0$. Thus, the process is isentropic, or one of constant entropy. To obtain an equation relating $p$, $V$ and $T$, we start with

$$p = f(S_{\mathrm{m}}, V_{\mathrm{m}}).$$

Then

$$\mathrm{d}p = \left(\frac{\partial p}{\partial V_{\mathrm{m}}}\right)_{S_{\mathrm{m}}} \mathrm{d}V_{\mathrm{m}} + \left(\frac{\partial p}{\partial S_{\mathrm{m}}}\right)_{V_{\mathrm{m}}} \mathrm{d}S_{\mathrm{m}}.$$

With $dS_m = 0$, we get

$$dp = \left(\frac{\partial p}{\partial V_m}\right)_{S_m} dV_m. \tag{3.70}$$

From the properties of an exact differential [see equation (1.37)], we get

$$\left(\frac{\partial p}{\partial V_m}\right)_{S_m} = -\frac{\left(\frac{\partial S_m}{\partial V_m}\right)_p}{\left(\frac{\partial S_m}{\partial p}\right)_{V_m}}$$

$$= -\frac{\left(\frac{\partial S_m}{\partial T}\right)_p \left(\frac{\partial T}{\partial V_m}\right)_p}{\left(\frac{\partial S_m}{\partial T}\right)_{V_m} \left(\frac{\partial T}{\partial p}\right)_{V_m}}.$$

Substituting

$$\left(\frac{\partial S_m}{\partial T}\right)_p = \frac{C_{p,m}}{T} \quad \text{and} \quad \left(\frac{\partial S_m}{\partial T}\right)_{V_m} = \frac{C_{V,m}}{T}$$

into the above equation gives

$$\left(\frac{\partial p}{\partial V_m}\right)_{S_m} = -\frac{C_{p,m}\left(\frac{\partial T}{\partial V_m}\right)_p}{C_{V,m}\left(\frac{\partial T}{\partial p}\right)_{V_m}}. \tag{3.71}$$

Substituting equation (3.71) into equation (3.70) gives

$$dp = -\frac{C_{p,\mathrm{m}}}{C_{V,\mathrm{m}}} \frac{\left(\dfrac{\partial T}{\partial V_\mathrm{m}}\right)_p}{\left(\dfrac{\partial T}{\partial p}\right)_{V_\mathrm{m}}} dV_\mathrm{m}. \tag{3.72}$$

Equation (3.72) is a general equation that relates a change in pressure to a change in volume for any reversible adiabatic process. It can be integrated by knowing how $C_{p,\mathrm{m}}$, $C_{V,\mathrm{m}}$, $(\partial T/\partial V_\mathrm{m})_p$, and $(\partial T/\partial p)_{V_\mathrm{m}}$ are related to $p$, $V_\mathrm{m}$, $T$. For an ideal gas

$$T = \frac{pV_\mathrm{m}}{R},$$

$$\left(\frac{\partial T}{\partial V_\mathrm{m}}\right)_p = \frac{p}{R},$$

$$\left(\frac{\partial T}{\partial p}\right)_{V_\mathrm{m}} = \frac{V_\mathrm{m}}{R}.$$

Substitution of these derivatives into equation (3.72) gives

$$\frac{dp}{p} = -\frac{C_{p,\mathrm{m}}}{C_{V,\mathrm{m}}} \frac{dV_\mathrm{m}}{V_\mathrm{m}}. \tag{3.73}$$

Integration assuming constant heat capacities gives

$$\ln p = -\frac{C_{p,\mathrm{m}}}{C_{V,\mathrm{m}}} \ln V_\mathrm{m} + \ln k_\mathrm{ad}$$

where $k_\mathrm{ad}$ is a constant of integration. Combining logarithms gives

$$\ln pV_\mathrm{m}^{C_{p,\mathrm{m}}/C_{V,\mathrm{m}}} = \ln k_\mathrm{ad}$$

or

$$pV_{\mathrm{m}}^{C_{p,\mathrm{m}}/C_{V,\mathrm{m}}} = k_{\mathrm{ad}}. \tag{3.69}$$

Alternate forms of equation (3.69) that involve $T$ can be obtained by substituting for $p$ or $V_{\mathrm{m}}$ in equation (3.69) and making use of the relationship for the ideal gas that $C_{p,\mathrm{m}} - C_{V,\mathrm{m}} = R$. The results are

$$TV_{\mathrm{m}}^{R/C_{V,\mathrm{m}}} = k'_{\mathrm{ad}} \tag{3.74}$$

$$\frac{T}{p^{R/C_{p,\mathrm{m}}}} = k''_{\mathrm{ad}} \tag{3.75}$$

where $k'_{\mathrm{ad}}$ and $k''_{\mathrm{ad}}$ are constants ($k'_{\mathrm{ad}} = k_{\mathrm{ad}}/R$ and $k''_{\mathrm{ad}} = k_{\mathrm{ad}}^{C_{V,\mathrm{m}}/C_{p,\mathrm{m}}}/R$). Instead of the indefinite integration, which yields equation (3.69), equation (3.73) could have been integrated between limits and substitutions made to obtain the relationships

$$\frac{1}{R}\ln\frac{T_2}{T_1} = \frac{1}{C_{V,\mathrm{m}}}\ln\frac{V_{\mathrm{m},1}}{V_{\mathrm{m},2}} = \frac{1}{C_{p,\mathrm{m}}}\ln\frac{p_2}{p_1}. \tag{3.76}$$

It is useful to compare the reversible adiabatic and reversible isothermal expansions of the ideal gas. For an isothermal process, the ideal gas equation can be written

$$pV_{\mathrm{m}} = k_{\mathrm{it}}, \tag{3.77}$$

where $k_{\mathrm{it}} = RT$ is a constant for the isothermal expansion. Figure 3.1 compares the pressure and volume change starting from the same initial conditions for the reversible isothermal [equation (3.77)] and reversible adiabatic [equation (3.69)] expansion of ideal gas with constant heat capacity. It can be seen that the pressure drops more rapidly in the adiabatic than in the isothermal expansion. This is as one would expect, since the temperature of the gas decreases during the adiabatic expansion.

Figure 3.2 compares a series of reversible isothermal expansions for the ideal gas starting at different initial conditions. Note that the isotherms are parallel. They cannot intersect since this would give the gas the same pressure and volume at two different temperatures. Figure 3.3 shows a similar comparison for a series of reversible adiabatic expansions. Like the isotherms, the adiabats cannot intersect. To do so would violate the Carathéodory principle and the Second Law of Thermodynamics, since the gas would have two different entropies at the same temperature, pressure, and volume.

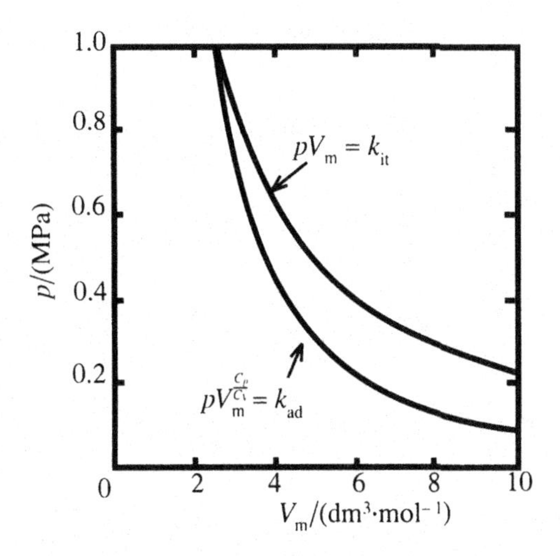


**Figure 3.1** Comparison of reversible isothermal and adiabatic ($C_V = \frac{3}{2}R$) ideal gas expansions.

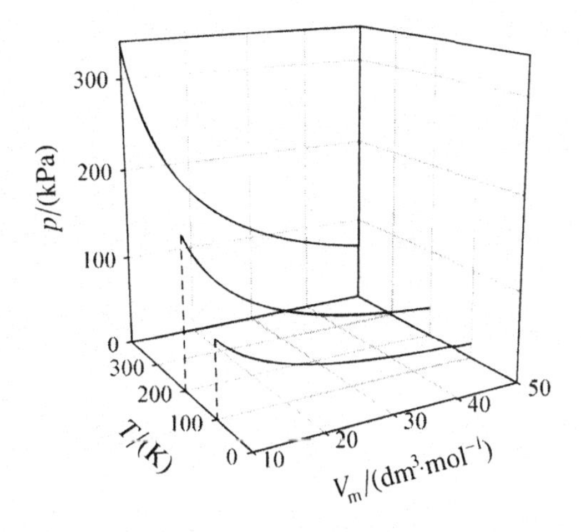


**Figure 3.2** Reversible isotherms of an ideal gas starting with the same initial and final volumes, but with different pressures.

## 3.3d The Carnot Cycle

In Chapter 2 (Section 2.2a) we qualitatively described the Carnot cycle, but were not able to quantitatively represent the process on a $p-V$ diagram because we did not know the pressure–volume relationship for a reversible adiabatic process. We now know this relationship (see section 3.3c), and in Figure 3.3, we compare a series of $p-V$ adiabats with different starting temperatures for an

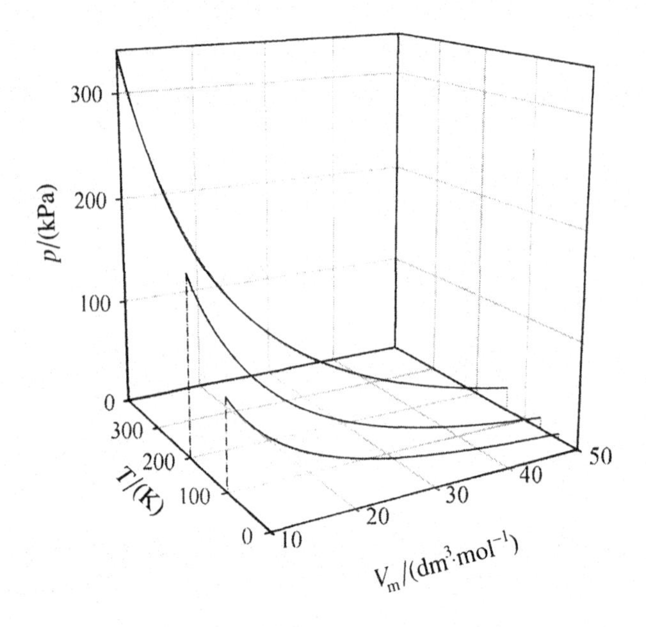


**Figure 3.3** Reversible adiabats of a monatomic ideal gas with $C_{V,\mathrm{m}} = 3/2R$, starting with the same initial and final volumes, but with different pressures.

ideal gas with constant heat capacity. As we described in Chapter 2, the Carnot cycle consists of a series of alternating reversible isothermal and adiabatic expansions and compressions put together in a cyclic process that returns the system to the initial state. This cycle is illustrated in Figure 3.4 for the ideal gas with constant heat capacity.

The four steps of the process can be summarized as follows:

(1) A reversible isothermal expansion of the ideal gas is made from an initial volume $V_1$ to a volume $V_2$ at an absolute (ideal gas) temperature $T_2$. The amount of pressure–volume work $w_2$ done by the system is obtained by substituting into Equation (2.16). The result is

$$w_2 = -nRT_2 \ln \frac{V_2}{V_1}. \tag{3.78}$$

To maintain isothermal conditions during this process, a quantity of heat $q_2$ is absorbed from a high-temperature heat reservoir operating at $T_2$. Since $\Delta U = 0$ for this isothermal expansion, $q_2 = -w_2$, so that

$$q_2 = nRT_2 \ln \frac{V_2}{V_1}. \tag{3.79}$$

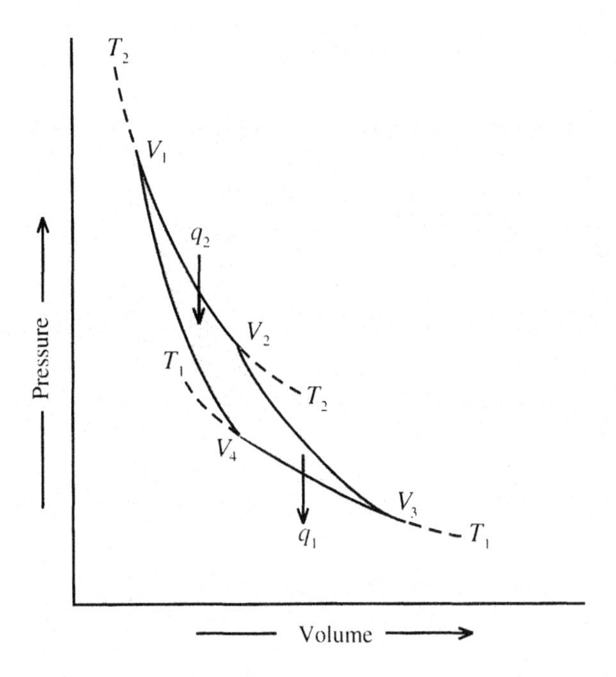


**Figure 3.4** Carnot cycle for the expansion and compression of an ideal gas. Isotherms alternate with adiabats in a reversible closed path. The shaded area enclosed by the curves gives the net work in the cyclic process.

(2) An adiabatic expansion of the ideal gas is made from $V_2$ to a volume $V_3$. Work is done by the system, and the temperature drops to a lower temperature $T_1$, since no heat is added to the system. In this adiabatic process, $q = 0$, and

$$\Delta U = w_{\text{adia}(1)} = C_V(T_1 - T_2). \tag{3.80}$$

(3) An isothermal compression of the fluid is made at $T_1$ from $V_3$ to $V_4$. The work $w_1$ added to the system is given by

$$w_1 = -nRT_1 \ln\frac{V_4}{V_3}. \tag{3.81}$$

To maintain isothermal conditions, a quantity of heat $q_1$ given by

$$q_1 = nRT_1 \ln\frac{V_4}{V_3} \tag{3.82}$$

is removed from the system and absorbed in a heat reservoir maintained at the temperature $T_1$.

(4) An adiabatic compression of the fluid returns the volume to $V_1$. Since work is added to the system and heat is not allowed to escape, the temperature increases to the initial temperature $T_2$. The amount of work is given by

$$w_{\text{adia}(2)} = C_V(T_2 - T_1). \tag{3.83}$$

The net work produced, which is the shaded area within the cycle, is the sum of equations (3.78), (3.80), (3.81), and (3.83). The result is

$$w = -nR\left( T_2 \ln\frac{V_2}{V_1} + T_1 \ln\frac{V_4}{V_3}\right). \tag{3.84}$$

Note that the sum of the adiabatic work in steps (2) and (4) is zero. Also, since the two adiabats span the same temperature range, equation (3.76) can be used to show that

$$\frac{V_4}{V_3} = \frac{V_1}{V_2}.$$

Substitution for $V_4/V_3$ from this equation into equation (3.84) and dividing by $q_2$ given by equation (3.79) gives the efficiency $\eta$ of conversion of heat into work for the Carnot cycle as

$$\frac{-w}{q_2} = \eta = \frac{T_2 - T_1}{T_2}. \tag{3.85}$$

But this is exactly the same as equation (2.82) that we obtained earlier

$$\frac{-w}{q_2} = \eta = \frac{T_2 - T_1}{T_2}. \tag{2.82}$$

The difference is that the temperature in equation (3.85) is the absolute or ideal gas temperature, while the temperature in equation (2.82) is the thermodynamic or Kelvin temperature. The conclusion we reach when comparing the two equations is that the absolute and Kelvin temperatures must be proportional to one another. That is

$$T(\text{Kelvin}) = \text{constant} \times T(\text{Absolute}).$$

The proportionality constant is arbitrary and can be set at any number. We set it equal to one so that the Kelvin and absolute temperature scales are the same.

In summary, the Carnot cycle can be used to define the thermodynamic temperature (see Section 2.2b), show that this thermodynamic temperature is an integrating denominator that converts the inexact differential $\delta q$ into an exact differential of the entropy d$S$, and show that this thermodynamic temperature is the same as the absolute temperature obtained from the ideal gas. This hypothetical engine is indeed a useful one to consider.

### 3.3e The Joule–Thomson Expansion

In a Joule–Thomson expansion, a gas under pressure flows through an orifice and expands at a lower pressure. The process can be represented schematically as shown in Figure 3.5. A gas with volume $V_1$, pressure $p_1$, and temperature $T_1$ flows through a porous membrane, pushed by a piston at pressure $p_1$. The gas expands against a piston at a lower pressure $p_2$ until all of the gas has been transferred to the other side of the membrane with a final volume $V_2$ and temperature $T_2$. Adiabatic shields keep heat from flowing into or out of the apparatus, and the flow rate is so slow that kinetic and potential energy changes associated with the flow are negligible. Under these condition, $q = 0$ and

$$\Delta U = w.$$

Both the compression and expansion are isobaric processes; hence, the total work is given by

$$\begin{aligned} w &= w_1 + w_2 = -p_1(0 - V_1) - p_2(V_2 - 0) \\ &= p_1 V_1 - p_2 V_2. \end{aligned}$$

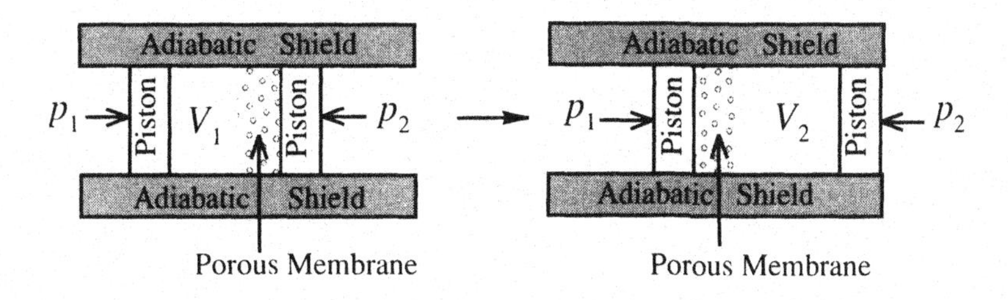


**Figure 3.5** In the Joule–Thomson expansion, a volume of gas $V_1$, is pushed through a porous plug by a piston at pressure $p_1$. The gas expands to a volume $V_2$ against a second piston at a pressure $p_2$.

Therefore,

$$\Delta U = U_2 - U_1 = p_1 V_1 - p_2 V_2,$$

or

$$U_2 + p_2 V_2 = U_1 + p_1 V_1.$$

Since

$$H = U + pV,$$

$$H_2 = H_1$$

or

$$\Delta H = 0.$$

Thus, the Joule–Thomson expansion is an isenthalpic process.

Figure 3.6 shows how pressure and temperature are related for a series of isenthalpic (Joule–Thomson) expansions. For example, if we start at the

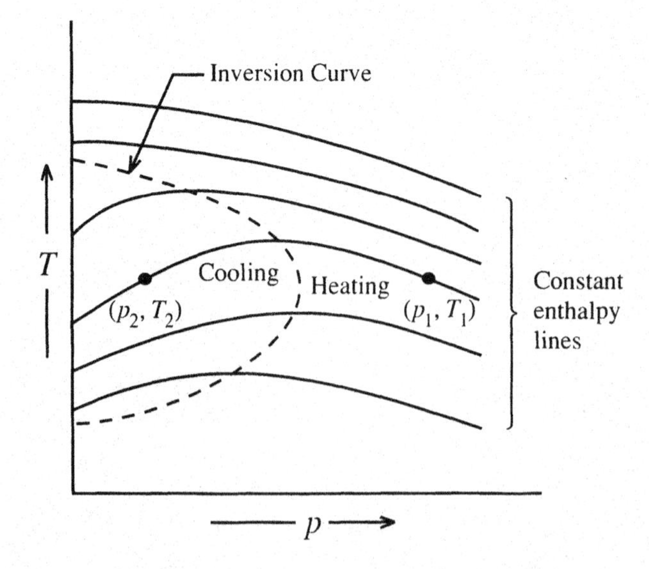


**Figure 3.6** Plot of $T$ versus $p$ along curves of constant enthalpy.

point $(p_1, T_1)$ in the figure and expand the gas, the pressure and temperature follow the isenthalpic line containing $(p_1, T_1)$. If the final pressure is $p_2$, then the final temperature will be $T_2$ as shown by point $(p_2, T_2)$ in the figure.

Since the Joule–Thomson process is isenthalpic, the slope of each line can be represented as $(\partial T/\partial p)_H$. This quantity is referred to as the Joule–Thomson coefficient, $\mu_{\mathrm{J.T.}}$. Thus[i]

$$\mu_{\mathrm{J.T.}} = \left(\frac{\partial T}{\partial p}\right)_H. \tag{3.86}$$

Figure 3.6 shows that $\mu_{\mathrm{J.T.}}$ is negative at high temperatures and pressures. Therefore, a gas heats up as it expands under these conditions. At lower temperatures, the gas continues to increase in temperature if the expansion occurs at high pressures. However, at lower pressures, the slope, and hence, $\mu_{\mathrm{J.T.}}$, becomes positive, and the gas cools upon expansion. Intermediate between these two effects is a pressure and temperature condition where $\mu_{\mathrm{J.T.}} = 0$. This temperature is known as the Joule–Thomson inversion temperature $T_{\mathrm{i}}$. Its value depends upon the starting pressure and temperature (and the nature of the gas). The dashed line in Figure 3.6 gives this inversion temperature as a function of the initial pressure. Note that when Joule–Thomson inversion temperatures occur, they occur in pairs at each pressure.[j]

Values for the Joule–Thomson coefficient can be obtained from equations of state. To do so, one starts with the relationship between exact differentials given by equation (1.37) to write (using molar quantities)

$$\mu_{\mathrm{J.T.}} = \left(\frac{\partial T}{\partial p}\right)_{H_{\mathrm{m}}} = -\frac{\left(\dfrac{\partial H_{\mathrm{m}}}{\partial p}\right)_T}{\left(\dfrac{\partial H_{\mathrm{m}}}{\partial T}\right)_p}.$$

[i] We will show later that $\mu_{\mathrm{J.T.}} = 0$ for an ideal gas. Thus the change in temperature resulting from a Joule–Thomson expansion is associated with the non-ideal behavior of the gas.

[j] The Joule–Thomson expansion can be used to liquify gases. An expansion at pressure and temperature conditions inside the dashed line envelope where $\mu_{\mathrm{J.T.}} < 0$ cools the gas. This gas is used to precool the incoming gas so that the expansion occurs at still lower temperatures. Continuing this process eventually cools the gas until it liquifies.

But

$$\left(\frac{\partial H_m}{\partial T}\right)_p = C_{p,m}$$

and

$$\left(\frac{\partial H_m}{\partial p}\right)_T = V_m - T\left(\frac{\partial V_m}{\partial T}\right)_p.$$

Hence,

$$\mu_{J.T.} = -\frac{1}{C_{p,m}}\left[V_m - T\left(\frac{\partial V_m}{\partial T}\right)_p\right]. \tag{3.87}$$

For the ideal gas

$$V_m = \frac{RT}{p}$$

$$\left(\frac{\partial V_m}{\partial T}\right)_p = \frac{R}{p}$$

$$\mu_{J.T.} = -\frac{1}{C_{p,m}}\left[V_m - \frac{TR}{p}\right] = 0.$$

Thus, $\mu_{J.T.} = 0$ for the ideal gas, and the change in temperature during the Joule–Thomson expansion depends upon non-ideal behavior of the gas.

Figure 3.7(a) shows experimental values of $\mu_{J.T.}$ obtained for $N_2$ gas, while Figure 3.7(b) shows how the Joule–Thomson coefficient for $N_2$ gas changes with pressure and temperature.[2]

Note, that in Figure 3.7(a), $\mu_{J.T.} > 0$ to the left (inside) of the curve, unless the temperature becomes very high (well above ambient), or very low (well below ambient), or the pressure becomes very high. Figure 3.7(b) shows that for pressures below 10 MPa, $\mu_{J.T.}$ is greater than zero in the temperature range from

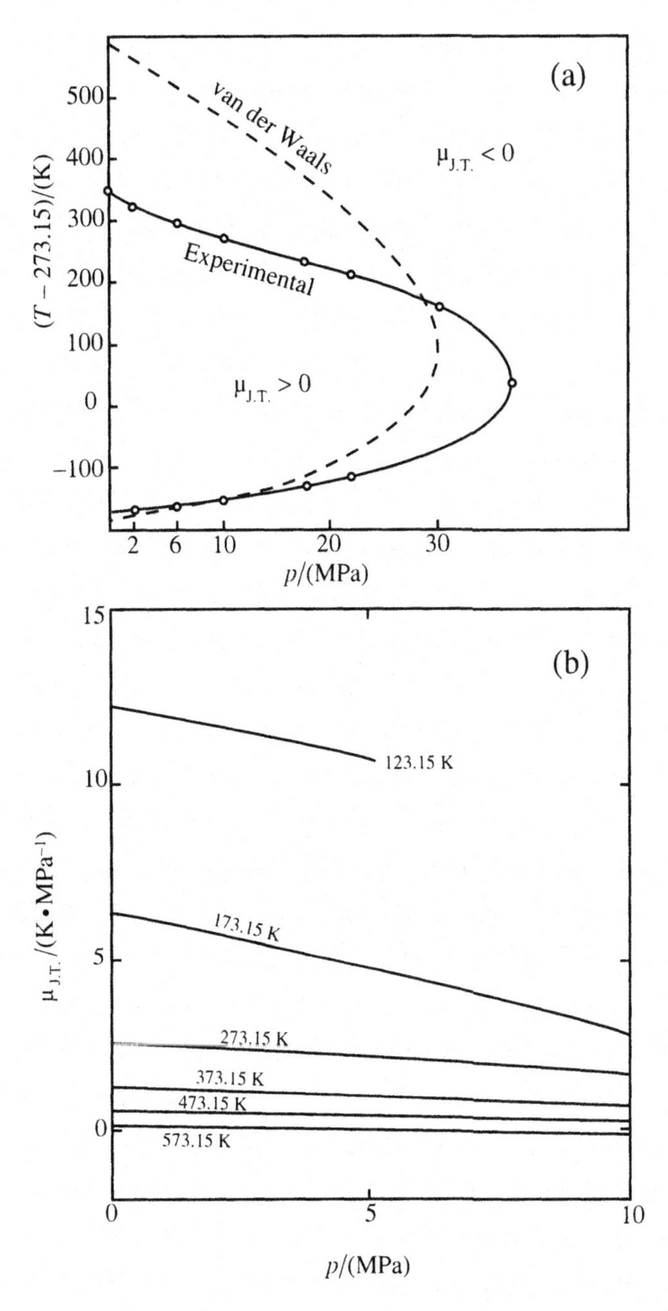


**Figure 3.7** (a) Joule–Thomson inversion curve ($\mu_{J.T.} = 0$) for nitrogen. (b) The Joule–Thomson coefficient of nitrogen gas. At the lowest temperature, 123.15 K, nitrogen liquifies; hence the curve for the gas terminates at the vapor pressure.

150 K to 500 K. A Joule–Thomson expansion in this range of pressure and temperature will cool the gas and can be used to liquify $N_2$.[k]

Equations of state can be used to predict $\mu_{\text{J.T.}}$ and $T_{\text{i}}$.

**Example 3.9:** Use the van der Waals equation to derive an equation relating $\mu_{\text{J.T.}}$ and $T_{\text{i}}$ to $p$, $T$, $C_{p,\text{m}}$ and the van der Waals constants $a$ and $b$.

**Solution:**

$$\mu_{\text{J.T.}} = -\frac{1}{C_{p,\text{m}}}\left[V_{\text{m}} - T\left(\frac{\partial V_{\text{m}}}{\partial T}\right)_p\right]. \tag{3.87}$$

For the van der Waals gas

$$\left(p + \frac{a}{V_{\text{m}}^2}\right)(V_{\text{m}} - b) = RT.$$

It is difficult to combine these two equations. The van der Waals equation is cubic in volume and equation (3.87) requires explicit values for the volume. Approximations must be made to the van der Waals equation before substitutions for $V_{\text{m}}$ and $(\partial V_{\text{m}}/\partial T)_p$ are made in equation (3.87).

The van der Waals equation can be written in the form.

$$V_{\text{m}} = \frac{RT}{p} - \frac{a}{pV_{\text{m}}} + b + \frac{ab}{pV_{\text{m}}^2}.$$

As a first approximation, $V_{\text{m}}$ may be replaced by the ideal gas equivalent in the correction terms $a/pV_{\text{m}}$ and $ab/pV_{\text{m}}^2$ to give

$$V_{\text{m}} = \frac{RT}{p} - \frac{a}{RT} + b + \frac{abp}{R^2T^2}. \tag{3.88}$$

[k] Air behaves much like $N_2$. Air is routinely liquified by using Joule–Thomson expansions starting at room temperature.

Differentiation of equation (3.88) gives

$$\left(\frac{\partial V_{\mathrm{m}}}{\partial T}\right)_p = \frac{R}{p} + \frac{a}{RT^2} - \frac{2abp}{R^2T^3}. \tag{3.89}$$

Substitution of equations (3.88) and (3.89) into equation (3.87) for $V_{\mathrm{m}}$ and $(\partial V_{\mathrm{m}}/\partial T)_p$ and simplifying gives

$$\mu_{\mathrm{J.T.}} = \frac{1}{C_{p,\mathrm{m}}}\left(\frac{2a}{RT} - b - \frac{3abp}{R^2T^2}\right). \tag{3.90}$$

At the inversion temperature, $\mu_{\mathrm{J.T.}} = 0$, which requires that

$$\frac{2a}{RT_{\mathrm{i}}} - b - \frac{3abp}{R^2T_{\mathrm{i}}^2} = 0.$$

Rearranging gives

$$T_{\mathrm{i}}^2 - \frac{2a}{Rb}T_{\mathrm{i}} + \frac{3ap}{R^2} = 0. \tag{3.91}$$

Figure 3.7(a) compares the experimental inversion curve for nitrogen gas with the van der Waals prediction. Considering the approximations involved, it is not surprising that the quantitative prediction of the van der Waals equation is not very good. Equation (3.91) is quadratic in $T_{\mathrm{i}}$ and hence, predicts two values for the inversion temperature, which is in qualitative agreement with the experimental observation.[1]

## 3.4 Relationship Between Free Energy and Work

Under certain $p$, $V$, $T$ conditions we can relate the Helmholtz free energy and the Gibbs free energy to work done in the process. To find the relationship between $\mathrm{d}A$ and $\delta w$, we write

$$A = U - TS.$$

[1] Errors can become especially large when derivatives of equations of state are involved, as is the case in this derivation.

Differentiating gives

$$dA = dU - T\,dS - S\,dT.$$

From the First Law

$$dU = \delta q + \delta w.$$

From the Second Law, for the reversible process

$$\delta q = T\,dS.$$

Combining equations gives

$$dA = \delta w - S\,dT.$$

At constant temperature, $dT = 0$ and

$$dA = \delta w. \tag{3.92}$$

Thus, in an isothermal reversible process, $dA$ equals the reversible work. Note that $\delta w$ in equation (3.92) is the total work. It includes pressure–volume work and any other forms, if present.[m]

A relationship between $dG$ and work can also be obtained. To do so we divide the work for a reversible process into pressure–volume work $(-p\,dV)$ and $\delta w'$, the work other than pressure–volume work that may occur, and write

$$\delta w = -p\,dV + \delta w'. \tag{3.93}$$

where $\delta w$ is the total work.

To obtain the relationship between $dG$ and $\delta w'$ we write

$$G = U + pV - TS.$$

Differentiating gives

$$dG = dU + p\,dV + V\,dp - T\,dS - S\,dT.$$

[m] In earlier times, $A$ was referred to as the work function because of the relationship of $dA$ or $\Delta A$ to the total work as given by equation (3.92).

At constant $p$ and $T$, $\mathrm{d}p$ and $\mathrm{d}T$ are zero and the differential expression becomes

$$\mathrm{d}G = \mathrm{d}U + p\,\mathrm{d}V - T\,\mathrm{d}S. \tag{3.94}$$

For the reversible process

$$\begin{aligned}\mathrm{d}U &= \delta q + \delta w \\ &= T\,\mathrm{d}S - p\,\mathrm{d}V + \delta w'.\end{aligned} \tag{3.95}$$

Substitution of equation (3.95) into (3.94) gives

$$\mathrm{d}G = \delta w'. \tag{3.96}$$

Thus, in a reversible process that is both isothermal and isobaric, $\mathrm{d}G$ equals the work other than pressure–volume work that occurs in the process.[n] Equation (3.96) is important in chemistry, since chemical processes such as chemical reactions or phase changes, occur at constant temperature and constant pressure. Equation (3.96) enables one to calculate work, other than pressure–volume work, for these processes. Conversely, it provides a method for incorporating the variables used to calculate these forms of work into the thermodynamic equations.

## Exercises

E3.1 (a) Prove that the variation of $C_{V,\mathrm{m}}$ with volume is given by the relationship

$$\left(\frac{\partial C_{V,\mathrm{m}}}{\partial V_\mathrm{m}}\right)_T = T\left(\frac{\partial^2 p}{\partial T^2}\right)_{V_\mathrm{m}}$$

(b) Prove that $C_{V,\mathrm{m}}$ for a van der Waals gas is independent of the volume of the gas.

E3.2 The expansion of a gas into a vacuum is a constant energy process, since $q = 0$ and $w = 0$, and hence $\Delta U = q + w = 0$.

(a) Show that the temperature does not change for the constant energy expansion of an ideal gas.

(b) Derive an expression for the constant energy expansion of a van der Waals gas with constant $C_{V,\mathrm{m}}$, relating the temperature change

[n] The name free energy, for $A$ and $G$ comes from the relationships of $\mathrm{d}A$ and $\mathrm{d}G$ to $\delta w$ as given in equations (3.92) and (3.96). Thus, $\Delta A$ and $\Delta G$ give the energy available (or free) to do work.

$(T_2 - T_1)$ to the initial and final volumes $V_{m,1}$ and $V_{m,2}$, the heat capacity $C_{V,m}$ and the van der Waals constants $a$ and $b$.

E3.3 A particular gas with $C_{V,m} = \frac{3}{2}R$ and independent of temperature obeys the equation of state

$$\left(p + \frac{a}{V_m}\right)V_m = RT,$$

where $a$ is a constant.

(a) Prove that $C_{p,m}$ for this gas varies with temperature by the expression

$$C_{p,m} = \frac{3}{2}R + \frac{R}{1 - a/RT}.$$

(b) Prove that the change in internal energy for an isothermal expansion of this gas from $V_{m,1}$ to $V_{m,2}$ is given by

$$\Delta U_m = a \ln \frac{V_{m,2}}{V_{m,1}}.$$

(c) One mole of this gas with $a = 0.304\ \text{dm}^3\cdot\text{MPa}\cdot\text{mol}^{-1}$ is taken at 300 K and 1.00 MPa pressure and expanded isothermally and reversibly to a pressure of 0.100 MPa. Calculate $w$, $q$, $\Delta U_m$, and $\Delta H_m$ for the process.

E3.4 A gas obeys the equation of state

$$pV_m = RT + Ap + BpT$$

where $A$ and $B$ are constants. Show that for this gas

$$C_{p,m} - C_{V,m} = R\left(1 + \frac{Bp}{R}\right)^2.$$

E3.5 A gas obeys the equation of state

$$PV_m = RT + \left(\frac{Bp}{T^2}\right)$$

where $B = 2.0 \times 10^4$ $dm^3 \cdot K^2 \cdot mol^{-1}$ is a constant. One mole of this gas is expanded isothermally at 300 K from 5.0 MPa to 0.100 MPa.
Calculate (in Joules)
(a) $\Delta G_m$
(b) $\Delta H_m$.

E3.6 A gas has the following properties

$$pV_m = RT + Bp$$

$$C_{p,m} = \alpha + \beta T$$

where $\alpha$, $\beta$ and $B$ are constants,
(a) Prove that $C_{p,m}$ is independent of $p$.
(b) Show that $\Delta S_m$ for the expansion of one mole of the gas from $(p_1, T_1)$ to $(p_2, T_2)$ is given by

$$\Delta S_m = \alpha \ln\left(\frac{T_2}{T_1}\right) + \beta(T_2 - T_1) - R \ln\left(\frac{p_2}{p_1}\right).$$

E3.7 A block of copper weighing 50 g is placed in 100 g of $H_2O$ for a short time. The copper is then removed from the liquid, with no adhering drops of water, and separated from it adiabatically. Temperature equilibrium is then established in both the copper and water. The entire process is carried out adiabatically at constant pressure. The initial temperature of the copper was 373 K and that of the water was 298 K. The final temperature of the copper block was 323 K. Consider the water and the block of copper as an isolated system and assume that the only transfer of heat was between the copper and the water. The specific heat of copper at constant pressure is 0.389 $J \cdot K^{-1} \cdot g^{-1}$ and that of water is 4.18 $J \cdot K^{-1} \cdot g^{-1}$. Calculate the entropy change in the isolated system.

E3.8 A sample of Ne gas contains 90 mole% $^{20}Ne$ and 10 mole% $^{22}Ne$. Calculate $\Delta G_m$ and $\Delta S_m$ at 300 K for the separation of one mole of this gas into two isotopically pure fractions, one containing only $^{20}Ne$ and the other $^{22}Ne$. Assume ideal gas behavior so that the isotopes behave independently in the separation, and that the pressure is 0.10 MPa.

E3.9 Use heat capacity coefficients in Table 2.1 of Chapter 2 to calculate $q$, $\Delta H_m$, and $\Delta S_m$ when one mole of HCl(g) is heated from 300 K to 1000 K.

E3.10 In the following table are presented $\alpha$ (coefficient of expansion), $\kappa$ (compressibility), and $\rho$ (density) for liquid water and liquid $CCl_4$. Use

| | $\alpha/(\mathrm{K}^{-1})$ | $\kappa/(\mathrm{MPa}^{-1})$ | $\rho/(\mathrm{g{\cdot}cm}^{-3})$ |
|---|---|---|---|
| water | $2.57 \times 10^{-4}$ | $4.51 \times 10^{-4}$ | 0.9971 |
| $CCl_4$ | $12.4 \times 10^{-4}$ | $10.56 \times 10^{-4}$ | 1.5940 |

this data to calculate $C_{p,\mathrm{m}} - C_{V,\mathrm{m}}$ for these two substances at 298.15 K and 0.101 MPa.

E3.11 Given $\Delta S_{\mathrm{m}} = 30.1\ \mathrm{J{\cdot}K^{-1}{\cdot}mol^{-1}}$ at $T = 300$ K for the isothermal change:

$$\text{monatomic gas } (p = 3.03 \text{ MPa}) = \text{monatomic gas } (p = 0.101 \text{ MPa}).$$

Calculate the final temperature if this gas expands adiabatically and reversibly from $T = 300$ K and $p = 3.03$ MPa to $p = 0.101$ MPa. Assume ideal behavior at $p = 0.101$ MPa, with $C_{V,\mathrm{m}} = \frac{3}{2}R$.

E3.12 An ideal gas with $C_{p,\mathrm{m}} = \frac{7}{2}R$ is to be compressed adiabatically and reversibly from 100 kPa to 2500 kPa, the initial temperature being 298.15 K. The compression may be carried out in one stage, or in two stages — from 100 kPa to 500 kPa, and then from 500 kPa to 2500 kPa, with the gas being allowed to regain its original temperature between the two stages. Which of the two processes can be carried out with the smaller expenditure of energy?

E3.13 A gas obeys the equation of state $PV_{\mathrm{m}} = RT + Bp$ and has a heat capacity $C_{V,\mathrm{m}}$ that is independent of temperature. Derive an expression relating $T$ and $V_{\mathrm{m}}$ in an adiabatic reversible expansion.

E3.14 The variation of the Joule–Thomson coefficient of air with the temperature $T$ at $p = 0.101$ MPa is given by

$$\mu_{\mathrm{J.T.}}/(\mathrm{K{\cdot}MPa^{-1}}) = -1.949 + \frac{1365}{T} - \frac{3150}{T^2}.$$

The dependence of $C_{p,\mathrm{m}}$ on temperature at 0.101 MPa is given approximately by

$$C_{p,\mathrm{m}}/(\mathrm{J{\cdot}K^{-1}{\cdot}mol^{-1}}) = 27.2 + 4.18 \times 10^{-3}T.$$

Determine the rate of change of $C_{p,\mathrm{m}}$ with pressure (in $\mathrm{J{\cdot}K^{-1}{\cdot}MPa^{-1}{\cdot}mol^{-1}}$) for air in the region of $T = 300$ K and $p = 0.101$ MPa.

## Problems

P3.1 A certain gas obeys the equation of state

$$PV_m = RT\left(1 + \frac{B}{V_m}\right)$$

where

$$B/(\text{dm}^3\cdot\text{mol}^{-1}) = -11.0 + 0.020T.$$

The heat capacity of the gas at two different molar volumes is given by

$$C_{V,m}/(\text{J}\cdot\text{K}^{-1}\cdot\text{mol}^{-1}) = 12.6 + 1.1 \times 10^{-2}\ T \quad (\text{at } 20\ \text{dm}^3\cdot\text{mole}^{-1})$$
$$C_{V,m}/(\text{J}\cdot\text{K}^{-1}\cdot\text{mol}^{-1}) = 12.6 + 2.2 \times 10^{-2}\ T \quad (\text{at } 60\ \text{dm}^3\cdot\text{mole}^{-1})$$

Given the following four processes (each for one mole):

(i) A(g, $V_m$ = 20 dm$^3$ · mol, $T$ = 300 K) = A(g, $V_m$ = 60 dm$^3$ · mol, $T$ = 300 K)

(ii) A(g, $V_m$ = 60 dm$^3$ · mol, $T$ = 300 K) = A(g, $V_m$ = 60 dm$^3$ · mol, $T$ = 500 K)

(iii) A(g, $V_m$ = 20 dm$^3$ · mol, $T$ = 300 K) = A(g, $V_m$ = 20 dm$^3$ · mol, $T$ = 500 K)

(iv) A(g, $V_m$ = 20 dm$^3$ · mol, $T$ = 500 K) = A(g, $V_m$ = 60 dm$^3$ · mol, $T$ = 500 K)

(a) Calculate $q$, $w$, $\Delta U_m$, $\Delta S_m$, and $\Delta H_m$ for the isothermal, reversible expansion of the gas for the changes of state represented by (i) and (iv).

(b) Calculate $q$, $w$, $\Delta U_m$, $\Delta S_m$, and $\Delta H_m$ for the isochoric, reversible processes represented by (ii) and (iii).

(c) If the direction of the two steps (i) and (ii) are reversed so that the combination of the four steps results in a cycle, calculate $q$, $w$, $\Delta U_m$, $\Delta S_m$, and $\Delta H_m$ for the overall process.

(d) Calculate $\Delta A_m$ and $\Delta G_m$ for the isothermal, reversible expansion of the gas for the changes of state represented by (i) and (iv).

(e) Consider the cycle composed of the four changes of state shown. What can be said about $\Delta A_m$ and $\Delta G_m$ for the changes of state represented by (ii) and (iii)?

P3.2 The volume of a liquid can be approximated by the equation of state

$$V_m = V_{0,m}[1 + \alpha(T - T_0) - \kappa(p - 1)]$$

where

$V_m$ = the molar volume at a pressure $p$ and temperature $T$

$V_{0,m}$ = the molar volume at $T = 273.15$ K and $p = 0.101$ MPa

$T_0 = 273.150$ K

$$\alpha = \frac{1}{V_m}\left(\frac{\partial V_m}{\partial T}\right)_p \quad \text{(coefficient of expansion)}$$

$$\kappa = -\frac{1}{V_m}\left(\frac{\partial V_m}{\partial p}\right)_T \quad \text{(compressibility).}$$

(a) Assume $\alpha$ and $\kappa$ are constants and show that

$$\left(\frac{\partial S_m}{\partial p}\right)_T = -V_{0,m}\alpha$$

$$\left(\frac{\partial H_m}{\partial p}\right)_T = V_{0,m}[1 - \alpha T_0 - \kappa(p - 1)]$$

(b) Calculate $\Delta H_m$ when one mole of water is compressed at $T = 293.15$ K from 0.10 to 2.50 MPa.
Given for $H_2O$:

$V_{0,m} = 18.018\ \text{cm}^3\cdot\text{mole}^{-1}$

$\alpha = 2.1 \times 10^{-4}\ \text{K}^{-1}$

$\kappa = 4.9 \times 10^{-4}\ \text{MPa}^{-1}$

P3.3 Molecular iodine is a solid at 298.15 K. The normal melting point is 386.8 K and the normal boiling point is 457.7 K. Given the following thermodynamic information for $I_2$ at $p = 0.101$ MPa:

$$C_{p,\mathrm{m}}(\mathrm{solid})/(\mathrm{J{\cdot}K^{-1}{\cdot}mol^{-1}}) = 54.68 + 1.34 \times 10^{-3}\,(T - 298.15)$$

$$[T = 298.15\ \mathrm{K\ to\ 386.8\ K}]$$

$$\Delta_{\mathrm{fus}}H^{\circ}_{\mathrm{m}}/(\mathrm{kJ{\cdot}mol^{-1}}) = 15.65 \qquad [T = 368.8\ \mathrm{K}]$$

$$C^{\circ}_{p,\mathrm{m}}(\mathrm{liquid})/(\mathrm{J{\cdot}K^{-1}{\cdot}mol^{-1}}) = 81.6 \qquad [T = 386.8\ \mathrm{K\ to\ 457.7\ K}]$$

$$\Delta_{\mathrm{vap}}H^{\circ}_{\mathrm{m}}/(\mathrm{kJ{\cdot}mol^{-1}}) = 25.5 \qquad [T = 457.7\ \mathrm{K}]$$

Calculate $\Delta S_{\mathrm{m}}$ for the process

$$(\mathrm{I_2, s, 298.15\ K, 0.101\ MPa}) \rightarrow (\mathrm{I_2, g, 457\ K, 0.101\ MPa}).$$

## References

1. The enthalpy of fusion and heat capacity of ice were taken from W. F. Giauque and J. W. Stout, "The Entropy of Water and the Third Law of Thermodynamics. The Heat Capacity of Ice from 15 to 237 °K", *J. Am. Chem. Soc.*, **58**, 1144–1150 (1936).
2. Experimental results of $\mu_{\mathrm{J.T.}}$ for $N_2$ are from J. R. Roebuck and H. Osterberg, "The Joule–Thomson Effect in Nitrogen", *Phys. Rev.*, **48**, 450–457 (1935).

## Chapter 4

# The Third Law and Absolute Entropy Measurements

In the previous chapter, we saw that entropy is the subject of the Second Law of Thermodynamics, and that the Second Law enabled us to calculate changes in entropy $\Delta S$. Another important generalization concerning entropy is known as the **Third Law of Thermodynamics**. It states that:

> *Every substance has a finite positive entropy, but at the absolute zero of temperature the entropy may become zero, and does so become in the case of a perfect crystalline substance.*[1]

As with the first and second laws, the Third Law is based on experimental measurements, not deduction. It is easy, however, to rationalize such a law. In a perfectly ordered[a] crystal, every atom is in its proper place in the crystal lattice. At $T = 0$ Kelvin, all molecules are in their lowest energy state. Such a configuration would have perfect order; and since entropy is a measure of the disorder in a system, perfect order would result in an entropy of zero.[b] Thus, the Third Law gives us an absolute reference point and enables us to assign values to $S$ and not just to $\Delta S$ as we have been restricted to do with $U$, $H$, $A$, and $G$.

---

[a] We will give examples later of crystals where disorder remains at zero Kelvin. In such substances, the entropy does not become zero at absolute zero.

[b] In Chapter 1 we wrote the Boltzmann equation relating entropy to the thermodynamic probability $W$ as

$$S = k \ln W.$$

In a completely ordered system, $W = 1$.

To obtain $S$, we start with equation (3.15) from the previous chapter

$$\left(\frac{\partial S}{\partial T}\right)_p = \frac{C_p}{T}. \tag{3.15}$$

Separating variables and integrating gives

$$\int_{S_0}^{S_T} \mathrm{d}S = \int_0^T \frac{C_p}{T}\,\mathrm{d}T. \tag{4.1}$$

According to the Third Law, $S_0 = 0$ and equation (4.1) becomes

$$S_T = \int_0^T \frac{C_p}{T}\,\mathrm{d}T. \tag{4.2}$$

Equation (4.2) can be used to determine the entropy of a substance. A pure crystalline sample is placed in a cryogenic calorimeter and cooled to low temperatures. Increments of heat, $q$, are added and the temperature change, $\Delta T$, is measured, from which the heat capacity can be calculated from the relationship

$$C_p = \frac{q}{\Delta T}. \tag{4.3}$$

Equation (4.3) is exactly true only if $q$ is an infinitesimal amount of heat, causing an infinitesimal temperature rise, $\mathrm{d}T$. However, unless the heat capacity is increasing rapidly and nonlinearly with temperature, equation (4.3) gives an accurate value for $C_p$ at the average temperature of the measurement.[c] Continued addition of heat gives the heat capacity as a function of temperature. The results of such measurements for glucose are shown in Figure 4.1.[2]

---

[c] Strictly speaking, the heat capacity at constant pressure is not measured in the cryogenic calorimeter. Instead, $C_{sat}$, the heat capacity at the saturated vapor pressure of the solid or liquid is obtained, and this pressure increases with increasing temperature. Usually, a small amount of helium gas is present in the calorimeter to add in the heat exchange. Small corrections can be made for the effect of the exchange gas and, unless the vapor pressure of the substance in the calorimeter is large, the difference between $C_{sat}$ and $C_p$ is negligible. We will use $C_p$ to represent this heat capacity.

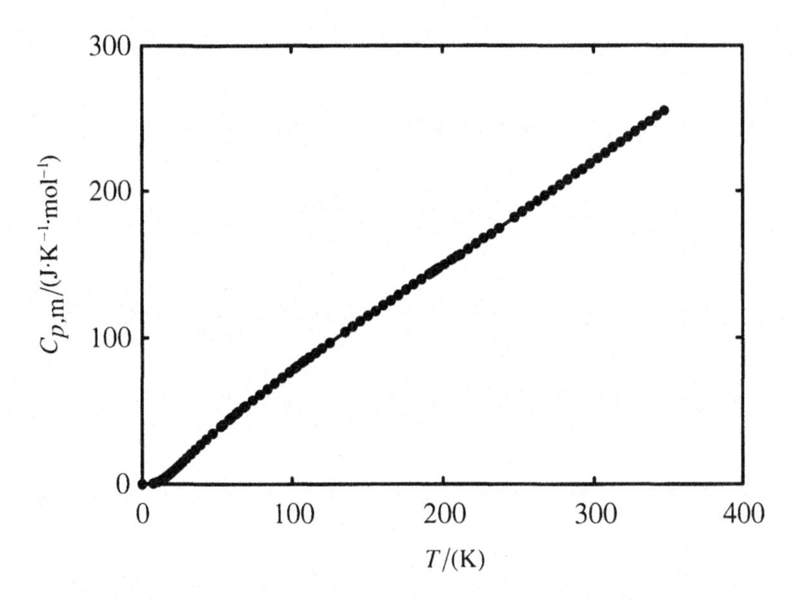


**Figure 4.1** Graph of $C_{p,m}$ against $T$ for glucose. The circles represent the experimental data points. The heat capacity becomes zero at 0 K.

The heat capacity of a solid is zero at 0 Kelvin and increases with temperature. If phase transitions are present, discontinuities occur in $C_p$, and an increase in entropy results from the phase change. We will consider an example of that type a little later, but for systems such as glucose with no phase changes, equation (4.2) gives the entropy $S_m$, which is obtained from an integration of a graph of $C_{p,m}/T$ against $T$ as shown in Figure 4.2. Determination of the area under the curve from 0 Kelvin to a temperature given by the dashed line (in this example, 298.15 K) gives $S_m$ at this temperature.

Equation (4.2) requires that the total area above 0 Kelvin be obtained, but heat capacity measurements cannot be made to the absolute zero of temperature. The lowest practical limit is usually in the range from 5 K to 10 K, and heat capacity below this temperature must be obtained by extrapolation. In the limit of low temperatures, $C_p$ for most substances follows the Debye low-temperature heat capacity relationship[d] given by equation (4.4)

$$C_{p,m} = aT^3. \tag{4.4}$$

[d] In Chapter 10, we will derive the Debye heat capacity equation and describe its properties in more detail.

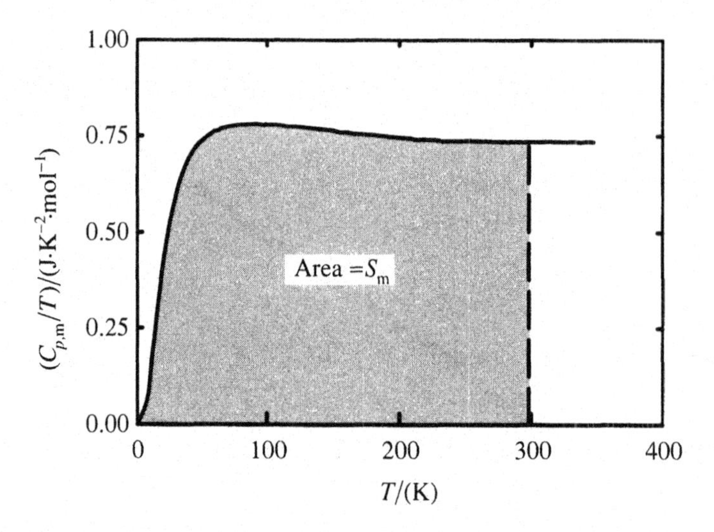


**Figure 4.2** Graph of $C_{p,\mathrm{m}}/T$ against $T$ for glucose. The area under the curve to the selected temperature gives the entropy at that temperature (in this example, 298.15 K).

Equation (4.4) indicates that at low temperatures, a graph of $C_{p,\mathrm{m}}/T$ against $T^2$ should be a straight line with an intercept of zero at $T = 0$. Figure 4.3 shows such a graph for glucose, where the lowest measurement is at approximately 7.5 K ($T^2 = 56\ \mathrm{K}^2$). A straight-line relationship is obtained below $T \cong 14$ K

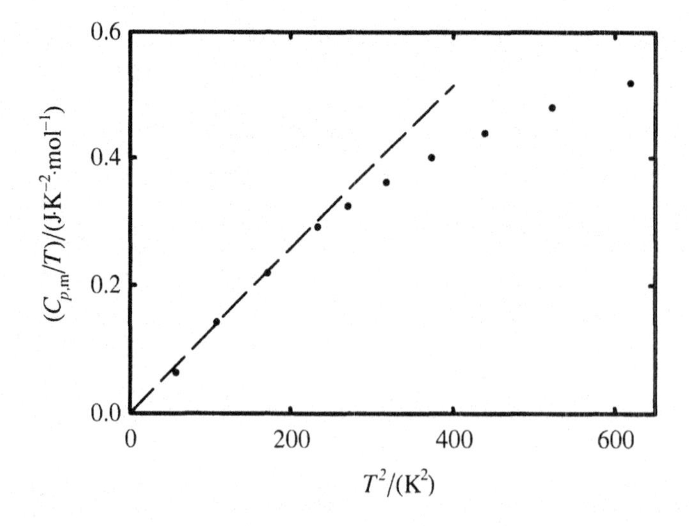


**Figure 4.3** Graph of $C_{p,\mathrm{m}}/T$ against $T^2$ for glucose. The extrapolation to 0 K is a straight line that can be used to calculate the entropy below the temperature of lowest measurement.

($T^2 \cong 200$ K$^2$). A graph such as the one shown in Figure 4.3 can be used to extrapolate $C_{p,\mathrm{m}}/T$ to 0 Kelvin. Values can be taken from the graph and included in the graphical integration.

Extrapolations are always subject to error, but fortunately the contribution to the entropy resulting from the extrapolation is a small part of the total. In glucose, for example, $S^\circ_{298} = 219.2 \pm 0.4$ J·K$^{-1}$·mol$^{-1}$, but the entropy contribution at 10 K obtained from the Debye extrapolation is only 0.28 J·K$^{-1}$·mol$^{-1}$. Well-designed cryogenic calorimeters are able to produce $C_p$ measurements of high accuracy; hence, the Third Law entropy obtained from the $C_p$ measurements can also be of high accuracy.

The $T^3$ relationship for heat capacity represented by equation (4.4) applies to most solids at low temperatures, but there are exceptions. Summarized in the table below are the different types of temperature dependencies on $C_p$ that have been observed at low temperatures.

| Temperature dependence | Type of solid | |
|---|---|---|
| $C_{p,\mathrm{m}} = aT^3$ | most nonmagnetic, nonmetallic crystals | (4.4) |
| $C_{p,\mathrm{m}} = bT^2$ | 2-dimensional layered crystals such as graphite and boron nitride | (4.5) |
| $C_{p,\mathrm{m}} = aT^3 + \gamma T$ | 3-dimensional metals | (4.6) |
| $C_{p,\mathrm{m}} = aT^3 + BT^{3/2}$ | ferromagnetic crystals at temperatures below the magnetic transition temperature if it occurs at low $T$ | (4.7) |
| $C_{p,\mathrm{m}} = aT^3$ | (antiferromagnetic crystals below the magnetic transition temperature if it occurs at low $T$ | (4.8) |

The linear term in $C_{p,\mathrm{m}}$ for metals results from the contribution to the heat capacity of the free electrons. It can become important at very low temperatures where the $T^3$ relationship becomes very small. For example, the electronic contribution to the heat capacity of Cu metal is 1.2% at 30 K, but becomes 80% of the total at 2 K.[e]

As indicated earlier, when phase transitions occur, $\Delta S$ for the phase transition must be included in the Third Law calculation of the entropy. For example, Figure 4.4 summarizes the heat capacity of $N_2$ as a function of temperature, up

[e] For metals, an extrapolation of $C_{p,\mathrm{m}}/T$ against $T^2$ to $T = 0$ does not give a value of zero {see equation (4.6)}. Instead, a value of $\gamma$ is obtained.

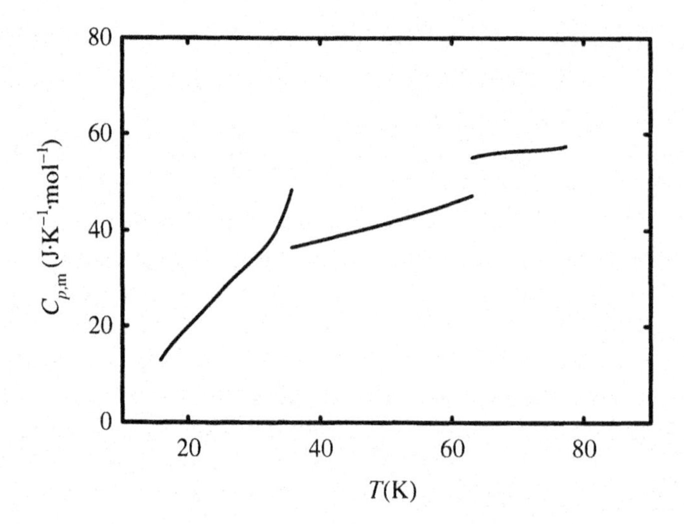


**Figure 4.4** Heat capacity of $N_2$ as a function of temperature. A solid phase transition occurs at 35.62 K, the melting temperature is 63.15 K, and the normal boiling temperature is 77.33 K.

to the formation of the gas phase.[3] A solid phase transition occurs at 35.62 K and the melting point (actually, triple point) is 63.15 K. The cryogenic calorimeter used to measure $C_p$ can also be used to measure the vapor pressure of the $N_2$ as a function of temperature, from which the normal boiling point[f] is determined to be 77.33 K. The enthalpy changes for the transition, for melting, and for vaporization, are also measured in the calorimeter, from which $\Delta S_m$ for the phase change can be calculated from

$$\Delta S_{\mathrm{m}} = \frac{\Delta H_{\mathrm{m}}}{T} \tag{4.9}$$

since the phase changes are reversible. A calculation of the entropy of $N_2$ gas at the normal boiling point is obtained as follows:

| | $\Delta S_{\mathrm{m}}/(\mathrm{J{\cdot}K^{-1}{\cdot}mol^{-1}})$ |
|---|---|
| 0–10 K, Debye extrapolation | 1.916 |
| 10–35.62 K, Graphical integration of $C_{p,\mathrm{m}}/T$ | 25.25 |
| 35.62 K, Transition: $\Delta S_{\mathrm{m}} = \Delta H_{\mathrm{m}}/T = 228.91/35.62$ | 6.43 |
| 35.62–63.15 K, Graphical integration of $C_{p,\mathrm{m}}/T$ | 23.38 |

[f] Temperature where the vapor pressure is 1 atm or 101.325 kPa.

| | |
|---|---|
| 63.15 K, Fusion: $\Delta S_m = \Delta H_m/T = 720.9/63.15$ | 11.42 |
| 63.15–77.33 K, Graphical integration of $C_{p,m}/T$ | 11.41 |
| 77.33 K, Vaporization: $\Delta S_m = \Delta H_m/T = 5577/77.33$ | 72.12 |
| Entropy of $N_2$ (g) at 77.32 K and 1 atm (101.325 kPa) | |
| | $151.93 \pm 0.4$ J·K$^{-1}$·mol$^{-1}$ |

The entropies of gases are often corrected to the ideal gas condition.[g] The correction $\Delta S_m^{corr}$ is made by following the cycle shown in Figure 4.5 with

$$\Delta S_m^{corr} = \Delta S_{m,1} + \Delta S_{m,2} + \Delta S_{m,3} = S_{m,i} - S_{m,r}$$

where $S_{m,i}$ is the entropy of the gas acting ideally and $S_{m,r}$ is the entropy of the real gas.

The pressure $p^*$ is taken to be low enough that the real gas behaves like an ideal gas, and $\Delta S_{m,2} = 0$. To calculate $\Delta S_{m,1}$ and $\Delta S_{m,3}$, we start with

$$\left(\frac{\partial S_m}{\partial p}\right)_T = -\left(\frac{\partial V_m}{\partial T}\right)_p.$$

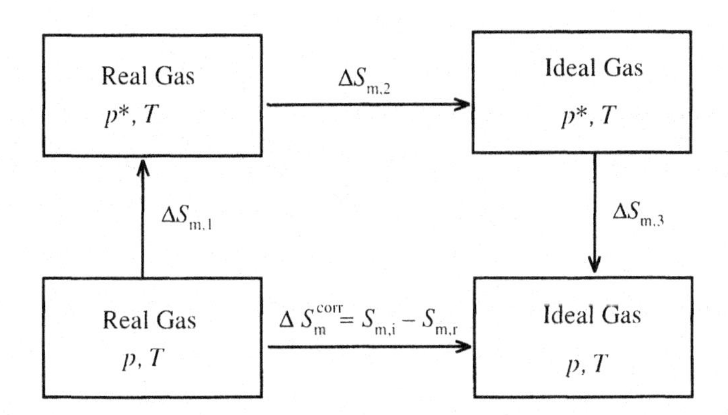


**Figure 4.5** The correction to ideal behavior $\Delta S_m^{corr} = S_{m,i} - S_{m,r}$ is given by $\Delta S_m^{corr} = \Delta S_{m,1} + \Delta S_{m,2} + \Delta S_{m,3}$ since $S$ is a state function. ($p^*$ is a very low pressure.)

---

[g] There are several reasons for correcting the Third Law entropy to ideal gas behavior. In Chapter 6, we will see that the ideal gas is the usual choice for the standard state of a gas. Thus, the correction to ideal gas behavior is necessary for obtaining the standard state value. In Chapter 10, we will see that the entropy of the ideal gas is the one calculated from molecular parameters. Correction of the Third Law entropy to the ideal gas condition allows for a direct comparison between the two values.

Separating variables and integrating gives

$$\Delta S_{\mathrm{m,\,1\ or\ 3}} = -\int_{p_1}^{p_2} \left(\frac{\partial V_{\mathrm{m,\,1\ or\ 3}}}{\partial T}\right)_p \mathrm{d}p \qquad (4.10)$$

To calculate $\Delta S_{\mathrm{m,\,3}}$, we use the ideal gas equation to calculate $(\partial V_{\mathrm{m,\,3}}/\partial T)_p$

$$V_{\mathrm{m}} = \frac{RT}{p}$$

$$\left(\frac{\partial V_{\mathrm{m,\,3}}}{\partial T}\right)_p = \frac{R}{p}$$

$$\Delta S_{\mathrm{m,\,3}} = -\int_{p^*}^{p} \frac{R}{p}\,\mathrm{d}p = -R\ln\frac{p}{p^*} \qquad (4.11)$$

We need an equation of state for a real gas to calculate $\Delta S_{\mathrm{m,\,1}}$. The modified Berthelot equation is often used for pressures near ambient and was used in the original reference to calculate the correction to ideal behavior for $N_2$ gas.[h] This equation is as follows

$$pV_{\mathrm{m}} = RT\left[1 + \frac{9pT_{\mathrm{c}}}{128p_{\mathrm{c}}T}\left(1 - \frac{6T_{\mathrm{c}}^2}{T^2}\right)\right] \qquad (4.12)$$

where $p_{\mathrm{c}}$ and $T_{\mathrm{c}}$ are the critical temperature and pressure. Differentiating gives

$$\left(\frac{\partial V_{\mathrm{m}}}{\partial T}\right)_p = \frac{R}{p} + \frac{27RT_{\mathrm{c}}^3}{32T^3p_{\mathrm{c}}}. \qquad (4.13)$$

[h] See Appendix 3 for a summary of the properties of several equations of state for gases.

Substitution of equation (4.13) into (4.10) gives

$$\Delta S_{\mathrm{m},1} = -\int_{p}^{p^*} \left( \frac{R}{p} + \frac{27RT_c^3}{32T^3 p_c} \right) \mathrm{d}p$$

$$\Delta S_{\mathrm{m},1} = -R \ln \frac{p^*}{p} - \frac{27RT_c^3}{32T^3 p_c}(p^* - p). \tag{4.14}$$

Combining equations (4.11) and (4.14) for $\Delta S_{\mathrm{m},3}$ and $\Delta S_{\mathrm{m},1}$ along with $\Delta S_{\mathrm{m},2} = 0$, gives

$$S_{\mathrm{m,i}} - S_{\mathrm{m,r}} = \frac{27RT_c^3 p}{32T^3 p_c}. \tag{4.15}$$

In equation (4.15) we have written $p - p^* = p$, since $p^*$ is very small. Substituting the critical pressure and temperature of $N_2$ into equation (4.15) with $p = 101.3$ kPa and $T = 77.33$ K gives $S_i - S_r = 0.92\ \mathrm{J \cdot K^{-1} \cdot mol^{-1}}$. Thus, the entropy of ideal $N_2$ gas at this pressure and temperature would be $151.93 + 0.92 = 152.8 \pm 0.4\ \mathrm{J \cdot K^{-1} \cdot mol^{-1}}$.

## 4.1 Verification of the Third Law

As mentioned earlier, the Third Law is a law of experience. But how does one go about verifying the law? As with so much else during the early days of science, the formulation and substantiation of the Third Law rested upon the shoulders of a group of exceptional scientists. From a comprehensive study of chemical reactions using electrochemical cells, T. W. Richards[4] demonstrated that $\Delta G$ and $\Delta H$ for a number of chemical reactions approach one another as shown in Figure 4.6.

Enthalpy and Gibbs free energy are related by

$$G = H - TS.$$

In a chemical reaction at a given temperature, this relationship can be used to give

$$\Delta G = \Delta H - T\Delta S. \tag{4.16}$$

Figure 4.6 demonstrates that

$$\lim_{T \to 0} (\Delta G - \Delta H) = \lim_{T \to 0} (T\Delta S) = 0$$

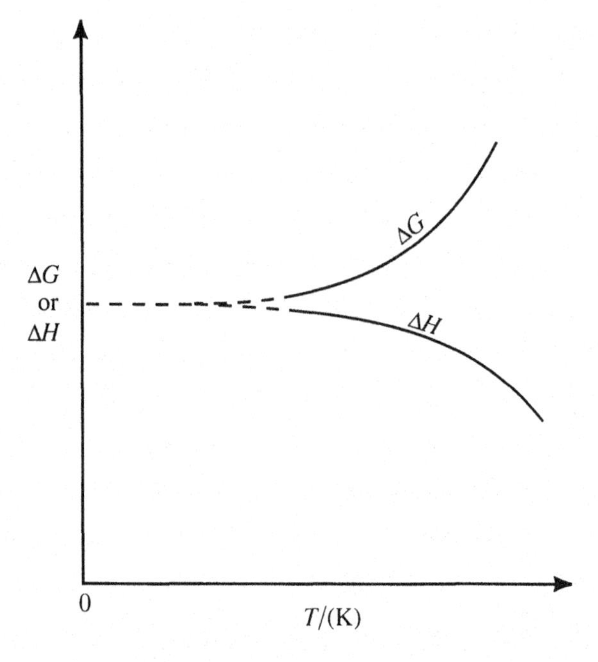


**Figure 4.6** For a chemical reaction, $\Delta H$ and $\Delta G$ approach the same value, with slopes $(\partial\Delta H/\partial T)_p$ and $(\partial\Delta G/\partial T)_p$ becoming zero as $T$ approaches zero.

which requires only that $\Delta S$ remain finite so that $T\Delta S$ will go to zero as $T$ goes to zero, and this conclusion is not particularly useful. However, the fact that $\Delta G$ approaches zero with a limiting slope of zero leads to the conclusion that

$$\lim_{T\to 0}\left(\frac{\partial\Delta G}{\partial T}\right)_p = \lim_{T\to 0}(-\Delta S) = 0.$$

This relationship led to an early formulation of the Third Law known as the **Nernst heat theorem**, which states that for any isothermal process

$$\lim_{T\to 0}\Delta S_T = 0. \qquad (4.17)$$

The conclusion that can be reached from the Nernst heat theorem is that the total entropy of the products and the reactants in a chemical reaction must be the same at 0 Kelvin. But nothing in the statement requires that the entropy of the individual substances in the chemical reaction be zero, although a value of zero for all reactants and products is an easy way to achieve the result of equation (4.17).

Further support for the conclusion that $S_0 = 0$ is obtained by comparing the entropy of a substance at a temperature $T$ by following two different paths from low temperatures to measure $S_T$. The process involves measuring the entropy change from 0 Kelvin for two different crystalline forms. For example, monoclinic sulfur reversibly converts to rhombic sulfur at the transition temperature $T_{tr} = 368.5$ K, but the conversion process is slow, and both rhombic and monoclinic sulfur can be cooled to low temperatures and the heat capacities determined for each as a function of temperature. Eastman and McGavock[5] have carefully measured $C_{p,m}$ of rhombic and of supercooled monoclinic sulfur from 13 to 365 K and West[6] has measured $C_{p,m}$ of rhombic sulfur from 298 K to 368.5 K and the enthalpy of transition $\Delta_{tr}H_m$ between the two phases. Combining these results enables one to calculate the entropy of monoclinic sulfur at the transition temperature by two different paths as follows.

- **The rhombic sulfur path**

| Path | $\Delta S_m/(\text{J}\cdot\text{K}^{-1}\cdot\text{mol}^{-1})$ |
|---|---|
| S(rhombic, $T = 0$ K)$\rightarrow$S(rhombic, $T = 368.5$ K): | $\displaystyle\int_0^{368.5} \frac{C_{p,m}(\text{rh})}{T}\,dT = 36.8$ |
| S(rhombic, $T = 368.5$ K)$\rightarrow$ S(monoclinic, $T = 368.5$ K): | $\displaystyle\frac{\Delta_{tr}H_m}{T_{tr}} = \frac{401.7}{368.5} = 1.09$ |
| S(rhombic, $T = 0$)$\rightarrow$S(monoclinic, $T = 368.5$ K): | $36.8 + 1.09 = 37.9 \pm 0.2$ |

- **The monoclinic sulfur path**

S(monoclinic, $T = 0$)$\rightarrow$ S(monoclinic, $T = 368.5$ K) $\qquad \displaystyle\int_0^{368.5} \frac{C_{p,m}(\text{mono})}{T}\,dT = 37.8 \pm 0.4$

Within experimental error, $S_{0,m}$ (monoclinic) $= S_{0,m}$ (rhombic), and it seems more probable that they both have zero entropy at 0 K than that both have the same non-zero entropy.

An even better example of this type of comparison is taken from measurements of Giauque[i] of the heat capacity on phosphine $PH_3$ shown in Figure 4.7.[7] Cooling $PH_3$ (g) causes it to condense to liquid at 185.41 K {point

[i] W. F. Giauque was one of those pioneers whose work led to the verification of the Third Law. He received the Nobel Prize in 1949 for his work.

(b)} when the pressure is 1 atm (101.33 kPa). Cooling results in solidification to solid I at 139.38 K, with a phase transition to form solid II occurring at 88.12 K. A transition occurs at 49.44 K {point (a)}, with solid III the stable form below this temperature. However, solid II can be supercooled to lower temperatures without conversion, until it undergoes an abrupt change in heat capacity at 35.67 K to solid IV and a transition to solid V at 30.29 K.

In a procedure similar to that described earlier for $N_2$, the entropy at $T = 49.44$ K {point (a)} can be calculated from the heat capacities and the enthalpies of transition, by following two different paths. $S_{m,49.44}$ was found to be 34.06 $J \cdot K^{-1} \cdot mol^{-1}$ following the solid III$\rightarrow$solid II path and $S_{m,49.44} =$ 34.02 $J \cdot K^{-1} \cdot mol^{-1}$ following the (solid V$\rightarrow$solid IV$\rightarrow$solid II) path. The two results agree well within experimental error, which requires that the entropy of solid III and solid V be the same at 0 K, an occurrence that is unlikely unless $S_{m,0} = 0$ for both forms.

The procedures described so far imply that $S_0 = 0$, but do not rigorously prove that this is so. The final proof comes from a comparison of $S_T$ for the ideal gas, obtained from the integration of $C_p$ data assuming the Third Law is valid combined with the entropies of transition, with values obtained from a calculation of $S_T$ for the ideal gas by statistical methods. The procedure, to be described in detail in Chapter 10, starts with the Boltzmann equation

$$S = k \ln W$$

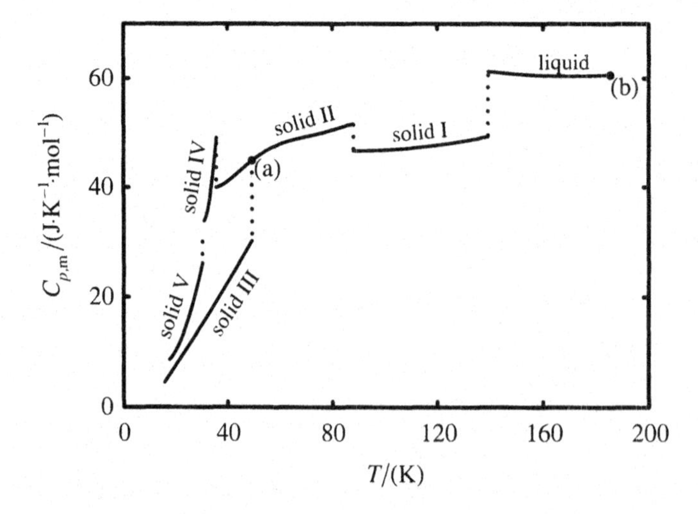


**Figure 4.7** Heat capacity and phase transitions in phosphine. The same entropy at point (a) is obtained by going the stable (lower) route or by going the metastable (upper) route. Point (b) is the normal boiling temperature.

**Table 4.1** Comparison of Third Law and statistical calculations of $S°$, the entropy of the ideal gas at 101.33 kPa ($J{\cdot}K^{-1}{\cdot}mol^{-1}$) at $T = 298.15$ K.

| Substance | $S = \int_0^T C_p \frac{dT}{T}$ | $S = k \ln W$ |
|---|---|---|
| $N_2$ | 192.0 | 191.6 |
| $O_2$ | 205.4 | 205.1 |
| $Cl_2$ | 223.1 | 223.0 |
| $H_2S$ | 205.4 | 205.6 |
| $CO_2$ | 213.8 | 213.7 |
| $NH_3$ | 192.2 | 192.2 |
| $SO_2$ | 247.9 | 248.0 |
| HCl | 186.2 | 186.6 |
| HBr | 199.2 | 198.7 |
| HI | 207.1 | 206.7 |

and leads to an equation for $S_T$ that involves knowledge of the molecular mass, moments of inertia, and vibrational frequencies of the molecule. Its derivation is completely independent of the Third Law. Entropy values of high accuracy can be calculated by this process.[j]

Comparison and agreement with the calorimetric value verifies the assumption that $S_0 = 0$. For example, we showed earlier that the entropy of ideal $N_2$ gas at the normal boiling point as calculated by the Third Law procedure had a value of $152.8 \pm 0.4$ $J{\cdot}K^{-1}{\cdot}mol^{-1}$. The statistical calculation gives a value of 152.37 $J{\cdot}K^{-1}{\cdot}mol^{-1}$, which is in agreement within experimental error. For $PH_3$, the Third Law and statistical values at $p = 101.33$ kPa and $T = 185.41$ K are $194.1 \pm 0.4$ $J{\cdot}K^{-1}{\cdot}mol^{-1}$ and 194.10 $J{\cdot}K^{-1}{\cdot}mol^{-1}$ respectively, an agreement that is fortuitously close. Similar comparisons have been made for a large number of compounds and agreement between the calorimetric (Third Law) and statistical value is obtained, all of which is verification of the Third Law. For example, Table 4.1 shows these comparisons for a number of substances.

## 4.2 Exceptions to the Third Law

The Third Law requires that a perfectly crystalline solid of a pure material be present at 0 Kelvin for $S_0$ to equal zero. Exceptions to the Third Law occur when this is not the case. For example, AgCl(s) and AgBr(s) mix to form a

[j] This calculation is one of the most satisfying in science. The values of the thermodynamic properties of the ideal gas calculated from molecular parameters are usually more accurate than the same thermodynamic results obtained from experimental measurements.

continuous set of solid solutions. Eastman and Milner[8] made a solid solution of AgBr with AgCl in which the mole fraction of AgBr was 0.728, and used this mixture to measure $\Delta_{mix}G_m$ and $\Delta_{mix}H_m$ for the mixing process

$$0.728 \text{ moles AgBr(s)} + 0.272 \text{ moles AgCl(s)} = 1.000 \text{ moles solid solution.}$$

They used electrochemical cell measurements[k] to determine $\Delta_{mix}G_m$, and solution calorimetric measurements to determine $\Delta_{mix}H_m$. The results they obtained at $T = 298.15$ K are $\Delta_{mix}G_m = 1060 \text{ J·mol}^{-1}$ and $\Delta_{mix}H_m = 340 \text{ J·mol}^{-1}$.

From these results $\Delta_{mix}S_m$ for the mixing process can be calculated. To do so, we start with equation (4.16),

$$\Delta G = \Delta H - T\Delta S, \tag{4.16}$$

from which $\Delta_{mix}S_m$ for the mixing process[l] can be calculated at $T = 298.15$ K from

$$\Delta_{mix}S_m = \frac{\Delta_{mix}H_m - \Delta_{mix}G_m}{T}$$

$$= \frac{340 + 1060}{298.15}$$

$$= 4.66 \text{ J·K}^{-1}\text{·mol}^{-1}.$$

Eastman and Milner also measured the heat capacities of the solid AgBr, solid AgCl, and the solid solution, from low temperatures to $T \cong 290$ K. Within experimental error, they found that $\Delta_{mix}C_{p,m} = 0$ for the mixing process over the entire temperature range. Hence, the entropy change remains essentially constant at $4.66 \text{ J·K}^{-1}\text{·mol}^{-1}$ over this temperature range, since

$$\left(\frac{\partial \Delta_{mix}S_m}{\partial T}\right)_p = \frac{\Delta_{mix}C_{p,m}}{T}.$$

[k] In Chapter 9 we will describe electrochemical cells and obtain the equations for calculating $\Delta G$ from cell emf measurements.

[l] This result is nearly equal to $4.87 \text{ J·K}^{-1}\text{·mol}^{-1}$, the value that would be calculated for the entropy of mixing to form an ideal solution. We will show in Chapter 7 that the equation to calculate $\Delta_{mix}S_m$ for the ideal mixing process is the same as the one to calculate the entropy of mixing of two ideal gases. That is, $\Delta_{mix}S_m = -R\sum x_i \ln x_i$.

The conclusion is that $\Delta_{\text{mix}}S_{\text{m}} = 4.66\ \text{J·K}^{-1}\text{·mol}^{-1}$ for the solution process at 0 Kelvin. If one assumes that the entropies of the AgBr and AgCl are zero at 0 Kelvin, then the solid solution must retain an amount of entropy that will give this entropy of mixing.

Supercooled liquids (glasses[m]) also have residual entropy at 0 Kelvin. As an example, glycerol supercools badly so that a glass is usually obtained at a low temperature. A crystalline solid can also be obtained if the liquid is cooled in a certain manner to initiate crystalization. Gibson and Giauque[9] started with

---

[m] In Figure 4.8, the broad step-like change in heat capacity for liquid glycerol at approximately 180 K is an example of a **glass transition**. The so-called **glass transition temperature**, $T_{\text{g}}$, is often taken as the mid-point of this step. The glass transition is not an equilibrium phase change where the liquid changes to a solid, but rather represents a temperature range where the heat capacity, and other physical properties such as viscosity, change rapidly with temperature. The name glass transition comes from the high-viscosity glass-like material that forms when the substance is cooled below $T_{\text{g}}$.

The most common type of glass transition is one that occurs for many liquids when they are cooled quickly below their freezing temperature. With rapid cooling, eventually a temperature region is reached where the translational and rotational motion associated with the liquid is lost, but the positional and orientational order associated with a crystal has not been achieved, so that the disorder remains frozen in. The loss of both translational and rotational motion leads to a large increase in viscosity and a large decrease in heat capacity.

Glass transitions occur in substances that supercool badly. Substances where hydrogen bonding occurs, such as glycerol, ethanol, cyclohexanol, and simple sugars, are common examples. Polymers and high-temperature ceramic materials also often have glass transitions, a change that is important in determining many of the desirable physical properties of these substances.

The glassy state does not represent a true equilibrium phase. Below the transition into a glass phase, the material is regarded as being in a metastable state. If one holds the substances at temperatures somewhat below the glass transition temperature, heat evolution can often be observed over time as the molecules slowly orient themselves into the lower energy, stable crystalline phase.

Glass transitions are always associated with the departure of a system from thermodynamic equilibrium in one or more of its degrees of freedom. In the liquid to glass transition described above, the material freezes into a state in which there is neither translational nor orientational equilibrium. In some materials, a translationally ordered but orientationally disordered, crystal can be prepared. Glass transitions have been observed in these materials under certain cooling conditions. Ice undergoes this type of transition at approximately 110 K. In this case, all translational and most rotational motions are absent from the crystal at temperatures above $T_{\text{g}}$. The glass transition results from the freezing in of a rearrangement of hydrogen-bonds associated with the reorientations of water molecules near defects in the crystal. Later in this chapter, we will describe this defect in ice in more detail. As one might expect, the change in heat capacity associated with the glass transition in ice is much smaller than that observed when translational and rotational motions are quenched in a liquid to glass transition. (See H. Suga, *Ann. N.Y. Acad. Sci.*, **484**, 248–263 (1986), for a discussion of the thermodynamic aspects of glassy crystals.)

liquid glycerol at the melting temperature of 291.05 K, measured the enthalpy of fusion at this temperature, the heat capacity of the solid down to 70 K, and the heat capacity of the supercooled liquid (glass) down to the same temperature.[n] The results they obtained are summarized in Figure 4.8.

Gibson and Giauque calculated the entropy difference between the solid and the supercooled liquid at 70 K. They found that at this temperature, the entropy of the supercooled liquid was $23.4 \pm 0.4\ \text{J·K}^{-1}\text{·mol}^{-1}$ larger then the entropy of the solid. Below 140 K, the solid and supercooled liquid have very nearly the same heat capacity. Thus, this entropy difference should not be much different at 0 Kelvin, and the liquid has an $S_{\text{m},0}$ of approximately $23.4\ \text{J·K}^{-1}\text{·mol}^{-1}$. This, of course, is a reflection of the lack of order in the supercooled liquid.

For most substances, the Third Law and statistical calculations of the entropy of the ideal gas are in agreement, but there are exceptions, some of which are summarized in Table 4.2. The difference results from residual entropy, $S_0$, left in the solid at 0 Kelvin because of disorder so that $S_T - S_0$ calculated from $\int_0^T C_p/T\,\mathrm{d}T$ is less than the $S_T$ calculated from statistical methods. In carbon monoxide the residual disorder results from a random arrangement of the CO molecules in the solid. Complete order in the solid can be represented schematically (in two-dimensions) by

[n] Giauque's report of the procedures he followed to obtain solid glycerol is one of the interesting stories of science. Initially, he experienced extreme difficulty in obtaining solid glycerol. To quote from Giauque:

> *The problem of crystallizing glycerol proved to be of some interest. A tube of glycerol was kept with one end in liquid air, the other at room temperature for a period of several weeks without results. Seeding with various organic crystals of similar structure was also tried. In fact, the artifices ordinarily used for starting crystallization in the absence of seed crystals were all tried without success.*
>
> *The few references in the literature indicated that glycerol had only been obtained in the crystalline state by chance. Inquiry among places storing large quantities of glycerol finally revealed some crystals at the plant of the Giant Powder Company at Manoose Bay, BC.*
>
> *After the seed crystals had arrived it was found that crystallization practically always occurred when amounts of 100 g of any laboratory sample were slowly warmed over a period of a day, after cooling to liquid-air temperatures. This occurred even when great precautions were taken to exclude the presence of seeds. However, it was found readily possible, by temperature manipulation alone, to produce crystalline or supercooled glycerol at will.*

What Giauque does not tell are the trials encountered in transporting the glycerol seed crystals by dog sled, boat, etc. from Manoose Bay, BC to Berkeley, California, while trying to keep the sample cold so that the seed crystals would not melt.

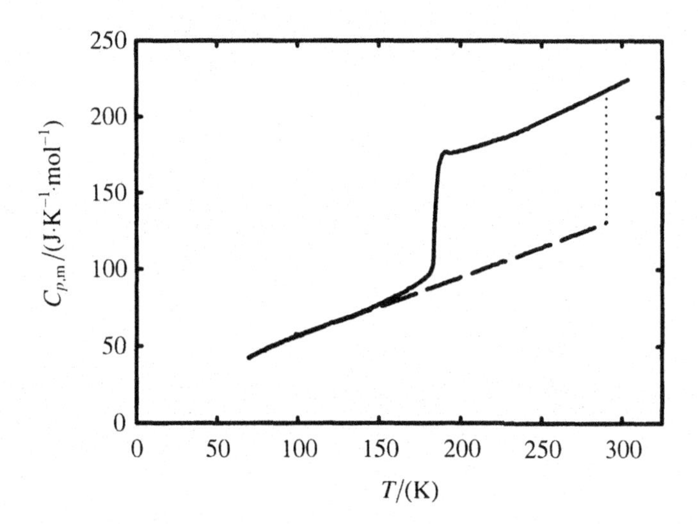


**Figure 4.8** Heat capacity of glycerol as a function of temperature. The solid line indicates $C_{p,\mathrm{m}}$ for the liquid and glassy phase. The dashed line represents $C_{p,\mathrm{m}}$ for the solid. The dotted line at the melting temperature of 291.05 K indicates the change in heat capacity upon melting. A glass transition occurs in the supercooled liquid at approximately 185 K. The heat capacities of the solid and the glass approach one another as the temperature is lowered; they are almost identical below 140 K.

**Table 4.2** Comparison of the ideal gas entropy in $\mathrm{J{\cdot}K^{-1}{\cdot}mol^{-1}}$ at a pressure of 101.32 kPa as calculated from the Third Law and from the statistical equations

| Substance | $T/(\mathrm{K})$ | $S_T$ (statistical) | $S_T$ (Third Law) | Difference | Reference |
|---|---|---|---|---|---|
| CO | 298.1 | 198.0 | 193.3 | 4.7 | a |
| $N_2O$ | 298.1 | 220.0 | 215.2 | 4.8 | b |
| NO | 121.4 | 183.1 | 179.9 | 3.2 | c |
| $ClO_3F$ | 226.48 | 261.9 | 251.8 | 10.1 | d |
| $H_2O$ | 298.1 | 188.7 | 185.3 | 3.4 | e |

[a] J. O. Clayton and W. F. Giauque, "The Heat Capacity and Entropy of Carbon Monoxide. Heat of Vaporization. Vapor Pressure of Solid and Liquid. Free Energy to 5000 °K from Spectroscopic Data," *J. Am. Chem. Soc.*, **54**, 2610–2626 (1932).
[b] R. W. Blue and W. F. Giauque, "The Heat Capacity and Vapor Pressure of Solid and Liquid Nitrous Oxide. The Entropy from its Band Spectrum," *J. Am Chem Soc.*, **57**, 991–997 (1935).
[c] H. L. Johnston and W. F. Giauque, "The Heat Capacity of Nitric Oxide from 14 °K to the Boiling Point and the Heat of Vaporization. Vapor Pressure of Solid and Liquid Phases. The Entropy from Spectroscopic Data," *J. Am. Chem. Soc.*, **51**, 3194–3214 (1929).
[d] J. K. Koehler and W. F. Giauque, "Perchloryl Fluoride. Vapor Pressure, Heat Capacity, Heats of Fusion and Vaporization. Failure of the Crystal to Distinguish O and F", *J. Am. Chem. Soc.*, **80**, 2659–2662 (1958).
[e] W. F. Giauque and J. W. Stout, "The Entropy of Water and the Third Law of Thermodynamics. The Heat Capacity of Ice from 25 to 273 °K" *J. Am. Chem. Soc.*, **58**, 1144–1150 (1936).

the arrangement:

C – O   C – O   C – O   C – O   C – O
O – C   O – C   O – C   O – C   O – C
C – O   C – O   C – O   C – O   C – O
O – C   O – C   O – C   O – C   O – C

But the two ends of the molecule are so similar that when the solid forms, the ends get mixed up and some molecules go in backwards to give an arrangement like the following:

C – O   O – C   C – O   C – O   O – C
C – O   O – C   C – O   O – C   O – C
C – O   O – C   C – O   O – C   O – C

Cooling the solid freezes in the disorder, and it persists to 0 Kelvin.
The entropy resulting from the disorder can be calculated from

$$S = k \ln W,$$

where $W$ is the number of possible orientations in the crystalline phase. For CO, each molecule has two possible arrangements, and for $N$ molecules $W = 2^N$, so that[o]

$$S = k \ln 2^N = Nk \ln 2.$$

For one mole, $N = N_A$ (Avogadro's number) and $N_A\, k = R$ so that

$$S_{m,0} = R \ln 2 = 5.76\ \mathrm{J{\cdot}K^{-1}{\cdot}mol^{-1}}.$$

This calculated number is larger than the 4.7 $\mathrm{J{\cdot}K^{-1}{\cdot}mol^{-1}}$ obtained from the experimental measurements, indicating that some residual order is present in the solid.

---

[o] An alternate way of calculating this entropy discrepancy is to attribute it to the mixing of two forms of CO that differ by having C and O reversed. The entropy of mixing 1/2 mole of each form is $S_0$ and can be calculated from

$$\begin{aligned} S_{m,0} &= -R[x_1 \ln x_1 + x_2 \ln x_2] \\ &= -R[0.5 \ln 0.5 + 0.5 \ln 0.5] \\ &= R \ln 2. \end{aligned}$$

Nitrous oxide is a linear molecule with the structure N—N—O. The Third Law entropy discrepancy for this compound results from the same disorder as in carbon monoxide. Again, the value obtained of 4.8 $J{\cdot}K^{-1}{\cdot}mol^{-1}$ is somewhat less than $R\ln 2$, indicating some residual order. In small molecules, discrepancies of this type seem to be largely restricted to interchange of the elements C, N, O and F of the second period of the Period Table.[p] In $ClO_3F$, three oxygen atoms and a fluorine atom are arranged tetrahedrally around the central chlorine. The fluorine atom can be found at each of the four corners of the tetrahedron, resulting in an entropy discrepancy of $R\ln 4 = 11.52$ $J{\cdot}K^{-1}{\cdot}mol^{-1}$. The observed value of 10.1 $J{\cdot}K^{-1}{\cdot}mol^{-1}$ is close to this amount, but somewhat less, indicating again some residual order in the solid.

Nitric oxide exists as a dimer in the solid state. This dimer can occupy two positions as follows:

```
N—O         O—N
|  |   or   |  |
O—N         N—O
```

The entropy discrepancy resulting from mixing the two different orientations of the dimer is $\frac{1}{2}R\ln 2 = 2.88$ $J{\cdot}K^{-1}{\cdot}mol^{-1}$. The factor of $\frac{1}{2}$ results from the fact that only $\frac{1}{2}$ mole of $N_2O_2$ is present. The experimental difference of 3.2 $J{\cdot}K^{-1}{\cdot}mol^{-1}$ is within experimental error of the expected value.

The entropy discrepancy in ice is attributed to a random arrangement of hydrogen bonds.[q] Figure 4.9 shows the arrangement of the H and O atoms in ice.[r] Each oxygen is surrounded by four other oxygens in a tetrahedral arrangement, with the structure held together by hydrogen bonds. Hydrogen atoms lie between each pair of oxygen atoms and are covalently bonded to one oxygen and hydrogen-bonded to the other. The O—H covalent bond is stronger and shorter than the O—H hydrogen bond so that the hydrogen is not equidistant between the two oxygens. When ice forms, the hydrogen bonds form randomly, leading to a disordered structure at low temperatures. To calculate

[p] Some residual disorder has been found in a metastable form of $COCl_2$ resulting from interchange of Cl and O. The residual entropy of 0.8 $J{\cdot}K^{-1}{\cdot}mol^{-1}$ is small and corresponds to only 1.75% of the oxygens in $COCl_2$ exchanging with the chlorines. See W. F. Giauque and J. B. Ott, "The Three Melting Points and Heats of Fusion of Phosgene. Entropy of Solids I and II and Atomic Exchange Disorder in Solid II," *J. Am. Chem. Soc.*, **82**, 2689–2695 (1960).

[q] The explanation is attributed to Linus Pauling, resulting from a conversation with W. F. Giauque.

[r] The positions of the covalent bonds and the hydrogen bonds are continually changing, so that Figure 4.9 represents only an instantaneous picture of the atomic arrangement.

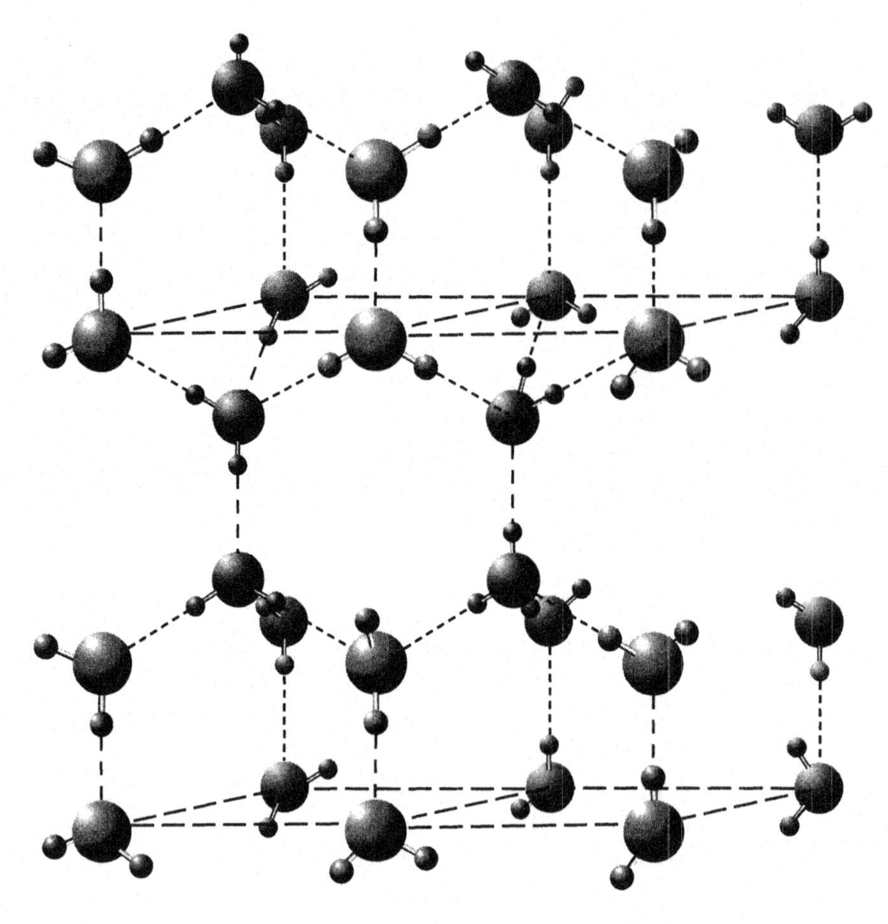

**Figure 4.9** Instantaneous crystal structure of ice. From: *General Chemistry*, Pauling and Pauling © 1970, 1953 and 1947 by W. H. Freeman. Reproduced with permission.

this disorder, we consider a $H_2O$ molecule at the center of a tetrahedron. Each of its covalently bonded protons can be directed to any one of three different oxygens from adjacent water molecules, with the second covalently bonded proton held in a fixed orientation. There are a total of six such possibilities. However, each of the adjacent oxygen atoms also has two protons. Each has a chance of $\frac{1}{2}$ of contributing the hydrogen bond to the central oxygen. For the two protons this chance becomes $(\frac{1}{2})(\frac{1}{2}) = \frac{1}{4}$. This cuts down the number of possible arrangements around the central oxygen by $\frac{1}{4}$ and $W = (\frac{6}{4})^{N_A}$. Hence,

$$S = k\ln(\tfrac{6}{4})^{N_A} = R\ln\tfrac{6}{4} = 3.37\ \text{J}\cdot\text{K}^{-1}\cdot\text{mol}^{-1}.$$

This number is in good agreement with the observed difference of $3.4 \text{ J}\cdot\text{K}^{-1}\cdot\text{mol}^{-1}$.[s]

It must have been perplexing to the scientists attempting to prove the Third Law to find that simple molecules such as CO, NO, $N_2O$ and $H_2O$ show a Third Law discrepancy, but satisfying when the explanations were finally obtained. Molecular hydrogen $H_2$ is a final example we want to consider. The proton nuclei in $H_2$ have nuclear spins of $\frac{1}{2}$. These spins couple together to form two energy levels that are often designated by the labels "parallel" and "antiparallel". Hydrogen with molecules in the parallel state is ortho-hydrogen. It is higher in energy than para-hydrogen, in which the spins couple in an antiparallel fashion.

Quantum mechanics predicts that a diatomic molecule has a set of rotational energy levels $\varepsilon_J$ given by

$$\varepsilon_J = \frac{h^2 J(J+1)}{8\pi^2 I} \tag{4.17}$$

where $h$ is Planck's constant, $I$ is the moment of inertia of the molecule and $J$ is the rotational quantum number with values of $J = 0, 1, 2, \ldots \infty$. A homonuclear diatomic such as $H_2$ has rotational energy levels that correspond to either even or odd values of $J$, but not both. It is found that the nuclear spin states determine which set of $J$ values are permitted. For para-hydrogen $J = 0, 2, 4, \ldots$ and for ortho-hydrogen $J = 1, 3, 5, \ldots$ The statistical methods we will develop in Chapter 10 enable us to calculate the equilibrium ratio of ortho-hydrogen to para-hydrogen. This ratio is shown in Figure 4.10. At high temperatures (which includes ambient temperature), equilibrium hydrogen is 75% ortho. Hydrogen with this ratio is known as normal hydrogen. With decreasing temperature, the equilibrium ratio of ortho to para decreases, until at very low temperatures, all of the hydrogen is para.

The conversion of ortho-hydrogen to para-hydrogen is slow in the absence of a catalyst. Therefore, as one cools room-temperature hydrogen to low temperatures, the ortho : para ratio remains at 3 : 1, and entropy is present that results from the mixing of these two different types of hydrogen.

The residual entropy can be calculated. For the para-hydrogen, $J = 0$ at $T = 0$ K. Since one-fourth of the hydrogen is para, the contribution to the

[s] There are other systems with residual entropy that we will not discuss here. For example, planar molecules such as thiophene are disordered in the solid due to a random arrangement (rotation) of the sulfur atoms. Ammonia, $NH_3$, provides another interesting example. Earlier statistical calculations neglected a vibration due to an umbrella inversion mode. The result in this case was that the statistical value was smaller than the Third Law value.

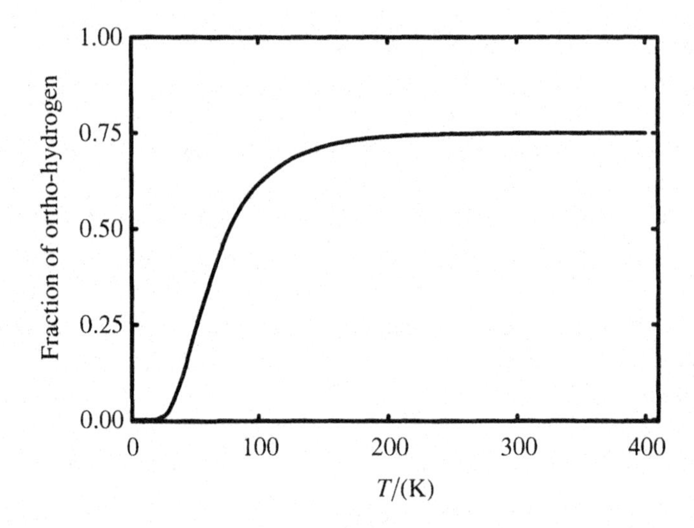


**Figure 4.10** Fraction of equilibrium molecular hydrogen that is ortho. At 0 K hydrogen is all para. The high temperature limit is 75% ortho and 25% para.

entropy of mixing is $\frac{1}{4}\ln\frac{1}{4}$. For the ortho-hydrogen, $J = 1$ at $T = 0$ K. Coupling of the rotational angular momentum of this spin with that of the nuclear spin gives nine kinds of ortho-hydrogen. These nine kinds each have a mole fraction of $\frac{1}{12}$. Hence, the contribution of the ortho is $\frac{9}{12}\ln\frac{1}{12}$. The total entropy of mixing, $S_{m,0}$, is given by

$$S_{m,0} = -R[\tfrac{1}{4}\ln\tfrac{1}{4} + \tfrac{9}{12}\ln\tfrac{1}{12}]$$

$$= 18.4\ J{\cdot}K^{-1}{\cdot}mol^{-1}.$$

This result is in excellent agreement with the value of $S_{m,0} = 18.1\ J{\cdot}K^{-1}{\cdot}mol^{-1}$ obtained from the heat capacity measurements.[10]

Catalysts such as charcoal can be used to maintain the equilibrium ratio of ortho-hydrogen to para-hydrogen with decreasing temperature.[t] When this happens, heat capacity measurements give the equilibrium value for the entropy of hydrogen.

[t] The energy difference between ortho- and para-hydrogen at 0 Kelvin is 1410 $J{\cdot}mol^{-1}$. The enthalpy of vaporization of normal hydrogen at the normal boiling point of 20.39 K is 902.9 $J{\cdot}mol^{-1}$. If one liquifies normal hydrogen and then inserts a catalyst to convert to equilibrium hydrogen, much of the hydrogen will boil off because the energy of conversion is larger than the energy of vaporization.

Many nuclei other than hydrogen have nuclear spins, and interaction with rotational energy levels similar to those given by equation (4.8) apply.[u] Hydrogen is unique in that it has a much smaller moment of inertia than other molecules. Since the spacing between the energy levels is inversely related to the moment of inertia, the rotational energy levels for most other substances are very close. Therefore, differences in macroscopic properties such as heat capacity, entropy, and energy, between species with even and odd $J$ values, are very small, the molecules are indistinguishable and thus the ortho–para entropy discrepancy is unique to $H_2$ and its isotopes.

There are still other forms of disorder that we have ignored in the Third Law entropies we have calculated. Most substances are a mixture of isotopes. For example, HCl is a mixture of $^{1}H^{35}Cl$, and $^{1}H^{37}Cl$ with small amounts of $^{2}H^{35}Cl$ and $^{2}H^{37}Cl$, and the entropy we calculate ignores an entropy of mixing of these different species. However, products and reactants in a chemical process have the same isotopic composition. This entropy of mixing would cancel in a calculation of $\Delta S$ for the process, and hence, is ignored. Nuclei with non-zero nuclear spins have nuclear moments that are randomly oriented in the solid. The alignment of these nuclei occurs only at very low temperatures, and the entropy change due to this alignment is not included in the extrapolation to 0 Kelvin. As with the isotope effect, these nuclear spin contributions cancel out in calculations of $\Delta S$ and are ignored.

In summary, the absolute entropies we calculate and tabulate are, in fact, not so absolute, since they do not include isotopic entropies of mixing nor nuclear spin alignment entropies. The entropies we tabulate are sometimes called **practical absolute entropies**. They can be used to correctly calculate $\Delta S$ for a chemical process, but they are not "true" absolute entropies.

## 4.3 Implications and Applications of the Third Law

### 4.3a Attainment of Perfect Order at Low Temperatures

Experience indicates that the Third Law of Thermodynamics not only predicts that $S_0 = 0$, but produces a potential to drive a substance to zero entropy at 0 Kelvin. Cooling a gas causes it to successively become more ordered. Phase changes to liquid and solid increase the order. Cooling through equilibrium solid phase transitions invariably results in evolution of heat and a decrease in entropy. A number of solids are disordered at higher temperatures, but the disorder decreases with cooling until perfect order is obtained. Exceptions are

[u] Equation (4.18) applies only to a diatomic or linear polyatomic molecule. Similar kinds of rotational energy levels are present in more complicated molecules. We will describe the various kinds in more detail in Chapter 10.

of the type we have discussed earlier. Substances such as CO try to achieve order, but cooling freezes in most of the disorder. Ordering a CO molecule in the solid at low temperatures requires that the molecule overcomes an energy of activation barrier as it flips over, and not enough energy is available to allow this to occur. Supercooled liquids attempt to order themselves by forming solids. Glasses persist to low temperature when large molecules are present in the liquid that are difficult to arrange in ordered structures. Hydrogen bonding produces hydrogen-bonded structures that are hard to align. Liquids such as alcohols, sugars (and even water) supercool more than most other substances, and often form glasses. It is interesting to note that most solid hydrate salts form perfect crystals. The presence of the anions and cations line up the water molecules and prevent random arrangements of the hydrogen bonds. An exception is $Na_2SO_4 \cdot 10H_2O$. Pitzer and Coulter[11] have shown that this hydrate, which has large amounts of water of hydration, has an $S_{m,0}$ of approximately $R\ln 2$. They attributed this disorder to the random arrangement of hydrogen bonds in the crystal from the two structures shown in Figure 4.11.

Helium is an interesting example of the application of the Third Law. At low temperatures, normal liquid helium converts to a superfluid with zero viscosity. This superfluid persists to 0 Kelvin without solidifying. Figure 4.12 shows how the entropy of He changes with temperature. The conversion from normal to superfluid occurs at what is known as the $\lambda$ transition temperature. Figure 4.12 indicates that at 0 Kelvin, superfluid He with zero viscosity has zero entropy, a condition that is hard to imagine.[v]

The Third Law predicts that isotopic species should separate at very low temperatures. This does not happen because the energy necessary to overcome

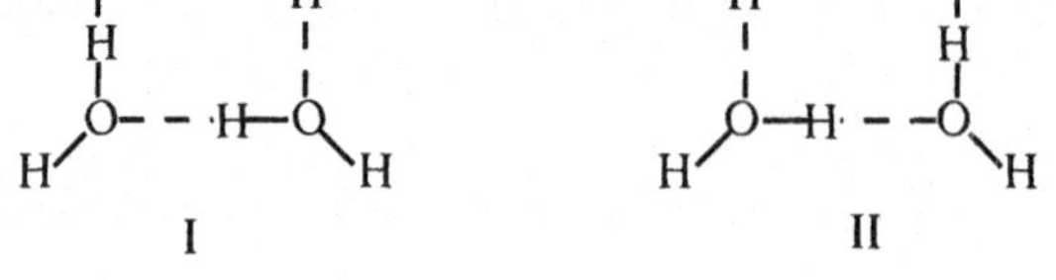

**Figure 4.11** Hydrogen bonded structures of the water of hydration in $Na_2SO_4 \cdot 10H_2O$.

[v] Note that there is no entropy change at the $\lambda$ point where the He changes from the normal fluid to the superfluid. The $\lambda$ transition is a continuous phase transition for which $\Delta S$ and $\Delta H$ are added over a temperature range without an abrupt change in enthalpy or entropy occurring at any particular temperature.

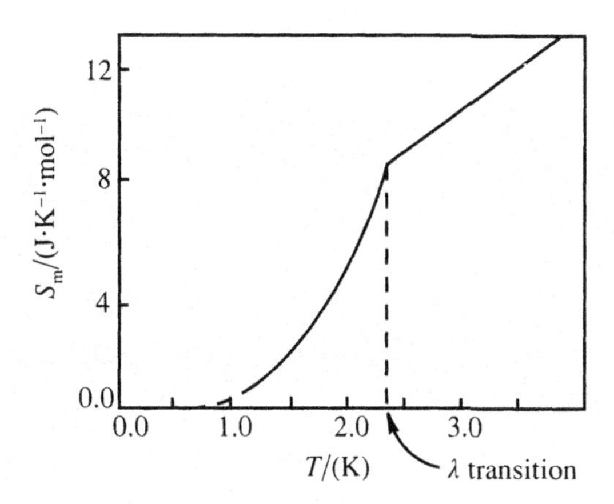


**Figure 4.12** Entropy of liquid helium near absolute zero.

the energy barrier for atoms to migrate through the solid as they segregate is not available. An exception is found in mixtures of liquid $^3He$ and $^4He$ at low temperatures where separation does occur because migration in a superfluid requires much less energy than in a solid.

Paramagnetic substances have a magnetic entropy associated with the random orientation of the spin of the unpaired electrons as shown in (a) below:

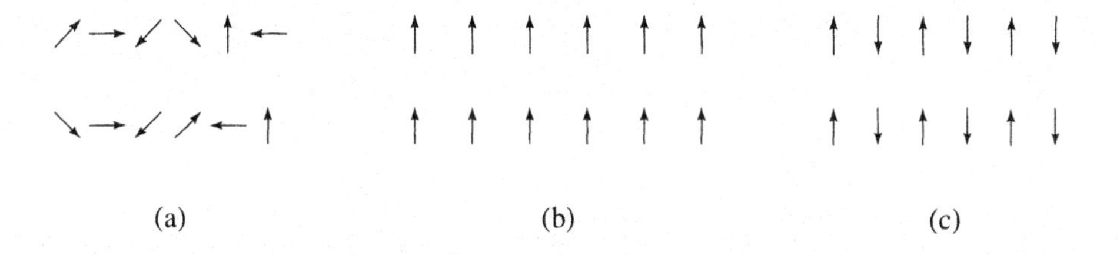


If a paramagnetic salt is cooled, eventually a temperature is reached where the magnetic moments of the electrons line up. This temperature is known as the **Curie point** or the **Neél point**, depending upon whether the spins line up parallel below this temperature as in (b), with the moments pointing in the same direction to reinforce one another and produce ferromagnetism, or the moments line up in opposite directions as in (c) so that they cancel and antiferromagnetism occurs.

Figure 4.13 shows the heat capacity of $NiCl_2$ as a function of temperature.[12] The heat capacity peak at 52 K occurs at the Neél point. Below this temperature, $NiCl_2$ becomes antiferromagnetic and the electron spins become increasingly ordered, so that at 0 Kelvin, $S_0 = 0$. Busey and Giauque[13] were able

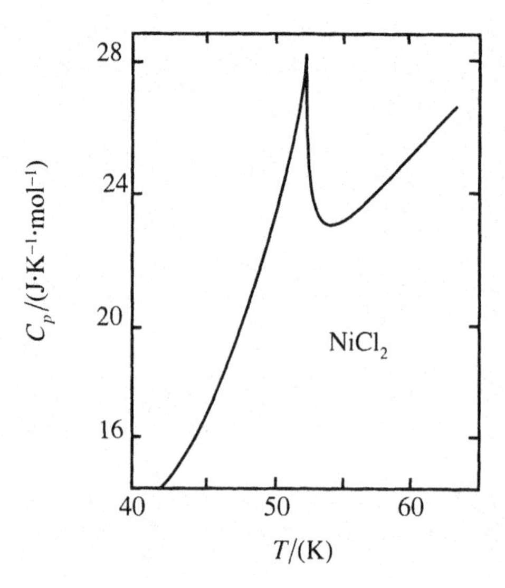


**Figure 4.13** The heat capacity of $NiCl_2$(cr). Above the peak at 52.5 K, $NiCl_2$ is paramagnetic, while below the peak it becomes antiferromagnetic.

to show this by studying the reaction

$$NiCl_2(s) + H_2(g) = Ni(s) + 2HCl(g)$$

at high temperatures. They obtained the entropy of $NiCl_2$ from their measurements and found that it gave the correct $\Delta S$ for this reaction, when combined with the entropies of $H_2$, Ni, and HCl.

Figure 4.14 shows a heat capacity curve[14] for $Ce_2Mg_3(NO_3)_{12}{\cdot}24H_2O$. Again, a maximum occurs in the $C_p$ curve, but this time the maximum is at a very low temperature. In a crystal of $Ce_2Mg_3(NO_3)_{12}{\cdot}24H_2O$, the unpaired electrons are diluted by the water of crystallization so that the interactions are very weak. The Third Law predicts that cooling this salt to a very low temperature should eventually cause the magnetic moments of the electrons to line up so that $S_0 = 0$. Because of the weak interactions, a magnetic field can align the electrons at temperatures somewhat above the very low temperatures required for spontaneous alignment. This magnetic alignment process forms the basis for **adiabatic demagnetization**, a procedure for obtaining very low temperatures that we will describe later in this section. The entropy associated with this alignment could easily be missed from an extrapolation to zero temperature from 10 K. The $C_{p,\mathrm{m}}$ peak is not large, but the area under the $C_{p,\mathrm{m}}/T$ against $T$ graph is significant because of the low $T$,

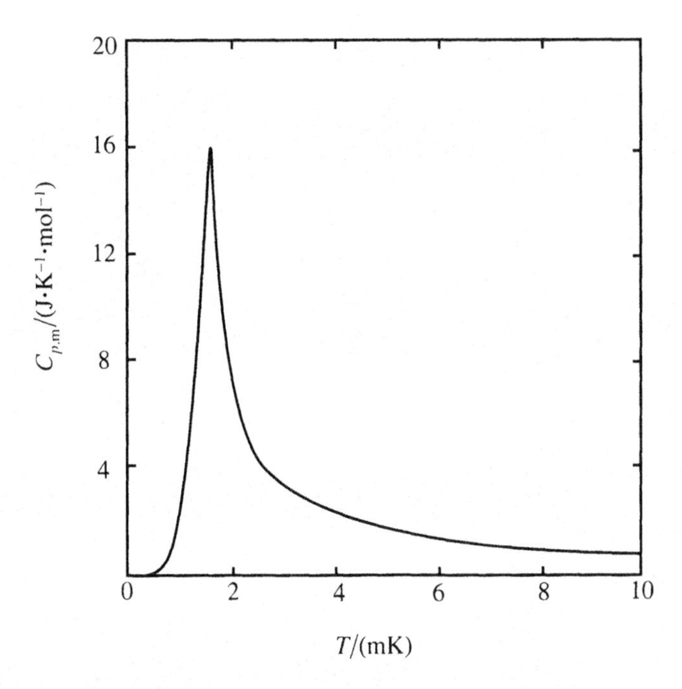


**Figure 4.14** The heat capacity of $Ce_2Mg_3(NO_3)_{12} \cdot 24H_2O$ at very low temperatures.

and large amounts of entropy can be tied up in such electron alignment processes.[w]

In summary, the Third Law predicts that ordering processes are favored as the temperature is lowered, so that eventually perfect order should be obtained in any solid as its temperature approaches 0 K. But kinetic effects are such that the equilibration times needed to achieve this order are sometimes very long.

---

[w] Measuring temperature at these very low temperatures is a major problem. In the paper reporting the $Ce_2Mg_3(NO)_3 \cdot 24H_2O$ measurements, W. F. Giauque and colleagues used an ingenious procedure to determine $T$ in which they measured the effect on the entropy of adding heat in a constant magnetic field $\mathcal{H}$. It can be shown that the thermodynamic temperature is given by

$$T = \left( \frac{\delta q}{\mathrm{d}S} \right)_{\mathcal{H}}.$$

The procedure is beyond the scope of this book. The reader is referred to the paper for details.

## 4.3b Limiting Values for Thermal Properties at Zero Kelvin

The Third Law predicts that $\lim_{T \to 0} S_T = 0$. It follows that $S_T$ must become independent of $p$, $V$, and $T$ at 0 Kelvin, so that derivatives of $S_T$ must also go to zero in the limit. Hence,

$$\lim_{T \to 0} \left( \frac{\partial S_T}{\partial p} \right)_T = 0 \tag{4.18}$$

$$\lim_{T \to 0} \left( \frac{\partial S_T}{\partial V} \right)_T = 0 \tag{4.19}$$

$$\lim_{T \to 0} \left( \frac{\partial S_T}{\partial T} \right)_V = 0 \tag{4.20}$$

$$\lim_{T \to 0} \left( \frac{\partial S_T}{\partial T} \right)_p = 0. \tag{4.21}$$

These relationships can be used to calculate limiting values for several thermal properties as the temperature approaches zero Kelvin.

**Coefficient of Expansion:** The change in entropy with pressure is related to the coefficient of expansion by

$$\left( \frac{\partial S}{\partial p} \right)_T = - \left( \frac{\partial V}{\partial T} \right)_p = -\alpha V.$$

From equation (4.18) given above

$$\lim_{T \to 0} \left( \frac{\partial S}{\partial p} \right)_T = \lim_{T \to 0} (-\alpha V) = 0.$$

Since $V$ remains finite at 0 Kelvin, $\alpha$ must go to zero. Thus,

$$\lim_{T \to 0} \alpha = 0.$$

**Temperature Gradient of Pressure:** From equation (4.19)

$$\lim_{T\to 0}\left(\frac{\partial S}{\partial V}\right)_T = \lim_{T\to 0}\left(\frac{\partial p}{\partial T}\right)_V = 0.$$

Thus, a solid confined to constant volume at 0 Kelvin does not exert an increase in pressure on the container as the temperature is increased.

**Heat Capacity:** From equation (4.20)

$$\lim_{T\to 0}\left(\frac{\partial S}{\partial T}\right)_V = \lim_{T\to 0}\frac{C_V}{T} = 0.$$

For $C_V/T$ to approach zero as $T$ approaches zero, $C_V$ must go to zero at a rate at least proportional to $T$. Earlier, we summarized the temperature dependence of $C_V$ on $T$ for different substances and showed that this is true. For example, most solids follow the Debye low-temperature heat capacity equation of low $T$ for which

$$C_{V,\mathrm{m}} = aT^3$$

so that

$$\frac{C_{V,\mathrm{m}}}{T} = aT^2$$

and

$$\lim_{T\to 0}\frac{C_{V,\mathrm{m}}}{T} = \lim_{T\to 0} T^2 = 0.$$

In equation (3.68), we found that $C_{p,\mathrm{m}}$ and $C_{V,\mathrm{m}}$ are related by

$$C_{p,\mathrm{m}} - C_{V,\mathrm{m}} = \frac{\alpha^2 V_\mathrm{m} T}{\kappa} \qquad (3.68)$$

where $\kappa$ is the compressibility.

Using the equation, we get

$$\lim_{T\to 0}(C_{p,\mathrm{m}} - C_{V,\mathrm{m}}) = \lim_{T\to 0}\frac{\alpha^2 V_\mathrm{m} T}{\kappa}.$$

Since $\alpha^2$ and $T$ go to zero as $T \to 0$, $\alpha^2 V_\mathrm{m} T/\kappa \to 0$ as $T \to 0$, and

$$\lim_{T\to 0}(C_{p,\mathrm{m}} - C_{V,\mathrm{m}}) = 0.$$

Thus $C_{p,\mathrm{m}}$ and $C_{V,\mathrm{m}}$ differ little from one another at low temperatures. The Debye low-temperature heat capacity equation (and other low-temperature relationships) we have summarized calculates $C_{p,\mathrm{m}}$, as well as $C_{V,\mathrm{m}}$, without significant error.

**$G_0$ and $H_0$:** Gibbs free energy and enthalpy are related by

$$G = H - TS.$$

At 0 Kelvin, both $T$ and $S$ are zero. Hence $G_0 = H_0$. This relationship will become important as we work with thermodynamic functions described at the end of this chapter.

## 4.4 Production of Low Temperatures and the Inaccessibility of Absolute Zero

### 4.4a Production of Low Temperatures

A number of techniques have been used to cool substances to low temperatures. Often, the process is done in steps: one procedure is used to cool to a certain temperature, then a second cools to still lower temperatures, and so on.

**Joule–Thomson Expansion and Evaporation Techniques** Air can be cooled through a Joule–Thomson expansion or similar processes until it liquifies. Distillation separates the oxygen and other minor components that are present in the liquid air to give liquid $N_2$, which has a normal boiling point of 77 K. Substances can easily be cooled to this temperature by immersing them in liquid $N_2$. Pumping on the liquid $N_2$ causes cooling until the $N_2$ solidifies at approximately 63 K. Further pumping cools the solid $N_2$ to 50–55 K (depending upon the conditions), but the vapor pressure of solid $N_2$ becomes too low below these temperatures to produce more cooling by evaporation (sublimation) of solid $N_2$.

In the next cooling step, helium gas is cooled to liquid nitrogen temperatures to get it below its Joule–Thomson inversion temperature. Compressions and expansions of this cold He eventually produce liquid helium at temperatures near its normal boiling point of 4.2 K. Immersing a sample in the liquid helium will cool it to this temperature. By pumping on liquid helium, temperatures as low as 1 K can be obtained in the liquid. Below this temperature the vapor pressure of the helium becomes very low and pumping to produce lower temperatures becomes ineffective. $^3$He has a higher vapor pressure than $^4$He. Pumping on $^3$He can produce temperatures as low as 0.3 K.

**Adiabatic Demagnetization** A process known as adiabatic demagnetization can be used to produce still lower temperatures. A schematic of the apparatus is shown in Figure 4.15. The sample contains a paramagnetic salt in which the interactions between the unpaired electrons are very weak. An example is the $Ce_2Mg_3(NO_3)_{12}{\cdot}24H_2O$ described earlier. Gadolinium salts such as $Gd(PMo_{12}O_{40})_3{\cdot}30H_2O$ can also be used.[15] In each case, alignment of the electron spins at zero magnetic field occurs at a temperature below 1 K.

The paramagnetic salt, along with the rest of the sample, is in thermal contact with liquid helium, and cooled by it to about 1 K. A strong magnetic field is then applied to the sample, causing the magnetic moments of the electrons to line-up. The result is a large decrease in entropy. The subsequent evolution of heat associated with the ordering process is absorbed by the liquid helium.

The sample is then isolated from the helium bath, usually by disengaging a mechanical thermal switch. After this thermal isolation of the sample, the magnetic field is removed, and the magnetic moments of the electrons resume a random arrangement. During the randomization process, the electronic system

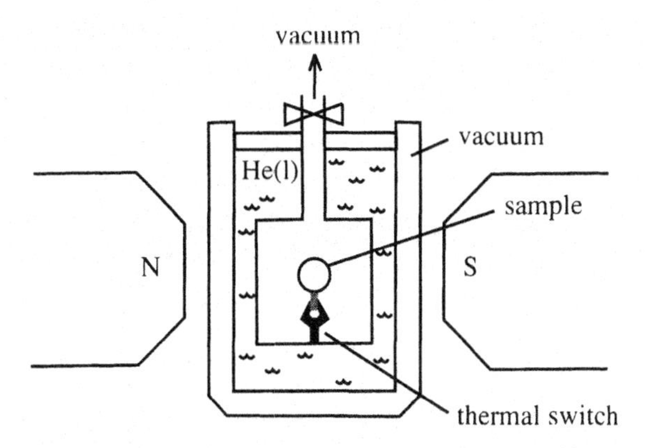


**Figure 4.15** Schematic of the experimental setup used in an adiabatic demagnetization experiment.

absorbs heat from the crystal lattice of the solid, causing a large increase in entropy, and a significant lowering of the temperature of the sample. Temperatures as low as $10^{-3}$ K can be obtained by this process.

**Nuclear Alignment** Alignment of nuclear spins can produce still lower temperatures, although of a specialized nature. The process, which may be called nuclear adiabatic demagnetization, starts with a sample containing nuclei with non-zero nuclear spins. The sample is cooled to $10^{-3}$ K in a magnetic field by adiabatic demagnetization of the paramagnetic salt as described above. This causes alignment of the nuclear spins. Isolation of the sample and removal of the magnetic field produces nuclei with temperatures corresponding to $10^{-6}$ K.[x] Hakonen, Lounasmaa and Oja[16] review the history of the experiments leading to these very low temperatures.

**Laser Cooling** The 1997 Nobel Prize in physics was shared by Steven Chu of Stanford University, William D. Phillips of the National Institute of Science and Technology, and Claude N. Cohen-Tannoudje of the Collège de France for their development and theoretical explanation of laser cooling, a process that can lower the temperature of a gas to a very low value.

In the process of laser cooling, an atomic beam of gaseous atoms (sodium and cesium have most often been used in neutral atom experiments) are directed into a region of intense laser radiation. Six laser beams arranged as three orthogonal pairs of counter-propagating beams are brought together.[y] The lasers are tuned to emit photons of a frequency slightly less than the resonance frequency for absorption by the atoms in the beam. For an atom moving toward the beam, a Doppler shift of the laser frequency occurs, bringing the frequency of the laser photon into resonance, where it is absorbed. Because the photon has momentum, the absorption slows the atom, and a deceleration specific for the direction of motion occurs. But since laser beams come from the six orthogonal directions, the overall effect is to slow fast moving atoms moving in any direction. The term "optical molasses" has been used to describe this region of intense laser light, because it has the effect of a viscous medium in slowing down the motion of the beam. No matter which direction the atom moves the molasses holds it back.

The absorbed radiation is quickly emitted by the atom, either through stimulated emission or spontaneous emission. Stimulated emission occurs in the

---

[x] It is only the temperature of the nuclear spin system that is at this very low temperature. The system is not in thermal equilibrium. Energy will immediately begin to flow from the lattice vibrations to the spin system, so that the spin temperature increases rapidly until it is at the same value as the lattice temperature (approximately $10^{-3}$ K).

[y] Another way to visualize this arrangement is to imagine that the beams are directed inward from the six faces of a cube.

same direction as the absorption, and the recoil effect accelerates the atom back to its original velocity so that the absorption and emission effects cancel. Spontaneous emission, on the other hand, results in a random emission of a photon (and hence recoil of the atom) in any direction, with the net effect that, on average, the atom slows down in the direction of the absorption.[z]

For a typical sodium atom, the initial velocity in the atomic beam is about 1000 $m{\cdot}s^{-1}$ and the velocity change per photon absorbed is 3 $cm{\cdot}s^{-1}$. This means that the sodium atom must absorb and spontaneously emit over $3 \times 10^4$ photons to be stopped. It can be shown that the maximum rate of velocity change for an atom of mass $m$ with a photon of frequency $\nu$ is equal to $h\nu/2mc\tau$ where $h$ and $c$ are Planck's constant and the speed of light, and $\tau$ is the lifetime for spontaneous emission from the excited state. For sodium, this corresponds to a deceleration of about $10^6$ $m{\cdot}s^{-2}$. This should be sufficient to stop the motion of 1000 $m{\cdot}s^{-1}$ sodium atoms in a time of approximately 1 ms over a distance of 0.5 m, a condition that can be realized in the laboratory.

In more recent experiments using a cesium atomic beam, temperatures as low as 2.5 μK have been reported.[aa] Enthusiastic workers in the field predict that the temperature may be decreased by as much as another factor of $10^6$ before the final minimum temperature is realized, and perhaps temperatures lower than 2.5 μK have been claimed by now.[bb]

---

[z] With spontaneous emission, the excited atom has an equal probability of emitting a photon in the direction of the beam, or opposite to the direction of the beam. The deceleration resulting from the absorption of the photon cancels when the emission is in the direction of the beam, but double the deceleration occurs when the emission is in the opposite direction. The net result is a slowing down of the atom with repeated absorptions.

[aa] Many interesting experiments can be performed with a sample cooled to this low a temperature, but methods must be devised to contain the sample to keep the atoms from "drifting off". One way is to switch on a magnetic field of the appropriate geometry so that when the laser is switched off, the sample is effectively confined in a magnetic box. Only small magnetic fields are required (unlike those required to contain hydrogen plasma at high temperatures where nuclear fusion could occur).

[bb] The question needs to be asked, how does one measure the final temperature after laser cooling? The temperature is not zero, but the molecular velocities have been reduced to values corresponding to very low temperatures. Time of flight measurements on the cooled beam can be used to determine the final distribution of the speeds of the atoms, and temperature can be calculated from the average energy above 0 Kelvin. The relationship (obtained from kinetic-molecular theory) is

$$\frac{U - U_0}{N} = \frac{3}{2} kT, \tag{4.22}$$

where $U - U_0$ is the total thermal energy of the atoms above zero Kelvin, $N$ is the number of atoms, and $k$ is the Boltzmann constant.

## 4.4b Inaccessibility of Absolute Zero

An alternate statement of the Third Law is the 1912 statement by W. Nernst: "Absolute zero is unattainable." To show the equivalence of the two statements of the Third Law consider the process

$$A(\mathcal{H} = 0, T = T') \rightarrow B(\mathcal{H} > 0, T = T'') \tag{4.23}$$

where $\mathcal{H}$ is the magnetic field strength. That is, a substance in state A in a zero field is changed to state B by turning on a magnetic field of strength $\mathcal{H}$ in an adiabatic enclosure. In the process the temperature changes from $T'$ to $T''$. The molar entropies $S^{\mathrm{A}}_{\mathrm{m},\,T'}$ and $S^{\mathrm{B}}_{\mathrm{m},\,T''}$ are given by

$$S^{\mathrm{A}}_{\mathrm{m},\,T'} = S^{\mathrm{A}}_{\mathrm{m},\,0} + \int_0^{T'} \frac{C^{\mathrm{A}}_{\mathrm{m}}}{T}\,\mathrm{d}T$$

and

$$S^{\mathrm{B}}_{\mathrm{m},\,T''} = S^{\mathrm{B}}_{\mathrm{m},\,0} + \int_0^{T''} \frac{C^{\mathrm{B}}_{\mathrm{m}}}{T}\,\mathrm{d}T,$$

where $C_{\mathrm{m}}$ is the molar heat capacity.

From the Third Law, $S^{\mathrm{A}}_{\mathrm{m},\,0} = S^{\mathrm{B}}_{\mathrm{m},\,0} = 0$, so that

$$S^{\mathrm{A}}_{\mathrm{m},\,T'} = \int_0^{T'} \frac{C^{\mathrm{A}}_{\mathrm{m}}}{T}\,\mathrm{d}T \tag{4.24}$$

$$S^{\mathrm{B}}_{\mathrm{m},\,T''} = \int_0^{T''} \frac{C^{\mathrm{B}}_{\mathrm{m}}}{T}\,\mathrm{d}T. \tag{4.25}$$

In the process represented by equation (4.23), the temperature changes from $T'$ to $T''$. What we must investigate is the possibility that $T''$ can equal 0 Kelvin. The Second Law predicts that for this process

$$\Delta S = S^{\mathrm{A}}_{\mathrm{m},\,T'} - S^{\mathrm{B}}_{\mathrm{m},\,T''} \geqslant 0$$

where the equality applies to the reversible process and the inequality to the spontaneous process. We will select the reversible process $[\Delta S(\mathrm{A} \rightarrow \mathrm{B}) = 0]$, since $T''$ will be the lowest when this occurs because

$$\int_0^{T''} \frac{C^{\mathrm{B}}_{\mathrm{m}}}{T}\,\mathrm{d}T$$

is the smallest when the process is reversible. In other words, the possibility of reaching absolute zero is best when the process is reversible.

In order that $\Delta S$ for this process be equal to zero, $S^{\mathrm{A}}_{\mathrm{m},T'}$ must equal $S^{\mathrm{B}}_{\mathrm{m},T''}$. For this to be so, equations (4.24) and (4.25) require that,

$$\int_0^{T'} \frac{C^{\mathrm{A}}_{\mathrm{m}}}{T}\,\mathrm{d}T = \int_0^{T''} \frac{C^{\mathrm{B}}_{\mathrm{m}}}{T}\,\mathrm{d}T. \qquad (4.26)$$

Since the heat capacities $C^{\mathrm{A}}_{\mathrm{m}}$ and $C^{\mathrm{B}}_{\mathrm{m}}$ cannot be less than zero, equation (4.26) can be true only if both $T'$ and $T''$ are greater than zero or $T'$ and $T''$ are both equal to zero (a trivial case). Hence, $T''$ cannot be zero if $T'$ is finite and it is not possible to devise a process that will produce a temperature of absolute zero. A quote from P. A. Rock[17] is appropriate at this point,

> *The predictions of the Third Law have been verified in a sufficiently large number of cases that experimental attempts to reach absolute zero are now placed in the same class as attempts to devise perpetual motion machines — which is to say there are much more productive ways to spend one's time. Much experimental work is carried out, however, at very low temperatures, because the behavior of matter under these conditions has produced many surprises and led to the uncovering of a great deal of new knowledge and the development of useful new devices, such as superconducting magnets.*[cc]

## 4.5 Thermodynamic Functions

A cryogenic calorimeter measures $C_{p,\mathrm{m}}$ as a function of temperature. We have seen that with the aid of the Third Law, the $C_{p,\mathrm{m}}$ data (along with $\Delta H_{\mathrm{m}}$ for phase changes) can be integrated to give the absolute entropy

$$S_{\mathrm{m},T} = \int_0^{T} \frac{C_{p,\mathrm{m}}}{T}\,\mathrm{d}T + \sum \frac{\Delta_{\text{phase change}}\mathrm{H}_{\mathrm{m}}}{T}.$$

Integrating to successively higher $T$ gives $S_{\mathrm{m}}$ values as a function of $T$. $S_{\mathrm{m},T}$ also depends upon $p$, especially if the substance is in the gas phase at the temperature. The entropy values are usually calculated and tabulated at a

[cc] W. F. Giauque was fond of saying that in the study of the effects of temperature on substances, a logarithmic temperature scale would be more appropriate in many ways. That is, as many interesting phenomena occur between 10 and 100 K as between 100 and 1000 K. Equally as important are the temperature ranges from 1 to 10 K, 0.1 to 1 K, 0.01 to 0.1 K, etc.

pressure of 0.1 MPa (1 bar). If the substance is present in the gas phase, corrections are made as described earlier to give the entropy of the ideal gas. The condition of 0.1 MPa pressure (and ideal gas behavior if a gas is involved) is referred to as the standard state and the superscript zero is used to represent this state. Thus, $S^\circ_m$ designates the entropy in the standard state.[dd]

Heat capacities can also be used to calculate enthalpy differences. To find the relationship we start with the equation

$$\left(\frac{\partial H_m}{\partial T}\right)_p = C_{p,m}.$$

Separating variables and integrating gives

$$\int_{H_{m,0}}^{H_{m,T}} dH_m = \int_0^T C_{p,m}\, dT$$

$$H_{m,T} - H_{m,0} = \int_0^T C_{p,m}\, dT + \sum \Delta_{\text{phase change}} H_m \tag{4.27}$$

where $H_{m,0}$ is the enthalpy at 0 Kelvin.[ee] Again, the results can be obtained for the standard state to give values for $H^\circ_{m,T} - H^\circ_{m,0}$. A value of particular interest is $H^\circ_{m,298} - H^\circ_{m,0}$. It can be obtained by integrating from 0 K to 298.15 K.

$$H^\circ_{m,298} - H^\circ_{m,0} = \int_0^{298} C^\circ_{p,m}\, dT + \sum \Delta_{\text{phase change}} \mathrm{H}^\circ_{m,}. \tag{4.28}$$

Integration of equation (4.27) at a series of temperatures gives a set of $H^\circ_{m,T} - H^\circ_{m,0}$ values. These values are known as enthalpy functions and can be tabulated along with $C^\circ_{p,m,T}$ and $S^\circ_{m,T}$.

Free energy functions, $(G^\circ_{m,T} - H^\circ_{m,0})$, can also be obtained by starting with the equation

$$\left(\frac{\partial G^\circ_{m,T}}{\partial T}\right)_p = -S^\circ_m$$

---

[dd] In the next chapter we will describe standard states in detail.

[ee] Equation (4.27) indicates that we must also add the enthalpy differences for any isothermal phase changes that may occur in the temperature interval from 0 Kelvin to $T$.

and integrating. The result is

$$\int_{G^\circ_{\mathrm{m},0}}^{G^\circ_{\mathrm{m},T}} \mathrm{d}G_\mathrm{m} = -\int_0^T S^\circ_\mathrm{m}\,\mathrm{d}T$$

$$G^\circ_{\mathrm{m},T} - G^\circ_{\mathrm{m},0} = -\int_0^T S^\circ_\mathrm{m}\,\mathrm{d}T. \tag{4.29}$$

But $G^\circ_{\mathrm{m},0} = H^\circ_{\mathrm{m},0}$ so that equation (4.29) can be written as

$$G^\circ_{\mathrm{m},T} - H^\circ_{\mathrm{m},0} = -\int_0^T S^\circ_{\mathrm{m},T}\,\mathrm{d}T. \tag{4.30}$$

Thus, values for $C^\circ_{p,\mathrm{m},T}$, $S^\circ_{\mathrm{m},T}$, $(H^\circ_{\mathrm{m},T} - H^\circ_{\mathrm{m},0})$ and $(G^\circ_{\mathrm{m},T} - H^\circ_{\mathrm{m},0})$ can be obtained as a function of temperature and tabulated. Figure 4.16 summarizes values for these four quantities as a function of temperature for glucose, obtained from the low-temperature heat capacity data described earlier. Note that the enthalpy and Gibbs free energy functions are graphed as $(H^\circ_{\mathrm{m},T} - H^\circ_{\mathrm{m},0})/T$ and $(G^\circ_{\mathrm{m},T} - H^\circ_{\mathrm{m},0})/T$. This allows all four functions to be plotted on the same scale. Figure 4.16 demonstrates the almost linear nature of the $(G^\circ_{\mathrm{m},T} - H^\circ_{\mathrm{m},0})/T$ function. This linearity allows one to easily interpolate between tabulated values of this function to obtain the value at the temperature of choice.

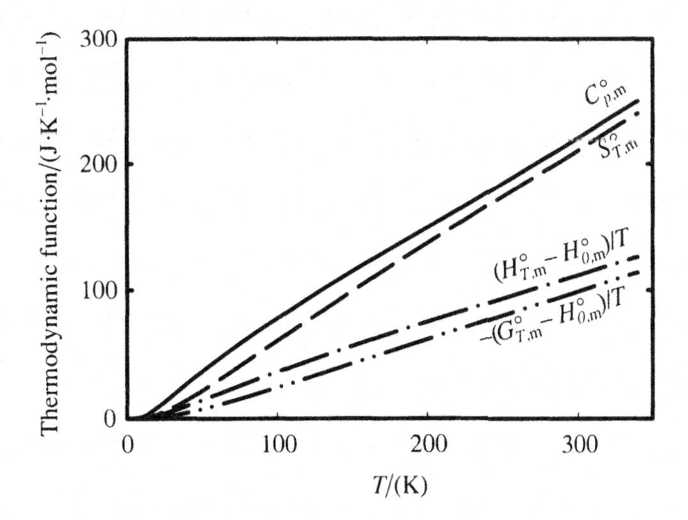


**Figure 4.16** Comparison of the thermodynamic functions for glucose. ———, $C^\circ_{p,\mathrm{m}}$; — — —, $S^\circ_{T,\mathrm{m}}$; —·—·—·—, $(H^\circ_{T,\mathrm{m}} - H^\circ_{0,\mathrm{m}})/T$; —··—··—, $-(G^\circ_{T,\mathrm{m}} - H^\circ_{0,\mathrm{m}})/T$.

Enthalpy and free energy functions can also be tabulated using $T = 298.15$ K as a reference temperature by making use of the relationships

$$(H^\circ_{m,T} - H^\circ_{m,298}) = (H^\circ_{m,T} - H^\circ_{m,0}) - (H^\circ_{m,298} - H^\circ_{m,0})$$

$$(G^\circ_{m,T} - H^\circ_{m,298}) = (G^\circ_{m,T} - H^\circ_{m,0}) - (H^\circ_{m,298} - H^\circ_{m,0}).$$

The free energy function can also be calculated from $S^\circ_{m,T}$ and $(H^\circ_{m,T} - H^\circ_{m,298})$ directly from the relationship

$$(G^\circ_{m,T} - H^\circ_{m,298}) = (H^\circ_{m,T} - H^\circ_{m,298}) - TS^\circ_{m,T}.$$

Values for the thermodynamic functions as a function of temperature for condensed phases are usually obtained from Third Law measurements. Values for ideal gases are usually calculated from the molecular parameters using the statistical mechanics procedures to be described in Chapter 10. In either case, $C^\circ_{p,m,T}$, $S^\circ_{m,T}$, $(H^\circ_{m,T} - H^\circ_{m,0})$ or $(H^\circ_{m,T} - H^\circ_{m,298})$, and $(G^\circ_{m,T} - H^\circ_{m,0})$ or $(G^\circ_{m,T} - H^\circ_{m,298})$ have been determined for a large number of chemical substances and are tabulated. The most comprehensive tabulation can be found in the JANAF tables.[18] Values are tabulated in those tables as $T/\mathrm{K}$, $C^\circ_p$, $S^\circ$, $-[G^\circ_T - H^\circ(T_r)]/T$, $H^\circ_T - H^\circ(T_r)$, and $\Delta_f H^\circ$, $\Delta_f G^\circ$ and $\log K_f$ (quantities that we will describe later). In the table $T_r$ is the reference temperature, which is almost always 298.15 K.

Figure 4.16 demonstrates that thermodynamic functions are not very exciting. That is, no unusual changes or effects occur. But these functions are very useful in calculating the thermodynamic quantities $\Delta_r S$, $\Delta_r H$, $\Delta_r G$, ... for a chemical reaction. When the reaction is written as[ff]

$$\sum_i \nu_i \mathrm{A_i} = 0$$

the heat capacity and entropy changes for the reaction at a particular temperature can be calculated from

$$\Delta_r C^\circ_{p,T} = \sum_i \nu_i C^\circ_{p,m,i,T} \qquad (4.31)$$

$$\Delta_r S^\circ_T = \sum_i \nu_i S^\circ_{m,i,T}. \qquad (4.32)$$

---

[ff] See equation (1.2).

The enthalpy and Gibbs free energy change can be calculated from the functions involving $H^\circ_{m,298}$ by

$$\Delta_r H^\circ_T = \sum_i \nu_i (H^\circ_{m,T} - H^\circ_{m,298})_i + \Delta_r H^\circ_{m,298} \tag{4.33}$$

$$\Delta_r G^\circ_T = \sum_i \nu_i (G^\circ_{m,T} - H^\circ_{m,298})_i + \Delta_r H^\circ_{298}. \tag{4.34}$$

When $H^\circ_{m,0}$ values are used, the equations become

$$\Delta_r H^\circ_T = \sum_i \nu_i (H^\circ_{m,T} - H^\circ_{m,0})_i + \Delta_r H^\circ_0 \tag{4.35}$$

$$\Delta_r G^\circ_T = \sum_i \nu_i (G^\circ_{m,T} - H^\circ_{m,0})_i + \Delta_r H^\circ_0. \tag{4.36}$$

Table 4.3 summarizes values taken from the JANAF tables for the Gibbs free energy functions and standard enthalpies of formation for a few common substances. The JANAF tables provide a more complete tabulation.

We will see in a later chapter how to calculate $\Delta_r H^\circ$ from enthalpies of formation, from which $\Delta_r G^\circ$ can be calculated using Table 4.3 and equations (4.34) or (4.36).

**Example 4.1:** Use the Gibbs free energy functions from Table 4.3 to calculate $\Delta_r G^\circ$ at 1000 K for the reaction

$$2H_2O(g) = 2H_2(g) + O_2(g),$$

given that $\Delta_r H^\circ_{298} = 483.652$ kJ, a value that we will later see how to obtain.

**Solution:** From Table 4.3, we obtain the following values at 1000 K.

| Substance | $-[(G^\circ_T - H^\circ_{298})/T]/$ $(J \cdot K^{-1} \cdot mol^{-1})$ |
|---|---|
| $H_2(g)$ | 145.536 |
| $O_2(g)$ | 220.875 |
| $H_2O(g)$ | 206.738 |

**Table 4.3** Thermodynamic functions*†

$T_r = 298.15$ K; standard state pressure = 0.1 MPa

| Substance | $\{-[G^\circ_T - H^\circ(T_r)]/T\}$ (J·K$^{-1}$·mol$^{-1}$) | | | | | | | $\Delta_f H^\circ(T_r)$/ (kJ·mol$^{-1}$) |
|---|---|---|---|---|---|---|---|---|
| | 100 K | 200 K | 298.15 K | 500 K | 1000 K | 1500 K | 2000 K | |
| C(g) | 176.684 | 160.007 | 158.100 | 160.459 | 168.678 | 175.044 | 179.996 | 716.670 |
| C(graphite) | 10.867 | 6.407 | 5.740 | 6.932 | 12.662 | 18.216 | 23.008 | 0 |
| $CCl_4$(g) | 374.168 | 317.054 | 309.809 | 319.840 | 357.595 | 388.320 | 412.725 | −95.981 |
| CH(g) | 208.978 | 185.708 | 183.040 | 186.349 | 198.062 | 207.595 | 215.440 | 594.128 |
| $CHCl_3$(g) | 345.889 | 301.259 | 295.620 | 303.641 | 335.240 | 362.168 | 384.180 | −103.177 |
| $CH_2$(g) | 223.967 | 197.055 | 193.931 | 197.957 | 212.971 | 225.773 | 236.489 | 386.392 |
| $CH_3$(g) | 226.465 | 197.609 | 194.170 | 198.783 | 216.836 | 232.885 | 246.636 | 145.687 |
| $CH_3F$(g) | 253.738 | 226.124 | 222.843 | 227.540 | 248.097 | 267.919 | 285.399 | −234.304 |
| $CH_4$(g) | 216.485 | 189.418 | 186.251 | 190.614 | 209.370 | 227.660 | 244.057 | −74.873 |
| CO(g) | 223.539 | 200.317 | 197.653 | 200.968 | 212.848 | 222.526 | 230.342 | −110.527 |
| $CO_2$(g) | 243.568 | 217.046 | 213.795 | 218.290 | 235.901 | 251.062 | 263.574 | −393.522 |
| $C_2$(g) | 235.431 | 203.335 | 199.382 | 204.042 | 219.054 | 230.338 | 239.145 | 837.737 |
| $C_2H_2$(g) | 234.338 | 204.720 | 200.958 | 206.393 | 227.984 | 246.853 | 262.733 | 226.731 |
| $C_2H_4$(g) | 252.466 | 222.975 | 219.330 | 224.879 | 249.742 | 273.827 | 295.101 | 52.467 |
| $C_3O_2$(g) | 328.013 | 281.870 | 276.071 | 284.216 | 316.075 | 343.406 | 365.899 | −93.638 |
| Cl(g) | 184.104 | 167.161 | 165.189 | 167.708 | 176.615 | 183.475 | 188.749 | 121.302 |
| $Cl_2$(g) | 251.696 | 226.120 | 223.079 | 227.020 | 241.203 | 252.438 | 261.277 | 0 |
| D(g) | 141.830 | 125.251 | 123.350 | 125.705 | 133.916 | 140.278 | 145.225 | 221.720 |
| $D_2$(g) | 170.973 | 147.630 | 144.960 | 148.272 | 159.943 | 169.311 | 176.885 | 0 |
| F(g) | 178.805 | 160.833 | 158.750 | 161.307 | 170.038 | 176.673 | 181.778 | 79.390 |
| $F_2$(g) | 229.549 | 205.596 | 202.789 | 206.452 | 219.930 | 230.839 | 239.531 | 0 |
| H(g) | 133.197 | 116.618 | 114.716 | 117.072 | 125.282 | 131.644 | 136.591 | 217.999 |
| HBr(g) | 224.595 | 201.363 | 198.699 | 202.006 | 213.737 | 223.228 | 230.905 | −36.443 |
| HCl(g) | 212.797 | 189.566 | 186.901 | 190.205 | 201.857 | 211.214 | 218.769 | −92.312 |
| HCN(g) | 230.911 | 204.966 | 201.828 | 206.125 | 222.619 | 236.724 | 248.441 | 135.143 |
| HF(g) | 199.679 | 176.445 | 173.780 | 177.082 | 188.631 | 197.733 | 205.011 | −272.546 |
| $HNO_3$(g) | 306.248 | 270.926 | 266.400 | 273.121 | 300.937 | 325.627 | 346.243 | −134.306 |
| $H_2$(g) | 155.408 | 133.284 | 130.680 | 133.973 | 145.536 | 154.652 | 161.943 | 0 |
| $H_2O$(g) | 218.534 | 191.896 | 188.834 | 192.685 | 206.738 | 218.520 | 228.374 | −241.826 |

| | | | | | | | | |
|---|---|---|---|---|---|---|---|---|
| $H_2O(l,g)$ | | | 69.950(l) | 104.712 | 162.736 | 189.184 | – | −285.830 |
| $H_2S(g)$ | 235.330 | 208.586 | 205.757 | 209.726 | 224.599 | 237.375 | 248.098 | −20.502 |
| $H_2SO_4(cr,l)$ | 300.474 | 181.887 | 156.895 | 173.938 | 245.666 | – | – | −813.989 |
| $H_2SO_4(g)$ | 360.452 | 305.914 | 298.796 | 309.259 | 351.592 | 388.416 | 418.890 | −735.129 |
| Hg(cr,l) | 113.592 | 80.284 | 76.028 | 79.158 | – | – | – | 0 |
| Hg(g) | 193.450 | 176.871 | 174.970 | 177.325 | 185.536 | 191.898 | 196.845 | 61.380 |
| HgO(cr) | 105.148 | 74.128 | 70.270 | 75.556 | 95.855 | – | – | −90.789 |
| I(g) | 199.267 | 182.688 | 180.786 | 183.142 | 191.352 | 197.717 | 202.679 | 106.762 |
| $I_2(g)$ | 292.808 | 264.038 | 260.685 | 264.891 | 279.680 | 291.251 | 300.396 | 62.421 |
| $I_2(cr)$ | 162.472 | 121.036 | 116.142 | – | – | – | – | 0 |
| KCl(cr,l) | 125.871 | 87.169 | 82.554 | 88.494 | 110.276 | 136.664 | 156.662 | −436.684 |
| N(g) | 171.780 | 155.201 | 153.300 | 155.655 | 163.866 | 170.228 | 175.175 | 472.683 |
| $NH_3(g)$ | 223.211 | 195.962 | 192.774 | 197.021 | 213.849 | 229.054 | 242.244 | −45.898 |
| NO(g) | 237.757 | 213.501 | 210.758 | 214.145 | 226.307 | 236.217 | 244.199 | 90.291 |
| $N_2(g)$ | 217.490 | 194.272 | 191.609 | 194.917 | 206.708 | 216.277 | 224.006 | 0 |
| $N_2O(g)$ | 250.829 | 223.335 | 219.957 | 224.613 | 242.694 | 258.120 | 270.769 | 82.048 |
| $N_2O_3(g)$ | 360.299 | 314.268 | 308.539 | 316.466 | 347.364 | 373.706 | 395.230 | 82.843 |
| $N_2O_4(cr,l)$ | 392.476 | 237.881 | 209.198 | 226.598 | – | – | – | −19.564 |
| $N_2O_4(g)$ | 363.098 | 311.015 | 304.376 | 313.907 | 352.127 | 385.251 | 412.484 | 9.079 |
| $N_2O_5(g)$ | | 354.795 | 346.548 | 358.466 | 405.784 | 445.871 | 478.290 | 11.297 |
| NaCl(cr,l) | 114.140 | 76.643 | 72.115 | 77.966 | 99.428 | 125.061 | 144.561 | −411.120 |
| NaCl(g) | 260.320 | 233.022 | 229.793 | 233.906 | 248.503 | 259.969 | 268.959 | −181.418 |
| O(g) | 181.131 | 163.085 | 161.058 | 163.511 | 171.930 | 178.390 | 183.391 | 249.173 |
| OH(g) | 210.980 | 186.471 | 183.708 | 187.082 | 198.801 | 208.043 | 215.446 | 38.987 |
| $O_2(g)$ | 231.094 | 207.823 | 205.147 | 208.524 | 220.875 | 231.002 | 239.160 | 0 |
| $O_3(g)$ | 271.040 | 242.401 | 238.932 | 243.688 | 262.228 | 277.866 | 290.533 | 142.674 |
| P(cr-white) | 62.415 | 43.173 | 41.077 | – | – | – | – | 0 |
| $P_4(g)$ | 331.278 | 285.819 | 279.992 | 288.032 | 317.923 | 341.996 | 361.020 | 58.907 |
| S(cr-rhoubic) | 49.744 | 34.038 | 32.056 | – | – | – | – | 0 |
| $SO_2(g)$ | 281.199 | 251.714 | 248.212 | 252.979 | 271.339 | 286.816 | 299.383 | −296.842 |
| $SO_3(g)$ | 295.976 | 261.145 | 256.769 | 262.992 | 287.768 | 309.041 | 326.421 | −395.765 |
| $S_2(g)$ | 255.684 | 231.072 | 228.165 | 231.958 | 245.760 | 256.832 | 265.657 | 128.600 |

* Taken from M. W. Chase, C. A. Davies, J. R. Downey, Jr., D. J. Frurip, R. A. McDonald, and A. N. Syverud, "JANAF Thermochemical Tables, Third Edition", *J. Phys. Chem. Ref. Data*, **14**, Supplement No. 1, 1985.

† Enthalpies of formation are included in Table 4.1 so that this table can be used independently to calculate $\Delta_r G^\circ$.

For the reaction $2H_2O(g) = 2H_2(g) + O_2(g)$, using equation (4.34) gives

$$\Delta_r G^\circ = \left\{ (2)\left(\frac{G_T^\circ - H_{298}^\circ}{T}\right)(H_2) + \left(\frac{G_T^\circ - H_{298}^\circ}{T}\right)(O_2) - (2)\left(\frac{G_T^\circ - H_{298}^\circ}{T}\right)(H_2O) \right\} \cdot T + \Delta_r H_{298}^\circ$$

$$= -[(2)(145.536) + (220.875) - (2)(206.738)](1000) + 483{,}652$$

$$= 385{,}181 \text{ J}.$$

The determination of $\Delta_r G^\circ$ for a chemical reaction is very useful in predicting the course of the reaction. Qualitatively, we will show in Chapter 5 that with $\Delta_r G^\circ < 0$, the reaction is spontaneous, at least when products and reactants are in their standard state condition. Quantitatively, we will see in Chapter 9 that $\Delta_r G^\circ$ can be used to calculate the equilibrium constant for the reaction, from which the final equilibrium conditions can be determined.

## Exercises

E4.1 Show that at very low temperatures where the Debye low temperature heat capacity equation applies that the entropy is one third of the heat capacity.

E4.2 Solid phases $\alpha$ and $\beta$ are in equilibrium for a pure solid substance at 12 K. Below 12 K, the heat capacities of $\alpha$ and $\beta$ vary with temperature according to the equations

$$C_{p,\text{m}}(\alpha)/(\text{J}\cdot\text{K}^{-1}\cdot\text{mol}^{-1}) = 8.0 \times 10^{-4} T^3$$

$$C_{p,\text{m}}(\beta)/(\text{J}\cdot\text{K}^{-1}\cdot\text{mol}^{-1}) = 5.0 \times 10^{-4} T^3.$$

(a) Calculate $\Delta_{\text{trans}} H_{\text{m}}$ and $\Delta_{\text{trans}} S_{\text{m}}$ at 12 K for the phase transition

$$\alpha(\text{s}) = \beta(\text{s}).$$

(b) Use the Second Law to determine which is the stable form of the substance below 12 K.

## Problems

P4.1 Given the following heat capacity information for magnesium metal as a function of temperature

| $T/(\text{K})$ | $C_{p,\text{m}}/(\text{J·K}^{-1}\text{·mol}^{-1})$ | $T/(\text{K})$ | $C_{p,\text{m}}/(\text{J·K}^{-1}\text{·mol}^{-1})$ |
|---|---|---|---|
| 12 | 0.067 | 130 | 18.94 |
| 14 | 0.109 | 140 | 19.74 |
| 16 | 0.176 | 150 | 20.40 |
| 18 | 0.272 | 160 | 20.97 |
| 20 | 0.360 | 170 | 21.48 |
| 25 | 0.787 | 180 | 21.91 |
| 30 | 1.427 | 190 | 22.30 |
| 35 | 2.301 | 200 | 22.67 |
| 40 | 3.360 | 210 | 22.96 |
| 45 | 4.502 | 220 | 23.22 |
| 50 | 5.720 | 230 | 23.48 |
| 60 | 8.171 | 240 | 23.71 |
| 70 | 10.45 | 250 | 23.93 |
| 80 | 12.47 | 260 | 24.12 |
| 90 | 14.24 | 270 | 24.31 |
| 100 | 15.70 | 280 | 24.49 |
| 110 | 16.95 | 290 | 24.67 |
| 120 | 18.02 | 298.15 | 24.81 |

Use a graphical method to calculate $S_\text{m}$ and $(H_\text{m} - H_{0,\text{m}})/T$ for Mg at each of the temperatures and make a graph as a function of $T$ comparing these two quantities.

P4.2 The organic molecule quinoline has the structure shown below.

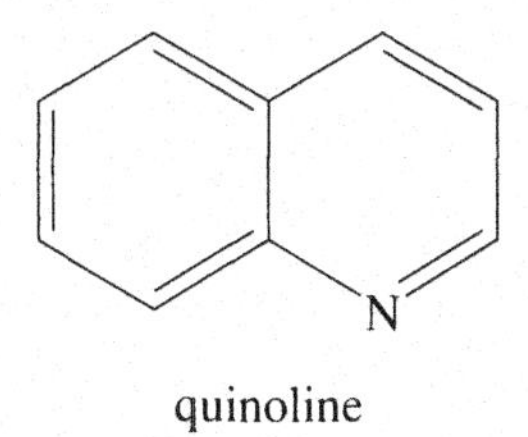


quinoline

(a) Heat capacities have been measured for this compound from 4 K to 450 K. The results below 10 K are summarized in the table below.[19]

| $T/(\text{K})$ | $C_{p,\text{m}}/(\text{J}{\cdot}\text{K}^{-1}{\cdot}\text{mol}^{-1})$ | $T/(\text{K})$ | $C_{p,\text{m}}/(\text{J}{\cdot}\text{K}^{-1}{\cdot}\text{mol}^{-1})$ |
|---|---|---|---|
| 4.813 | 0.407 | 7.725 | 1.596 |
| 6.018 | 0.831 | 8.749 | 2.262 |
| 6.847 | 1.056 | 9.761 | 2.952 |

Graph the above data in the form $C_{p,\text{m}}/T$ against $T^2$ to test the validity of the Debye low-temperature heat capacity relationship [equation (4.4)] and find a value for the constant in the equation.

(b) The heat capacity study also revealed that quinoline undergoes equilibrium phase transitions, with enthalpies as follows:

| | |
|---|---|
| crystal I = crystal II at 220.00 K | $\Delta_{\text{trans}}H_{\text{m}} = 68.179\ \text{J}{\cdot}\text{mol}^{-1}$ |
| crystal II = liquid at 258.369 K | $\Delta_{\text{fus}}H_{\text{m}} = 10.663\ \text{kJ}{\cdot}\text{mol}^{-1}$ |
| liquid = vapor ($p$ = 0.0112 kPa) at 298.15 K | $\Delta_{\text{vap}}H_{\text{m}} = 59.316\ \text{kJ}{\cdot}\text{mol}^{-1}$ |

The heat capacity of the different one-phase regions obtained in this study can be integrated to give the following values.

$$\int_{10.00}^{220.00} \frac{C_{p,\text{m}}(\text{crystal I})}{T}\,\text{d}T = 128.725\ \text{J}{\cdot}\text{K}^{-1}{\cdot}\text{mol}^{-1}$$

$$\int_{220.00}^{258.369} \frac{C_{p,\text{m}}(\text{crystal II})}{T}\,\text{d}T = 21.410\ \text{J}{\cdot}\text{K}^{-1}{\cdot}\text{mol}^{-1}$$

$$\int_{258.369}^{298.15} \frac{C_{p,\text{m}}(\text{liquid})}{T}\,\text{d}T = 26.889\ \text{J}{\cdot}\text{K}^{-1}{\cdot}\text{mol}^{-1}.$$

Use the Third Law to calculate the standard entropy, $S^\circ_{\text{m}}$, of quinoline (g) ($p = 0.101325$ MPa) at $T = 298.15$ K. (You may assume that the effects of pressure on all of the condensed phases are negligible, and that the vapor may be treated as an ideal gas at a pressure of 0.0112 kPa, the vapor pressure of quinoline at 298.15 K.)

(c) Statistical mechanical calculations have been performed on this molecule and yield a value for $S^\circ_{\text{m}}$ of quinoline gas at 298.15 K of 344 $\text{J}{\cdot}\text{K}^{-1}{\cdot}\text{mol}^{-1}$. Assuming an uncertainty of about $\pm 1\ \text{J}{\cdot}\text{K}^{-1}{\cdot}\text{mol}^{-1}$ for both your calculation in part (b) and the statistical calculation, discuss the agreement of the calorimetric value with the statistical result. Does this compound possess a residual entropy at 0 K? Can you rule out a frozen-in disorder arising from the random mixing of molecules with one of the two orientations shown below?

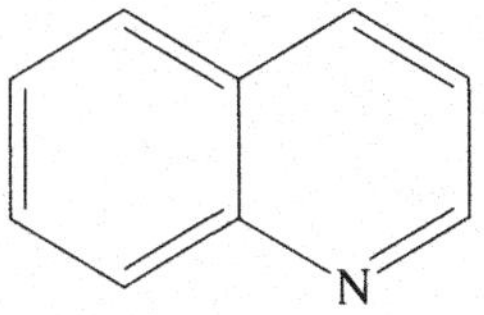

N or N

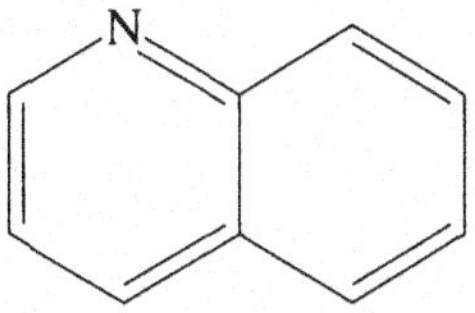

P4.3 Given the following data for perchloryl fluoride ($ClO_3F$)

| $T$/(K) | $C_{p,m}(s)$/ (J·K$^{-1}$·mol$^{-1}$) | $T$/(K) | $C_{p,m}(s)$/ (J·K$^{-1}$·mol$^{-1}$) | $T$/(K) | $C_{p,m}(l)$/ (J·K$^{-1}$·mol$^{-1}$) |
|---|---|---|---|---|---|
| 14.98 | 7.24 | 56.95 | 44.43 | 130.51 | 87.95 |
| 16.55 | 10.33 | 60.14 | 46.74 | 136.41 | 87.99 |
| 18.10 | 11.38 | 64.42 | 49.20 | 142.92 | 88.12 |
| 19.63 | 13.51 | 69.08 | 50.96 | 149.31 | 88.49 |
| 21.75 | 16.23 | 73.31 | 52.22 | 155.38 | 88.95 |
| 24.08 | 19.46 | 78.84 | 53.97 | 162.14 | 89.20 |
| 26.21 | 21.97 | 88.39 | 57.28 | 169.26 | 89.75 |
| 28.48 | 24.56 | 93.06 | 59.08 | 176.16 | 90.29 |
| 30.90 | 27.20 | 97.91 | 61.30 | 183.24 | 91.04 |
| 33.45 | 29.71 | 102.48 | 63.47 | 190.35 | 91.67 |
| 36.62 | 32.38 | 107.67 | 66.07 | 197.27 | 92.55 |
| 40.13 | 34.98 | 113.25 | 70.21 | 204.40 | 93.22 |
| 43.59 | 37.20 | 118.88 | 75.48 | 211.39 | 94.18 |
| 46.97 | 39.04 | 123.31 | 80.37* | 218.22 | 95.10 |
| 50.20 | 40.50 | | | 223.34 | 95.81 |

* High value due to premelting.

$\Delta_{fus}H_m = 3834$ J·mol$^{-1}$ at the triple point temperature, $T = 125.41$ K

$\Delta_{vap}H_m = 19326$ J·mol$^{-1}$ at the normal boiling point temperature, $T = 226.48$ K

(At the normal boiling point, $p = 1$ atm or 101.328 kPa)

Vapor pressure

$$\ln(p/\text{kPa}) = \frac{-3804.72}{T} - 8.62625 \ln T + 0.0106145T + 65.79108$$

Critical constants

$$T_c = 368.3 \text{ K}, \ p_c = 5.37 \text{ bar}$$

(a) Assume the Third Law is valid and calculate the entropy of ideal gaseous $ClO_3F$ at the normal boiling point. You may use the trapezoidal rule to perform the graphical integration on the $C_{p,m}$ data. Use the Debye low-temperature heat capacity relationship to extrapolate the $C_{p,m}$ data below 15 K. Assume $ClO_3F(g)$ obeys the modified Berthelot equation to calculate the correction to ideal gas behavior for the $ClO_3F(g)$.

(b) Statistical calculations of $S_m$ for $ClO_3F$(ideal gas) at the normal boiling point give a value of 261.88 $J \cdot K^{-1} \cdot mol^{-1}$ (see problem P10.2). Explain the difference between the two values.

P4.4 The heat capacity of $Na_2SO_4 \cdot 10H_2O(s)$ has been measured from 15 K to 300 K, and $S_m^\circ$ at $T = 298.15$ K, which was computed from[20]

$$\int_{0\text{ K}}^{298.15\text{ K}} \frac{C_{p,m}}{T} \, dT,$$

was found to be 585.55 $J \cdot K^{-1} \cdot mol^{-1}$. The following thermodynamic data are also known for the hydration reaction at $T = 298.15$ K:

$$Na_2SO_4(s) + 10H_2O(g) = Na_2SO_4 \cdot 10H_2O(s)$$

$$\Delta_r G^\circ = -91{,}190 \text{ J} \cdot \text{mol}^{-1}$$

$$\Delta_r H^\circ = -521{,}950 \text{ J} \cdot \text{mol}^{-1}$$

In addition, the entropies $S_m^\circ$ at $T = 298.15$ K for anhydrous $Na_2SO_4$ and for water vapor are 149.39 $J \cdot K^{-1} \cdot mol^{-1}$ and 188.72 $J \cdot K^{-1} \cdot mol^{-1}$, respectively. Is $Na_2SO_4 \cdot 10H_2O$ a perfect crystal at 0 K?

## References

1. G. N. Lewis and M. Randall, "Thermodynamics and the Free Energy of Chemical Substances", First Edition, McGraw-Hill Book Company, New York, 1923, p. 448.

2. J. Boerio-Goates, "Heat-Capacity Measurements and Thermodynamic Functions of Crystalline α-D-Glucose at Temperatures from 0 K to 350 K", *J. Chem. Thermodyn.*, **23**, 403–409 (1991).
3. W. F. Giauque and J. O. Clayton, "The Heat Capacity and Entropy of Nitrogen. Heat of Vaporization. Vapor Pressures of Solid and Liquid. The Reaction $\frac{1}{2}N_2 + O_2 = NO$ from Spectroscopic Data", *J. Am. Chem. Soc.*, **55**, 4875–4889 (1993).
4. T. W. Richards, "The Significance of Changing Atomic Volume III. The Relation of Changing Heat Capacity to Change of Free Energy, Heat of Reaction, Change of Volume, and Chemical Affinity", *Z. Physik. Chem.* **42**, 129–154 (1902).
5. E. D. Eastman and W. C. McGavock, "The Heat Capacity and Entropy of Rhombic and Monoclinic Sulfur", *J. Am. Chem. Soc.*, **59**, 145–151 (1937).
6. E. D. West, "The Heat Capacity of Sulfur from 25 to 450°, the Heats and Temperatures of Transition and Fusion", *J. Am. Chem. Soc.*, **81**, 29–37 (1959).
7. C. C. Stephenson and W. F. Giauque, "A Test of the Third Law of Thermodynamics by Means of Two Crystalline Forms of Phosphine. The Heat Capacity, Heat of Vaporization and Vapor Pressure of Phosphine. Entropy of the Gas", *J. Chem. Phys.*, **5**, 149–158 (1937).
8. E. D. Eastman and R. T. Milner, "The Entropy of a Crystalline Solution of Silver Bromide and Silver Chloride in Relation to the Third Law of Thermodynamics", *J. Chem. Phys.*, **1**, 444–456 (1933).
9. G. E. Gibson and W. F. Giauque, "The Third Law of Thermodynamics. Evidence from the Specific Heats of Glycerol that the Entropy of a Glass Exceeds that of a Crystal at the Absolute Zero", *J. Am. Chem. Soc.*, **45**, 93–104 (1923).
10. W. F. Giauque, "The Entropy of Hydrogen and the Third Law of Thermodynamics. The Free Energy and Dissociation of Hydrogen", *J. Am. Chem. Soc.*, **52**, 4816–4831 (1930).
11. K. S. Pitzer and L. V. Coulter, "The Heat Capacities, Entropies and Heats of Solution of Anhydrous Sodium Sulfate and of Sodium Sulfate Decahydrate. The Application of the Third Law of Thermodynamics to Hydrated Crystals", *J. Am. Chem. Soc.*, **60**, 1310–1313 (1938).
12. R. H. Busey and W. F. Giauque, "The Heat Capacity of Anhydrous $NiCl_2$ from 15 to 300 °K. The Antiferromagnetic Anomaly near 52 °K. Entropy and Free Energy", *J. Am. Chem. Soc.*, **74**, 4443–4446 (1952).
13. R. H. Busey and W. F. Giauque, "The Equilibrium Reaction $NiCl_2 + H_2 = Ni + 2HCl$. Ferromagnetism and the Third Law of Thermodynamics", *J. Am. Chem. Soc.*, **75**, 1791–1794 (1953).
14. R. A. Fisher, E. W. Hornung, G. E. Brodale and W. F. Giauque, "Magnetothermodynamics of $Ce_2Mg_3(NO_3)_{12}{\cdot}24H_2O$. II. The Evaluation of Absolute Temperature and other Thermodynamic Properties of CMN to 0.6 m° K", *J. Chem. Phys.*, **58**, 5584–5604 (1973).
15. See W. F. Giauque and D. P. MacDougall, "Experiments Establishing the Thermodynamic Temperature Scale below 1 °K. The Magnetic and Thermodynamic Properties of Gadolinium Phosphomolybdate as a Function of Field and Temperature", *J. Am. Chem. Soc.*, **60**, 376–388 (1938).
16. P. Hakonen, O. V. Lounasmaa and A. Oja, "Spontaneous Nuclear Magnetic Ordering in Copper and Silver at Nano- and Picokelvin Temperatures", *J. Magn. Magn. Mater.*, **100**, 394–412 (1991).
17. P. A. Rock, *Chemical Thermodynamics, University Science Books*, Mill Valley, California, (1983), p. 157.
18. The reference to the JANAF Tables is as follows: W. W. Chase Jr., C. A. Davies, J. R. Downey Jr., D. J. Frurip, R. A. McDonald and A. N. Syverud, "JANAF Thermodynamical Tables, Third Edition", *J. Phys. Chem. Ref. Data*, **14**, Supplement No. 1, American Chemical Society, Washington, DC (1985).

19. W. V. Steele, D. G. Archer, R. D. Chirico, W. B. Collier, I. A. Hossenlopp, A. Nguyen, N. K. Smith and B. E. Gammon, "The Thermodynamic Properties of Quinoline and Isoquinoline", *J. Chem. Thermodyn.*, **20**, 1233–1264 (1988).
20. G. Brodale and W. F. Giauque, "The Heat of Hydration of Sodium Sulfate. Low Temperature Heat Capacity and Entropy of Sodium Sulfate Decahydrate", *J. Am. Chem. Soc.*, **80**, 2042–2044 (1958).

## Chapter 5

# The Chemical Potential and Equilibrium

## 5.1 Composition as a Variable

So far our discussion of chemical thermodynamics has been limited to systems in which the chemical composition does not change. We have dealt with pure substances, often in molar quantities, but always with a fixed number of moles, $n$. The Gibbs equations

$$dU = T\,dS - p\,dV,$$

$$dH = T\,dS + V\,dp,$$

$$dA = -S\,dT - p\,dV,$$

and

$$dG = -S\,dT + V\,dp$$

assume constant $n$, as do all of the differential expressions derived in Chapter 3.

But much of chemistry involves mixtures, solutions, and reacting systems in which the number of moles or mole number, $n_i$, of each species present can be variable. When this happens, the extensive properties, $Z = V, S, U, H, A$ or $G$ become functions of the composition variables, as well as two of the state variables as described earlier.[a] We can express this mathematically as

$$Z = f(X, Y, n_1, n_2, \ldots),$$

---

[a] The most common state variables chosen are the intensive variables $p$ and $T$, although $V$ and $S$ are also sometimes used.

where $X$ and $Y$ represent the state variables, and each $n_i$ gives the mole number for the $i$th component. The total differential d$Z$ is then given by

$$\mathrm{d}Z = \left(\frac{\partial Z}{\partial X}\right)_{Y,n} \mathrm{d}X + \left(\frac{\partial Z}{\partial Y}\right)_{X,n} \mathrm{d}Y + \left(\frac{\partial Z}{\partial n_1}\right)_{X,Y,n_{j\neq 1}} \mathrm{d}n_1 + \left(\frac{\partial Z}{\partial n_2}\right)_{X,Y,n_{j\neq 2}} \mathrm{d}n_2 + \cdots + \left(\frac{\partial Z}{\partial n_i}\right)_{X,Y,n_{j\neq i}} \mathrm{d}n_i + \cdots, \tag{5.1}$$

where the subscripts on the partial derivatives $(\partial Z/\partial n_i)_{X,Y,n_{j\neq i}}$ mean that variables $X$, $Y$, and all mole numbers except that for the $i$th species, stay constant. Thus, for a mixture, $Z$ is a function of the two variables, $X$ and $Y$ as before, but with the addition of a variable mole number $n_i$, for each component in the system.

## 5.2 The Chemical Potential

For the Gibbs free energy, the total differential given above as equation (5.1) can be written as

$$\mathrm{d}G = \left(\frac{\partial G}{\partial T}\right)_{p,n} \mathrm{d}T + \left(\frac{\partial G}{\partial p}\right)_{T,n} \mathrm{d}p + \sum_i \left(\frac{\partial G}{\partial n_i}\right)_{T,p,n_{j\neq i}} \mathrm{d}n_i, \tag{5.2}$$

where the first two partial derivatives (in front of d$T$ and d$p$) are taken at constant composition, and those inside the summation are taken with constant $T$ and $p$, and only one composition variable changing at a time.

The quantity $(\partial G/\partial n_i)_{T,p,n_{j\neq i}}$ is called the chemical potential of the $i$th component and given the symbol $\mu_i$ so that

$$\mu_i = \left(\frac{\partial G}{\partial n_i}\right)_{T,p,n_{j\neq i}}. \tag{5.3}$$

Substitution of equation (5.3) into equation (5.2) gives

$$\mathrm{d}G = \left(\frac{\partial G}{\partial T}\right)_{p,n} \mathrm{d}T + \left(\frac{\partial G}{\partial p}\right)_{T,n} \mathrm{d}p + \sum_i \mu_i \,\mathrm{d}n_i. \tag{5.4}$$

We know from Chapter 3 that

$$\left(\frac{\partial G}{\partial T}\right)_{p,n} = -S$$

and

$$\left(\frac{\partial G}{\partial p}\right)_{T,n} = V.$$

Substitution into equation (5.4) gives

$$dG = -S\,dT + V\,dp + \sum_i \mu_i\,dn_i. \tag{5.5}$$

Equation (5.5) is an extended form of the fourth Gibbs equation that applies when the mole numbers become variables. Under the condition of constant $T$ and $p$, it becomes

$$dG = \sum \mu_i\,dn_i\,. \tag{5.6}$$

We can find alternate, but equivalent, statements that define $\mu_i$. Starting with

$$U = f(S, V, n_1, n_2, \ldots)$$

we can write

$$dU = \left(\frac{\partial U}{\partial S}\right)_{V,n} dS + \left(\frac{\partial U}{\partial V}\right)_{S,n} dV + \sum_i \left(\frac{\partial U}{\partial n_i}\right)_{S,V,n_{j\neq i}} dn_i. \tag{5.7}$$

With constant $n$'s, the first Gibbs equation can be written

$$dU = T\,dS - p\,dV,$$

and equation (5.7) becomes

$$dU = \left(\frac{\partial U}{\partial S}\right)_{V,n} dS + \left(\frac{\partial U}{\partial V}\right)_{S,n} dV.$$

Comparing coefficients gives

$$\left(\frac{\partial U}{\partial S}\right)_{V,n} = T$$

$$\left(\frac{\partial U}{\partial V}\right)_{S,n} = -p.$$

Substituting into equation (5.7) gives

$$dU = T\,dS - p\,dV + \sum_i \left(\frac{\partial U}{\partial n_i}\right)_{S,\,V,\,n_{j\neq i}} dn_i. \tag{5.8}$$

If we add $d(pV - TS)$ to both sides of equation (5.8) while recognizing that $dG = dU + d(pV - TS)$, we get

$$dG = T\,dS - p\,dV$$
$$+ \sum_i \left(\frac{\partial U}{\partial n_i}\right)_{S,\,V,\,n_{j\neq i}} dn_i + p\,dV + V\,dp - T\,dS - S\,dT.$$

Cancelling terms gives

$$dG = -S\,dT + V\,dp + \sum_i \left(\frac{\partial U}{\partial n_i}\right)_{S,\,V,\,n_{j\neq i}} dn_i. \tag{5.9}$$

Comparing the terms in equations (5.5) and (5.9) enables us to write

$$\mu_i = \left(\frac{\partial U}{\partial n_i}\right)_{S,\,V,\,n_{j\neq i}}. \tag{5.10}$$

Similar derivations involving $H$ and $A$ enable us to write[b] that

$$\mu_i = \left(\frac{\partial G}{\partial n_i}\right)_{T,\,p,\,n_{j\neq i}} = \left(\frac{\partial A}{\partial n_i}\right)_{T,\,V,\,n_{j\neq i}} = \left(\frac{\partial H}{\partial n_i}\right)_{S,\,p,\,n_{j\neq i}} = \left(\frac{\partial U}{\partial n_i}\right)_{S,\,V,\,n_{j\neq i}}. \tag{5.11}$$

We can derive very simply an important interpretation of the chemical potential for a pure substance, for which we can write

$$G = nG_{\mathrm{m}}, \tag{5.12}$$

where $G_{\mathrm{m}}$ is the molar free energy. Differentiating equation (5.12) with respect to $n$ gives

$$\left(\frac{\partial G}{\partial n}\right)_{T,\,p} = G_{\mathrm{m}} = \mu. \tag{5.13}$$

Hence, for a pure substance, the chemical potential is a measure of its molar Gibbs free energy. We next want to describe the chemical potential for a component in a mixture, but to do so, we first need to define and describe a quantity known as a partial molar property.

---

[b] The significance of equation (5.11) can perhaps be better understood by considering a system consisting of $n_1$ moles of component 1, $n_2$ moles of component 2, ..., $n_i$ moles of component $i$, and so forth. The state of this system can be fixed by specifying a point on a multidimensional $(G, T, p, n_1, n_2, ..., n_i, ...)$ surface. The chemical potential $\mu_i$ in this state is simply the slope of a tangent to the surface at this point that is drawn in a direction such that $p$, $T$, and all $n_{j\neq i}$ are constant.

But the same state of the system can equally well be specified by a point on a $(U, S, V, n_1, n_2, ..., n_i, ...)$ surface. Equation (5.11) requires that the slope of the tangent at this point drawn in the direction of constant $S$, $V$, and all $n_{j\neq i}$ have the same slope as the tangent on the Gibbs free energy surface, since this slope is also $\mu_i$. Similar arguments can be used to equate the slopes of the tangent on a $(H, S, p, n_1, n_2, ..., n_i, ...)$ surface in the direction of constant $S$, $p$, and $n_{j\neq i}$, and the tangent on a $(A, T, V, n_1, n_2, ..., n_i, ...)$ surface in the direction of constant $T$, $V$, $n_{j\neq i}$ to the above, since these slopes are also $\mu_i$ if these points represent the same state of the system.

We see from equation (5.11) that the chemical potential is the rate of change of an energy quantity associated with a mass change when the process takes place under a specific set of reaction conditions. For the (rare) process that proceeds at constant $S$ and $V$, $\mu_i$ is a measure of the rate of change of $U$ with $n_i$. For the more common example of a process taking place at constant $T$ and $p$, $\mu_i$ is the rate of change of $G$ with respect to $n_i$. Later in this chapter, we will see that the chemical potential is the driving force that causes the flow of mass in a chemical process, and since these processes usually occur under the condition of constant $T$ and $p$, the expression relating $\mu_i$ to $G$ and $n_i$ is the one we will almost always use.

## 5.3 Partial Molar Properties

As chemists, we are most often concerned with reactions proceeding under conditions in which the temperature and pressure are the variables we control. Therefore, it is useful to have a set of properties that describe the effect of a change in concentration on the various thermodynamic quantities under conditions of constant temperature and pressure. We refer to these properties as the **partial molar quantities**.

The partial molar quantity $\bar{Z}_i$ is defined as

$$\bar{Z}_i = \left(\frac{\partial Z}{\partial n_i}\right)_{T,p,n_{j \neq i}}, \tag{5.14}$$

where $Z$ can be any of the extensive thermodynamic properties ($Z = U, H, S, A, G$ or $V$). Note that the subscripted variables indicate that the partial derivative is to be taken at constant values of $T$ and $p$.

To understand the significance of the partial molar property, we start with

$$Z = f(p, T, n_1, n_2, \ldots),$$

so that the total differential of $Z$ can be written as

$$\mathrm{d}Z = \sum_i \left(\frac{\partial Z}{\partial n_i}\right)_{T,p,n_{j \neq i}} \mathrm{d}n_i \tag{5.15}$$

for a process at constant $T$ and $p$. Substituting the notation from equation (5.14) into equation (5.15) gives

$$\mathrm{d}Z = \sum_i \bar{Z}_i \, \mathrm{d}n_i. \tag{5.16}$$

Equation (5.16) can be integrated. We expect the partial molar properties to be functions of composition, and of temperature and pressure. For a system at constant temperature and constant pressure, the partial molar properties would be functions only of composition. We will start with an infinitesimal quantity of material, with the composition fixed by the initial amounts of each component present, and then increase the amounts of each component but always in that same fixed ratio so that the composition stays constant. When we do this, $\bar{Z}_i$ stays constant, and the integration of equation

(5.16) gives

$$Z = \sum_i \bar{Z}_i n_i. \tag{5.17}$$

The same result can be obtained from an application of Euler's theorem, explained in more detail in Appendix 1. The thermodynamic quantities, $Z$, are homogeneous functions of degree one with respect to mole numbers.[c] At constant $T$ and $p$, one can use Euler's theorem to write an expression for $Z$ in terms of the mole numbers and the derivatives of $Z$ with respect to the mole numbers. The result is[d]

$$n_1\left(\frac{\partial Z}{\partial n_1}\right)_{n_j \neq 1} + n_2\left(\frac{\partial Z}{\partial n_2}\right)_{n_j \neq 2} + n_3\left(\frac{\partial Z}{\partial n_3}\right)_{n_j \neq 3} + \cdots = Z. \tag{5.18}$$

Because $p$ and $T$ are held constant, and only a single composition variable is changed in each derivative, the partial derivatives in equation (5.18) are equivalent to those in equation (5.14), and equation (5.18) becomes the same as equation (5.17).

By either a direct integration in which $\bar{Z}_i$ is held constant, or by using Euler's theorem, we have accomplished the integration of equation (5.16), and are now prepared to understand the physical significance of the partial molar property. For a one-component system, $Z = nZ_{\mathrm{m}}$, where $Z_{\mathrm{m}}$ is the molar property. Thus, $Z_{\mathrm{m}}$ is the contribution to $Z$ for a mole of substance, and the total $Z$ is the molar $Z_{\mathrm{m}}$ multiplied by the number of moles. For a two-component system, equation (5.17) gives

$$Z = n_1\bar{Z}_1 + n_2\bar{Z}_2.$$

This equation suggests that $\bar{Z}_1$ is the effective contribution per mole of component 1 to the total $Z$. Thus, when $\bar{Z}_1$ is multiplied by $n_1$, the number of moles of component 1, the product gives the total contribution of component 1 to $Z$. Likewise $\bar{Z}_2$ is the contribution per mole of component 2 and $n_2\bar{Z}_2$ is the total contribution of component 2 to $Z$. As an example, for a binary mixture, the total volume is given by $V = n_1\bar{V}_1 + n_2\bar{V}_2$, with $\bar{V}_1$ and $\bar{V}_2$ as the partial

[c] A homogeneous function of degree $n$ ($n$ is not the mole number) is one for which $f(kx, ky, kz) = k^n f(x, y, z)$. See Appendix 1 for more details.

[d] $Z$ is not homogeneous in either $T$ or $p$. But those variables are held constant and do not enter into the Euler expression for Z.

molar volumes, or the volume per mole that each component contributes to the total volume $V$.[e]

In summary, we have defined a partial molar property $\bar{Z}_i$ as

$$\bar{Z}_i = \left(\frac{\partial Z}{\partial n_i}\right)_{T,p,n_{j\neq i}}. \tag{5.14}$$

It is related to the total property $Z$ by

$$Z = \sum_i n_i \bar{Z}_i. \tag{5.17}$$

We want to be able to measure $\bar{Z}_i$. But before we do so, we want to note a useful property of $\bar{Z}_i$, describe a particular application, and derive an important equation.

First, we note that all of the thermodynamic equations that we have derived for the total extensive variables apply to the partial molar properties. Thus, if

$$\left(\frac{\partial H}{\partial p}\right)_T = V - T\left(\frac{\partial V}{\partial T}\right)_p \tag{5.19}$$

then

$$\left(\frac{\partial \bar{H}_i}{\partial p}\right)_T = \bar{V}_i - T\left(\frac{\partial \bar{V}_i}{\partial T}\right)_p. \tag{5.20}$$

Equation (5.20) can be proved by differentiating equation (5.19)

$$\left[\frac{\partial}{\partial n_i}\left(\frac{\partial H}{\partial p}\right)_{T,n}\right]_{T,p,n_{j\neq i}} = \left(\frac{\partial V}{\partial n_i}\right)_{T,p,n_{j\neq i}} - T\left[\frac{\partial}{\partial n_i}\left(\frac{\partial V}{\partial T}\right)_{p,n}\right]_{T,p,n_{j\neq i}}.$$

[e] In some circumstances, $\bar{Z}_i$ can be negative. For example, $\bar{V}_2$ can be negative if component 2 fits into the structure of component 1 and causes a collapse of the structure when it is added so that the total volume decreases with the addition of component 2.

But the order or differentiation does not matter so that

$$\left[\frac{\partial}{\partial n_i}\left(\frac{\partial H}{\partial p}\right)_{T,n}\right]_{T,p,n_{j\neq i}} = \left[\frac{\partial}{\partial p}\left(\frac{\partial H}{\partial n_i}\right)_{T,p,n_{j\neq i}}\right]_{T,n},$$

and

$$\left[\frac{\partial}{\partial n_i}\left(\frac{\partial V}{\partial T}\right)_{p,n}\right]_{T,p,n_{j\neq i}} = \left[\frac{\partial}{\partial T}\left(\frac{\partial V}{\partial n_i}\right)_{T,p,n_{j\neq i}}\right]_{p,n}.$$

The result is

$$\left(\frac{\partial \bar{H}_i}{\partial p}\right)_{T,n} = \bar{V}_i - T\left(\frac{\partial \bar{V}_i}{\partial T}\right)_{p,n}.$$

A similar proof can be used for applying any of our thermodynamic equations to partial molar properties. For example, if

$$\left(\frac{\partial S}{\partial T}\right)_p = \frac{C_p}{T}$$

then

$$\left(\frac{\partial \bar{S}_i}{\partial T}\right)_p = \frac{\bar{C}_{p,i}}{T}.$$

If

$$\left(\frac{\partial G}{\partial p}\right)_T = V$$

then

$$\left(\frac{\partial \overline{G}_i}{\partial p}\right)_T = \overline{V}_i.$$

The last equation can be put into a different form by noting that

$$\overline{G}_i = \left(\frac{\partial G}{\partial n_i}\right)_{T,p,n_{j\neq i}} = \mu_i$$

so that

$$\left(\frac{\partial \mu_i}{\partial p}\right)_T = \overline{V}_i.$$

This last equation is only one of a number that we can write that apply to $\mu_i$, after recognizing the relationship between $\mu_i$ and $\overline{G}_i$. For example,

$$\left(\frac{\partial \mu_i}{\partial T}\right)_p = -\overline{S}_i$$

$$\left(\frac{\partial \mu_i/T}{\partial T}\right)_p = -\frac{\overline{H}_i}{T^2}$$

$$\mathrm{d}\mu_i = -\overline{S}_i\,\mathrm{d}T + \overline{V}_i\,\mathrm{d}p$$

$$G = \sum_i n_i\mu_i. \tag{5.21}$$

Equation (5.21) is especially important in that it indicates that $\mu_i$, the chemical potential of the *i*th component in the mixture is the contribution (per mole) of that component to the total Gibbs free energy.[f]

---

[f] In summary, for a pure substance, the chemical potential is the molar Gibbs free energy of that substance, while in a mixture, the chemical potential of a component is the partial molar Gibbs free energy of that component.

Before leaving our discussion of partial molar properties, we want to emphasize that only the partial molar Gibbs free energy is equal to $\mu_i$. The chemical potential can be written as $(\partial A/\partial n_i)_{T, V, n_{j\neq i}}$ or $(\partial H/\partial n_i)_{S, p, n_{j\neq i}}$ or $(\partial U/\partial n_i)_{S, V, n_{j\neq i}}$, but these quantities are not $\overline{A}_i$, $\overline{H}_i$, and $\overline{U}_i$, and we cannot substitute these partial molar quantities for $\mu_i$ into equations such as those given above.

## 5.4 The Gibbs–Duhem Equation

Equations (5.16) and (5.18)

$$\mathrm{d}Z = \sum_i \overline{Z}_i \, \mathrm{d}n_i \tag{5.16}$$

$$Z = \sum_i n_i \overline{Z}_i \tag{5.18}$$

can be used to derive an important equation relating partial molar properties. Differentiating equation (5.18) gives

$$\mathrm{d}Z = \sum_i n_i \, \mathrm{d}\overline{Z}_i + \sum_i \overline{Z}_i \, \mathrm{d}n_i. \tag{5.22}$$

Comparing terms in equations (5.16) and (5.22) requires that

$$\sum_i n_i \, \mathrm{d}\overline{Z}_i = 0. \tag{5.23}$$

Equation (5.23) is known as the **Gibbs–Duhem equation**. It relates the partial molar properties of the components in a mixture. Equation (5.23) can be used to calculate one partial molar property from the other. For example, solving for $\mathrm{d}\overline{Z}_1$ gives

$$\mathrm{d}\overline{Z}_1 = -\frac{n_2}{n_1} \, \mathrm{d}\overline{Z}_2.$$

Integration gives

$$\int_{\overline{Z}_1^*}^{\overline{Z}_1} \mathrm{d}\overline{Z}_1 = -\int_{n_2=0}^{n_2} \frac{n_2}{n_1} \, \mathrm{d}\overline{Z}_2$$

or

$$\overline{Z}_1 - \overline{Z}_1^* = -\int_{n_2=0}^{n_2} \frac{n_2}{n_1}\,d\overline{Z}_2 \tag{5.24}$$

where $\overline{Z}_1^* = Z_{m,1}^*$ is the molar property for pure component 1 ($n_2 = 0$). Equation (5.24) can be expressed in terms of mole fractions instead of number of moles by noting that

$$\frac{x_2}{x_1} = \frac{\left(\dfrac{n_2}{n_1 + n_2}\right)}{\left(\dfrac{n_1}{n_1 + n_2}\right)} = \frac{n_2}{n_1}.$$

Substitution into equation (5.24) gives

$$\overline{Z}_1 - \overline{Z}_1^* = -\int_{x_2=0}^{x_2} \frac{x_2}{x_1}\,d\overline{Z}_2. \tag{5.25}$$

By knowing $\overline{Z}_2$ as a function of composition, $\overline{Z}_1$ can be calculated from equations (5.24) and (5.25).

A particularly useful form of equation (5.23) is the one that relates chemical potentials

$$\sum_i n_1\,d\mu_i = 0, \tag{5.26}$$

or, in a binary mixture,

$$n_1\,d\mu_1 + n_2\,d\mu_2 = 0. \tag{5.27}$$

We will return to this equation later.

## 5.5 Determination of Partial Molar Properties

Obtaining partial molar properties involves the determination of the derivative as given by equation (5.14)

$$\overline{Z}_i = \left(\frac{\partial Z}{\partial n_i}\right)_{T,p,n_{j \neq i}} . \tag{5.14}$$

A variety of procedures can be used to determine $\overline{Z}_i$ as a function of composition.[g] Care must be taken if reliable values are to be obtained, since the determination of a derivative or a slope is often difficult to do with high accuracy. A number of different techniques are employed, depending upon the accuracy of the data that is used to calculate $\overline{Z}_i$, and the nature of the system. We will now consider several examples involving the determination of $\overline{V}_i$ and $\overline{C}_{p,i}$, since these are the properties for which absolute values for the partial molar quantity can be obtained. Only relative values of $\overline{H}_i$ and $\mu_i$ can be obtained, since absolute values of $H$ and $G$ are not available. For $\overline{H}_i$ and $\mu_i$ we determine $\overline{H}_i - \overline{H}_i^\circ$ or $\mu_i - \mu_i^\circ$, where $\overline{H}_i^\circ$ and $\mu_i^\circ$ are values for $\overline{H}_i$ and $\mu_i$ in a reference or standard state. We will delay a discussion of these quantities until we have described standard states.

The following examples demonstrate commonly used techniques for determining $\overline{C}_{p,i}$ and $\overline{V}_i$.

## 5.5a Numerical Methods

Figure 5.1 is a graph of the specific heat capacity $c_p$ (heat capacity per gram) of aqueous sulfuric acid solutions at $T = 298.15$ K against $A$, the ratio of moles of water to moles of sulfuric acid. The values plotted were obtained from a very

[g] Chemical composition will usually be expressed in terms of mole fraction $x_i$ given by

$$x_i = \frac{n_i}{\sum_i n_i}$$

or molality $m$ given by

$$m = \frac{\text{moles solute}}{\text{kg of solvent}}$$

or molarity $c$ given by

$$c = \frac{\text{moles solute}}{\text{dm}^3 \text{ of solution}}$$

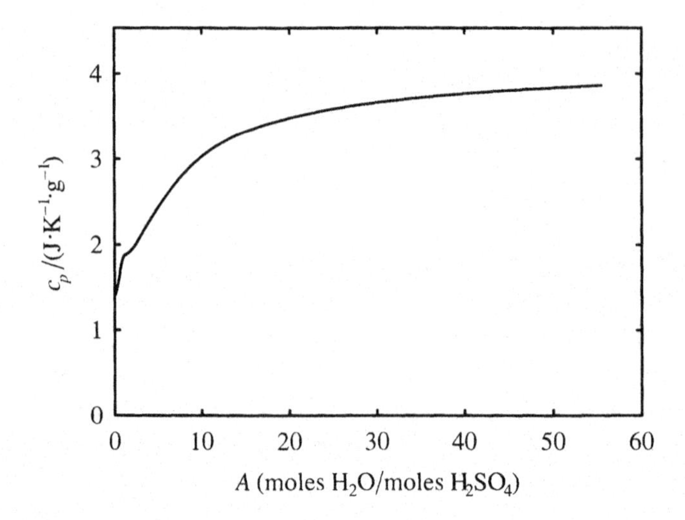


**Figure 5.1** Specific heat capacity of aqueous sulfuric acid solutions at 298.15 K as a function of $A$, the ratio of moles of $H_2O$ to moles of $H_2SO_4$.

careful study of the thermodynamic properties of sulfuric acid solutions.[1] The curve in Figure 5.1 was generated from over a hundred heat capacity measurements, determined with high accuracy. These experimental specific heat capacities can be used to calculate the partial molar heat capacities.[h] To calculate $\overline{C}_{p,1}$, we start with an amount of solution that contains $n_1 = A$ moles of $H_2O$ with molecular weight $M_1$ (in g·mol$^{-1}$) and $n_2 = 1$ mole of $H_2SO_4$ with molecular weight $M_2$ (again in g·mol$^{-1}$). The total heat capacity $C_p$ of this solution at composition $A$ is given by $C_p = c_p(M_1 A + M_2)$. This heat capacity can be used to calculate $\overline{C}_{p,1}$ using the relationship $\overline{C}_{p,1} = (\partial C_p/\partial n_1)_{n_2} = (\partial C_p/\partial A)_{n_2}$. When high quality heat capacity data are available, such as those shown in Figure 5.1 for the (sulfuric acid + water) system, one can substitute smoothed $(\Delta C_p/\Delta A)$ values for $(\partial C_p/\partial A)$ so that $\overline{C}_{p,1} = (\Delta C_p/\Delta A)_{n_2}$. Reversing the procedure by holding the moles of $H_2O$ constant while varying the amount of $H_2SO_4$ gives $\overline{C}_{p,2}$. The $\overline{C}_{p,1}$ and $\overline{C}_{p,2}$ values obtained can then be checked for thermodynamic consistency[i] with the Gibbs–Duhem equation.

---

[h] In aqueous solutions, the subscript 1 is nearly always used to represent water. In general, the subscript 1 is usually used to represent the component designated as the solvent while 2, 3, ... represent the solutes.

[i] Thermodynamic variables are related through a number of different thermodynamic equations. Experimentalists who measure thermodynamic properties by one method often check the results using other relationships. This test checks for thermodynamic consistency, which must follow if the results are to be trusted.

The $\overline{C}_{p,1}$ and $\overline{C}_{p,2}$ results for this system as summarized in Figure 5.2, are interesting to interpret. The changes in $\overline{C}_{p,1}$ and $\overline{C}_{p,2}$ in Figure 5.2 can be attributed to changes in the hydration of ionized sulfuric acid in the water. The negative $\overline{C}_{p,1}$ at high $x_2$ results from ionization of $H_2SO_4$ as $H_2O$ is added. The ionized $H_2SO_4$ has a lower heat capacity than unionized $H_2SO_4$. Thus, adding water to pure $H_2SO_4$ decreases the total heat capacity and the water makes a negative contribution so that $\overline{C}_{p,1}$ is negative. Abrupt changes in $\overline{C}_{p,1}$ and $\overline{C}_{p,2}$ at other mole fractions can be attributed to the formation of various hydrates in the solutions.

## 5.5b Analytical Methods Using Molality

The volume of a solution is sometimes expressed as a function of composition and the partial molar volume is then obtained by differentiation. For example, Klotz and Rosenburg[2] have expressed the volume of aqueous sodium chloride solutions at 298.15 K and ambient pressure as a function of the molality $m$ of the solution by the equation[j]

$$V/(\text{cm}^3) = 1001.38 + 16.6253m + 1.77738m^{3/2} + 0.1194m^2. \tag{5.28}$$

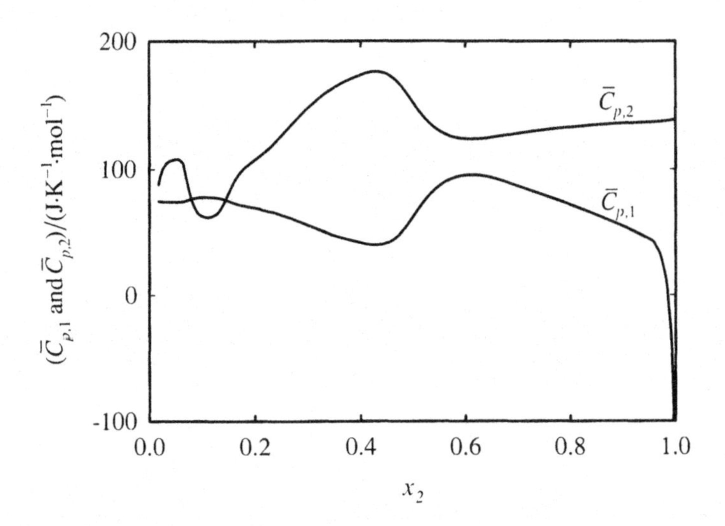


**Figure 5.2** Partial molar heat capacities of $H_2O$ ($\overline{C}_{p,1}$) and $H_2SO_4$ ($\overline{C}_{p,2}$) against $x_2$, the mole fraction of $H_2SO_4$.

[j] Molality $m$ is often used to express composition for electrolyte solutions, with the thermodynamic property a function of $m^{1/2}$, $m^{3/2}$ etc. The Debye–Hückel theory, which we will describe later, will help in understanding the reasons for this concentration dependence.

The value of $\overline{V}_2$ can be obtained by differentiation with respect to $m$, since $m$ is the moles of solute per constant number of moles of solvent ($n_1 = 1/M_1$, where $M_1$ is the molecular weight of the solvent in kg·mol$^{-1}$). Differentiating equation (5.28) gives

$$\overline{V}_2 = \left(\frac{\partial V}{\partial m}\right)_{T,p,n_1} = 16.6253 + 2.6607m^{1/2} + 0.2388m. \tag{5.29}$$

Substituting values of $m$ into equation (5.29) gives $\overline{V}_2$ at different concentrations.

An expression for $\overline{V}_1$ can be obtained from equation (5.29) by integration of the Gibbs–Duhem equation. Starting with the Gibbs–Duhem equation {equation (5.23)} applied to volume gives

$$n_1\,\mathrm{d}\overline{V}_1 + n_2\,\mathrm{d}\overline{V}_2 = 0.$$

We use this relationship with $n_2 = m$ so that $n_1 = 1/M_1$. In aqueous solution, $M_1 = 0.018015$ kg·mol$^{-1}$, in which case $n_1 = 1/0.018015 = 55.51$, and the above equation becomes

$$55.51\,\mathrm{d}\overline{V}_1 + m\,\mathrm{d}\overline{V}_2 = 0. \tag{5.30}$$

Differentiating equation (5.29) gives

$$\mathrm{d}\overline{V}_2 = (1.3304m^{-1/2} + 0.2388)\,\mathrm{d}m. \tag{5.31}$$

Combining equations (5.30) and (5.31) gives

$$55.51\,\mathrm{d}\overline{V}_1 + m(1.3304m^{-1/2} + 0.2388)\,\mathrm{d}m = 0.$$

Solving for $\mathrm{d}\overline{V}_1$ and integrating gives

$$\int_{V^*_{\mathrm{m},1}}^{\overline{V}_1} \mathrm{d}\overline{V}_1 = -\int_{m=0}^{m} \frac{m}{55.51}(1.3304m^{-1/2} + 0.2388)\,\mathrm{d}m$$

$$\overline{V}_1 - V^*_{\mathrm{m},1} = -\frac{0.8869}{55.51}m^{3/2} - \frac{0.1194}{55.51}m^2,$$

where $V^*_{m,1}$ is the molar volume of the pure solvent. Solving for $\overline{V}_1$ with $V^*_{m,1} = 18.08\ \text{cm}^3\cdot\text{mol}^{-1}$ (the molar volume of pure water at 298.15 K), gives

$$\overline{V}_1/(\text{cm}^3\cdot\text{mol}^{-1}) = 18.08 - 0.015977m^{3/2} - 0.002151m^2. \tag{5.32}$$

### 5.5c Analytical Methods Using Mole Fractions

It can be shown that $\overline{V}_1$ and $\overline{V}_2$ are related to the mole fractions $x_1$ and $x_2$ by[k]

$$\overline{V}_1 = V^*_{m,1} + \frac{V^E_m}{x_1} + x_1 x_2 \left[\frac{\partial(V^E_m/x_1)}{\partial x_1}\right]_{n_2} \tag{5.33}$$

$$\overline{V}_2 = V^*_{m,2} + \frac{V^E_m}{x_2} + x_1 x_2 \left[\frac{\partial(V^E_m/x_2)}{\partial x_2}\right]_{n_1}, \tag{5.34}$$

where $V^E_m$ is the volume change upon mixing to form one mole of solution at constant temperature and pressure. It is known as the excess molar volume,[l] and can be calculated from

$$V^E_m = \frac{V - n_1 V^*_{m,1} - n_2 V^*_{m,2}}{(n_1 + n_2)} \tag{5.35}$$

where $V$ is the volume of the solution containing $n_1$ moles of component 1 and $n_2$ moles of component 2, and $V^*_{m,1}$ and $V^*_{m,2}$ are the molar volumes of the pure components. The derivation is as follows.

Equation (5.35) can be solved for $V$ to give

$$V = n_1 V^*_{m,1} + n_2 V^*_{m,2} + n_1 \frac{V^E_m}{x_1} \tag{5.36}$$

---

[k] Equations (5.33) and (5.34) can be applied equally as well to the other thermodynamic variables such as $\overline{C}_{p,i}$. For enthalpy, we find $(\overline{H}_i - H^*_{m,i}) = \overline{L}_i$, the relative partial molar enthalpy since absolute values of $\overline{H}_i$ cannot be obtained. In Chapter 7, we will describe relative partial molar properties in detail.

[l] In Chapter 7, we will describe excess molar properties in detail.

where

$$x_1 = \frac{n_1}{n_1 + n_2}.$$

Differentiating equation (5.36) (with $T$ and $p$ constant) gives

$$\left(\frac{\partial V}{\partial n_1}\right)_{n_2} = V^*_{\mathrm{m},1} + \frac{V^{\mathrm{E}}_{\mathrm{m}}}{x_1} + n_1\left(\frac{\partial(V^{\mathrm{E}}_{\mathrm{m}}/x_1)}{\partial n_1}\right)_{n_2}. \tag{5.37}$$

But

$$\left(\frac{\partial(V^{\mathrm{E}}_{\mathrm{m}}/x_1)}{\partial n_1}\right)_{n_2} = \left(\frac{\partial(V^{\mathrm{E}}_{\mathrm{m}}/x_1)}{\partial x_1}\right)_{n_2}\left(\frac{\partial x_1}{\partial n_1}\right)_{n_2}. \tag{5.38}$$

and

$$\left(\frac{\partial x_1}{\partial n_1}\right)_{n_2} = \frac{(n_1 + n_2) - n_1}{(n_1 + n_2)^2} = \frac{n_2}{(n_1 + n_2)^2}$$

$$= \frac{n_1 n_2}{n_1(n_1 + n_2)^2} = \frac{x_1 x_2}{n_1}. \tag{5.39}$$

Combining equations (5.37), (5.38), and (5.39) gives

$$\bar{V}_1 = V^*_{\mathrm{m},1} + \frac{V^{\mathrm{E}}_{\mathrm{m}}}{x_1} + x_1 x_2\left(\frac{\partial(V^{\mathrm{E}}_{\mathrm{m}}/x_1)}{\partial x_1}\right)_{n_2}. \tag{5.33}$$

A similar derivation gives equation (5.34). To use equations (5.33) and (5.34) to calculate $\bar{V}_1$ and $\bar{V}_2$, $V^{\mathrm{E}}_{\mathrm{m}}$ must be known as a function of $x_1$ or $x_2$. An expression often used is the Redlich–Kister equation, given by

$$Z^{\mathrm{E}}_{\mathrm{m}} = x_1(1 - x_1)\sum_{j=0}^{n} a_j(2x_1 - 1)^j, \tag{5.40}$$

where $Z$ is any of our thermodynamic variables, and enough coefficients $a_j$ are used to give a good fit to the experimental results. Expressed in terms of volume, equation (5.40) becomes

$$V_m^E = x_1(1 - x_1) \sum_{j=0}^{n} a_j(2x_1 - 1)^j.$$

Dividing this equation by $x_1$ gives

$$\frac{V_m^E}{x_1} = (1 - x_1) \sum_{j=0}^{n} a_j(2x_1 - 1)^j.$$

Differentiation and substitution into equation (5.33) gives

$$\bar{V}_1 = V_{m,1}^* + \frac{V_m^E}{x_1} - x_1(1 - x_1)$$

$$\times \left\{ a_0 + \sum_{j=1}^{n} a_j(2x_1 - 1)^{j-1}[2(j+1)x_1 - (2j+1)] \right\}. \tag{5.41}$$

The equivalent equation for $\bar{V}_2$ is[m]

$$\bar{V}_2 = V_2^* + \frac{V_m^E}{x_2} - x_2(1 - x_2)$$

$$\times \left\{ a_0 + \sum_{j=1}^{n} a_j(1 - 2x_2)^{j-1}[2(j+1) - 2(j+1)x_2] \right\}. \tag{5.42}$$

Figure 5.3 shows $\bar{V}_1$ and $\bar{V}_2$ for the (benzene + cyclohexane) system as a function of mole fraction, obtained in this manner.[3] Shown on the graph are $V_{m,1}^*$ and $V_{m,2}^*$, the partial molar volumes (which are the molar volumes) of the pure benzene and pure cyclohexane. The opposite ends of the curves gives $\bar{V}_1^\infty$ and $\bar{V}_2^\infty$, the partial molar volumes in an infinitely dilute solution. We note that

---

[m] A similar procedure can be used to obtain $(\bar{H}_1 - \bar{H}_1^*)$ and $(\bar{H}_2 - \bar{H}_2^*)$ from $H_m^E$, the excess molar enthalpy. These differences are relative partial molar enthalpies that we will describe in more detail in Chapter 7.

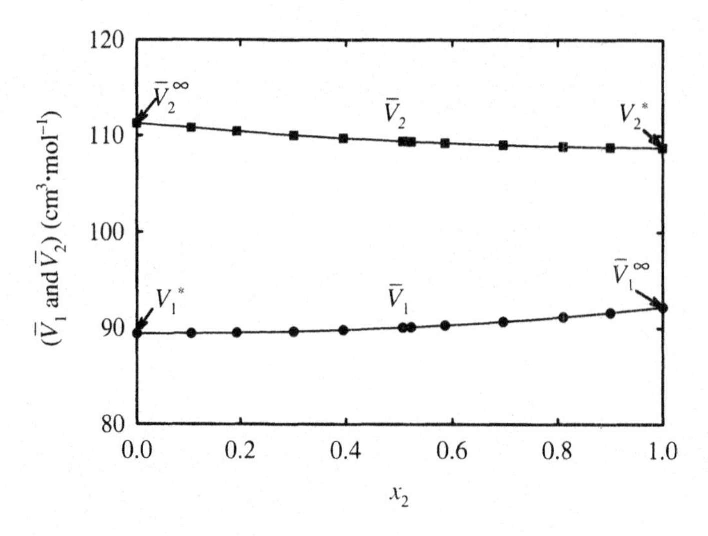


**Figure 5.3** Partial molar volumes $\overline{V}_1$ and $\overline{V}_2$ at 298.15 K for $(x_1C_6H_6 + x_2c\text{-}C_6H_{12})$ against $x_2$. The circles and squares represent the experimental data points.

the volume occupied by a mole of benzene (or a benzene molecule) is larger in the infinitely dilute solution where it is surrounded only by cyclohexane molecules, and the volume occupied by a cyclohexane molecule is larger in the infinitely dilute solution where it is surrounded only by benzene molecules.[n]

## 5.5d Calculations of Partial Molar Properties From Apparent Molar Properties

A method for determining partial molar properties, most often applied to electrolyte solutions, involves using the apparent molar property $\phi Z$, defined as

$$\phi Z = \frac{Z - n_1 Z^*_{\text{m},1}}{n_2}. \tag{5.43}$$

---

[n] We cannot generalize from this example. In some systems $\overline{V}_1^{\infty} < V^*_{\text{m},1}$ while $\overline{V}_2^{\infty} > V^*_{\text{m},2}$. In other examples $\overline{V}_1^{\infty} > V^*_{\text{m},1}$ while $\overline{V}_2^{\infty} < V^*_{\text{m},2}$. The thing that must be true from the Gibbs–Duhem equation is that

$$x_1 \frac{\mathrm{d}\overline{V}_1}{\mathrm{d}x_2} = -x_2 \frac{\mathrm{d}\overline{V}_2}{\mathrm{d}x_2}.$$

Thus, in the infinitely dilute solution, if $\mathrm{d}\overline{V}_1^{\infty}/\mathrm{d}x_2 > 0$ then $\mathrm{d}V_2^*/\mathrm{d}x_2 < 0$. In other words, as one approaches infinite dilution, if $\overline{V}_1$ is increasing then $\overline{V}_2$ must be decreasing. Conversely, if $\overline{V}_1$ is decreasing, $\overline{V}_2$ must be increasing.

The significance of equation (5.43) can be seen by referring to Figure 5.4 where the total volume of an aqueous sodium acetate solution containing a kg of water is graphed against $m$ (the molality) moles of sodium acetate.[4] When $m$ is the molality, the mass of the water is exactly one kilogram and its volume is held constant at 1002.94 cm$^3$. The volume increase in the solution as moles of sodium acetate are added is the apparent molar volume of the sodium acetate multiplied by $m$, the number of moles of sodium acetate. It is the volume the sodium acetate would occupy if the volume of the water does not change. The apparent molar volume is not the partial molar volume, since the volume of the water does change with concentration in the solution, but $\phi V$ should not differ too much from $\bar{V}_2$, which can be determined from $\phi V$. We will now see how this is done by starting with equation (5.43) and solving for $Z$ to obtain

$$Z = n_2\phi Z + n_1 Z^*_{\mathrm{m},1}.$$

Differentiating gives

$$\left(\frac{\partial Z}{\partial n_2}\right)_{T,p,n_1} = \bar{Z}_2$$

$$= \phi Z + n_2\left(\frac{\partial \phi Z}{\partial n_2}\right)_{T,p,n_1}. \tag{5.44}$$

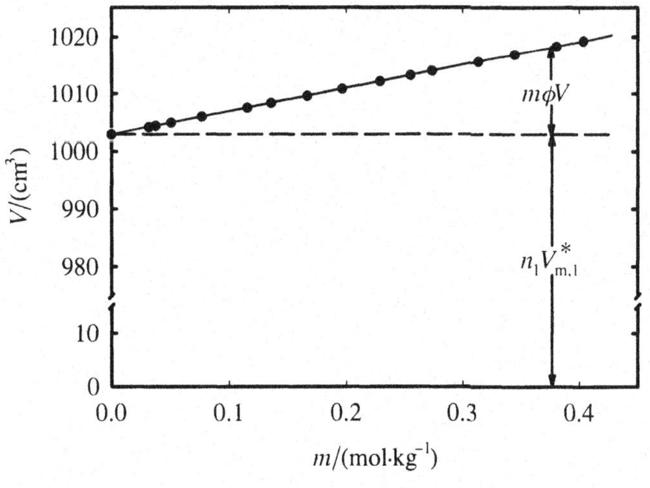

**Figure 5.4** Significance of the apparent molar volume $\phi V$ using aqueous sodium acetate solutions as an example. The dots represent the experimental results at $T = 298.15$ K with $m$ the molality of the sodium acetate.

In our example, $Z = V$, $n_2 = m$, $n_1 = 1\ \text{kg}/0.018016\ \text{kg}\cdot\text{mol}^{-1} = 55.506$ moles, and $\overline{V}_2$ is given by

$$\overline{V}_2 = \phi V + m\left(\frac{\partial \phi V}{\partial m}\right)_{T,p,n_1}. \tag{5.45}$$

An advantage of using equation (5.44) to calculate $\overline{Z}_2$ is that $\phi Z$ often varies almost linearly with $n_2$ (or $m$) so that the derivative can be accurately determined. Furthermore, the derivative multiplied by $n_2$ (or $m$) is often small when compared with $\phi Z$ so that errors in determining the derivative do not affect the accuracy of $\overline{Z}_2$ as much as if $\overline{Z}_2$ is calculated completely from a derivative, as was done in the determination of $\overline{C}_{p,1}$ and $\overline{C}_{p,2}$ for sulfuric acid described earlier.

**Example 5.1:** The apparent molar heat capacities at 298.15 K of $HNO_3$ in aqueous nitric acid solutions are given by the expression[5]

$$\phi C_p/(\text{J}\cdot\text{mol}^{-1}) = -72.1 + 28.95m^{1/2} + 46.1m$$

where $m$ is the molality of $HNO_3$. Calculate $\overline{C}_{p,2}$ for $HNO_3$ when $m = 0.100\ \text{mol}\cdot\text{kg}^{-1}$.

**Solution:**

$$\left(\frac{\partial \phi C_p}{\partial m}\right)_{n_1} = (\tfrac{1}{2})(28.95)m^{-1/2} + 46.1.$$

When $m = 0.100$,

$$\phi C_p = -72.1 + (28.95)(0.100)^{1/2} + (46.1)(0.100)$$

$$= -58.3\ \text{J}\cdot\text{K}^{-1}\cdot\text{mol}^{-1}$$

and

$$\left(\frac{\partial \phi C_p}{\partial m}\right)_{n_1} = (\tfrac{1}{2})(28.95)(0.100)^{-1/2} + 46.1$$

$$= 91.9.$$

$$\overline{C}_{p,2} = -58.3 + (0.100)(91.9)$$
$$= -49.1 \text{ J·K}^{-1}\text{·mol}^{-1}.$$

(Note that the $m(\partial\phi C_p/\partial m)_{n_2}$ term contributes about 19% to the total value of $\overline{C}_{p,2}$.) The negative $\overline{C}_{p,2}$ for $HNO_3$ is not unusual for strong electrolytes in dilute solution. It results because the charged ions break up the hydrogen bonded structure of the water and decrease the heat capacity of the solution over that of pure water. Thus, the contribution of $\overline{C}_{p,2}$ to $C_{p,m}$ is negative.

## 5.6 Criteria For Equilibrium

The prediction and understanding of a state of equilibrium constitutes one of the most important applications of thermodynamics. If we wait long enough, a system consisting of subsystems that are not at equilibrium will change until equilibrium is established. Heat will flow until all parts of the system are at the same temperature. Thus $T_1 = T_2 = T_3 = \cdots = T_i$, is a criterion for equilibrium. Volumes will adjust until all parts of the system are at the same pressure, unless some mechanical constraint keeps this from happening. Thus $p_1 = p_2 = p_3 \cdots = p_i$ is a second criterion for equilibrium.

In a chemical system, mass will flow until equilibrium is achieved. Consider, as an example, the approach to equilibrium between the liquid and gas phases of a substance. If a sample of liquid is introduced into an evacuated closed container, the liquid will begin to evaporate to form a gas phase. If one measures the pressure of this gas as a function of time, a graph similar to that shown in Figure 5.5 will be obtained.[o] The pressure at first increases rapidly. After a short time period, however, it levels off at the value represented by $p^*$.

Kinetic-molecular theory provides an explanation on a molecular level for this equilibrium. Evaporation from the liquid occurs as fast moving molecules on the surface escape from the liquid. In turn, molecules in the gas phase strike the liquid and condense. As the concentration (pressure) of gas molecules builds up in the gas phase, the rate of condensation increases. Eventually, a pressure is reached where the rate of condensation and rate of evaporation just balance, and equilibrium is achieved. The equilibrium pressure is denoted by $p^*$ and is known as the vapor pressure. The magnitude of $p^*$ depends upon the substance, composition of the liquid, and any two of our thermodynamic variables such as temperature and total pressure. The criteria for equilibrium that we will now derive provide the thermodynamic relationships that will help

[o] The liquid sample must be large enough that all does not evaporate before equilibrium is achieved. The condition of phase equilibrium requires that two (or more) phases be present.

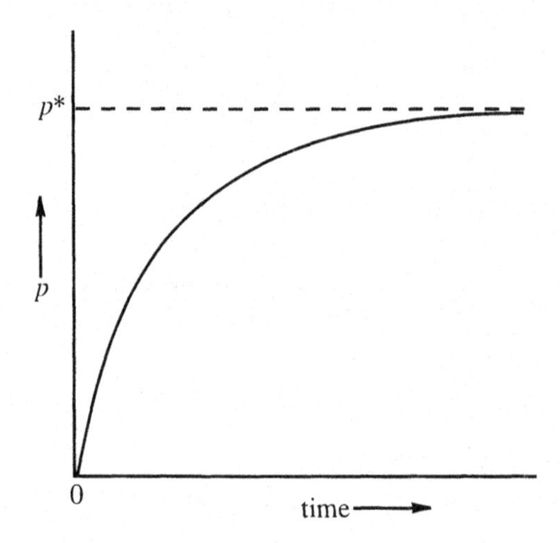


**Figure 5.5** Change of gas pressure with time as a liquid evaporates into a closed container. The pressure $p^*$ is the equilibrium vapor pressure.

us predict the effect of $p$, $T$, and composition on the point of equilibrium for phase changes, such as the one we have used as an example, and for chemical equilibrium.

To derive the condition for thermodynamic equilibrium, we start with an isolated system consisting of two subsystems as shown in Figure 5.6. Subsystem A is the one of primary interest in that it is the one in which the chemical process is occurring. Subsystem B is a reservoir in contact with subsystem A in such a way that energy in the form of heat or work can flow between the two subsystems. If left alone, the system will come to equilibrium. Energy will be transferred between the subsystems so that the temperature and pressure will be

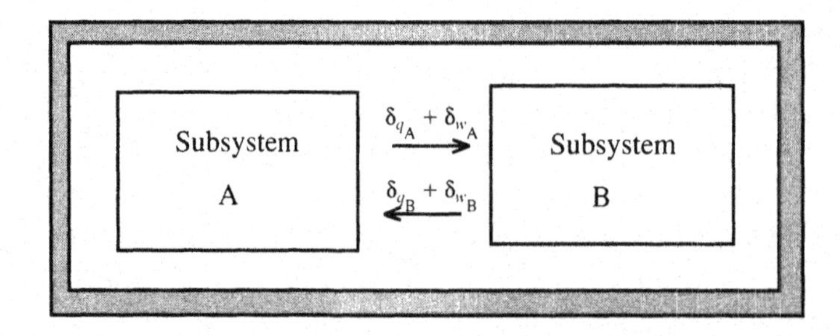


**Figure 5.6** An isolated system composed of subsystem A (the one we will eventually designate as the system) and subsystem B (the surroundings containing a heat reservoir). Heat and work will be exchanged until $T_A = T_B$, $p_A = p_B$, and equilibrium is established.

the same throughout, and mass will be transferred within subsystem A until equilibrium is achieved.[p] This process is spontaneous and hence, not reversible, and it requires that entropy be produced in the universe. Since the system is isolated, entropy cannot be produced outside the system, so that all of the entropy change must occur within the system.

We will take this isolated system and displace it from equilibrium by an infinitesimal amount in some manner. The system will then return to equilibrium and an entropy change will occur in the system given by

$$\mathrm{d}S = \mathrm{d}S_{\mathrm{A}} + \mathrm{d}S_{\mathrm{B}} \geqslant 0$$

where $\mathrm{d}S_{\mathrm{A}}$ is the entropy change in subsystem A, $\mathrm{d}S_{\mathrm{B}}$ is the entropy change in subsystem B, and $\mathrm{d}S$ is the total entropy change that, according to the Second Law,[q] will be greater than zero if the process is spontaneous and equal to zero if the process is reversible.

In the process that produces entropy, $\mathrm{d}S$, heat flows between the two subsystems so that

$$\delta q_{\mathrm{A}} = -\delta q_{\mathrm{B}}. \tag{5.46}$$

We will include in A any irreversibility in the flow of heat so that changes in B are reversible, and hence,

$$\mathrm{d}S_{\mathrm{B}} = \frac{\delta q_{\mathrm{B}}}{T}.$$

Substituting for $\delta q_{\mathrm{B}}$ from equation (5.46) gives

$$\mathrm{d}S_{\mathrm{B}} = -\frac{\delta q_{\mathrm{A}}}{T}$$

so that

$$\mathrm{d}S_{\mathrm{A}} - \frac{\delta q_{\mathrm{A}}}{T} \geqslant 0$$

[p] Subsystem A is a closed system with open phases so that mass can transfer within A, but not from A to B.

[q] See the discussion in Chapter 3 of the Second Law in terms of the Carathéodory principle.

or

$$T\mathrm{d}S_\mathrm{A} - \delta q_\mathrm{A} \geqslant 0 \tag{5.47}$$

since $T$ is always greater than zero.

Equation (5.47) gives the criterion for reversibility or spontaneity within subsystem A of an isolated system. The inequality applies to the spontaneous process, while the equality holds for the reversible process. Only when equilibrium is present can a change in an isolated system be conceived to occur reversibly. Therefore, the criterion for reversibility is a criterion for equilibrium, and equation (5.47) applies to the spontaneous or the equilibrium process, depending upon whether the inequality or equality is used.

Consider as an example, an isolated system in which subsystem B is a heat reservoir at $T = 273.15$ K and $p = 1$ atm (0.101 MPa) and subsystem A is a mixture of pure liquid water and solid ice. This temperature is the normal melting temperature of water, with solid and liquid in equilibrium at one atmosphere (0.101 MPa) pressure. We now slightly perturb subsystem A by changing infinitesimally its temperature or pressure. If we increase the temperature by $\mathrm{d}T$, an infinitesimal amount of ice melts, an amount of heat $\delta q$ flows into subsystem B, and the temperature decreases by $\mathrm{d}T$ and returns to the starting temperature. By repeating this process an infinite number of times, we can reversibly melt the solid while staying at equilibrium. In a similar manner, decreasing the temperature by $\mathrm{d}T$ causes the liquid to solidify reversibly. Thus, in a reversible phase change, the temperature never deviates more than an infinitesimal $\mathrm{d}T$ from the equilibrium value.

Changing the pressure will have a similar effect. If we increase $p$ by $\mathrm{d}p$, the solid melts. This process can be reversed at any time by decreasing the pressure by $\mathrm{d}p$. Note that at $p = 1$ atm (101.325 kPa), only at $T = 273.15$ K can the phase change be made to occur reversibly because this is the temperature where solid and liquid are in equilibrium at this pressure. If we tried to freeze liquid water at $p = 1$ atm and a lower temperature such as 263.15 K, the process, once started, would proceed spontaneously and could not be reversed by an infinitesimal change in $p$ or $T$.

The restriction of applying equation (5.47) to an isolated system seems to seriously limit the usefulness of this equation since we seldom work with isolated systems. But this is not so; the pre-eminent example of an isolated system is the universe, since neither mass nor energy can flow in or out of the universe. Thus, the isolated system shown in Figure 5.6 can be made the universe, with A the system of interest, and B the surroundings. When we designate the combined system as the universe, we can drop the subscript A in equation (5.47) and multiply by $-1$ to obtain

$$\delta q - T\,\mathrm{d}S \leqslant 0. \tag{5.48}$$

Equation (5.48) applies to any closed system of interest (in the universe).

Starting with equation (5.48) we can derive four equations that provide the criteria for spontaneity or equilibrium in a system under a specific set of conditions. To derive the first equation, we use the first law to write

$$\delta q = \mathrm{d}U - \delta w. \tag{5.49}$$

If we restrict the work in our process to pressure–volume work, then

$$\delta w = -p_{\mathrm{ext}}\,\mathrm{d}V. \tag{5.50}$$

Combining equations (5.48), (5.49), and (5.50) gives[r]

$$\mathrm{d}U - T\,\mathrm{d}S \leqslant -p_{\mathrm{ext}}\,\mathrm{d}V. \tag{5.51}$$

In a constant $S$, $V$ and $n$ (since the system is closed) process, $\mathrm{d}S = \mathrm{d}V = 0$ and equation (5.51) becomes

$$\mathrm{d}U_{S,V,n} \leqslant 0. \tag{5.52}$$

Equation (5.52) is the first of our criteria. The subscripts indicate that equation (5.52) applies to the condition of constant entropy, volume, and total moles, with the equality applying to the equilibrium process and the inequality to the spontaneous process.

The second criterion is obtained by writing

$$U = H - pV$$

and differentiating to get

$$\mathrm{d}U = \mathrm{d}H - p\,\mathrm{d}V - V\,\mathrm{d}p. \tag{5.53}$$

Equation (5.53) is substituted into equation (5.51) and rearranged to give

$$\mathrm{d}H - V\,\mathrm{d}p - T\,\mathrm{d}S \leqslant p\,\mathrm{d}V - p_{\mathrm{ext}}\,\mathrm{d}V. \tag{5.54}$$

---

[r] For the reversible (equilibrium) process, $p_{\mathrm{ext}} = p$ in equation (5.50). In the spontaneous process, the two pressures are not equal. In either case, $\delta w = -p_{\mathrm{ext}}\,\mathrm{d}V$.

In a constant pressure and entropy process, $p = p_{ext}$[s] and $dp$ and $dS$ are equal to zero so that equation (5.54) becomes

$$dH_{S,p,n} \leqslant 0. \tag{5.55}$$

Equation (5.55) gives the criterion for equilibrium or spontaneity in a constant pressure and entropy process.

The third criterion is obtained by writing

$$U = A + TS$$

and differentiating to obtain

$$dU = dA + T\,dS + S\,dT.$$

This equation is substituted into equation (5.51) to give

$$dA + T\,dS + S\,dT - T\,dS \leqslant -p_{ext}\,dV.$$

Cancelling terms gives

$$dA + S\,dT \leqslant -p_{ext}\,dV. \tag{5.56}$$

In a constant $T$ and $V$ process, $dT$ and $dV$ are equal to zero and equation (5.56) becomes

$$dA_{T,V,n} \leqslant 0. \tag{5.57}$$

Equation (5.57) gives the criterion for equilibrium and spontaneity in a constant temperature and volume process.

The final criterion is obtained from

$$H = G + TS.$$

Differentiating gives

$$dH = dG + T\,dS + S\,dT. \tag{5.58}$$

---

[s] In an isobaric process, $p = p_{ext}$ unless a mechanical constraint inside the system prevents pressure equilibrium. The criteria for equilibrium that we derive do not apply to this situation, but this is not a serious concern, since such a constraint is rarely encountered and never under circumstances that are of interest in chemical processes.

Substituting equation (5.58) into equation (5.54) and cancelling terms gives

$$\mathrm{d}G - V\,\mathrm{d}p + S\,\mathrm{d}T \leqslant p\,\mathrm{d}V - p_{\text{ext}}\,\mathrm{d}V.$$

At constant pressure and temperature $p_{\text{ext}} = p$ and $\mathrm{d}T$ and $\mathrm{d}p$ are equal to zero so that

$$\mathrm{d}G_{T,p,n} \leqslant 0. \tag{5.59}$$

This is the fourth criterion for equilibrium or spontaneity.

In summary the four criteria are[t]

$$\mathrm{d}U_{S,V,n} \leqslant 0 \tag{5.52}$$

$$\mathrm{d}H_{S,p,n} \leqslant 0 \tag{5.55}$$

$$\mathrm{d}A_{V,T,n} \leqslant 0 \tag{5.57}$$

$$\mathrm{d}G_{p,T,n} \leqslant 0, \tag{5.59}$$

where the equality applies to the equilibrium process and the inequality to the spontaneous process. The last of the four equations is of most interest in chemistry since most chemical processes occur at constant $p$ and $T$. We will now use this criterion to obtain the conditions that apply to phase equilibrium and chemical equilibrium.

### 5.6a Criterion for Phase Equilibrium

Under certain pressure and temperature conditions, a system can contain two or more phases in equilibrium. An example is the temperature and pressure where solid and liquid are in equilibrium. We refer to this condition as (solid + liquid) equilibrium, and the temperature as the melting temperature. This temperature changes with pressure and with composition. The melting temperature when the

---

[t] Again, we see that $S$ and $V$ become the variables of primary interest when working with $U$. With $H$ we work with $S$ and $p$, with $A$ it is $V$ and $T$, and with $G$ it is $p$ and $T$. We will use equation (5.59) to derive a criterion for equilibrium in a constant $T$ and $p$ process in terms of chemical potential defined earlier in this chapter as $\mu_i = (\partial G/\partial n_i)_{T,p,n_{j\neq i}}$. Equivalent relationships can be derived for a constant $T$ and $V$ process using equation (5.57) and $\mu_i = (\partial A/\partial n_i)_{T,V,n_{j\neq i}}$. For a constant $S$ and $p$ process, we would use equation (5.55) with $\mu_i = (\partial H/\partial n_i)_{S,p,n_{j\neq i}}$; and for a constant $S$ and $V$ process, we would use equation (5.52) with $\mu_i = (\partial U/\partial n_i)_{S,V,n_{j\neq i}}$. It seems safe to generalize and say that the chemical potential provides the criterion for equilibrium in any process in which we hold constant any two of the thermodynamic variables $S$ and $V$, or $S$ and $p$, or $T$ and $V$, or $T$ and $p$.

pressure is exactly one atmosphere (0.101325 MPa) is known as the **normal melting temperature**.

Liquid and vapor are in equilibrium when the pressure of the vapor phase is the vapor pressure. When the (vapor + liquid) equilibrium mixture is exposed to the atmosphere, the mixture will boil at a temperature where the vapor pressure equals the external (atmospheric) pressure. This temperature is known as the boiling temperature. At the **normal boiling** temperature, the substance has a vapor pressure of exactly one atmosphere (0.101325 MPa) and hence, boils at this external pressure.

The equilibrium pressure when (solid + vapor) equilibrium occurs is known as the **sublimation pressure**. (The sublimation temperature is the temperature at which the vapor pressure of the solid equals the pressure of the atmosphere.) A **normal sublimation temperature** is the temperature at which the sublimation pressure equals one atmosphere (0.101325 MPa). Two solid phases can be in equilibrium at a transition temperature (solid + solid) equilibrium, and (liquid + liquid) equilibrium occurs when two liquids are mixed that are not miscible and separate into two phases. Again, "normal" refers to the condition of one atmosphere (0.101325 MPa) pressure. Thus, the **normal transition temperature** is the transition temperature when the pressure is one atmosphere (0.101325 MPa) and at the **normal (liquid + liquid) solubility** condition, the composition of the liquid phases are those that are in equilibrium at an external pressure of one atmosphere (0.101325 MPa).

In summary, the following types of phase equilibrium involving two phases are commonly encountered[u]

solid = liquid

solid = gas

liquid = gas

(solid A = solid B)

(liquid A = liquid B).

More than two phases can exist at equilibrium. For example, solid ice, liquid water, and water vapor exist together at the triple point of water. Various combinations of solid, liquid, and vapor can exist together to give triple points.[v]

---

[u] Phase equilibrium involving liquids and gases at high pressure and at a temperature near the critical locus of the mixture is known as (fluid + fluid) equilibrium.

[v] We will see later that for a pure substance, only three phases can be in equilibrium at a time. With multiple components more than three phases can exist together at equilibrium. For example, in a binary mixture, quadruple points can occur.

At a particular temperature and pressure, equation (5.59) provides a criterion for determining when phases are in equilibrium and for predicting when one phase will convert to another. A more useful criterion however, is obtained from the chemical potential. Consider as an example the system shown below in Figure 5.7 that consists of a single pure substance with phases A and B in equilibrium. The system is closed so that mass cannot flow in or out of the system, but the phases are open so that mass can flow between the two phases. An infinitesimal (reversible) displacement in the system causes $\mathrm{d}n_\mathrm{A}$ moles to flow from A and $\mathrm{d}n_\mathrm{B}$ moles to flow from B. The total free energy change for this process is the sum of the free energy changes $\mathrm{d}G_\mathrm{A}$ and $\mathrm{d}G_\mathrm{B}$ in each phase. That is,

$$\mathrm{d}G_{T,p,n} = \mathrm{d}G_\mathrm{A} + \mathrm{d}G_\mathrm{B}. \tag{5.60}$$

But

$$\mathrm{d}G_\mathrm{A} = \mu_\mathrm{A}\,\mathrm{d}n_\mathrm{A}$$

and

$$\mathrm{d}G_\mathrm{B} = \mu_\mathrm{B}\,\mathrm{d}n_\mathrm{B}.$$

so that

$$\mathrm{d}G_{T,p,n} = \mu_\mathrm{A}\,\mathrm{d}n_\mathrm{A} + \mu_\mathrm{B}\,\mathrm{d}n_\mathrm{B}\,. \tag{5.61}$$

At equilibrium

$$\mathrm{d}G_{T,p,n} = 0 \tag{5.59}$$

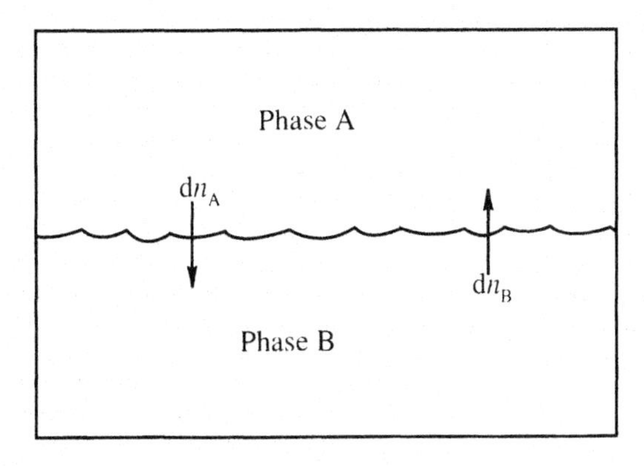


**Figure 5.7** Infinitesimal amounts of mass $\mathrm{d}n_\mathrm{A}$ and $\mathrm{d}n_\mathrm{B}$ are transferred in a system containing two phases A and B that are in equilibrium.

so that

$$\mu_A \, dn_A + \mu_B \, dn_B = 0. \tag{5.62}$$

Since the system is closed, the total number of moles $n$ given by

$$n = n_A + n_B$$

is constant. Hence,

$$dn = dn_A + dn_B = 0$$

or

$$dn_A = -dn_B.$$

Substitution into equation (5.62) gives

$$(\mu_A - \mu_B) \, dn_A = 0.$$

Since the differential $dn_A$ is unrestricted and can have any value, we conclude that at equilibrium

$$\mu_A = \mu_B.$$

Thus, the condition for equilibrium between two phases of a pure substance is given by

$$\mu_A = \mu_B \quad \text{(equilibrium)}. \tag{5.63}$$

The condition for spontaneity is obtained by repeating the derivation using the inequality instead of the equality in equation (5.59). The result is

$$\mu_B < \mu_A \quad \text{(spontaneous for A} \rightarrow \text{B)}. \tag{5.64}$$

The consequences of these equations are seen in Figure 5.8 in which $\mu_A$ and $\mu_B$ are plotted against temperature at a fixed pressure. At the temperature $T_0$, $\mu_A = \mu_B$ and the two phases are in equilibrium. For $T > T_0$, $\mu_A > \mu_B$ and B is the stable phase. For $T < T_0$, $\mu_B > \mu_A$ and A is the stable phase. It can be seen from these relationships that $\mu$ is a potential that drives the flow of mass in a phase change. Mass flows from the phase with high potential to the phase with low potential. When the two potentials are equal, equilibrium is established and there is no net flow of mass.

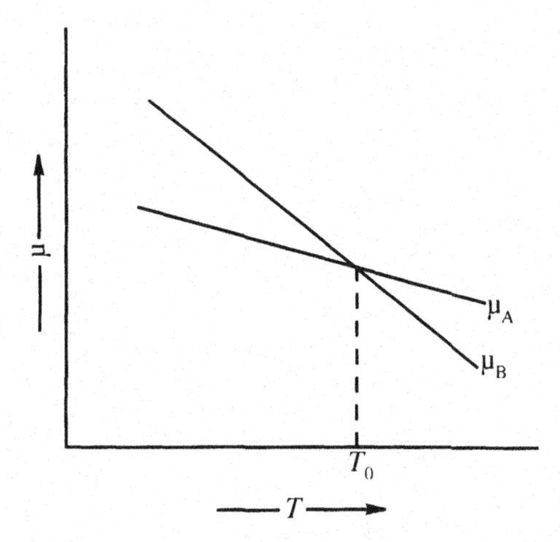


**Figure 5.8** At the equilibrium temperature $T_0$, $\mu_A = \mu_B$. For $T < T_0$, $\mu_B > \mu_A$ and B spontaneously changes to A. For $T > T_0$, $\mu_A > \mu_B$ and A spontaneously changes to B.

Equations (5.63) and (5.64) are actually more general than is apparent from the derivation. Consider a closed system at a given temperature and pressure with $n_1^0, n_2^0, n_3^0, \ldots$ moles of the components $1, 2, 3, \ldots$ distributed among the phases $A, B, C, \ldots$ For the flow of mass between the phases due to an infinitesimal reversible (equilibrium) displacement we can write

$$\mathrm{d}G_A = \mu_{A,1}\,\mathrm{d}n_{A,1} + \mu_{A,2}\,\mathrm{d}n_{A,2} + \mu_{A,3}\,\mathrm{d}n_{A,3} + \cdots$$

$$\mathrm{d}G_B = \mu_{B,1}\,\mathrm{d}n_{B,1} + \mu_{B,2}\,\mathrm{d}n_{B,2} + \mu_{B,3}\,\mathrm{d}n_{B,3} + \cdots$$

$$\mathrm{d}G_C = \mu_{C,1}\,\mathrm{d}n_{C,1} + \mu_{C,2}\,\mathrm{d}n_{C,2} + \mu_{C,3}\,\mathrm{d}n_{C,3} + \cdots$$

where $\mathrm{d}G_A$, $\mathrm{d}G_B$, and $\mathrm{d}G_C, \ldots$ are the changes in Gibbs free energy in each of the $A, B, C, \ldots$ phases.

At equilibrium, the total Gibbs free energy change is given by

$$\mathrm{d}G_{T,p,n} = \mathrm{d}G_A + \mathrm{d}G_B + \mathrm{d}G_C + \cdots = 0.$$

Substitution gives

$$\begin{aligned}&\mu_{A,1}\,\mathrm{d}n_{A,1} + \mu_{B,1}\,\mathrm{d}n_{B,1} + \mu_{C,1}\,\mathrm{d}n_{C,1} + \cdots\\ &\quad + \mu_{A,2}\,\mathrm{d}n_{A,2} + \mu_{B,2}\,\mathrm{d}n_{B,2} + \mu_{C,2}\,\mathrm{d}n_{C,2} + \cdots\\ &\quad + \mu_{A,3}\,\mathrm{d}n_{A,3} + \mu_{B,3}\,\mathrm{d}n_{B,3} + \mu_{C,3}\,\mathrm{d}n_{C,3} + \cdots = 0.\end{aligned} \tag{5.65}$$

The total moles of each component is given by

$$n_1^0 = n_{A,1} + n_{B,1} + n_{C,1} + \cdots$$

$$n_2^0 = n_{A,2} + n_{B,2} + n_{C,2} + \cdots$$

$$n_3^0 = n_{A,3} + n_{B,3} + n_{c,3} + \cdots \qquad (5.66)$$

Differentiation gives

$$dn_1^0 = 0 = dn_{A,1} + dn_{B,1} + dn_{C,1} + \cdots$$

$$dn_2^0 = 0 = dn_{A,2} + dn_{B,2} + dn_{C,2} + \cdots$$

$$dn_3^0 = 0 = dn_{A,3} + dn_{B,3} + dn_{C,3} + \cdots \qquad (5.67)$$

By using the equations in (5.67), $dn_{A,1}, dn_{A,2}, dn_{A,3}$ ... can be eliminated from equation (5.65) to obtain

$$(\mu_{B,1} - \mu_{A,1})\, dn_{B,1} + (\mu_{C,1} - \mu_{A,1})\, dn_{C,1} + \cdots$$
$$+ (\mu_{B,2} - \mu_{A,2})\, dn_{B,2} + (\mu_{C,2} - \mu_{A,2})\, dn_{C,2} + \cdots$$
$$+ (\mu_{B,3} - \mu_{A,3})\, dn_{B,3} + (\mu_{C,3} - \mu_{A,3})\, dn_{C,3} + \cdots$$
$$+ \cdots = 0. \qquad (5.68)$$

Since each $dn$ is independent, each coefficient must be zero. This requires that

$$\mu_{A,1} = \mu_{B,1} = \mu_{C,1} = \cdots$$

$$\mu_{A,2} = \mu_{B,2} = \mu_{C,2} = \cdots$$

$$\mu_{A,3} = \mu_{B,3} = \mu_{C,3} = \cdots$$

In words, for phase equilibrium to occur in a constant temperature and pressure process, the chemical potential of each component must have the same value in all of the phases that are in equilibrium.

Applying the inequality in equation (5.59) gives the conditions under which one phase will spontaneously convert to another. The result is simply that for any component, moles of that component will flow from a phase in which it has a higher chemical potential to a phase in which it has a lower chemical potential until the two chemical potentials are equal. This flow of mass will disrupt other equilibria that are present and mass will flow between phases from high potential to a low potential until each component has the same chemical potential in all phases.[w]

## 5.6b The Gibbs Phase Rule

Having phases together in equilibrium restricts the number of thermodynamic variables that can be varied independently and still maintain equilibrium. An expression known as the **Gibbs phase rule** relates the number of independent components $C$[x] and number of phases $P$ to the number of variables that can be changed independently. This number, known as the **degrees of freedom** $f$, is equal to the number of independent variables present in the system minus the number of equations of constraint between the variables.

The number of independent variables for each phase are the two that we have considered earlier among $p$, $T$, $V$, $U$, $H$, $A$, and $G$, plus $(C-1)$ composition variables.[y] Thus the total number of independent variables for $P$ phases is $\{(C-1+2)$ per phase$\} \cdot (P$ phases$) = (C+1)P$.

Equations of constraint can be written by noting that the temperature, pressure, and chemical potential of each component must be the same in all phases. Thus,

$$T_{\mathrm{A}} = T_{\mathrm{B}} = T_{\mathrm{C}} = \cdots = T_P$$

$$p_{\mathrm{A}} = p_{\mathrm{B}} = p_{\mathrm{C}} = \cdots = p_P$$

$$\mu_{\mathrm{A},1} = \mu_{\mathrm{B},1} = \mu_{\mathrm{C},1} = \cdots = \mu_{P,1}$$

$$\mu_{\mathrm{A},2} = \mu_{\mathrm{B},2} = \mu_{\mathrm{C},2} = \cdots = \mu_{P,2}$$

$$\vdots \qquad \vdots \qquad \vdots \qquad \qquad \vdots$$

$$\mu_{\mathrm{A},C} = \mu_{\mathrm{B},C} = \mu_{\mathrm{C},C} = \cdots = \mu_{P,C}.$$

From the above, the total number of equations of constraint is given by $(P-1)(C+2)$. Since the degrees of freedom $f$ is the number of independent

[w] This does not imply that all components will have the same chemical potential at equilibrium, only that the chemical potential for each component is the same in all phases.

[x] The independent components are those that are not related through chemical equilibrium reactions. See Section 1.2c for a discussion of components in a system.

[y] Only $(C-1)$ composition variables are independent. Thus, if we know the molarity of the solute in a binary solution, we know the concentration of the solvent. Similarly, if we know the mole fraction of two components in a ternary system, the composition of the third is known, since $x_1 + x_2 + x_3 = 1$.

variables minus the number of equations of restraint, we get

$$f = P(C+1) - (P-1)(C+2)$$

$$f = C - P + 2. \tag{5.69}$$

This relationship is the Gibbs phase rule.

**Example 5.3:** Predict the degrees of freedom for (a) pure liquid water and solid ice in equilibrium; (b) pure liquid water, solid ice, and water vapor in equilibrium, and (c) solid ice in equilibrium with a liquid mixture of (ethanol + water).

**Solution:**

(a) $C = 1$, $P = 2$,
$f = C - P + 2 = 1 - 2 + 2 = 1$.
Thus, one variable can be specified. If one fixes the melting temperature, the pressure is fixed. Conversely, if one fixes the pressure, the melting temperature is fixed. As we indicated earlier, the normal melting temperature of water is the temperature at which the pressure is exactly one atmosphere. This temperature is fixed at 273.150 K.

(b) $C = 1$, $P = 3$,
$f = 1 - 3 + 2 = 0$.
For a pure substance, having three phases in equilibrium results in a triple point that is invariant. When pure solid, liquid, and gaseous water are in equilibrium, the temperature is fixed at a value of 273.16 K, and the pressure of the gas is fixed at the vapor pressure value (0.6105 kPa).

(c) $C = 2$, $P = 2$,
$f = 2 - 2 + 2 = 2$.
Two variables can be varied independently. Specifying the pressure and temperature fixes the composition, specifying the pressure and composition fixes the temperature, or specifying the temperature and composition fixes the pressure.

## 5.6c The Clapeyron Equation

A useful relationship between the temperature and pressure of phases in equilibrium can be derived from the condition for equilibrium. We start with equilibrium between phases A and B written as

$$\mathrm{A} = \mathrm{B},$$

and write the condition of equilibrium for the $i$th component

$$\mu_{A,i} = \mu_{B,i}.$$

A pressure displacement $dp$ and a temperature displacement $dT$ are made on the system. This causes changes in the chemical potentials $d\mu_{A,i}$ and $d\mu_{B,i}$. If the phases are to remain in equilibrium, these changes must be equal so that

$$d\mu_{A,i} = d\mu_{B,i}.$$

The Gibbs equation applied to each component relates the change in chemical potential to the change in temperature and pressure.

$$d\mu_{A,i} = -\bar{S}_{A,i}\, dT + \bar{V}_{A,i}\, dp$$

$$d\mu_{B,i} = -\bar{S}_{B,i}\, dT + \bar{V}_{B,i}\, dp.$$

Equating these two equations and collecting terms gives

$$\frac{dp}{dT} = \frac{\bar{S}_{B,i} - \bar{S}_{A,i}}{\bar{V}_{B,i} - \bar{V}_{A,i}}$$

or

$$\frac{dp}{dT} = \frac{\Delta\bar{S}_i}{\Delta\bar{V}_i} \tag{5.70}$$

where $\Delta\bar{S}_i$ and $\Delta\bar{V}_i$ are the change in partial molar entropy and partial molar volume between the two phases.

Equation (5.70) is known as the **Clapeyron equation**. It is most often applied to a pure substance (although it is not restricted to do so), in which case $\bar{V} = V_m^*$ and $\bar{S} = S_m^*$ and

$$\frac{dp}{dT} = \frac{\Delta S_m}{\Delta V_m}, \tag{5.71}$$

where $\Delta S_m$ and $\Delta V_m$ are the molar changes in entropy or volume for the change in phase. We have shown in Chapter 2 {equation (2.64)} that for an equilibrium phase change

$$\Delta S_\mathrm{m} = \frac{\Delta H_\mathrm{m}}{T}.$$

Substitution of this relationship into equation (5.71) gives an alternate form of the Clapeyron equation.

$$\frac{\mathrm{d}p}{\mathrm{d}T} = \frac{\Delta H_\mathrm{m}}{T\Delta V_\mathrm{m}}. \tag{5.72}$$

Thus, for (solid + liquid) equilibrium

$$\frac{\mathrm{d}p}{\mathrm{d}T} = \frac{\Delta_\mathrm{fus} H_\mathrm{m}}{T\Delta_\mathrm{fus} V_\mathrm{m}}, \tag{5.73}$$

and for (vapor + liquid) equilibrium

$$\frac{\mathrm{d}p}{\mathrm{d}T} = \frac{\Delta_\mathrm{vap} H_\mathrm{m}}{T\Delta_\mathrm{vap} V_\mathrm{m}}, \tag{5.74}$$

where $\Delta_\mathrm{fus} H_\mathrm{m}$ and $\Delta_\mathrm{vap} H_\mathrm{m}$ are the molar enthalpy changes for fusion and vaporization, respectively, and $\Delta_\mathrm{fus} V_\mathrm{m}$ and $\Delta_\mathrm{vap} V_\mathrm{m}$ are the molar volume changes for fusion and vaporization. Values for the molar enthalpies of fusion and vaporization, along with the normal melting and boiling temperatures, are given for selected substances in Table 2.2 of Chapter 2.

### 5.6d Criterion for Chemical Equilibrium

Equation (5.59) is the starting point for deriving the condition for chemical equilibrium. We have written a generalized chemical reaction as

$$\nu_1 \mathrm{A}_1 + \nu_2 \mathrm{A}_2 + \cdots = \nu_m \mathrm{A}_m + \nu_{m+1} \mathrm{A}_{m+1}$$

or

$$\sum \nu_i \mathrm{A}_i = 0, \tag{5.75}$$

where the stoichiometric $\nu_i$ coefficients are positive for products and negative for reactants and the $A_i$ are the components. For example, for the reaction

$$3H_2 + N_2 = 2NH_3$$

$A_1 = H_2$, $\nu_1 = -3$, $A_2 = N_2$, $\nu_2 = -1$, and $A_3 = NH_3$, $\nu_3 = 2$.

During a chemical reaction, the change $\Delta n_i$ in the number of moles for each substance is proportional to $\nu_i$, with the proportionality constant the same for all components. This proportionality constant is given the symbol $\xi$ and is called the **extent of the reaction**. It is related to the number of moles reacted by

$$\Delta n_i = \nu_i \xi, \tag{5.76}$$

where

$$\Delta n_i = n_i - n_{i,0}, \tag{5.77}$$

with $n_{i,0}$ equal to the moles of the component before reaction. For example, in our reaction suppose that 2 moles of $N_2$ reacts so that $\Delta n_{N_2} = -2$. Then, from equation (5.76)

$$-2 \text{ moles} = (-1 \text{ moles})(\xi)$$

or

$$\xi = 2.$$

Using this value for $\xi$ gives

$$\Delta n_{H_2} = -(3)(\xi) = -(3)(2) = -6 \text{ moles}$$

$$\Delta n_{NH_3} = (2)(\xi) = (2)(2) = 4 \text{ moles.}$$

To obtain the criteria for equilibrium we combine equations (5.76) and (5.77) to give

$$n_i = n_{i,0} + \nu_i \xi.$$

Differentiation gives

$$dn_i = \nu_i \, d\xi$$

which when substituted into the equation

$$dG_{T,p,n} = \sum_i \mu_i \, dn_i \leqslant 0 \tag{5.78}$$

gives

$$dG_{T,p,n} = \sum_i \mu_i \nu_i \, d\xi \leqslant 0 \tag{5.79}$$

or

$$d\xi \sum_i \mu_i \nu_i \leqslant 0. \tag{5.80}$$

No reaction takes place if $d\xi = 0$, so that

$$\sum_i \mu_i \nu_i \leqslant 0. \tag{5.81}$$

Equation (5.81) gives us the criteria for equilibrium or spontaneity. For example, for equilibrium in the reaction leading to the formation of ammonia

$$2\mu_{NH_3} - 3\mu_{H_2} - \mu_{N_2} = 0.$$

For the spontaneous process in which $H_2$ and $N_2$ combine to form $NH_3$,

$$2\mu_{NH_3} - 3\mu_{H_2} - \mu_{N_2} < 0.$$

Note that by rearranging equation (5.79), we get

$$\frac{dG}{d\xi} = \sum_i \nu_i \mu_i.$$

At equilibrium, $dG/d\xi = 0$ so that a reaction proceeds until $dG/d\xi$ is a minimum. If we plot $G$ against $\xi$ we get a graph similar to that shown in Figure 5.9. The minimum in the curve gives the extent of the reaction when equilibrium is obtained.

In summary, we now have the tools for describing phase equilibrium for both pure materials and for mixtures, and for understanding chemical processes at equilibrium. We will rely upon the foundation developed in this chapter as we

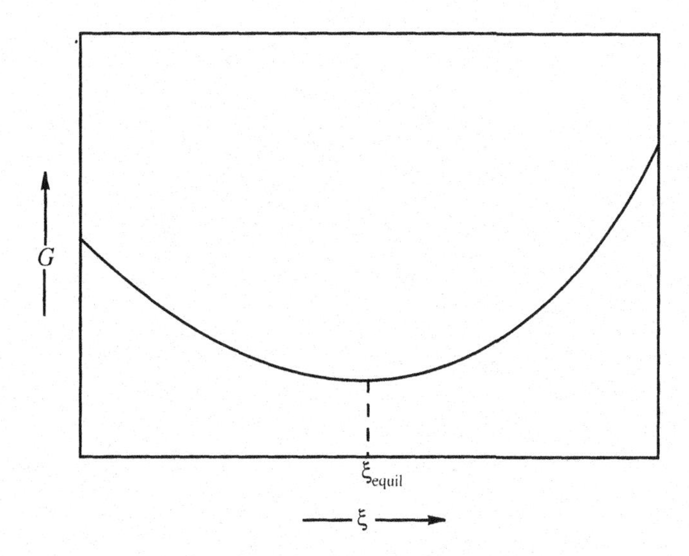


**Figure 5.9** Graph of Gibbs free energy against $\xi$, the extent of the reaction. The minimum in the curve gives the extent of the reaction at equilibrium.

apply our thermodynamic tools in subsequent chapters to phase changes and to chemical reactions.

## Exercises

E5.1 The densities of $\alpha$ and $\gamma$ iron are 7.571 and 7.633 g·cm$^{-3}$, respectively, at their transformation temperature of 1183 K under a pressure of 0.101 MPa. The enthalpy of the phase change, $\Delta_{trans}H_m$, from $\alpha$ to $\gamma$ is 900 J·mol$^{-1}$ at 1183 K, and 1665 J·mol$^{-1}$ at 1100 K. Assuming (a) that $\Delta_{trans}H_m$ is independent of pressure but varies linearly with temperature, and (b) the value of $V_m(\alpha) - V_m(\gamma)$ is constant, calculate the pressure under which both forms of iron coexist at 1100 K.

E5.2 The density of aqueous acetic acid solutions varies with composition $w_2$ (mass fraction of acetic acid) at 298 K as follows:

| $\rho$/(g·cm$^{-3}$) | $w_2$ | $\rho$/(g·cm$^{-3}$) | $w_2$ | $\rho$/(g·cm$^{-3}$) | $w_2$ |
|---|---|---|---|---|---|
| 0.9982 | 0 | 1.0488 | 0.40 | 1.0700 | 0.80 |
| 1.0125 | 0.10 | 1.0575 | 0.50 | 1.0661 | 0.90 |
| 1.0263 | 0.20 | 1.0642 | 0.60 | 1.0498 | 1.00 |
| 1.0384 | 0.30 | 1.0685 | 0.70 | | |

Plot an appropriate volume unit vs. mole fraction of acetic acid. Determine the partial molar volumes of water and acetic acid at $x_2 = 0$, 1, and several intermediate compositions (at least three). Plot $\overline{V}_1$ and $\overline{V}_2$ as a function of $x_2$.

E5.3 At $T = 291.15$ K the total volume of a solution formed from one kilogram of water mixed with $m$ moles of $MgSO_4$ is given by

$$V/(\text{cm}^3) = 1001.21 + 34.69\ (m - 0.07)^2.$$

The expression applies up to a molality of approximately $m = 1.10\ \text{mol·kg}^{-1}$. What are the partial molar volumes of (a) the $MgSO_4$ and (b) the water when $m = 0.050\ \text{moles·kg}^{-1}$?

E5.4 The apparent molar heat capacity of sucrose (2) in water (1) is given as a function of the molality, $m$ by the expression

$$\phi C_p/(\text{J·K}^{-1}\text{·mol}^{-1}) = 633.88 + 4.728m - 0.1948m^2$$

with $C^*_{p,1} = 75.40\ \text{J·K}^{-1}\text{·mol}^{-1}$.

(a) Derive expressions for $\overline{C}_{p,1}$ and $\overline{C}_{p,2}$ as a function of $m$.

(b) Calculate the heat capacity of a solution consisting of 0.3 moles of sucrose and 600 g of water.

E5.5 The following equation has been found to represent the volume of aqueous solutions of NaCl at $T = 298.15$ K and $p = 0.101$ MPa that contains 1.000 kg of water

$$V/(\text{cm}^3) = 1002.96 + 16.6253n + 1.7738n^{3/2} + 0.1194n^2$$

where $n$ is the moles of NaCl.

(a) Obtain expressions for $\overline{V}_1$ and $\overline{V}_2$.

(b) Calculate $\Delta V_m$ for each of the processes in which one mole of NaCl or one mole of water ends up in a solution with $m = 4\ \text{mol·kg}^{-1}$. That is,

$$\text{NaCl(s)} = \text{NaCl (aq}, m = 4\ \text{mol·kg}^{-1})$$

$$H_2O(l) = H_2O\ (\text{aq}, m = 4\ \text{mol·kg}^{-1})$$

where $m$ is the molality of NaCl. The density of solid NaCl is $2.165\ \text{g·cm}^{-3}$ at $T = 298.15$ K.

## Problems

P5.1 The specific heat $c_p$ and apparent molar heat capacities $\phi C_p$ of aqueous solutions of hydroxyacetamide ($HOCH_2CONH_2$, $M_2 = 0.07507\ \text{kg·mol}^{-1}$) at 298.15 K and a series of molalities $m$ are as follows[6]

| $m/(\text{mol}\cdot\text{kg}^{-1})$ | $c_p/(\text{J}\cdot\text{K}^{-1}\cdot\text{g}^{-1})$ | $\phi C_p/(\text{J}\cdot\text{K}^{-1}\cdot\text{mol}^{-1})$ |
|---|---|---|
| 0.0000 | 4.1840 | – |
| 0.2014 | 4.1515 | 150.2 |
| 0.4107 | 4.1187 | 150.6 |
| 0.7905 | 4.0630 | 152.00 |
| 1.2890 | 3.9943 | – |
| 1.7632 | 3.9350 | 154.14 |
| 2.6537 | 3.8353 | 156.52 |
| 4.3696 | 3.6777 | 160.21 |
| 4.3697 | 3.6777 | 160.21 |
| 6.1124 | 3.5518 | 163.22 |

(a) Calculate the value of $\phi C_p$ of a 1.2890 molal solution. Note that if we take 1 kg of solvent, then $m$ is the moles of solute and the total heat capacity is 1000 $c_p(1 + mM_2)$. In this solution the heat capacity of the pure solvent is 1000 $c_p^*$.

(b) Make a graph of $\phi C_p$ against molality. Determine enough values of $\overline{C}_{p,2}$ so that you can plot them on the graph of $\phi C_p$ and compare the values.

P5.2 Given the following densities $\rho$ at $T = 298.15$ K and $p = 15.0$ MPa for mixtures of $(x_1H_2O + x_2C_2H_5OH)$, where $x_1$ and $x_2$ are the mole fractions

| $x_2$ | $\rho/(\text{kg}\cdot\text{m}^{-3})$ | $x_2$ | $\rho/(\text{kg}\cdot\text{m}^{-3})$ |
|---|---|---|---|
| 0 | 1003.7 | 0.4029 | 888.3 |
| 0.0218 | 993.8 | 0.4623 | 875.9 |
| 0.0336 | 989.3 | 0.5067 | 867.5 |
| 0.0459 | 985.1 | 0.5557 | 858.7 |
| 0.0549 | 982.2 | 0.6099 | 849.7 |
| 0.0656 | 979.1 | 0.6392 | 845.0 |
| 0.0725 | 977.0 | 0.7030 | 835.8 |
| 0.0825 | 974.1 | 0.7378 | 830.9 |
| 0.0944 | 970.8 | 0.7747 | 825.8 |
| 0.1182 | 964.3 | 0.8140 | 820.8 |
| 0.1627 | 951.5 | 0.8559 | 815.4 |
| 0.2148 | 936.0 | 0.9007 | 810.1 |
| 0.2633 | 922.7 | 0.9486 | 804.2 |
| 0.3193 | 907.8 | 1 | 798.4 |
| 0.3673 | 896.2 | – | – |

(a) Obtain values for $\overline{V}_1$ and $\overline{V}_2$ at a series of mole fractions using equations (5.33) and (5.34). Make a plot of these values and $V_m$ against $x_2$ and compare the results.
(b) Calculate the volume of the solution that contains 1 kg of solvent ($H_2O$) for selected mole fractions $x_2$. Calculate the molality $m$ from $x_2$ for these solutions and then the apparent molar volume $\phi V$ for ethanol using equation (5.43).
(c) Use $\phi V$ to calculate $\overline{V}_1$ and $\overline{V}_2$ for these solutions {see equation (5.45)}. Add $\phi V$ and $\overline{V}_1$ and $\overline{V}_2$ obtained in this manner to the graph described in (a) so that all results can be compared.

## References


1. W. F. Giauque, E. W. Hornung, J. E. Kunzler, and T. R. Rubin, "The Thermodynamic Properties of Aqueous Sulfuric Acid Solutions and Hydrates from 15 to 300 °K", *J. Am. Chem. Soc.*, **82**, 62–70 (1960).
2. I. M. Klotz and R. M. Rosenburg, *Chemical Thermodynamics: Basic Theory and Methods*, W. A. Benjamin Inc. Menlo Park, California, 1972, p. 276.
3. J. R. Goates, J. B. Ott and J. F. Moellmer, "Determination of Excess Volumes in Cyclohexane + Benzene and + *n*-Hexane with a Vibrating Tube Densimeter", *J. Chem. Thermodyn.*, **9**, 249–257 (1977).
4. G. C. Allred and E. M. Woolley, "Heat Capacities of Aqueous Acetic Acid, Sodium Acetate, Ammonia, and Ammonium Chloride at 283.15, 298.15, and 313.15 K. $\Delta C_p^\circ$ for Ionization of Acetic Acid and for Dissociation of Ammonium Ion", *J. Chem. Thermodyn.*, **13**, 155–164 (1981).
5. O. Enea, P. P. Singh, E. M. Woolley, K. G. McCurdy, and L. G. Hepler, "Heat Capacities of Aqueous Nitric Acid, Sodium Nitrate, and Potassium Nitrate at 298.15 K: $\Delta C_p^\circ$ of Ionization of Water", *J. Chem. Thermodyn.*, **9**, 731–734 (1977).
6. F. T. Gucker, Jr., W. L. Ford, and C. E. Moser, "The Apparent and Partial Molal Heat Capacities and Volumes of Glycine and Glycolamide", *J. Phys. Chem.*, **43**, 153–168, (1939); F. T. Gucker, Jr. and W. L. Ford, "Apparent and Partial Molal Heat Capacities and Volumes of Glycerine and Glycolamide. II. Results for Concentrated Solutions of Glycolamide", **45**, 309–313 (1941).

# Chapter 6

# Fugacity, Activity, and Standard States

In Chapter 5, we considered systems in which composition becomes a variable, and defined and described the chemical potential. We showed that the chemical potential provides the condition for spontaneity or equilibrium. It is the potential that drives the flow of mass in a chemical process. A useful quantity related to the chemical potential is the fugacity. It can also be thought of as a measure of the flow of mass in a chemical process, and can be used to determine the point of equilibrium. It is often known as the **escaping tendency** since it can be used to describe the ease with which mass flows from one phase to another, particularly the flow from a solid or liquid phase to a gas phase.

Fugacity, like other thermodynamics properties, is a defined quantity that does not need to have physical significance, but it is nice that it does relate to physical quantities. Under some conditions, it becomes (within experimental error) the equilibrium gas pressure (vapor pressure) above a condensed phase. It is this property that makes fugacity especially useful. We will now define fugacity, see how to calculate it, and see how it is related to vapor pressure. We will then define a related quantity known as the activity and describe the properties of fugacity and activity, especially in solution.

## 6.1 Fugacity

### 6.1a Definition of Fugacity

The genesis of the idea for defining fugacity comes from the equation for calculating the free energy change at constant temperature for the ideal gas. Starting with the Gibbs equation for a pure substance

$$d\mu = -S_m\, dT + V_m\, dp \tag{6.1}$$

and specifying constant temperature ($\mathrm{d}T = 0$) and ideal gas behavior so that

$$V_m = \frac{RT}{p}$$

gives

$$\mathrm{d}\mu = \frac{RT}{p}\,\mathrm{d}p$$

or

$$\mathrm{d}\mu = RT\,\mathrm{d}\ln p. \qquad (6.2)$$

Equation (6.2) can be integrated to calculate $\Delta G_m$ for a change in pressure for the ideal gas. The result is

$$\Delta G_m = \mu_2 - \mu_1 = RT \ln\frac{p_2}{p_1}. \qquad (6.3)$$

The form of equation (6.2), which states that the change in chemical potential for an isothermal pressure change is proportional to the change in the natural logarithm of the ideal gas pressure, turns out to be very useful. However, this equation works only for changes involving ideal gases. G. N. Lewis defined a new quantity called the fugacity that maintains the form of this equality, but applies to non-ideal gases as well. The fugacity, $f$, is defined as

$$\mathrm{d}\mu = RT\,\mathrm{d}\ln f, \qquad (6.4)$$

with the added stipulation that

$$\lim_{p\to 0} \frac{f}{p} = 1. \qquad (6.5)$$

The limiting behavior ensures that the fugacities of real gases approach those of the ideal gas in the limit of low pressure. Since at low pressures the fugacity and pressure become the same, it should be clear that fugacities will be expressed in the same units as pressure, Pa, MPa, atm, Torr, etc.

The ratio $f/p$ is called the fugacity coefficient $\phi$,

$$\frac{f}{p} = \phi, \tag{6.6}$$

and equation (6.5) can be written as

$$\lim_{p \to 0} \phi = 1. \tag{6.7}$$

Figure 6.1 compares the fugacity and pressure of butane gas at $T = 500$ K as a function of pressure.[1] Note that as the pressure decreases, $f$ approaches $p$. Equation (6.5) ensures that this will happen. The manner in which $\phi$ deviates from unity as $p$ increases varies with the gas and the conditions of pressure and temperature. Figure 6.2 gives $\phi$ for butane gas as a function of pressure at the temperatures of 500 K and 700 K. Note that at 700 K, $\phi$ becomes greater than unity at high pressures. For some gases at high temperatures, $\phi$ is greater than one at all pressures.

## 6.1b Determination of Fugacities

So far we have defined fugacity for a single component gas. We will first see how fugacities are determined for a pure gas before we expand to include

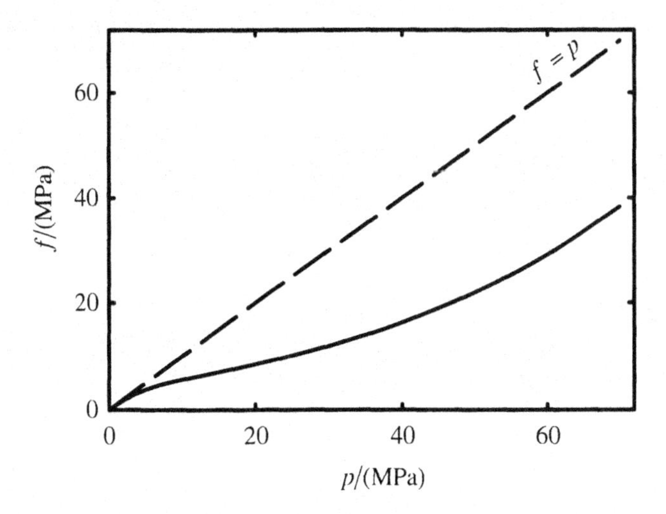


**Figure 6.1** Fugacity of n-butane gas as a function of pressure at $T = 500$ K. The dashed line represents the fugacity that would be obtained if n-butane behaved as an ideal gas at each pressure.

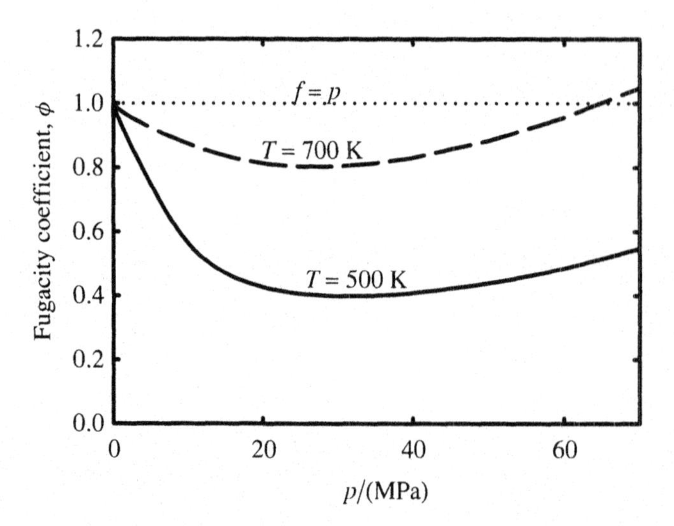


**Figure 6.2** Fugacity coefficient, $\phi$, as a function of pressure for n-$C_4H_{10}$ at $T = 500$ K and 700 K.

condensed phases and mixtures. A number of different equations have been developed to calculate $\phi$ for a gas from ($p$, $V$, $T$) data. One of the most useful involves the quantity, $\alpha$, defined as

$$\alpha = V_m^{\text{ideal}} - V_m \tag{6.8}$$

where $V_m$ is the molar volume of the gas and $V_m^{\text{ideal}}$ is the molar volume the gas would have if it were behaving ideally at the same $T$ and $p$. To relate $\alpha$ to $\phi$, we write

$$\phi = \frac{f}{p}.$$

Taking logarithms gives

$$\ln \phi = \ln f - \ln p.$$

Differentiating gives

$$\left(\frac{\partial \ln \phi}{\partial p}\right)_T = \left(\frac{\partial \ln f}{\partial p}\right)_T - \left(\frac{\partial \ln p}{\partial p}\right)_T. \tag{6.9}$$

From equation (6.4) we get

$$\mathrm{d}\ln f = \frac{\mathrm{d}\mu}{RT}. \tag{6.10}$$

Dividing equation (6.10) by d$p$ while including the condition of constant $T$, and remembering that $(\partial\mu/\partial p)_T = V_\mathrm{m}$ gives

$$\left(\frac{\partial \ln f}{\partial p}\right)_T = \frac{V_\mathrm{m}}{RT}. \tag{6.11}$$

Substituting equation (6.11) and $(\partial \ln p/\partial p)_T = 1/p$ into equation (6.9) gives

$$\left(\frac{\partial \ln \phi}{\partial p}\right)_T = \frac{V_\mathrm{m}}{RT} - \frac{1}{p}$$

or

$$\left(\frac{\partial \ln \phi}{\partial p}\right)_T = -\frac{1}{RT}\left[\frac{RT}{p} - V_\mathrm{m}\right]. \tag{6.12}$$

The quantity $RT/p$ in equation (6.12) is the ideal gas volume $V_\mathrm{m}^\mathrm{ideal}$. Hence, the quantity in the brackets is $\alpha$ and equation (6.12) can be written as

$$\left(\frac{\partial \ln \phi}{\partial p}\right)_T = -\frac{\alpha}{RT}. \tag{6.13}$$

The fugacity coefficient $\phi$ at a pressure $p$ can be obtained by separating variables and integrating equation (6.13). In setting the limits of the integration we remember that $\phi = 1$ when $p = 0$. Hence,

$$\int_{\ln\phi=0}^{\ln\phi} \mathrm{d}\ln\phi = -\frac{1}{RT}\int_0^p \alpha\, \mathrm{d}p \tag{6.14}$$

or

$$\ln\phi = -\frac{1}{RT}\int_0^p \alpha\, \mathrm{d}p. \tag{6.15}$$

The integral of $\alpha$ d$p$ can be obtained from experimental ($p$, $V$, $T$) data or from an equation of state. Figure 6.3 shows a graph of $\alpha$ against $p$ for butane gas at $T = 500$ K obtained from a fit of experimental data to the modified Benedict–Webb–Rubin (m-BWR) equation.[2] This 32 constant equation very accurately represents the ($p$, $V$, $T$) properties of n-butane. The area under the curve up to a pressure $p$ gives the $\int_0^p \alpha dp$ at this pressure. Both $V_m^{ideal}$ and $V_m$ approach infinity as $p$ approaches zero. Note, however, from Figure 6.3 that the difference between them ($\alpha$) does not go to infinity, and an extrapolation can be made to zero pressure. For n-butane at 500 K, the extrapolated value[a] is $2.10 \times 10^{-4} m^3 \cdot mol^{-1}$.

In Figure 6.3, the area obtained from the m-BWR equation gives a value for $\phi$ of 0.4386 for n-butane at $T = 500$ K and $p = 50$ MPa. Since the m-BWR equation represents the ($p$, $V$, $T$) properties of the gas with high accuracy, these $\phi$ values are reliable and can be used as a reference for comparing results obtained in other ways.

Other equations of state can be used to calculate $\phi$. For example, the virial equation[b] is given by

$$pV_m = RT + Bp + Cp^2 + Dp^3 + \cdots, \tag{6.16}$$

where $B$, $C$, $D$, etc. are the second, third, ... virial coefficients. These coefficients are functions of $T$ but constant with $p$ and $V$. Solving equation (6.16) for $V_m$ and substitution into equation (6.8) with $V_m^{ideal} = RT/p$ gives

$$\alpha = \frac{RT}{p} - \frac{RT}{p} - B - Cp - Dp^2 - \cdots \tag{6.17}$$

$$= -(B + Cp + Dp^2 + \cdots). \tag{6.18}$$

Substitution of equation (6.18) into equation (6.15) and integrating gives

$$\ln \phi = \frac{1}{RT}\left[Bp + \frac{C}{2}p^2 + \frac{D}{3}p^3 + \right] \cdots \tag{6.19}$$

---

[a] Equations of state, such as the virial equation, demonstrate that $\alpha$ remains finite as $p \to 0$.

[b] See Appendix 3 for a discussion of the virial and other equations of state.

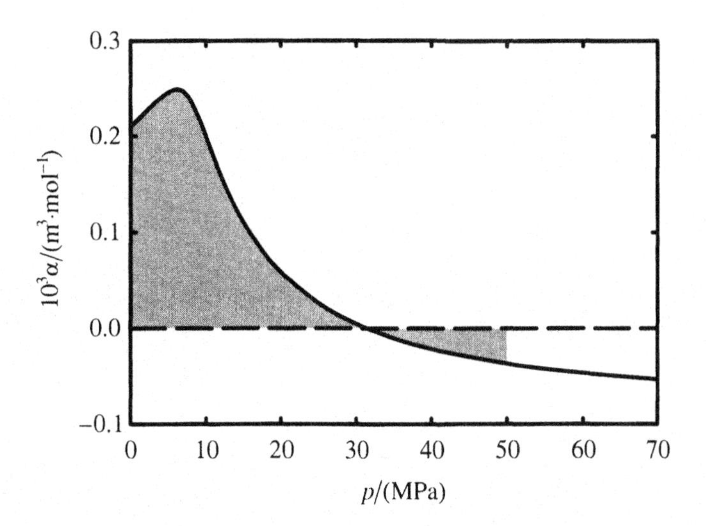


**Figure 6.3** Graph of $\alpha = (V_m^{\text{ideal}} - V_m)$ against $p$ for n-butane at $T = 500$ K. The shaded area shown between the curve and the dashed line ($\alpha = 0$) gives $-RT\ln\phi$ at $p = 50$ MPa.

**Example 6.1** For $T = 500$ K and $p \leqslant 10$ MPa, $pV_m$ for n-butane gas can be accurately represented by the equation

$$pV_m = 4155.8 - 209.9p - 5.186p^2 - 1.448p^3 + 0.2079p^4$$

where $p$ is the pressure in MPa and $V_m$ is the molar volume in $\text{m}^3\cdot\text{mol}^{-1}$. Calculate $\phi$ for n-butane at $p = 5.00$ MPa and $T = 500$ K.

**Solution:** This equation of state is in the form of equation (6.16) with B = −209.9, C = −5.186, D = −1.448, and E = 0.2079.

Substitution into equation (6.19) gives

$$\ln\phi = \frac{1}{(8.314)(500)}\left[-(209.9)(5.00) - \frac{(5.186)(5.00)^2}{2} - \left(\frac{1.448}{3}\right)(5.00)^3 + \left(\frac{0.2079}{4}\right)(5.00)^4\right]$$

$$= -0.275, \quad \text{so that } \phi = 0.760.$$

By comparison, the value obtained from the m-BWR equation is $\phi = 0.7492$.

At moderate to low pressures, the higher order terms in equation (6.19) can be neglected to give

$$\ln \phi = \frac{Bp}{RT}. \tag{6.20}$$

Under these conditions, other relationships can be used to calculate $\phi$. Equation (6.20) can be written in the form

$$\phi = \exp\left(\frac{Bp}{RT}\right).$$

Expanding the exponential and neglecting higher order terms gives

$$\phi = 1 + \frac{Bp}{RT} = \frac{RT + Bp}{RT}. \tag{6.21}$$

From the virial equation, after neglecting higher order terms we get

$$RT + Bp = pV_m.$$

Substitution into equation (6.21) gives

$$\phi = \frac{pV_m}{RT}. \tag{6.22}$$

If we substitute $p^{ideal}$ for $RT/V_m$, where $p^{ideal}$ is the pressure that an ideal gas would exert at the same volume, we obtain

$$\phi = \frac{p}{p^{ideal}}. \tag{6.23}$$

Thus, $\phi$ can be estimated from the observed pressure and the ideal pressure calculated from the molar volume and the ideal gas equation. Klotz and Rosenburg[3] report that the error in using equation (6.23) to calculate $\phi$ is less than 1% for $O_2$ up to a pressure of 10 MPa. For $CO_2$ (g) the error is 1% at 2.5 MPa and 4% at 5 MPa.

The modified Berthelot equation given by

$$pV_{\mathrm{m}} = RT + \frac{9RT_{\mathrm{c}}p}{128p_{\mathrm{c}}}\left(1 - 6\frac{T_{\mathrm{c}}^2}{T^2}\right), \tag{6.24}$$

where $p_{\mathrm{c}}$ and $T_{\mathrm{c}}$ are the critical pressure and temperature, is sometimes used to calculate $\phi$ at low to moderate pressures. It has the form of a truncated virial equation with

$$B = \frac{9RT_{\mathrm{c}}}{128p_{\mathrm{c}}}\left(1 - 6\frac{T_{\mathrm{c}}^2}{T^2}\right). \tag{6.25}$$

Substitution of equation (6.25) into equation (6.20) gives

$$\ln\phi = \frac{9T_{\mathrm{c}}p}{128p_{\mathrm{c}}T}\left(1 - 6\frac{T_{\mathrm{c}}^2}{T^2}\right). \tag{6.26}$$

**Example 6.2** Calculate $\phi$ for n-butane at $T = 500$ K and $p = 5.00$ MPa using the modified Berthelot equation of state.[c]

---

[c] A number of years ago, one of us {see J. B. Ott, J. R. Goates, and H. T. Hall Jr., "Comparisons of Equations of State in Effectively Describing PVT Relations," *J. Chem. Ed.*, **48**, 515–517 (1971).} suggested a further modification of the modified Berthelot equation. It has the form

$$pV_{\mathrm{m}} = RT + \frac{RT_{\mathrm{c}}p}{17p_{\mathrm{c}}}\left(1 - \frac{15T_{\mathrm{c}}^2}{2T^2}\right).$$

Using this form gives

$$\ln\phi = \frac{T_{\mathrm{c}}p}{17p_{\mathrm{c}}T}\left(1 - \frac{15T_{\mathrm{c}}^2}{2T^2}\right).$$

Using equation (6.28), one calculates $\ln\phi = -0.2919$ for n-butane at $T = 500$ K and $p = 5.00$ MPa. The resulting $\phi$ of 0.747 is in excellent agreement with the m-BWR result.

**Solution:** For n-butane, $p_c = 3.79$ MPa and $T_c = 425.10$ K. Substitution into equation (6.26) gives

$$\ln\phi = \frac{(9)(425.10\ \text{K})(5.00\ \text{MPa})}{(128)(3.79\ \text{MPa})(500\ \text{K})}\left[1-(6)\left(\frac{425.10}{500.00}\right)^2\right]$$

$$= -0.263.$$

$$\phi = 0.769.$$

This result can be compared with the m-BWR value of $\phi = 0.7492$.

The fugacity coefficient can be calculated from other equations of state such as the van der Waals, Redlick-Kwong, Peng-Robinson, and Soave,[d] but the calculation is complicated, since these equations are cubic in volume, and therefore they cannot be solved explicitly for $V_m$, as is needed to apply equation (6.12). Klotz and Rosenburg[4] have shown a way to get around this problem by eliminating $p$ from equation (6.12) and integrating over volume, but the process is not easy. For the van der Waals equation, they end up with the relationship

$$\ln f = \ln\frac{RT}{V_m - b} + \frac{b}{V_m - b} - \frac{2a}{RTV_m} \tag{6.27}$$

or

$$\ln\phi = \ln\frac{RT}{V_m - b} + \frac{b}{V_m - b} - \frac{2a}{RTV_m} - \ln p, \tag{6.28}$$

where $a$ and $b$ are the van der Waals constants. To obtain $\phi$ at a pressure $p$ and temperature $T$, $p$ and $T$ are substituted into the van der Waals equation, which is then solved for $V_m$. Substitution of $V_m$ and $p$ into equation (6.28) then gives $\ln\phi$ or $\phi$. Similar procedures could be applied to other equations of state that can not be solved explicitly for volume.

With modern high-speed computers, an easier method is to numerically integrate equation (6.15), written in the form

$$\ln\phi = -\frac{1}{RT}\int_0^p\left(\frac{RT}{p} - V_m\right)dp. \tag{6.15}$$

---

[d] The form and properties of these various equations of state are summarized in Appendix 3.

Values of $V_m$ as a function of $p$ are computed from an equation of state, substituted into equation (6.15), and numerically integrated. In principle, this method can be used to calculate $\phi$ for any well-behaved equation of state. (It is essentially the method used at the beginning of this section to obtain $\phi$ from the m-BWR equation.)

An alternate to equation (6.15) is an equation that relates $\ln\phi$ to $z$, the compressibility factor of the gas, that is defined as

$$z = \frac{pV_m}{RT}. \tag{6.29}$$

To relate $\ln\phi$ to $z$, equation (6.15) is written in terms of $z$ by noting that

$$\alpha = \frac{RT}{p} - V_m = \frac{RT}{p}\left(1 - \frac{pV_m}{RT}\right)$$

$$= \frac{RT}{p}(1 - z).$$

Hence

$$\mathrm{d}\ln\phi = -\frac{1}{RT}\frac{RT}{p}(1 - z)\,\mathrm{d}p,$$

which upon integration yields

$$\ln\phi = -\int_0^p \frac{1 - z}{p}\,\mathrm{d}p. \tag{6.30}$$

$\ln\phi$ can then be obtained as the area under the curve when $(1 - z)/p$ is graphed against $p$.

Equation (6.30) leads to a final method of obtaining an approximate value for $\ln\phi$ by making use of the **law of corresponding states**. This law states that all gases obey the same equation of state when expressed in terms of the reduced variables $T_r = T/T_c$, $p_r = p/p_c$, and $V_r = V/V_c$, where $T_c$, $p_c$, and $V_c$ are the critical temperature, pressure, and volume, respectively.

At temperatures greater than the critical temperature, the law of corresponding states is a reasonably good approximation for most gases. Goug-Jen Su[5] showed the correspondence by graphing the compressibility factor, $z$, of a number of gases as a function of $p_r$ and $T_r$. This graph, which is shown in Appendix 3 at the end of this book, demonstrates that corresponding states is a reasonably good approximation for a large number of gases, and by inference, should work for most gases.

Converting equation (6.30) to reduced variables gives

$$\ln \phi = -\int_0^{p_r} \frac{1-z}{p_r}\, dp_r . \tag{6.31}$$

The law of corresponding states indicates that all gases should show the same behavior in applying equation (6.31). This enables one to construct a chart showing $T_r$ isotherms of $z$ against $p_r$, such as that shown in Figure 6.4, from which $\phi$ can be estimated.

**Example 6.3** Use the reduced fugacity coefficient chart in Figure 6.4 to estimate $\phi$ for n-butane gas at $p$ = 5.00 MPa and $T$ = 500 K.

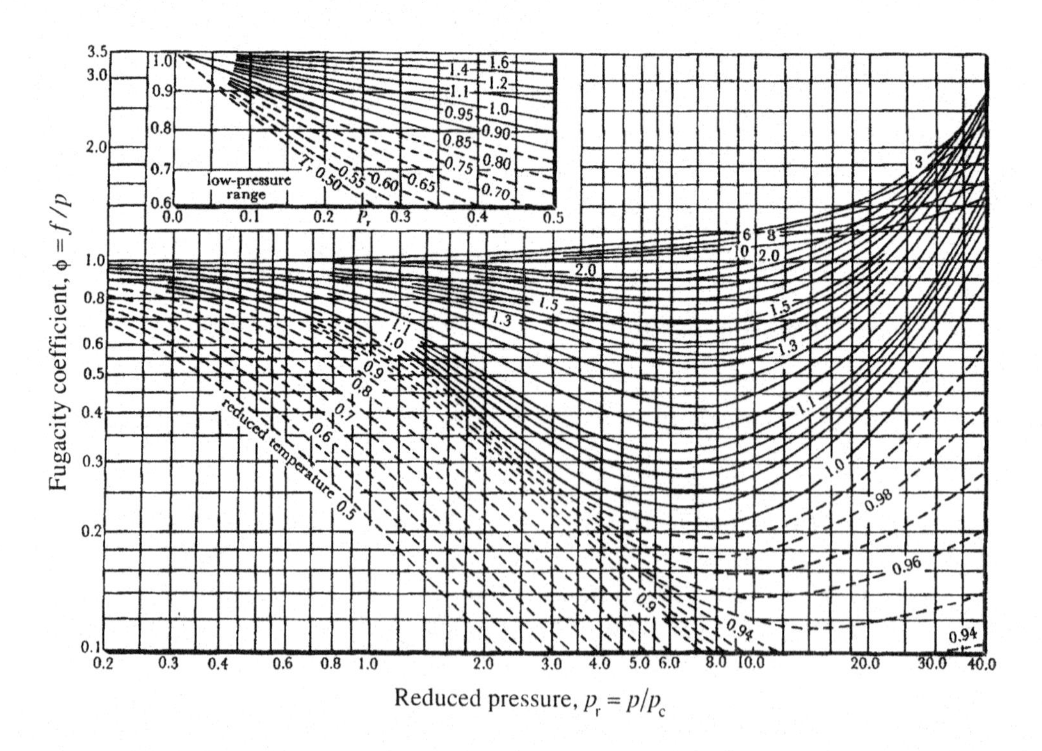


**Figure 6.4** Fugacity coefficients of gases in terms of the reduced pressure and temperature. Based on data taken from B. W. Gamson and K. M. Watson, *Natl. Petrol. News*, Tech. Sec. 36, R623 (Sept. 6, 1944).

**Solution:** For n-butane $T_c = 425.10$ K and $p_c = 3.79$ MPa. At $p = 5$ MPa and $T = 500$ K

$$T_r = \frac{500}{425.10} = 1.18.$$

$$p_r = \frac{5.00}{3.79} = 1.32.$$

Interpolating from the chart in Figure 6.4 gives $\phi = 0.73$ so that $f = 3.65$ MPa. This result can be compared with those referred to earlier.

## 6.1c Fugacity for Pure Condensed Phases

The defining equation for fugacity $f_c$ in a condensed phase (solid or liquid) is the same as in the gas phase

$$d\mu_c = RT\, d\ln f_c \tag{6.32}$$

$$\lim_{p \to o} \frac{f_c}{p} = 1, \tag{6.33}$$

where $\mu_c$ is the chemical potential in the condensed phase. The fugacity $f_c$ is related directly to $f_g$, the fugacity of the gas that is in equilibrium with the condensed phase. Applying the criteria for phase equilibrium derived in the previous chapter gives

$$d\mu_c = d\mu_g$$

where $d\mu_g$ is the chemical potential in the gas phase. Substitution of equations (6.4) (applied to the gas phase) and (6.32) into the above equality gives

$$RT\, d\ln f_c = RT\, d\ln f_g. \tag{6.34}$$

Equation (6.34) requires that the fugacity in the condensed phase be proportional to the fugacity in the gas phase

$$f_c = kf_g.$$

The second part of the definition for fugacity requires that $k = 1$, since, in theory, any solid or liquid taken to zero pressure would eventually be a gas for which $f = p$ so that $k$ equals one.

The conclusion we reach is that the **fugacity of a condensed phase is the fugacity of the gas in equilibrium with the solid or liquid**. This vapor fugacity can be related to a gaseous vapor pressure by the methods we have described earlier for the gas phase. In practice, the vapor pressures of the solids and liquids we work with are usually small enough that the difference between fugacity and pressure in the gas phase is small.[e] We will often equate vapor pressure with vapor fugacity for a solid or liquid and interchange the terms "vapor pressure" and "vapor fugacity" in our discussions, using the latter only when the vapor pressure is high, or the rigor of the discussion requires the use of fugacity.

### 6.1d Effect of Pressure and Temperature on the Vapor Fugacity

**Change of fugacity with pressure:** Equation (6.11) provides the starting place for calculating the effect of pressure on the vapor fugacity of a solid or liquid

$$\left(\frac{\partial \ln f}{\partial p}\right)_T = \frac{V_m}{RT}. \tag{6.11}$$

**Example 6.4** The vapor pressure of benzene is 12.69 kPa at $T = 298.15$ K and a total pressure equal to the vapor pressure. Calculate the vapor pressure of benzene at $T = 298.15$ K under He gas at a total pressure of 100 MPa.

**Solution:** We will assume that the vapor pressure of benzene equals the vapor fugacity so that equation (6.11) applies to the vapor pressure. We will also assume that a negligible amount of He dissolves in the benzene so that it remains pure. Integrating equation (6.11) gives

$$\int_{\ln p_1}^{\ln p_2} \mathrm{d} \ln p = \int_{p'_1}^{p'_2} \frac{V_m}{RT}\, \mathrm{d}p',$$

where $p_1$ and $p_2$ are the vapor pressures, and $p'_1$ and $p'_2$ are the total pressures. The molar volume of benzene at $T = 298.15$ K and 0.101 MPa is 89.40 $\text{cm}^3\cdot\text{mol}^{-1}$. Assuming that the liquid is incompressible gives

$$\ln \frac{p_2}{p_1} = \frac{V_m}{RT}(p'_2 - p'_1). \tag{6.22}$$

[e] As an example, $\phi$ for n-butane at $p = 0.101$ MPa and $T = 500$ K is 0.990. Usually, vapor pressures are less than 0.101 MPa and $\phi$ is very nearly one.

Substituting (while converting to SI units)

$$f_1 = p_1 = p'_1 = 12.69 \text{ kPa} \times 10^3 \text{ Pa·kPa}^{-1} = 1.269 \times 10^4 \text{ Pa},$$

$$p'_2 = 1.00 \text{ MPa} \times 10^6 \text{ Pa·MPa}^{-1} = 1.00 \times 10^6 \text{ Pa},$$

$$V_m = 89.40 \text{ cm}^3\text{·mol}^{-1} \times 10^{-6} \text{ m}^3\text{·cm}^{-3} = 8.940 \times 10^{-5} \text{ m}^3\text{·mol}^{-1}$$

gives

$$\ln \frac{p_2}{12.69 \text{ kPa}} = \frac{(8.940 \times 10^{-5})}{(8.314)(298.15)}(1.00 \times 10^6 \text{ Pa} - 1.269 \times 10^4 \text{ Pa})$$

$$= 3.56 \times 10^{-2}.$$

$$\frac{p_2}{12.61 \text{ kPa}} = 1.036.$$

$$p_2 = 13.07 \text{ kPa}.$$

To many, it seems counter intuitive that increasing the total pressure increases the vapor pressure, but such is the case. However, the change is small, and the effect of total pressure on vapor pressure can generally be ignored, unless large pressure changes are involved.

**Change of fugacity with temperature:** Consider an isothermal process in which a substance (solid, liquid, or gas) at a pressure, $p$, and fugacity, $f$, is converted to a gas at a very low pressure, $p^\ddagger$, where the fugacity is $f^\ddagger$. Starting with the defining equation for $f$ and integrating gives

$$\int_{\mu}^{\mu^\ddagger} d\mu = RT \int_{\ln f}^{\ln f^\ddagger} d \ln f$$

$$\mu^\ddagger - \mu = RT \ln \frac{f^\ddagger}{f}$$

or

$$\frac{\mu^\ddagger}{T} - \frac{\mu}{T} = R \ln f^\ddagger - R \ln f. \qquad (6.35)$$

Differentiation of equation (6.35) with respect to $T$ at constant $p$ gives

$$\left(\frac{\partial(\mu^{\ddagger}/T)}{\partial T}\right)_p - \left(\frac{\partial(\mu/T)}{\partial T}\right)_p = R\left(\frac{\partial \ln f^{\ddagger}}{\partial T}\right)_p - R\left(\frac{\partial \ln f}{\partial T}\right)_p. \tag{6.36}$$

In Chapter 3, we showed[f] that

$$\left(\frac{\partial(\mu^{\ddagger}/T)}{\partial T}\right)_p = -\frac{H_m^{\ddagger}}{T^2}$$

$$\left(\frac{\partial(\mu/T)}{\partial T}\right)_p = -\frac{H_m}{T^2}.$$

Also, $(\partial \ln f^{\ddagger}/\partial T)_p = 0$ since at $p^{\ddagger}, f^{\ddagger} = p^{\ddagger}$; hence, $f^{\ddagger}$ does not change when $p$ is held constant. Substitution of these relationships into equation (6.36) gives

$$\left(\frac{\partial \ln f}{\partial T}\right)_p = \frac{H_m^{\ddagger} - H_m}{RT^2}. \tag{6.37}$$

For a gas, $H_m^{\ddagger} - H_m$ is the change in enthalpy as the gas at pressure, $p$, is expanded into a vacuum. For a liquid (or solid), $H_m^{\ddagger} - H_m$ is the enthalpy change as the liquid (or solid) is vaporized (or sublimed) into a vacuum. It has been called the **ideal enthalpy of vaporization (or sublimation)** since it represents the enthalpy change as the liquid (or solid) becomes an ideal gas.

## 6.1e Fugacity in a Mixture

For a component in a mixture, the fugacity is defined by the same equation as for a pure substance, except that partial molar quantities are substituted for molar quantities. Thus,

$$d\mu_i = RT \, d \ln f_i \tag{6.38}$$

$$\lim_{p \to 0} \frac{f_i}{p_i} = 1, \tag{6.39}$$

---

[f] See Table 3.1 in Chapter 3.

and the effect of pressure and temperature is given by

$$\left(\frac{\partial \ln f_i}{\partial p}\right)_T = \frac{\overline{V}_i}{RT} \tag{6.40}$$

$$\left(\frac{\partial \ln f_i}{\partial T}\right)_p = \frac{\overline{H}_i^\ddagger - \overline{H}_i}{RT^2}. \tag{6.41}$$

These equations serve as the basis for determining $f_i$, the fugacity of the $i$th component in a mixture.

**Fugacity of a Component in a Gaseous Mixture:** One could guess that the determination of fugacities, $f_i$, for the individual components in a gaseous mixture can become complicated as one takes into account the different types of interactions that are present. The mathematical relationship that applies is obtained by starting with the defining equations

$$\mathrm{d}\mu_i = RT\,\mathrm{d}\ln f_i$$

with

$$\mathrm{d}\mu_i = \overline{V}_i\,\mathrm{d}p - \overline{S}_i\,\mathrm{d}T.$$

At constant temperature, these equations can be combined to give (after dividing by $RT$)

$$\mathrm{d}\ln f_i = \frac{\overline{V}_i}{RT}\,\mathrm{d}p, \tag{6.42}$$

where $\overline{V}_i$ is the partial molar volume of the component and $p$ is the total pressure. The fugacity coefficient, $\phi_i$ of the $i$th component in a mixture is defined as

$$\phi_i = \frac{f_i}{p_i} \tag{6.43}$$

where $p_i$ is the partial pressure given by Dalton's law,

$$p_i = y_i p, \tag{6.44}$$

with $p$ as the total pressure and $y_i$ as the mole fraction[g] in the gas. Taking the logarithms of both sides of equation (6.44) and differentiating while holding the composition ($y_i$) constant gives

$$\mathrm{d}\ln p_i = \mathrm{d}\ln p. \tag{6.45}$$

Subtracting equation (6.45) from equation (6.42) and substituting for $\phi_i$ from equation (6.43) gives

$$\mathrm{d}\ln \phi_i = \left(\frac{\overline{V}_i}{RT} - \frac{1}{p}\right)\mathrm{d}p. \tag{6.46}$$

Integration of equation (6.46) while remembering that $\phi = 1$ when $p = 0$ gives

$$\int_{\ln\phi=0}^{\ln\phi_i} \mathrm{d}\ln\phi_i = \int_0^p \left(\frac{\overline{V}_i}{RT} - \frac{1}{p}\right)\mathrm{d}p$$

or

$$\ln\phi_i = \int_0^p \left(\frac{\overline{V}_i}{RT} - \frac{1}{p}\right)\mathrm{d}p. \tag{6.47}$$

With sufficient data, equation (6.47) can be integrated to give $\phi_i$. However, adequate data are available for only a few mixtures, and approximate relationships are used to estimate $\phi_i$. The simplest approximation uses the Lewis and Randall rule given by[h]

$$f_i = y_i f_i^*, \tag{6.48}$$

[g] We will follow the convention of designating mole fraction in the vapor phase by $y_i$ and the mole fraction in the liquid phase as $x_i$ when we need to differentiate between the compositions of the two phases.

[h] Note the similarity between the Lewis and Randall rule

$$f_i = y_i f_i^*$$

and Dalton's law of partial pressure

$$p_i = y_i p^*.$$

Dalton's law is based on the assumption of ideal gases so that each behaves independently and exerts the same partial pressure as it would if alone in the container. The Lewis and Randall rule assumes that the fugacity of the gases is independent so that the gas has the same fugacity coefficient as it would have at the same total pressure when other gases were not present.

where $f_i$ is the fugacity of the component in the mixture and $f_i^*$ is the fugacity of the pure gaseous component at the same temperature and total pressure, $p$. Combining equations (6.43) and (6.48) gives

$$\phi_i = \frac{f_i}{p_i} = \frac{y_i f_i^*}{p_i}.$$

Using Dalton's law of partial pressures to express $p_i$ in terms of $p$ gives

$$\phi_i = \frac{y_i f_i^*}{y_i p} = \phi_i^*. \tag{6.49}$$

Thus, with the Lewis and Randall approximation, $\phi_i$ has the same value as the pure gas would have at the same temperature and total pressure.

The next level of approximation is valid to higher pressures. It assumes that the gas mixture obeys the virial equation of state, with the third, fourth and higher, virial coefficients equal to zero. Thus

$$pV = n_t(RT + pB_{\text{mix}}) \tag{6.50}$$

where $B_{\text{mix}}$ is the second virial coefficient for the mixture given by

$$B_{\text{mix}} = \sum_i \sum_j y_i y_j B_{ij} \tag{6.51}$$

and

$$n_t = \sum_i n_i. \tag{6.52}$$

Equations (6.51) and (6.52) can be substituted into equation (6.50), which can then be solved for $V$ and differentiated to obtain $\overline{V}_i$. This result can be substituted into equation (6.47) and integrated to give $\phi_i$. As an example, we will derive the equation for calculating $\phi_1$ in a binary mixture. The calculation of $\phi_2$ would be similar.

First, we solve equation (6.50) for $V$:

$$V = n_t \frac{RT}{p} + n_t B_{\text{mix}}. \tag{6.53}$$

Since $n_t = n_1 + n_2$,

$$\left(\frac{\partial n_t}{\partial n_1}\right)_{p,T,n_2} = 1,$$

then

$$\left(\frac{\partial V}{\partial n_t}\right)_{p,T,n_2} = \left(\frac{\partial V}{\partial n_t}\right)_{p,T,n_2}\left(\frac{\partial n_t}{\partial n_1}\right)_{p,T,n_2},$$

and

$$\left(\frac{\partial V}{\partial n_t}\right)_{p,T,n_2} = \left(\frac{\partial V}{\partial n_1}\right)_{p,T,n_2} = \bar{V}_1. \tag{6.54}$$

Differentiation of equation (6.53) with respect to $n_t$ at constant $T$, $p$ and $n_2$ and combining with equation (6.54) gives equation (6.55)

$$\bar{V}_1 = \frac{RT}{p} + B_{\text{mix}} + n_t\left(\frac{\partial B_{\text{mix}}}{\partial n_t}\right)_{p,T,n_2}. \tag{6.55}$$

For a binary mixture, equation (6.51) becomes

$$B_{\text{mix}} = y_1^2 B_{11} + 2y_1 y_2 B_{12} + y_2^2 B_{22} \tag{6.56}$$

where $B_{11}$, and $B_{22}$ are the second virial coefficients for the pure components, and $B_{12}$ is the cross virial term for the mixture. Differentiation of equation (6.56) with respect to $n_t$ at constant $n_2$ is obtained by noting that

$$\left(\frac{\partial y_1}{\partial n_t}\right)_{n_2} = \frac{n_t - n_1}{n_t^2}$$

$$\left(\frac{\partial y_2}{\partial n_t}\right)_{n_2} = -\frac{n_2}{n_t^2}$$

so that

$$\left(\frac{\partial B_{\text{mix}}}{\partial n_t}\right)_{p,T,n_2} = 2y_1\frac{(n_t - n_1)}{n_t^2}B_{11} + 2B_{12}\left[y_1\left(-\frac{n_2}{n_t^2}\right) + y_2\left(\frac{n_t - n_1}{n_t^2}\right)\right]$$

$$+ 2y_2\left(-\frac{n_2}{n_t^2}\right)B_{22}. \tag{6.57}$$

Substituting equation (6.57) into equation (6.55) and rearranging gives

$$\bar{V}_1 - \frac{RT}{p} = 2(y_1B_{11} + y_2B_{12}) - B_{\text{mix}}. \tag{6.58}$$

Substitution of equation (6.58) into equation (6.47) and integrating gives[i]

$$\ln\phi_1 = \frac{p}{RT}[2(y_1B_{11} + y_2B_{12}) - B_{\text{mix}}]. \tag{6.59}$$

Equation (6.60) extends equation (6.59) to calculate $\phi_i$ in a multicomponent mixture,

$$\ln\phi_i = \frac{p}{RT}\left(2\sum_j y_jB_{ij} - B_{\text{mix}}\right), \tag{6.60}$$

with $B_{\text{mix}}$ given by equation (6.51). More complete expressions include the introduction of the third virial coefficient defined as

$$C_{\text{mix}} = \sum_i\sum_j\sum_k y_iy_jy_kC_{ijk}. \tag{6.61}$$

We will limit our treatment of what can obviously become a very complex problem to what we have done.[6]

---

[i] Equation (6.59) can be rearranged to put it in the form

$$\ln\phi_1 = \frac{p}{RT}[(1 - y_2^2)B_{11} + y_2^2(2B_{12} - B_{22})].$$

Applying equation (6.59) presents problems in that $B_{12}$ is often not known. When components 1 and 2 are very similar in their molecular properties one can approximate $B_{12}$ as the mean of $B_{11}$ and $B_{22}$. That is,

$$B_{12} = \tfrac{1}{2}(B_{11} + B_{22}). \tag{6.62}$$

In this case, equation (6.59) simplifies to

$$\ln \phi_1 = \frac{p}{RT} B_{11}. \tag{6.63}$$

**Fugacity in Liquid Mixtures: Raoult's Law and Henry's Law:** Each component in a liquid mixture has an equilibrium vapor pressure, and hence, a vapor fugacity. These fugacities are functions of the composition and the nature of the components, with the total vapor fugacity equal to the sum of the fugacities of the components. That is,

$$f_{\text{total}} = \sum_i f_i. \tag{6.64}$$

Binary (vapor + liquid) equilibria studies involve the determination of $f_i$ as a function of composition, $x_i$, the mole fraction in the liquid phase. Of special interest is the dependence of $f_i$ on composition in the limit of infinite dilution. In the examples which follow, equilibrium vapor pressures, $p_i$, are measured and described. These vapor pressures can be corrected to vapor fugacities using the techniques described in the previous section. As stated earlier, at the low pressures involved in most experiments, the difference between $p_i$ and $f_i$ is very small, and we will ignore it unless a specific application requires a differentiation between the two.

**(a) Raoult's Law and the Ideal Solution:** Figure 6.5 summarizes the experimentally determined vapor pressures at $T = 308.15$ K for the binary mixture that we can represent as $\{x_1\text{c-C}_6\text{H}_{11}\text{CH}_3 + x_2\text{c-C}_6\text{H}_{12}\}$, where $\text{c-C}_6\text{H}_{12}$ is cyclohexane and $\text{c-C}_6\text{H}_{11}\text{CH}_3$ is methylcyclohexane.[7] The most striking feature of the diagram is the linear relationship between vapor pressure and mole fraction. The dashed lines are graphs of the equations

$$f_2 = x_2 f_2^*, \tag{6.65}$$

$$f_1 = x_1 f_1^* = (1 - x_2) f_1^*, \tag{6.66}$$

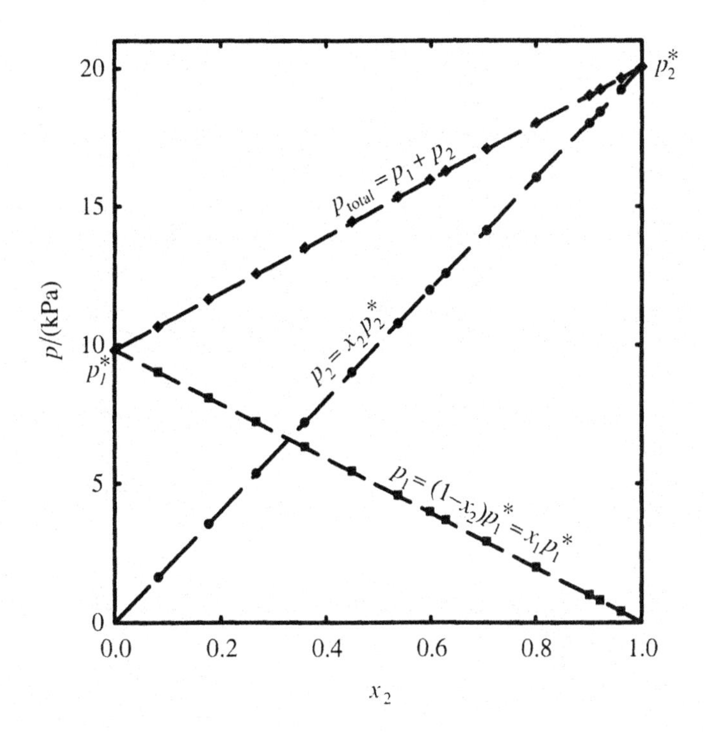


**Figure 6.5** Vapor pressures for $\{x_1\text{c-}C_6H_{11}CH_3 + x_2\text{c-}C_6H_{12}\}$ at $T = 308.15$ K. The symbols represent the experimental vapor pressures as follows: ●, vapor pressure of c-$C_6H_{12}$; ■, vapor pressure of c-$C_6H_{11}CH_3$; ◆, total vapor pressure. The dashed lines represent the ideal solution prediction.

and

$$f_{total} = f_1^* + x_2(f_2^* - f_1^*), \tag{6.67}$$

with the approximation that $f_i = p_i$. In equations (6.65) to (6.67), $f_1$ and $f_2$ ($p_1$ and $p_2$) are the vapor fugacities (pressures) of the two components above the liquid mixture, with $x_1$ and $x_2$ the mole fractions, and $f_1^*(p_1^*)$ and $f_2^*(p_2^*)$ are the vapor fugacities (pressures) of the pure substances.

Equations (6.65) and (6.66) are statements of Raoult's law. They form the basis for defining an ideal solution.

*An ideal solution is one for which all components obey Raoult's law*

$$f_i = x_i f_i^* \tag{6.68}$$

*over the entire range of composition at all pressures and temperatures.*

Just as no real gas has the exact behavior of an ideal gas, no real solution follows equation (6.68) exactly.[j] Deviations are always present. A careful examination of Figure 6.5 will show that the experimental results lie just slightly above the lines drawn from equations (6.65) to (6.67). However, Raoult's law is a good approximation when the components have similar molecular size, shape, and polarity. The more common examples of near-ideal mixtures are those in which the components consist of molecules with similar size and shape that have low or zero polarity, such as in the cyclohexane and methylcyclohexane solutions shown in Figure 6.5. However, polar mixtures can also be nearly ideal. Figure 6.6 summarizes the vapor pressures for $\{x_1CH_3CN + x_2CH_3NO_2\}$ at $T = 333.15$ K.[8] The acetonitrile and nitromethane molecules have similar sizes and shapes, and both have a high polarity. (The gas phase dipole moments are $13.2 \times 10^{-30}$ C·m and $11.5 \times 10^{-30}$ C·m for $CH_3CN$ and $CH_3NO_2$, respectively.)

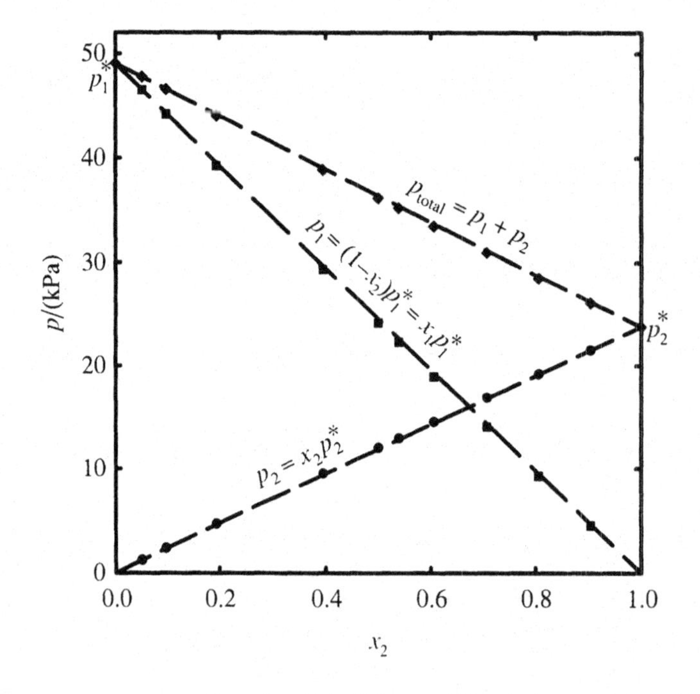


**Figure 6.6** Vapor pressures for $\{x_1CH_3CN + x_2CH_3NO_2\}$ at $T = 333.15$ K. The symbols represent the experimental vapor pressures as follows: ●, vapor pressure of $CH_3NO_2$; ■, vapor pressure of $CH_3CN$; ◆, total vapor pressure. The dashed lines represent the ideal solution prediction.

---

[j] A possible exception is a mixture of *d* and *l* optical isomers of a compound.

Mixtures of molecules with dissimilar shapes and sizes, or where strong interactions such as hydrogen bonding or change transfer are possible, usually show significant deviation from ideal behavior. Figure 6.7 shows the vapor pressures for $\{x_1CH_3CH_2OH + x_2H_2O\}$ at $T = 303.15$ K, where such behavior is observed.[9] The ideal solution predictions are shown as dashed lines, and it is evident that the vapor pressures are larger than would be expected for the ideal solution. A mixture in which the vapor pressure exceeds those expected for an ideal solution is said to show positive deviations from Raoult's law.

Deviations in which the observed vapor pressure are smaller than predicted for ideal solution behavior are also observed. Figure 6.8 gives the vapor pressure of $\{x_1(CH_3CH_2)_3N + x_2CHCl_3\}$ at $T = 283.15$ K, an example of such behavior.[10] This system is said to exhibit negative deviations from Raoult's law.

We repeat that ideal solutions, like ideal gases, do not exist. But like the ideal gas, the ideal solution has an application as a reference state, and it is important to know the conditions under which Raoult's law is a good

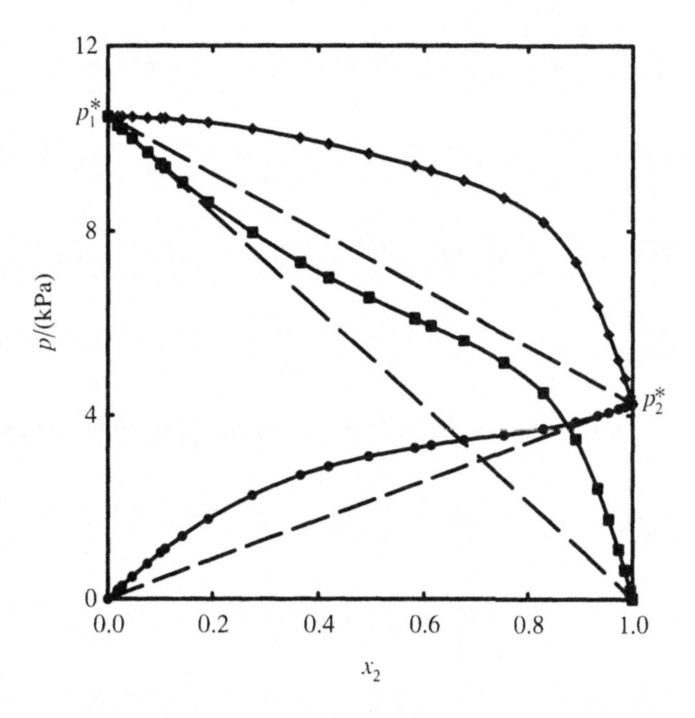


**Figure 6.7** Vapor pressures for $\{x_1CH_3CH_2OH + x_2H_2O\}$ at $T = 303.15$ K. The symbols represent the experimental vapor pressures as follows: ●, vapor pressure of $H_2O$; ■, vapor pressure of $CH_3CH_2OH$; ◆, total vapor pressure. The solid lines are the fits to the experimental data, and the dashed lines represent the ideal solution prediction.

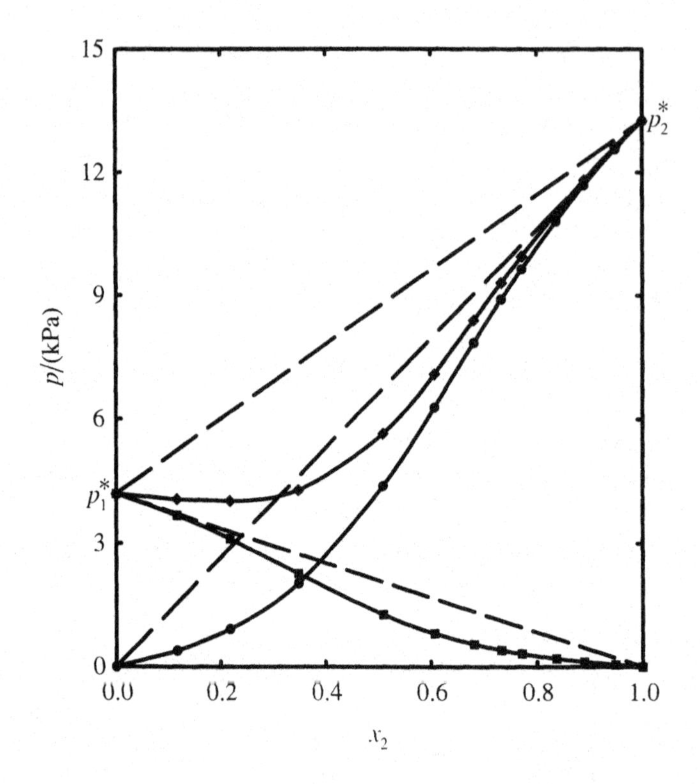


**Figure 6.8** Vapor pressures for $\{x_1(CH_3CH_2)_3N + x_2CHCl_3\}$ at $T = 283.15$ K. The symbols represent the experimental vapor pressures as follows: ●, vapor pressure of $CHCl_3$; ■, vapor pressure of $(CH_3CH_2)_3N$; ◆, total vapor pressure. The solid lines are the fits to the experimental data and the dashed lines represent the ideal solution prediction.

approximation for real solutions.[k] An examination of Figures 6.7 and 6.8 shows that in each case, a component's vapor pressure approaches Raoult's law behavior in the limit as the composition of the solution approaches the pure component. Put another way,

*In a dilute solution the solvent approaches Raoult's law behavior.*

---

[k] A gas approaches ideal behavior as the pressure approaches zero, with high temperatures aiding the process. Under these conditions, molecular interactions and molecular volumes become negligible, and no change in energy occurs when the distance between the molecules changes. A mixture approaches ideal solution behavior over the entire composition range when interactions between like and unlike molecules are the same so that no change in energy occurs when, during the mixing process, interactions between like molecules are replaced by interactions between unlike molecules. Having molecules of similar size aids this process.

By the solvent, we mean the component with the larger mole fraction. Mathematically, we can express this relationship as

$$p_1 \rightarrow x_1 p_1^* \quad \text{as } x_1 \rightarrow 1.$$

Or, more properly,

$$f_1 \rightarrow x_1 f_1^* \quad \text{as } x_1 \rightarrow 1.$$

**(b) Henry's Law:** It is apparent from Figures 6.7 and 6.8 that the solute (component with the smaller mole fraction) does not approach Raoult's law behavior as the solution becomes more dilute ($x_2 \rightarrow 0$). A straight line is approached, but it does not have the Raoult's law slope. This can be seen in Figure 6.9(a) , which is a graph of the vapor pressure, $p_2$, of ethanol against the mole fraction $x_2$ as $x_2 \rightarrow 0$, along with the Raoult's law prediction, for the system shown in Figure 6.7.

This linear relationship is known as Henry's law. It is represented mathematically by

$$f_2 = k_{H,x} x_2 \tag{6.69}$$

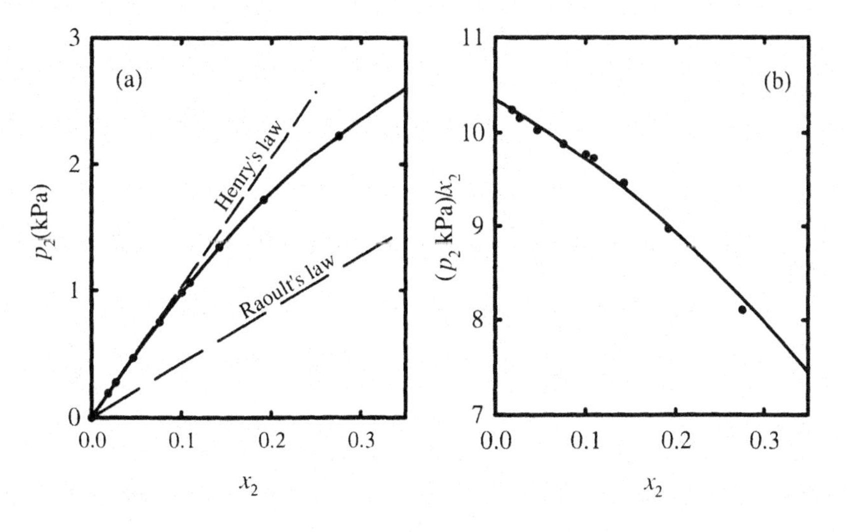


**Figure 6.9** (a) Graph of $x_2$ (at low $x_2$) against $p_2$(kPa) for ($x_1CH_3CH_2OH + x_2H_2O$) at $T = 303.15$ K. The long-dashed line is the Raoult's law prediction, while the short-dashed line is drawn with the Henry's law limiting slope as determined in (b), which is a graph of $x_2$ against $p_2/x_2$. Extrapolation of the line in (b) to $x_2 = 0$ gives a Henry's law constant for $H_2O$ in $CH_3CH_2OH$ of 10.35 kPa.

where $k_{H,x}$ is the Henry's law constant. Henry's law can also be written in terms of other concentration units. For a solute, mole fraction, $x_2$, molality, $m$, and molarity, $c$, are related by the equations

$$x_2 = \frac{M_1 m}{M_1 m + 1} = \frac{M_1 c}{c(M_1 - M_2) + 10^{-3}\rho},$$

where $M_1$ and $M_2$ are the molecular weights (kg·mol$^{-1}$) of the solvent and solute, and $\rho$ is the density of the solution (kg·m$^{-3}$). As $m \rightarrow 0$ or $c \rightarrow 0$, $\rho \rightarrow \rho_1^*$, the density of the pure solvent, and the terms in the denominator involving $m$ or $c$ become negligible so that $x_2 = M_1 m = M_1 c/10^{-3}\rho_1^*$. Substitution for $x_2$ in equation (6.69) gives

$$f_2 = k_{H,m} m \tag{6.70}$$

or

$$f_2 = k_{H,c} c, \tag{6.71}$$

where

$$k_{H,m} = M_1 k_{H,x} \tag{6.72}$$

and

$$k_{H,c} = \frac{M_1 k_{H,x}}{10^{-3}\rho_1^*}. \tag{6.73}$$

Equations (6.69), (6.70), and (6.71) are all expressions of Henry's law, which may be stated as follows:

> *The vapor fugacity of the solute in dilute solution is proportional to the concentration of solute. That is,* $f_2 = k_{H,x} x_2$ *or* $f_2 = k_{H,m} m$ *or* $f_2 = k_{H,c} c$.

Figure 6.9(b) demonstrates a procedure for determining $k_{H,x}$. The ratio $p_2/x_2$ is plotted against $x_2$ and the limit of $(p_2/x_2)$ as $x_2 \rightarrow 0$ is then taken as $k_{H,x}$. For the (ethanol + water) data shown in figure 6.9(b), the intercept of 10.35 at $x_2 = 0$ gives $k_{H,x}$ for this system.[1]

[1] The change in $(p_2/x_2)$ with $x_2$ shown in Figure 6.9(b) demonstrates that Henry's law is a limiting law that is rigorously true only at $x_2 = 0$.

In summary, in the limit as $x_2 \rightarrow 0$ and $x_1 \rightarrow 1$, $f_1 \rightarrow x_1 f_1^*$ and $f_2 \rightarrow x_2 k_{H,x}$. It can be shown from the Gibbs–Duhem equation that when the solute obeys Henry's law, the solvent obeys Raoult's law. To prove this, we start with the Gibbs–Duhem equation relating the chemical potentials

$$n_1 \, d\mu_1 + n_2 \, d\mu_2 = 0.$$

Dividing by $(n_1 + n_2)$ gives

$$x_1 \, d\mu_1 + x_2 \, d\mu_2 = 0.$$

Substitution of vapor fugacities for chemical potentials ($d\mu_i = RT \, d \ln f_i$) and dividing by $RT$ gives

$$x_1 \, d \ln f_1 + x_2 \, d \ln f_2 = 0. \tag{6.74}$$

When the solute obeys Henry's law,

$$f_2 = x_2 k_{H,x}$$

or

$$\ln f_2 = \ln x_2 + \ln k_{H,x}.$$

Differentiating at constant $T$ and $p$ ($k_{H,x}$ is a constant) gives

$$d \ln f_2 = d \ln x_2 = \frac{dx_2}{x_2}. \tag{6.75}$$

Substitution of equation (6.75) into equation (6.74) gives

$$x_1 \, d \ln f_1 + x_2 \frac{dx_2}{x_2} = 0.$$

Since $x_2 = 1 - x_1$, $dx_2 = -dx_1$. After substitution and rearranging the equation becomes

$$d \ln f_1 = d \ln x_1.$$

Integration gives

$$\int_{\ln f_1^*}^{\ln f_1} \mathrm{d}\ln f_1 = \int_{\ln x_1 = 0}^{\ln x_1} \mathrm{d}\ln x_1$$

$$\ln \frac{f_1}{f_1^*} = \ln x_1$$

or

$$f_1 = x_1 f_1^*.$$

**(c) The Duhem–Margules Equation:** The Gibbs–Duhem equation relates the slopes d ln $f_1/\mathrm{d}x_1$ and d ln $f_2/\mathrm{d}x_2$, from which we have shown that when the solute obeys Henry's law, the solvent obeys Raoult's law. The Gibbs–Duhem equation is also the starting place for deriving a more general relationship between the change of fugacity with composition for the two different components in a binary mixture. We start with equation (6.74) and divide by d$x_1$ to get

$$x_1 \frac{\mathrm{d}\ln f_1}{\mathrm{d}x_1} + x_2 \frac{\mathrm{d}\ln f_2}{\mathrm{d}x_1} = 0. \tag{6.76}$$

Substituting d$x_2 = -$d$x_1$ in the second term gives

$$x_1 \frac{\mathrm{d}\ln f_1}{\mathrm{d}x_1} - x_2 \frac{\mathrm{d}\ln f_2}{\mathrm{d}x_2} = 0.$$

Since $p$ and $T$ are constant, the equation is more properly written as

$$x_1 \left(\frac{\partial \ln f_1}{\partial x_1}\right)_{p,T} = x_2 \left(\frac{\partial \ln f_2}{\partial x_2}\right)_{p,T}. \tag{6.77}$$

Equation (6.77) is known as the Duhem–Margules equation. It can also be written as

$$\frac{x_1}{f_1}\left(\frac{\partial f_1}{\partial x_1}\right)_{p,T} = \frac{x_2}{f_2}\left(\frac{\partial f_2}{\partial x_2}\right)_{p,T}$$

or

$$\frac{(\partial f_1/\partial x_1)_{p,T}}{(\partial f_2/\partial x_2)_{p,T}} = \frac{f_1/x_1}{f_2/x_2}. \tag{6.78}$$

With the approximation of ideal vapors, equation (6.78) becomes

$$\frac{(\partial p_1/\partial x_1)_{p,T}}{(\partial p_2/\partial x_2)_{p,T}} = \frac{p_1/x_1}{p_2/x_2}. \tag{6.79}$$

Equation (6.79) relates the slopes of the vapor pressure lines in a graph of vapor pressure against mole fraction such as those shown in Figures 6.5 to 6.8. It requires that if one of the components in a binary mixture obeys Raoult's law over the entire composition range, then the other must do the same. This can be seen as follows. If component 1 obeys Raoult's law over the entire composition range, then the slope of the graph of $p_1$ against $x_1$ is

$$\text{Slope} = \frac{\partial p_1}{\partial x_1} = \frac{p_1 - 0}{x_1 - 0} = \frac{p_1}{x_1}.$$

Substitution into equation (6.79) gives

$$\left(\frac{\partial p_2}{\partial x_2}\right) = \frac{p_2}{x_2}. \tag{6.80}$$

Separating variables and integrating gives

$$\int_{p_2^*}^{p_2} \frac{\mathrm{d}\,p_2}{p_2} = \int_1^{x_2} \frac{\mathrm{d}\,x_2}{x_2}$$

$$\ln \frac{p_2}{p_2^*} = \ln x_2$$

or

$$p_2 = x_2 p_2^*.$$

Therefore, it is a sufficient condition for ideal solution behavior in a binary mixture that one component obeys Raoult's law over the entire composition range, since the other component must do the same.

When deviations from ideal solution behavior occur, the changes in the deviations with mole fraction for the two components are not independent, and the Duhem–Margules equation can be used to obtain this relationship. The allowed combinations[m] are shown in Figure 6.10 in which $p_1/p_1^*$ and $p_2/p_2^*$ are

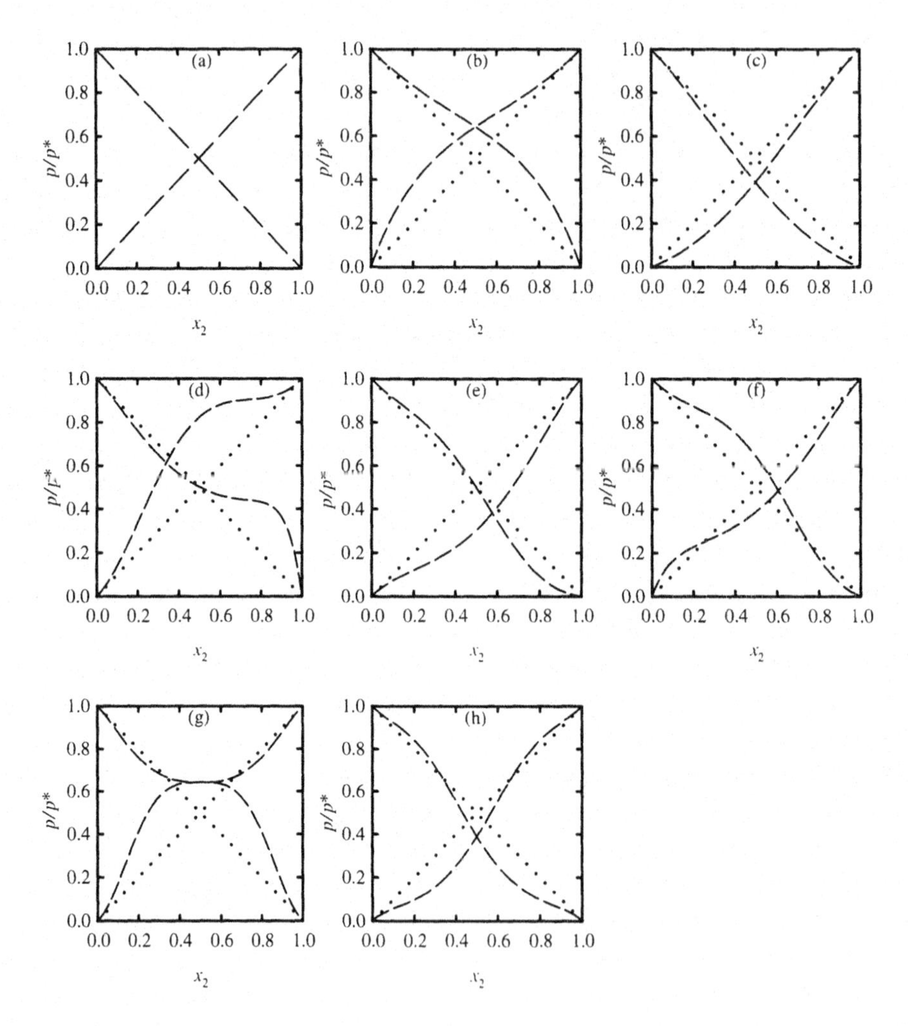


**Figure 6.10** Representative deviations from ideal solution behavior allowed by the Duhem–Margules equation. The dotted lines are the ideal solution predictions. The dashed lines give $p_2/p_2^*$ (lower left to upper right), and $p_1/p_1^*$ (upper left to lower right).

---

[m] For a detailed discussion of the use of the Duhem–Margules equation to show these possible combinations, see M. L. McGlashan, "Deviations from Raoult's Law," *J. Chem Ed.*, **40**, 516–518 (1963).

plotted against $x_2$. Graph (a) is the condition where both components obey Raoult's law. In graph (b) positive deviations from Raoult's law are present over the entire composition range for both components, while in (c), negative deviations occur at all concentrations for both components. In graph (d), component 2 shows positive deviations at all compositions while deviations for component 1 change from negative at low $x_2$ to positive at high $x_2$. In graph (e) the reverse is true. That is, component 2 shows negative deviations at all compositions while deviations for component 1 change from positive to negative with increasing $x_2$. In graph (f) both components change from positive to negative with increasing $x_2$. In graphs (g) and (h), deviations for one component change from positive to negative (or negative to positive) while the second changes from negative to positive (or positive to negative) with an increase in $x_2$. It is important to note (and can be seen in the figure) that in each example, the solvent obeys Raoult's law and the solute obeys Henry's law in the limit of infinitely dilute solution.

## 6.2 The Activity

Equation (6.38) defines fugacity in a mixture through the relationship

$$d\mu_i = RT \, d \ln f_i. \tag{6.38}$$

This equation can be integrated to give

$$\int_{\mu_i^\circ}^{\mu_i} d\mu_i = RT \int_{\ln f_i^\circ}^{\ln f_i} d \ln f_i \tag{6.81}$$

or

$$\mu_i = \mu_i^\circ + RT \ln \frac{f_i}{f_i^\circ}, \tag{6.82}$$

where $\mu_i^\circ$ and $f_i^\circ$ are the chemical potential and the fugacity in a reference or **standard state**.

The ratio $f_i/f_i^\circ$ is defined as the activity $a_i$. That is,

$$a_i = \frac{f_i}{f_i^\circ}, \tag{6.83}$$

and equation (6.82) becomes

$$\mu_i = \mu_i^\circ + RT \ln a_i. \tag{6.84}$$

Activity can be thought of as the quantity that corrects the chemical potential at some pressure and/or composition condition[n] to a standard or reference state. The concept of a standard state is an important one in thermodynamics. The choice of the pressure and composition conditions for the standard state are completely arbitrary, and unusual choices are sometimes made. The common choices are those of convenience. In the next section, we will describe and summarize the usual choices of standard states. But, first, we want to describe the effect of pressure and temperature on $a_i$.

## 6.2a Effect of Pressure on Activity

The effect of pressure on $a_i$ is obtained by differentiating equation (6.84)

$$\left(\frac{\partial \mu_i}{\partial p}\right)_T = \left(\frac{\partial \mu_i^\circ}{\partial p}\right)_T + RT\left(\frac{\partial \ln a_i}{\partial p}\right)_T.$$

But

$$\left(\frac{\partial \mu_i}{\partial p}\right)_T = \bar{V}_i$$

and

$$\left(\frac{\partial \mu_i^\circ}{\partial p}\right)_T = 0$$

since $\mu_i^\circ$ is the chemical potential in the standard state, which is defined for some fixed pressure. Hence,

$$\left(\frac{\partial \ln a_i}{\partial p}\right)_T = \frac{\bar{V}_i}{RT}. \tag{6.85}$$

It can be shown from equation (6.85) that pressure has only a small effect on the activity of liquids and solids.

---

[n] Temperature is not a part of the specification of a standard state. Thus $\mu_i^\circ$ changes with temperature, but not with pressure, since the pressure is specified for a standard state.

**Example 6.5** At 298.15 K the partial molar volumes[11] of benzene (1) and cyclohexane (2) in an equimolar mixture of the two are $\overline{V}_1 = 68.880\ \text{cm}^3\cdot\text{mol}^{-1}$ and $\overline{V}_2 = 65.817\ \text{cm}^3\cdot\text{mol}^{-1}$. Calculate the percent change in $a_2$ per MPa.

**Solution:**

$$\left(\frac{\partial \ln a_2}{\partial p}\right)_T = \frac{\overline{V}_2}{RT} = \frac{1}{a_2}\left(\frac{\partial a_2}{\partial p}\right)_T.$$

$$\left(\frac{\partial a_2}{\partial p}\right)_T = \frac{(65.817\ \text{cm}^3\cdot\text{mol}^{-1})(10^{-6}\text{m}^3\cdot\text{cm}^{-3})}{(8.314\ \text{J}\cdot\text{K}^{-1}\cdot\text{mol}^{-1})(298.15\ \text{K})}a_2$$

$$= (2.66 \times 10^{-8}\ \text{Pa}^{-1})(10^6\ \text{Pa}\cdot\text{MPa}^{-1})a_2$$

$$= 0.0266\ \text{MPa}^{-1}\ a_2.$$

A change of 1 MPa in pressure changes $a_2$ by 2.66%. Thus, the change of a few kPa in pressure around ambient pressure has only a small effect on $a_i$.

## 6.2b Effect of Temperature on Activity

To determine the effect of temperature on $a_i$, we start with equation (6.84) and divide by $T$ to get

$$\frac{\mu_i}{T} = \frac{\mu_i^\circ}{T} + R \ln a_i.$$

Differentiating gives

$$\left(\frac{\partial(\mu_i/T)}{\partial T}\right)_p - \left(\frac{\partial(\mu_i^\circ/T)}{\partial T}\right)_p = R\left(\frac{\partial \ln a_i}{\partial T}\right)_p. \tag{6.86}$$

We use the Gibbs–Helmholtz relationship to obtain the derivatives on the left

$$\left(\frac{\partial(\mu_i/T)}{\partial T}\right)_p = -\frac{\overline{H}_i}{T^2} \tag{6.87}$$

$$\left(\frac{\partial(\mu_i^\circ/T)}{\partial T}\right)_p = -\frac{\overline{H}_i^\circ}{T^2}. \tag{6.88}$$

We substitute equations (6.87) and (6.88) into equation (6.86) and rearrange to get

$$\left(\frac{\partial \ln a_i}{\partial T}\right)_p = -\frac{\overline{H}_i - \overline{H}_i^\circ}{RT^2}. \tag{6.89}$$

Equation (6.89) is often written as

$$\left(\frac{\partial \ln a_i}{\partial T}\right)_p = -\frac{\overline{L}_i}{RT^2} \tag{6.90}$$

where $\overline{L}_i = \overline{H}_i - \overline{H}_i^\circ$ is the **relative partial molar enthalpy**. It is the difference in partial molar enthalpies of the ith component in between the mixture of interest and the standard state mixture.[o]

## 6.3 Standard States

In defining the activity through equations (6.83) and (6.84), we have made no restrictions on the choice of a standard state except to note that specification of temperature is not a part of the standard state condition. We are free to choose standard states in whatever manner we desire.[p] However, choices are usually made that are convenient and simplify calculations involving activities. The usual choices differ for a gas, pure solid or liquid, and solvent or solute in solution. We will now summarize these choices of standard states and indicate the reasons. Before doing so, we note that activities for a substance with different choices of standard states are proportional to one another. This can be seen as follows:

With a particular choice of standard state

$$a_i = \frac{f_i}{f_i^\circ}$$

where $f_i^\circ$ is the standard state fugacity. With a different choice of standard state for the same system, we can write

$$a_i' = \frac{f_i}{f_i^{\circ\prime}}$$

---

[o] In Chapter 7, we will describe relative partial molar enthalpy in more detail.

[p] When working with activities, one must always be careful to specify the chosen standard state. Otherwise, confusion can, and often has, occurred.

where $a_i'$ is the activity when $f_i^{\circ\prime}$ is the standard state fugacity. Dividing one equation by the other gives

$$a_i' = a_i\left(\frac{f_i^{\circ}}{f_i^{\circ\prime}}\right). \tag{6.91}$$

The ratio $(f_i^{\circ}/f^{\circ\prime})$ is a constant that converts $a_i$ to $a_i'$ at any composition or pressure.

## 6.3a Choice of Standard States

**Standard State of a Gas:** The standard state for a gas is usually chosen as the gas behaving as ideal gas at a pressure of exactly 100 kPa (0.1 MPa). This choice of standard state is demonstrated in Figure 6.11. Note that when $p = 100$ kPa, the fugacity of the real gas would not be exactly 100 kPa. But on the $f = p$ (ideal gas) line, $f = 100$ kPa when $p = 100$ kPa, and this is the standard state fugacity. With this choice of standard state the activity is

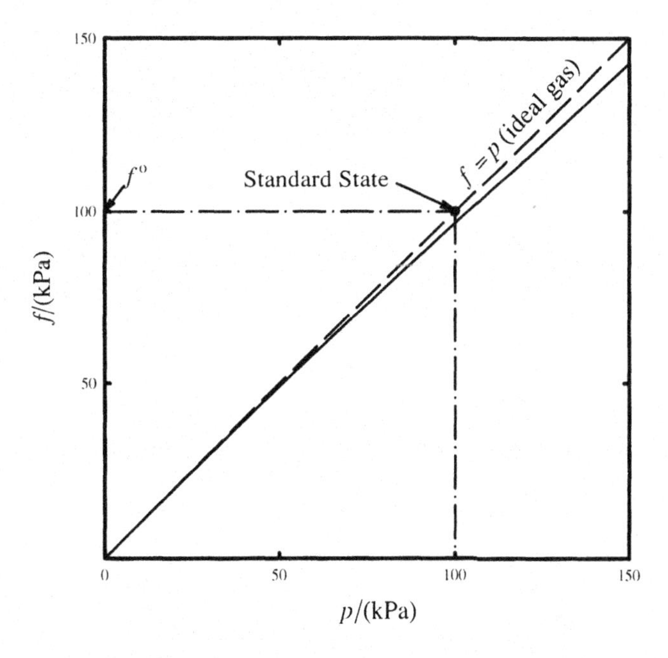


**Figure 6.11** The standard state for a gas is the ideal gas with a standard state fugacity $f^{\circ} = p^{\circ} = 100$ kPa (1 bar).

given by

$$a = \frac{f}{100 \text{ kPa}}. \tag{6.92}$$

Activity is a dimensionless quantity, and $f$ must be expressed in kPa with this choice of standard state. It is inconvenient to carry $f^\circ = 100$ kPa through calculations involving activity of gases. Choosing the standard state for a gas as we have described above creates a situation where SI units are not convenient. Instead of expressing the standard state as $f^\circ = 100$ kPa, we often express the pressure and fugacity in bars, since 1 bar = 100 kPa. In this case, $f^\circ = 1$ bar, and equation (6.92) becomes[q]

$$a = \frac{f}{1 \text{ bar}} \qquad \text{(gas)}. \tag{6.93}$$

With the standard state expressed in this manner, the activity of the gas becomes the fugacity expressed in bars. We will usually follow this convention as we work with activities of gases. An added convenience comes from being able to relate fugacity to pressure through the fugacity coefficient $\phi$,

$$f = \phi p.$$

Thus, when $f$ and $p$ are expressed in bars, activity is given by

$$a = \phi p \tag{6.94}$$

and the fugacity coefficient can be used to correct pressure to activity. We remember that

$$\lim_{p \to 0} \phi = 1. \tag{6.95}$$

At low pressures, it is a good approximation to replace the activity of gases by the pressure (in bars).

---

[q] The acitivity $a$ in equation (6.93) must be dimensionless. This happens in this equation only if we express $f$ in the units of bars.

Later, we will make equilibrium calculations that involve activities, and we will see why it is convenient to choose the ideal gas as a part of the standard state condition, even though it is a hypothetical state.[r] With this choice of standard state, equations (6.94) and (6.95) allow us to use pressures, corrected for non-ideality, for activities as we make equilibrium calculations for real gases.[s]

**Standard States for Pure Solids and Pure Liquids:** For condensed phases, the pure solid or liquid under 1 bar (100 kPa) external pressure is usually chosen as the standard state. With this choice, the standard state fugacity is given by

$$f^\circ = f^*(1 \text{ bar})$$

where $f^*(1 \text{ bar})$ is the vapor fugacity of the pure condensed phase under 1 bar total pressure. The activity is still defined as

$$a = \frac{f}{f^\circ} = \frac{f}{f^*(1 \text{ bar})}, \tag{6.96}$$

and for the pure material in its standard state, the activity equals one. As we saw in equation (6.85), the activity is a function of pressure, and a variation in the external pressure above the condensed phase will change its activity, even for a pure substance. It is convenient to define a quantity $\Gamma$ as the ratio of the activity under a given external pressure to that of standard state activity. That is,

$$\Gamma = \frac{a}{1}. \tag{6.97}$$

With this definition, $\Gamma$ is the numerical value of the activity for the substance under some pressure $p$. It is also the ratio of the fugacity of the pure condensed phase under pressure $p$ to that of the phase under 1 bar pressure.

---

[r] We will employ a similar procedure for choosing a standard state of a solute in solution by choosing a hypothetical solution as the standard state condition.

[s] In Chapter 10, we will calculate the thermodynamic properties of gases from the molecular parameters, and these calculations, which provide the standard state values, are most easily done for the ideal gas.

Integration of equation (6.85) yields $\ln \Gamma$:

$$\ln \Gamma = \int_{a=1}^{a} \mathrm{d}(\ln a) \tag{6.98}$$

or

$$\ln \Gamma = \int_{1\ \text{bar}}^{p} \frac{V_{\mathrm{m}}}{RT} \,\mathrm{d}p \tag{6.99}$$

If the pressure dependence of the molar volume is known as a function of $p$, the activity coefficient, and hence the activity, of the substance under pressure conditions that differ from the standard state can be calculated from equation (6.99). The following example illustrates that changes of the activity of condensed phases with pressure are small. It is only when pressures become large, for example, for those on a geologic scale, that significant changes in the activity from its standard state value are found. This observation is the justification for the common practice in many textbooks of setting the activity of pure solids and liquids equal to one, regardless of the pressure.

**Example 6.6** Calculate the activity $\Gamma$ at $T = 298.15$ K for $CaCO_3$(s) at $p = 10$ bars. The density of $CaCO_3$(s) at this temperature and $p = 1$ bar is $2.711 \times 10^3$ kg·m$^{-3}$.

**Solution:** At $p = 1$ bar

$$V_{\mathrm{m}} = \frac{0.1001\ \mathrm{kg{\cdot}mol^{-1}}}{2.711 \times 10^3\ \mathrm{kg{\cdot}m^{-3}}}$$

$$= 3.692 \times 10^{-5}\ \mathrm{m^3{\cdot}mol^{-1}}.$$

From equation (6.99)

$$\ln \Gamma = \int_{1}^{p} \frac{V_{\mathrm{m}}}{RT} \,\mathrm{d}p.$$

We will assume that $CaCO_3$ is incompressible so that $V_m$ is a constant[t] and integrate to get

$$\ln \Gamma = \frac{V_m}{RT}(p - 1)$$

$$= \frac{3.692 \times 10^{-5}\ \text{m}^3\cdot\text{mol}^{-1}}{(8.314\ \text{J}\cdot\text{K}^{-1}\cdot\text{mol}^{-1})(298.15\ \text{K})}(10\ \text{bar} - 1\ \text{bar})(10^5\ \text{Pa}\cdot\text{bar}^{-1})$$

$$\ln \Gamma = 1.340 \times 10^{-2}.$$

$$\Gamma = 1.013.$$

Thus, an increase in pressure by a factor of ten changes the activity of $CaCO_3$ by approximately 1%.

**Standard State of a Solvent in a Mixture:** The usual choice of a standard state for a solvent in a solution is the pure solvent at a pressure of 1 bar, the same convention as for a pure solid or liquid. Thus,[u]

$$f_1^\circ = f_1^*\ (1\ \text{bar})$$

[t] A calculation of the effect of pressure on the activity that does not involve the assumption of constant $V_m$ usually starts with the compressibility $\kappa$. Integration of equation (1.39) that relates $\kappa$ to $V$, while assuming that $\kappa$ is independent of pressure, gives the equation

$$V_m = V_m^\circ \exp\{-\kappa(p - 1)\},$$

where $p$ is the pressure in bars and $V_m^\circ$ is the molar volume at $p = 1$ bar. Expanding the exponential gives a simpler equation:

$$V_m = V_m^\circ \left\{ 1 - \kappa(p - 1) + \frac{\kappa^2(p - 1)^2}{2} - \cdots \right\}.$$

Either of these relationships can be substituted into equation (6.99) and integrated to give an equation for calculating the activity $a$ at a pressure $p$. For $CaCO_2$, one must go to a pressure of 3 kbar before the assumption of a constant $V_m$ causes a 1% error in the calculation of the activity. At this pressure, the activity of the $CaCO_3$ is calculated to be 86 if one assumes that $\kappa$ is constant at the ambient pressure value of $1.41 \times 10^{-2}\ \text{GPa}^{-1}$. This is an increase from $\Gamma = 1.16$ at $p = 0.10$ kbar, to 2.1 at 0.5 kbar, 4.4 at 1 kbar, and 20 at 2 kbars. The logarithmic relationship in equation (6.99) causes $\Gamma$ to increase exponentially with $p$, and $\Gamma$ can become large at high $p$.

[u] As indicated earlier, we will use the subscript 1 to refer to the solvent in solution, with subscripts 2, 3, ... referring to solutes.

and

$$a_1 = \frac{f_1}{f_1^*(1\ \text{bar})}, \tag{6.100}$$

where $f_1$ is the fugacity of the solvent in the solution under some external pressure and $f_1^*(1\ \text{bar})$ is the fugacity under 1 bar pressure. If the fugacity of the pure solvent is known at a total pressure other than 1 bar, the quantity $\Gamma_1$ defined in equation (6.98) can be used to correct it to the standard state fugacity. That is,

$$f_1^*(1\ \text{bar}) = \frac{f_1^*(p)}{\Gamma_1}, \tag{6.101}$$

so that the activity for the solvent under some total pressure $p$ other than 1 bar is given by

$$a_1(p) = \Gamma_1 \frac{f_1(p)}{f_1^*(p)}. \tag{6.102}$$

In the case of solutions, $\ln \Gamma_1$ would be determined from equation (6.85) using the partial molar volume of the solvent, $\overline{V}_1$, in the solution, rather than the molar volume of the pure solvent.

As we have seen, unless the pressure is considerably larger than 1 bar, $\Gamma_1$ is very nearly 1. Except for special circumstances, we will assume it is unity in future calculations and discussions of activity in solution, and we will drop the designation of (1 bar) pressure for the standard state pressure. That is, $f_1^*(1\ \text{bar})$ will be set equal to $f_1^*(p)$, the vapor fugacity at a pressure $p$, and both will be designated as $f_1^*$, so that equation (6.100) can be written as

$$a_1 = \frac{f_1}{f_1^*}.$$

Comparing this equation with Raoult's law

$$x_1 = \frac{f_1}{f_1^*}$$

shows that for this choice of standard state, the activity $a_1$ of a component in an ideal solution is equivalent to its concentration, $x_1$. For this reason, this standard state is often referred to as the **Raoult's law standard state**.

For solutions that do not follow Raoult's law, it is convenient to define a quantity called the activity coefficient,[v] $\gamma_{R,1}$, that is a measure of the deviation of the activity from the concentration, through the relationship

$$a_1 = \gamma_{R,1} x_1. \tag{6.103}$$

For ideal solutions, the activity coefficient will be unity, but for real solutions, $\gamma_{R,1}$ will differ from unity, and, in fact, can be used as a measure of the non-ideality of the solution. But we have seen earlier that real solutions approach ideal solution behavior in dilute solution. That is, the behavior of the solvent in a solution approaches Raoult's law as $x_1 \rightarrow 1$, and we can write for the solvent

$$\lim_{x_1 \rightarrow 1} \gamma_{R,1} = 1. \tag{6.104}$$

For liquid mixtures (especially when the components are nonelectrolytes) in which we work with solutions over the entire range of composition, we often choose the Raoult's law standard state for both components. Thus, for the second component

$$a_2 = \frac{f_2}{f_2^*} \tag{6.105}$$

and

$$a_2 = \gamma_{R,2} x_2. \tag{6.106}$$

$$\lim_{x_2 \rightarrow 1} \gamma_{R,2} = 1. \tag{6.107}$$

Figure 6.12 shows a graph of $a_1$ and $a_2$ as a function of mole fraction for mixtures of $\{x_1(C_4H_9)_2O + x_2CCl_4\}$[12] at $T = 308.15$ K. A Raoult's law standard state has been chosen for both components. The system shows negative deviation from Raoult's law over the entire range of composition, with $a_1$ less than $x_1$ and $a_2$ less than $x_2$, so that all $\gamma_{R,1}$ and $\gamma_{R,2}$ are less than 1.

---

[v] IUPAC suggests $f_i$ for the activity coefficient with a Raoult's law standard state. We will use $\gamma_R$ instead so as not to confuse the activity coefficient with the fugacity, which is also represented by the symbol $f_i$.

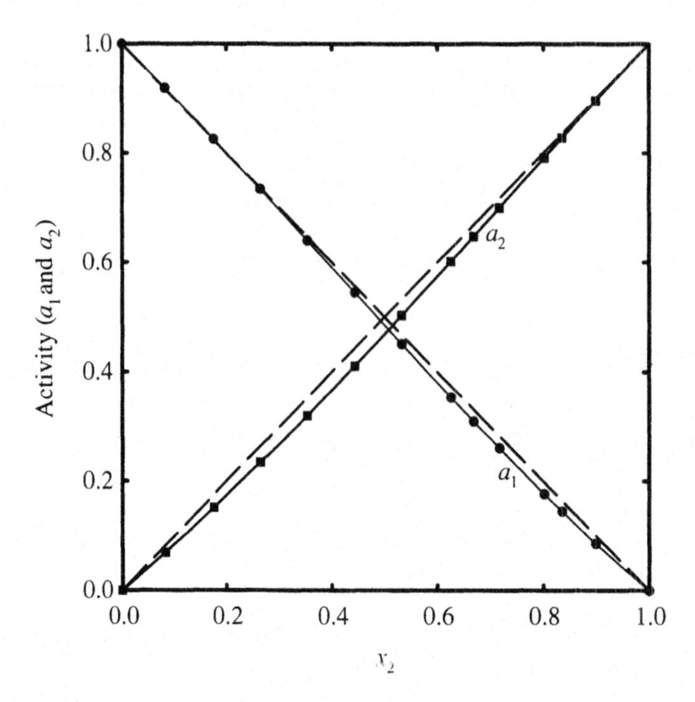


**Figure 6.12** Activity ($a_1$ and $a_2$) for $\{x_1(C_4H_9)_2O + x_2CCl_4\}$ at $T = 308.15$ K. The dashed lines are the ideal solution predictions.

**Standard States of Solutes in Solution:** For a solute, particularly in situations where only dilute solutions can or will be considered, the usual procedure is to define the standard state in terms of a hypothetical solution that follows Henry's law at either a concentration of $x_2 = 1$ or $m_2 = 1$. These standard states are known as **Henry's law standard states**. The standard state solutions are said to be hypothetical because real solutions at these high concentrations do not follow Henry's law.

When Henry's law is followed, we have

$$f_2 = k_{H,x}x_2$$

or

$$f_2 = k_{H,m}m,$$

depending upon our choice for expressing the concentration of the solute. Since the standard states are defined to follow Henry's law when the concentration is one, the numerical value of the standard state fugacity is equal to the appropriate Henry's law constant. That is

$$f_2^\circ = k_{H,x}(1)$$

or

$$f_2^\circ = k_{H,m}(1\ \text{mol}\cdot\text{kg}^{-1}).$$

With our definition of activity as the ratio of the fugacity of the component in the solution to that in the standard state, we find that

$$a_2 = \frac{f_2}{f_2^\circ} = \frac{f_2}{k_{H,x}} \tag{6.108}$$

or

$$a_2 = \frac{f_2}{f_2^\circ} = \frac{f_2}{k_{H,m}(1\ \text{mol}\cdot\text{kg}^{-1})}. \tag{6.109}$$

The factor of (1 mol·kg$^{-1}$) in the denominator of equation (6.109) is necessary to keep the activity dimensionless. It is inconvenient to write this factor in all of our expressions for activity. We usually leave it out of the expression and write the above equation as

$$a_2 = \frac{f_2}{k_{H,m}}, \tag{6.109}$$

while remembering that the dimension factor must be included to keep activity dimensionless.[w]

[w] The situation is similar to that for a gas where $a = f/f^\circ$. But $f^\circ = 1$ bar, and we write $a = f$ and remember that $f$ must be expressed in bars. An alternative to equation (6.109) that is followed by many authors is to write this equation as

$$a_2 = \frac{f_2}{k_{H,m}m^\circ},$$

with $m^\circ = 1$ mol·kg$^{-1}$. We find the use of $m^\circ$ inconvenient and unnecessary, and we will follow the convention used in equation (6.109) throughout the book.

Comparing equations (6.108) and (6.109) with the Henry's law equations written in the form

$$x_2 = \frac{f_2}{k_{H,x}}$$

or

$$m = \frac{f_2}{k_{H,m}}$$

shows that when Henry's law is obeyed, $a_2 = x_2$ or $a'_2 = m$, depending on which standard state is chosen. When Henry's law is not obeyed, $a_2$ and $x_2$, or $a'_2$ and $m$, are related through the activity coefficients $\gamma_{H,x}$ or $\gamma_{H,m}$ as given by equations (6.110) and (6.111)

$$a_2 = \gamma_{H,x} x_2 \tag{6.110}$$

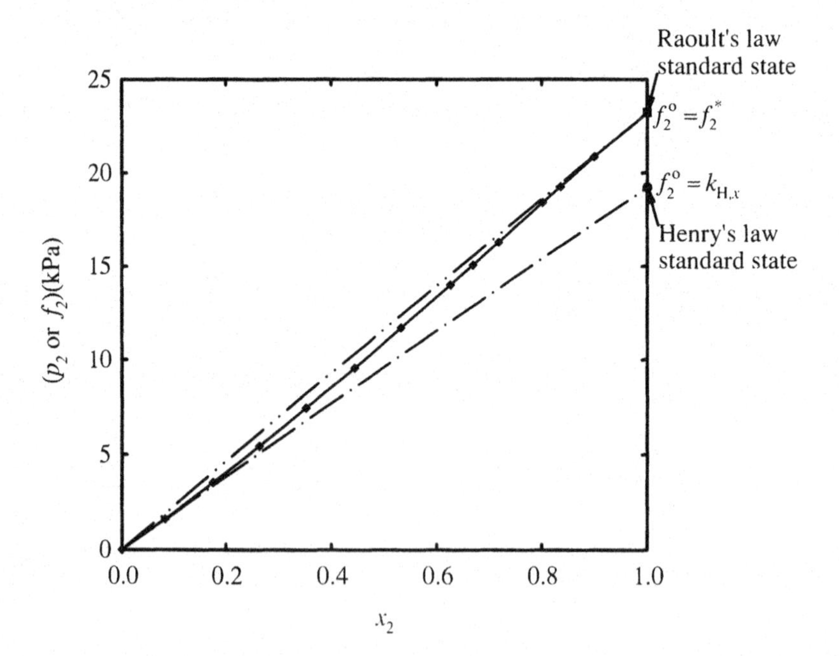


**Figure 6.13** Standard states for $CCl_4$ in $\{x_1(C_4H_9)_2O + x_2CCl_4\}$ at $T = 308.15$ K. The ■ represents the Raoult's law standard state with $f_2^\circ = f_2^* = 23.26$ kPa and the ● represents the Henry's law standard state with $f_2^\circ = k_{H,x} = 19.23$ kPa. The ◆ symbols represent the experimental results.

or

$$a'_2 = \gamma_{H,m} m. \tag{6.111}$$

The reason for choosing a Henry's Law standard state can be seen by referring to Figure 6.13, which compares the Henry's law and Raoult's law standard states for $CCl_4$ in $\{x_1(C_4H_9)_2O + x_2CCl_4\}$. At high $x_2$, Raoult's law

**Table 6.1** Choice of standard states ($a_i = f_i/f_i^\circ$)

| State of matter | Standard state | Standard state fugacity | Activity coefficient | Limiting relationship |
|---|---|---|---|---|
| Gas | Ideal gas at $p = 1$ bar | $f^\circ = 1$ bar | $a = \phi\, p$ | $\phi \rightarrow 1$ as $p \rightarrow 0$ |
| Pure solid or pure liquid | Pure substance at $p = 1$ bar | $f^\circ = f^*$ | $a = \Gamma$ | $\Gamma \rightarrow 1$ as $p \rightarrow 1$ |
| Solvent in mixture | (Raoult's law) Pure substance at $p = 1$ bar | $f_1^\circ = f_1^*$ | $a_1 = \gamma_{R,1} x_1$ | $\gamma_{R,1} \rightarrow 1$ as $x_1 \rightarrow 1$ |
| Solute* in mixture | (Raoult's law)* Pure substance at $p = 1$ bar | $f_2^\circ = f_2^*$ | $a_2 = \gamma_{R,2} x_2$ | $\gamma_{R,2} \rightarrow 1$ as $x_2 \rightarrow 1$ |
| | or (Henry's law) Hypothetical solution with $x_2 = 1$ that obeys Henry's law | $f_2^\circ = k_{H,x}$ | $a_2 = \gamma_{H,x} x_2$ | $\gamma_{H,x} \rightarrow 1$ as $x_2 \rightarrow 0$ |
| | or (Henry's law) Hypothetical solution with $m = 1$ that obeys Henry's law | $f_2^\circ = k_{H,m}$ | $a_2 = \gamma_{H,m} m$ | $\gamma_{H,m} \rightarrow 1$ as $m \rightarrow 0$ |

*A Raoult's law standard state for the solute is often chosen for nonelectrolyte mixtures that cover the entire concentration range from $x_2 = 0$ to $x_2 = 1$.

behavior is approximated for $CCl_4$ and a Raoult's law standard state allows one to represent $a_2$ by $x_2$. On the other hand, at low $x_2$ (or $m$), Henry's law is a better approximation, and a Henry's law standard state allows one to represent $a_2$ by $x_2$ (or $m$).[x]

In summary, the usual choice of standard states and the implications of these choices are shown in Table 6.1.

## 6.4 Activities of Electrolyte Solutions

Electrolytes are solutes that dissociate to produce ionic species when dissolved in a solvent. For example, an equilibrium is established in each of the following processes:

$$NaNO_3(aq) = Na^+(aq) + NO_3^-(aq)$$

$$HCl(aq) = H^+(aq) + Cl^-(aq)$$

$$H_3PO_4(aq) = H^+(aq) + H_2PO_4^-(aq)$$

$$HgCl_2(aq) = HgCl^+(aq) + Cl^-(aq).$$

As we apply thermodynamics to the dissociation process, we usually designate the solute as a **strong electrolyte** or a **weak electrolyte**. In a strong electrolyte, the extent of the reaction, $\xi$, is so far toward the products that essentially all of the electrolyte is dissociated, and very few undissociated species (known as ion pairs) are present. In a weak electrolyte, significant amounts of the undissociated species are present. Often the dissociation is incomplete to the point that only a small fraction is dissociated. Of the examples given above, $NaNO_3$ and HCl are usually considered to be strong electrolytes while $H_3PO_4$ and $HgCl_2$ are weak.

For electrolytes where dissociation is extensive, but not complete, the classification is somewhat arbitrary, and the electrolyte can be considered to be either strong or weak. Thermodynamics does not prevent us from treating an electrolyte either way, but we must be careful to designate our assignment because the choice of standard state is different for a strong electrolyte and a weak electrolyte. Assuming that an electrolyte is weak requires that we have some nonthermodynamic procedure for distinguishing clearly between the dissociated and undissociated species. For example, Raman spectroscopy

[x] We find that Henry's law is a valid limiting law only when the solute is a nonelectrolyte. When dissociation occurs, as in a strong electrolyte, the standard state is defined differently because of a different limiting law. We will discuss activity of electrolytes in the next section.

shows different characteristics in the vibrational spectra of $HNO_3$ and $NO_3^-$. The intensity of these spectral bands can be used to calculate the degree of dissociation, and $HNO_3$ can be treated as a weak electrolyte based on these measurements.[13] In dilute solution, the $HNO_3$ band disappears and we consider $HNO_3$ to be a strong electrolyte. Other techniques such as electrolytic conduction and freezing point lowering can be used instead, but they might give a different answer for the degree of dissociation. The strength of the strong electrolyte description is that it is not based on any such assumptions, and for this reason it is often the preferred method for treating electrolytes, a method that we now describe.[y]

## 6.4a Activities and Standard States of Strong Electrolytes

In Section 6.3a we saw that choosing a Henry's law standard state for a nonelectrolyte solute allows us to substitute mole fraction or molality for activity in dilute solution. A Henry's law standard state will not work for a strong electrolyte solute because Henry's law is not the limiting relationship in dilute solution. As an example,[z] Figure 6.14 summarizes the vapor fugacities as a function of mole fraction for HCl and $H_2O$ in aqueous solution[14] at 303.15 K. The measurements cover the mole fraction range[aa] from $x_1$ (mole fraction water) $= 0.78$ to $x_1 = 1$. Note that as $x_1$ approaches 1, $p_1$ approaches the Raoult's law value, represented by the dashed line in the figure, and a Raoult's law standard state is the logical choice for water (the solvent).[bb]

However, as can be seen in Figure 6.15, which is a graph of the fugacity of HCl against molality in dilute aqueous solutions of HCl (near $x_1 = 1$), $f_2$ approaches the $m$ axis with zero slope. This behavior would lead to a Henry's law constant, $k_{H,m} = 0$, given the treatment we have developed so far. Since the activity with a Henry's law standard state is defined as $a_2 = f_2/k_{H,m}$, this would yield infinite activities for all solutions.

Figure 6.15 also shows a graph of $f_2$ against $m^2$. Note that this plot has a non-zero slope and it becomes linear as $m^2 \rightarrow 0$. Therefore, the limiting law

---

[y] We will return to a discussion of equilibria in solutions containing weak electrolytes in Chapter 9.

[z] We use HCl as an example, but the analysis works equally well for any 1 : 1 electrolyte. Later in this section, we will see what must be done to expand the analysis to include electrolytes that dissociate into more than two ions.

[aa] When $x_1 = 0.78$, $m = 15.7$ mol·kg$^{-1}$, which is about the molality of concentrated hydrochloric acid.

[bb] Note the minimum in the total vapor pressure curve around $x_1 = 0.86$. The result is that aqueous HCl solutions form a maximum boiling azeotrope. We will describe azeotropes in more detail in a later chapter.

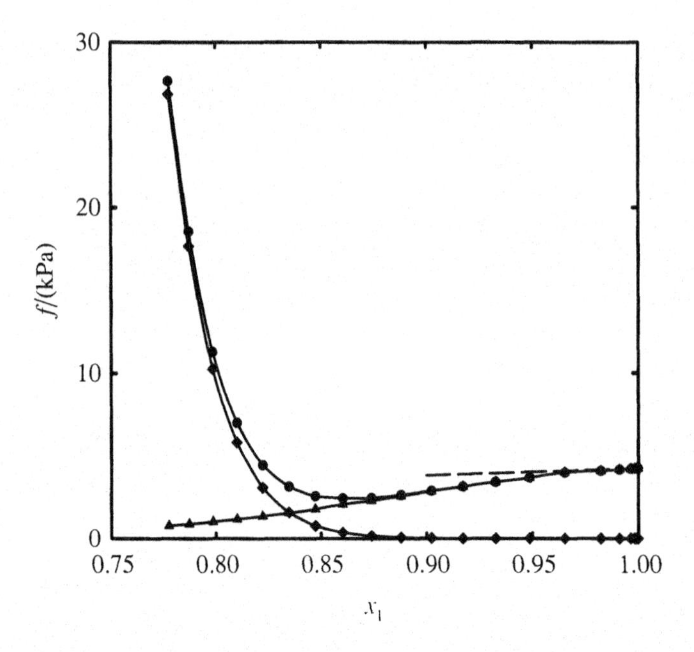


**Figure 6.14** Graph of vapor fugacity $f$ against $x_1$ for $\{x_1H_2O + x_2HCl\}$. The various curves are as follows: ▲, vapor fugacity of $H_2O$; ◆, vapor fugacity of HCl; ●, total vapor fugacity ($H_2O + HCl$). The dashed line gives the Raoult's law limiting values for the vapor fugacity of $H_2O$.

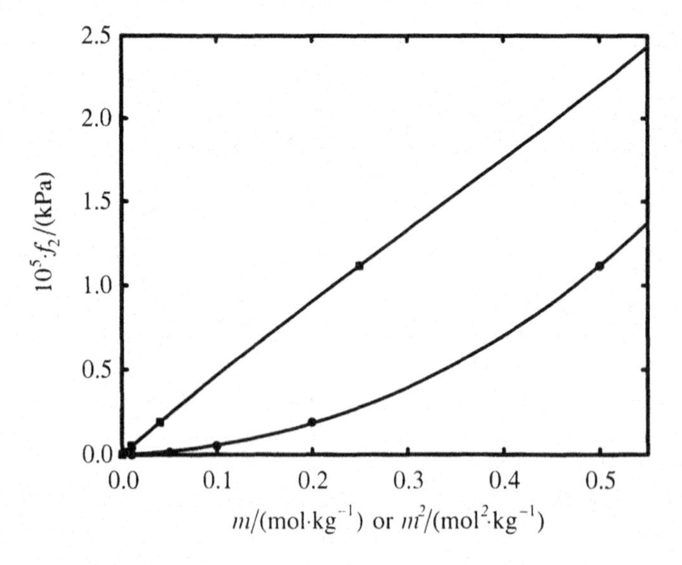


**Figure 6.15** Graph of vapor fugacity of HCl in aqueous solutions at $T = 303.15$ K, against $m$ (●) or $m^2$ (■).

behavior for HCl (and other 1 : 1 electrolytes as well) is given by

$$f_2 = k_2 m^2 \tag{6.112}$$

where $k_2$ is a constant.

This behavior suggests as a convenient standard state, a hypothetical solution that obeys equation (6.112) when $m = 1$. In this standard state

$$f_2^\circ = k_2(1)^2 = k_2$$

and

$$a_2 = \frac{f_2}{k_2}. \tag{6.113}$$

In the limit of infinitely dilute solutions, where equation (6.112) holds, $m^2 = f_2/k_2$. If we maintain this ratio as our definition of activity, $a_2$, then $a_2 = m^2$ in these solutions. For solutions which are not in the limiting region, we write[cc]

$$a_2 = \gamma_2 m^2, \tag{6.114}$$

where $\gamma_2$ is the activity coefficient, and $\gamma_2 \rightarrow 1$ as $m^2 \rightarrow 0$.

The limiting law dependence of $a_2$ on $m^2$ and a better understanding of the nature of $\gamma_2$ can be obtained by considering the equilibrium that is present when the electrolyte dissociates. For HCl, this equilibrium is

$$HCl(aq) = H^+(aq) + Cl^-(aq). \tag{6.115}$$

---

[cc] For reasons similar to those described earlier for the nonelectrolyte {equation (6.109)}, we should be writing equation (6.114) as

$$a_2 = \gamma_2 \frac{m^2}{(m^\circ)^2}$$

where $m^\circ = 1\ \text{mol·kg}^{-1}$ is the standard state molality. Again we will leave the $m^\circ$ out of our equations and remember that activity is dimensionless when $m$ has the units of $\text{mol·kg}^{-1}$.

It is the $H^+$ and $Cl^-$ that exist as independent species in solution. They can each be assigned a Henry's law standard state. That is

$$a_+ = \gamma_+ m_+ \tag{6.116}$$

$$a_- = \gamma_- m_- \tag{6.117}$$

where $a_+$, $a_-$, $\gamma_+$, $\gamma_-$, $m_+$, and $m_-$ represent the activity, activity coefficient, and molality of the ionic species.

The activity of the HCl (aq) is obtained from the equilibrium process represented by equation (6.115). The free energy change for this reaction is given by

$$\Delta_r G = \mu_+ + \mu_- - \mu_2 \tag{6.118}$$

where $\mu_+$ and $\mu_-$ are the chemical potentials of $H^+$ and $Cl^-$ and $\mu_2$ is the chemical potential for HCl (aq); but we can write (with a Henry's law standard state)

$$\mu_+ = \mu_+^\circ + RT \ln a_+$$

$$\mu_- = \mu_-^\circ + RT \ln a_-$$

$$\mu_2 = \mu_2^\circ + RT \ln a_2.$$

Substitution into equation (6.118) gives

$$\Delta_r G = (\mu_+^\circ + \mu_-^\circ) - \mu_2^\circ + RT \ln \frac{a_+ a_-}{a_2}. \tag{6.119}$$

At equilibrium, $\Delta_r G = 0$, and equation (6.119) becomes

$$(\mu_+^\circ + \mu_-^\circ) - \mu_2^\circ = -RT \ln \frac{a_+ a_-}{a_2}, \tag{6.120}$$

where $a_+$, $a_-$, and $a_2$ are now the equilibrium activities in the solution. Equation (6.120) can be rewritten as

$$\Delta_r G^\circ = -RT \ln \frac{a_+ a_-}{a_2} \tag{6.121}$$

where $\Delta_r G^\circ$, the standard free energy change for the dissociation reaction, is given by

$$\Delta_r G^\circ = (\mu_+^\circ + \mu_-^\circ) - \mu_2^\circ. \tag{6.122}$$

Equation (6.121) can be solved for the activity ratio to give

$$\frac{a_+ a_-}{a_2} = \exp(-\Delta_r G^\circ / RT). \tag{6.123}$$

For a strong electrolyte such as HCl we assume that all of the electrolyte is dissociated to form the ionic species. In this case, it is appropriate to consider the chemical potential of the HCl as the sum of the chemical potentials of the dissociated species. That is[dd]

$$\mu_2^\circ = \mu_+^\circ + \mu_-^\circ. \tag{6.124}$$

This equation is the basis of the strong electrolyte standard state. Substitution into equation (6.122) gives $\Delta_r G^\circ = 0$, which when substituted into equation (6.123) gives

$$\frac{a_+ a_-}{a_2} = 1 \tag{6.125}$$

---

[dd] In Chapter 9, we will define the equilibrium constant $K$ for a chemical reaction represented by

$$\sum_i \nu_i A_i = 0$$

as

$$K = \prod_i a_i^{\nu_i}.$$

For reactions such as (6.115), $K$ becomes

$$K = \frac{a_+ a_-}{a_2}.$$

Comparing this equation with equation (6.125) demonstrates that $K = 1$ for the dissociation of an electrolyte when the strong electrolyte standard state is chosen.

or

$$a_2 = a_+ a_- . \tag{6.126}$$

Equation (6.126) is the starting point for relating activity to molality for a 1 : 1 electrolyte such as HCl. It leads to the correct limiting law expression as we will now see. We start by substituting equations (6.116) and (6.117) into equation (6.126) to obtain

$$\begin{aligned} a_2 &= (\gamma_+ m_+)(\gamma_- m_-) \\ &= (\gamma_+ \gamma_-) m^2 \end{aligned} \tag{6.127}$$

where $m_+ = m_- = m$.

It is not possible to experimentally measure values for $\gamma_+$ or $\gamma_-$. A solution cannot be formed that contains only the cation or the anion. Therefore, what we do measure is the geometrical mean, $\gamma_\pm$, given by

$$\gamma_\pm = (\gamma_+ \gamma_-)^{1/2} \tag{6.128}$$

or

$$\gamma_\pm^2 = \gamma_+ \gamma_- . \tag{6.129}$$

Substitution of equation (6.129) into equation (6.127) gives

$$a_2 = \gamma_\pm^2 m^2 . \tag{6.130}$$

It is easy to show that equation (6.130) leads to the correct limiting law expression. With

$$a_2 = \frac{f_2}{k_2} = \gamma_\pm^2 m^2 ,$$

in the limit as $m \to 0$, $\gamma_\pm \to 1$, and $a_2 \to m^2$. Hence,

$$\frac{f_2}{k_2} = m^2 .$$

Other equations can be derived for 1 : 1 electrolytes that relate mean ionic quantities. We define the mean ionic activity as

$$a_\pm = (a_+ a_-)^{1/2} ; \tag{6.131}$$

the mean ionic molality as

$$m_{\pm} = (m_{+}m_{-})^{1/2}; \tag{6.132}$$

and combine these equations with equations (6.126) and (6.130) to obtain

$$a_{\pm} = \gamma_{\pm}m_{\pm}. \tag{6.133}$$

In summary, for a 1 : 1 strong electrolyte

limiting law: $f_2 = km^2$

activity: $a_2 = \gamma_{\pm}^2 m^2$

mean ionic activity: $a_{\pm} = \gamma_{\pm}m_{\pm}$

with $a_{\pm} = (a_{+}a_{-})^{1/2}$

$\gamma_{\pm} = (\gamma_{+}\gamma_{-})^{1/2}$

$m_{\pm} = (m_{+}m_{-})^{1/2}$.

### 6.4b Activities of Strong Unsymmetrical Electrolytes

So far we have considered only symmetrical 1 : 1 electrolytes such as HCl, KCl, or $MgSO_4$. For unsymmetrical electrolytes, the limiting law takes a different form, and different relationships between activity, molality and activity coefficient are obtained. For example, for the 2 : 1 electrolyte, $Na_2SO_4$, the dissociation reaction is

$$Na_2SO_4\ (aq) = 2Na^{+}(aq) + SO_4^{2-}(aq). \tag{6.134}$$

Experimental measurements show that the limiting law is

$$f_2 = k_3 m^3. \tag{6.135}$$

The standard state is chosen as a hypothetical 1 molar equation that obeys equation (6.135), in which case

$$f_2^{\circ} = k_3,$$

$$a_2 = \frac{f_2}{k_3}, \tag{6.136}$$

and

$$a_2 = a_+^2 a_-. \tag{6.137}$$

The activity coefficients are given by

$$a_+ = \gamma_+ m_+$$

$$a_- = \gamma_- m_-$$

so that

$$a_2 = \gamma_+^2 m_+^2 \gamma_- m_-.$$

The geometrical means are given by

$$\gamma_\pm = (\gamma_+^2 \gamma_-)^{1/3} \tag{6.138}$$

$$m_\pm = (m_+^2 m_-)^{1/3}. \tag{6.139}$$

But $m_+ = 2m$ and $m_- = m$ so that

$$m_\pm = 4^{1/3} m. \tag{6.140}$$

Combining all of this together gives

$$a_2 = 4\gamma_\pm^3 m^3. \tag{6.141}$$

It is also easy to show that once again

$$a_\pm = \gamma_\pm m_\pm.$$

In general, for an electrolyte that dissociates as follows

$$M_{\nu^+} X_{\nu^-} = \nu_+ M^{z+} + \nu_- X^{z-} \tag{6.142}$$

the limiting law is given by

$$f_2 = k_\nu m^{(\nu_+ + \nu_-)} \tag{6.143}$$

**Table 6.2** Activity coefficient relationships for electrolyte solutions (single electrolyte)

| Limiting law | Nonelectrolyte Sucrose | 1 : 1 NaCl $MgSO_4$ | 2 : 1 or 1 : 2 $Na_2SO_4$ $CaCl_2$ | 3 : 1 or 1 : 3 $AlCl_3$ $Na_3PO_4$ | 3 : 2 or 2 : 3 $La_2(SO_4)_3$ $Ca_3(PO_4)_2$ | General $A_{\nu+} B_{\nu-}$ |
|---|---|---|---|---|---|---|
| $f_2 =$ | $km$ | $km^2$ | $km^3$ | $km^4$ | $km^5$ | $km^{(\nu_+ + \nu_-)}$ |
| $a_2 =$ | $a$ | $(a_+)(a_-)$ or $[\gamma_\pm^2 m^2]$ | $(a_+)^2(a_-)$ or $(a_+)(a_-)^2$ or $[4\gamma_\pm^3 m^3]$ | $(a_+)(a_-)^3$ or $(a_+)^3(a_-)$ or $[27\gamma_\pm^4 m^4]$ | $(a_+)^2(a_-)^3$ or $(a_+)^3(a_-)^2$ or $[108\gamma_\pm^5 m^5]$ | $(a_+)^{\nu+}(a_-)^{\nu-}$ or $[(\nu_+)^{\nu_+}(\nu_-)^{\nu_-}][\gamma_\pm m]^{(\nu_+ + \nu_-)}$ |
| $a_\pm =$ | — | $[(a_+)(a_-)]^{1/2}$ | $[(a_+)^2(a_-)]^{1/3}$ or $[(a_+)(a_-)^2]^{1/3}$ | $[(a_+)(a_-)^3]^{1/4}$ or $[(a_+)^3(a_-)]^{1/4}$ | $[(a_+)^2(a_-)^3]^{1/5}$ or $[(a_+)^3(a_-)^2]^{1/5}$ | $[(a_+)^{\nu+}(a_-)^{\nu-}]^{1/(\nu_+ + \nu_-)}$ |
| $m_\pm =$ | — | $m$ | $2^{2/3}m$ | $3^{3/4}m$ | $4^{4/5}m$ | $[(\nu_+)^{\nu+}(\nu_-)^{\nu-}]^{1/(\nu_+ + \nu_-)}m$ |
| $\gamma_\pm =$ | — | $\frac{a_\pm}{m_\pm}$ | $\frac{a_\pm}{m_\pm}$ | $\frac{a_\pm}{m_\pm}$ | $\frac{a_\pm}{m_\pm}$ | $\frac{a_\pm}{m_\pm}$ |

with

$$a_2 = \frac{f_2}{k_\nu} \tag{6.144}$$

so that

$$a_2 = (a_+)^{\nu+}(a_-)^{\nu-} \tag{6.145}$$

$$\gamma_\pm = [(\gamma_+)^{\nu+}(\gamma_-)^{\nu-}]^{1/\nu} \tag{6.146}$$

$$\begin{aligned} m_\pm &= [(m_+)^{\nu+}(m_-)^{\nu-}]^{1/\nu} \\ &= [(\nu_+ m)^{\nu+}(\nu_- m)^{\nu-}]^{1/\nu} \\ &= m[(\nu_+)^{\nu+}(\nu_-)^{\nu-}]^{1/\nu} \end{aligned} \tag{6.147}$$

$$a_\pm = [(a_+)^{\nu+}(a_-)^{\nu-}]^{1/\nu} \tag{6.148}$$

$$a_2 = \gamma_\pm^\nu m^\nu [(\nu_+)^{\nu+}(\nu_-)^{\nu-}] \tag{6.149}$$

$$a_\pm = \gamma_\pm m_\pm \tag{6.150}$$

where

$$\nu = \nu_+ + \nu_-. \tag{6.151}$$

Table 6.2 summarizes the activity relationships for different types of strong electrolytes.

## 6.5 Determination Of Activity

A number of techniques are employed to determine activities or activity coefficients, depending upon the nature and type of the system. Often, procedures are used that are very specific for the system being studied. We will describe several commonly encountered examples.

### 6.5a Activity from Vapor Pressure Measurements

For a mixture containing volatile components, the activities of both components (usually based on a Raoult's law standard state) can be obtained by measuring the total vapor pressure $p$ above the mixture, the composition of the mixture in the liquid phase ($x_i$) and in the vapor phase ($y_i$), and the vapor pressures of the pure components. The vapor pressure of an individual component in the mixture is obtained from[ee]

$$p_i = y_i p. \tag{6.152}$$

Although the difference is small, $p_i$ and $p_i^*$ are usually corrected to $f_i$ and $f_i^*$ using an equation of state, and $\gamma_{R,i}$ is obtained from

$$a_i = \gamma_{R,i} x_i = \frac{f_i}{f_i^*}. \tag{6.153}$$

## 6.5b Activities from Freezing Point and Boiling Point Measurements

The activity of a solvent can be determined in a solution from the change in the freezing temperature or the boiling temperature due to the addition of a solute. Consider a process in which a solution is cooled until solid (component 1) crystallizes from solution. An equilibrium is established so that

$$\mu_1 = \mu_1(\text{s}) \tag{6.154}$$

where $\mu_1$ and $\mu_1(\text{s})$ are the chemical potentials of component 1 (the solvent) in the solution and in the solid phases, respectively. Often, the solid that crystallizes from solution is pure so that $\mu_1(\text{s}) = \mu_1^*(\text{s})$, and the condition for equilibrium becomes

$$\mu_1 = \mu_1^*(\text{s}). \tag{6.155}$$

---

[ee] With the aid of the computer, $a_i$ can be obtained from total pressure measurements without the necessity of sampling the liquid and vapor phases to determine composition, if accurate total vapor pressure data are available. Known quantities of the two components are mixed in a cell of known volume and the total vapor pressure is determined over the composition range. An iterative method is then used that minimizes the deviations of the individual pressures from the total while maintaining thermodynamic consistency for the results, corrects the composition for evaporation into the vapor phase, and corrects vapor pressures to vapor fugacities. Ewing and Marsh {M. B. Ewing and K. N. Marsh, *J. Chem. Thermodyn.*, **5**, 651–657 (1973)} have detailed the procedure to be used to calculate $\gamma_{R,i}$ from the total pressure measurements, based on a process originally proposed by Barker {J. A. Barker, *Aust. J. Chem.*, **6**, 207–210 (1953)}.

This procedure is commonly used to calculate vapor pressures and activities for volatile mixtures. For example, it was used to determine the vapor pressures for the (ethanol + water) system shown in Figure 6.7.

The chemical potential in the solution is related to the activity of the solvent $a_1$ by

$$\mu_1 = \mu_1^\circ + RT \ln a_1. \tag{6.156}$$

A Raoult's law standard state is chosen for the solvent so that $\mu_1^\circ = \mu_1^*$. Substituting this equality and equation (6.155) into equation (6.156) and rearranging gives

$$R \ln a_1 = \frac{\mu_1^*(\mathrm{s})}{T} - \frac{\mu_1^*}{T}. \tag{6.157}$$

To determine the effect of temperature on $a_1$, we differentiate equation (6.157) to obtain

$$R\left(\frac{\partial \ln a_1}{\partial T}\right)_{p,x} - \left(\frac{\partial[\mu_1^*(\mathrm{s})/T]}{\partial T}\right)_p \left(\frac{\partial[\mu_1^*/T]}{\partial T}\right)_p \tag{6.158}$$

$$= \frac{-H_1^*(\mathrm{s})}{T^2} + \frac{H_1^*}{T^2}, \tag{6.159}$$

or

$$\left(\frac{\partial \ln a_1}{\partial T}\right)_{p,x} = \frac{H_1^* - H_1^*(\mathrm{s})}{RT^2}. \tag{6.160}$$

The difference $H_1^* - H_1^*(\mathrm{s})$ is the enthalpy of fusion of the pure solvent $\Delta_{\mathrm{fus}} H_{\mathrm{m},1}$, and equation (6.160) becomes

$$\left(\frac{\partial \ln a_1}{\partial T}\right)_{p,x} = \frac{\Delta_{\mathrm{fus}} H_{\mathrm{m},1}}{RT^2}. \tag{6.161}$$

We must remember that $T$ in equation (6.161) is the equilibrium melting temperature. Integration of this equation will give an equation that relates melting temperature to activity. Separating variables and integrating gives

$$\int_{\ln a_1 = 0}^{\ln a_1} \mathrm{d} \ln a_1 = \int_{T_1^*}^{T_1} \frac{\Delta_{\mathrm{fus}} H_{\mathrm{m},1}}{RT^2} \mathrm{d}T.$$

We have chosen the pure solvent as the lower limit of the integration. Under this condition $a_1 = 1$ (Raoult's law standard state), and $T = T_1^*$, the melting temperature of the pure solvent. Integration of the left side of the equation gives

$$R \ln a_1 = \int_{T_1^*}^{T_1} \frac{\Delta_{\mathrm{fus}} H_{\mathrm{m},1}}{T^2} \mathrm{d}T. \tag{6.162}$$

To integrate the right side of the equation, we must know how $\Delta_{\mathrm{fus}} H_{\mathrm{m},1}$ changes with temperature. To find this temperature relationship, we write

$$\left( \frac{\partial \Delta_{\mathrm{fus}} H_{\mathrm{m},1}}{\partial T} \right)_p = \Delta_{\mathrm{fus}} C_{p,\mathrm{m},1}. \tag{6.163}$$

We have seen earlier that $C_{p,\mathrm{m},1}$ can be expressed as a function of temperature by an equation of the type

$$C_{p,\mathrm{m},1}(\mathrm{l}) = a + bT + cT^{-2} + dT^2 + \cdots$$

$$C_{p,\mathrm{m},1}(\mathrm{s}) = a' + b'T + c'T^{-2} + d'T^2 + \cdots$$

so that

$$\Delta_{\mathrm{fus}} C_{p,\mathrm{m},1} = \Delta a + \Delta bT + \Delta cT^{-2} + \Delta\ dT^2 + \cdots, \tag{6.164}$$

where $\Delta a = a - a'$, $\Delta b = b - b'$ $\Delta c = c - c'$, ... As many coefficients are included as are justified by the accuracy of the $C_{p,\mathrm{m},1}$ data, or as are needed to give the desired accuracy in the activity equation. Substitution of equation (6.164) into equation (6.163) and integrating as an indefinite integral gives

$$\Delta_{\mathrm{fus}} H_{\mathrm{m},1} = \Delta H_1 + \Delta aT + \frac{\Delta b}{2} T^2 - \Delta c \left( \frac{1}{T} \right) + \frac{\Delta d}{3} T^3 + \cdots \tag{6.165}$$

where $\Delta H_{\mathrm{I}}$ is a constant of integration that can be obtained by substituting into equation (6.165), the known value of $\Delta_{\mathrm{fus}}H_{\mathrm{m,1}}$ at $T_1^*$, the melting temperature of the pure solvent. Substitution of equation (6.165) into equation (6.162) and integrating gives[ff]

$$R \ln a_1 = \Delta H_{\mathrm{I}}\left(\frac{1}{T^*} - \frac{1}{T}\right) + \Delta a \ln \frac{T}{T^*} + \frac{\Delta b}{2}(T - T^*) + \frac{\Delta c}{2}\left(\frac{1}{(T)^2} - \frac{1}{(T^*)^2}\right) + \frac{\Delta d}{6}[(T)^2 - (T^*)^2] + \cdots \tag{6.166}$$

To determine the activity coefficient, $\gamma_{\mathrm{R,1}}$, one measures the melting temperature, $T$, of a solution with mole fraction, $x_1$, and calculates $a_1$ from equation (6.166). The activity coefficient is then obtained from

$$\gamma_{\mathrm{R,1}} = \frac{a_1}{x_1}.$$

This procedure gives $\gamma_{\mathrm{R,1}}$ at the melting temperature of the solution, a temperature that varies with $x_1$. These activity coefficients are usually corrected to a common reference temperature such as 298.15 K by integrating equation (6.90) derived earlier.

$$\left(\frac{\partial \ln a_1}{\partial T}\right)_p = -\frac{\overline{L}_1}{RT^2} \tag{6.90}$$

where $\overline{L}_1 = \overline{H}_1 - \overline{H}_1^*$ is the relative partial molar enthalpy (with a Raoult's law standard state). In Section 5.5c of Chapter 5, we briefly described a procedure that can be used for calculating $\overline{L}_1$ from excess enthalpy values.[15] In Chapter 7, we will return to a more detailed description of procedures for calculating and applying $\overline{L}_i$.

---

[ff] For ideal solutions, $a_1 = x_1$, and equation (6.166) can be used to calculate the freezing points, and hence, the (solid + liquid) phase diagram for an ideal solution. We will return to this calculation in Chapter 8.

Freezing point methods are often applied to the measurement of activities of electrolytes in dilute aqueous solution because the freezing point lowering, $\theta = T_1 - T_1^*$, can be determined with high accuracy, and the solute does not dissolve in the solid to any appreciable extent. Equations can be derived[gg] relating $a_1$ to $\theta$ instead of $T_1$ and $T_1^*$. The detailed expressions can be found in the literature.[16]

Activities can also be calculated from boiling point measurements. Either pressure is held constant and the boiling temperature is determined, or the boiling temperature is held constant and the total pressure is measured. In both instances, the compositions of the distillate and condensate are determined. From this information, partial pressures and hence, activities can be calculated at the boiling temperature in a manner similar to that described earlier for determining activities from total vapor pressure measurements. As an example, Scatchard and coworkers assembled a (vapor + liquid) distillation apparatus and measured the pressure required to keep mixtures boiling at a constant temperature.[17] They used the apparatus to measure vapor pressures, and hence, activities for a number of mixtures, including the (trichloroethane + ethanol)[18] system. The vapor pressures and activities at $T = 328.15$ K that they obtained for this system are shown in Figure 6.16.

Boiling point measurements of sufficient accuracy to obtain reliable activities are not easy to make. It is difficult to ensure that equilibrium conditions are achieved in the still. As a result, boiling point measurements, unlike freezing point measurements, are not often used to determine these quantities.

## 6.5c Activity from Isopiestic Methods

The activity of a volatile solvent in a solution that contains a nonvolatile solute can be obtained from an experimental technique known as the "**isopiestic method**".[19] An apparatus is constructed similar to that shown in Figure 6.17. The mixture in container A is a solution of a nonvolatile solute in a solvent in which $a_1$, the activity of the solvent, has been accurately determined in other experiments as a function of concentration. Containers B and C hold solutions of other nonvolatile solutes in the same solvent. These are the solutions for which the activity of the solvent is to be determined.

The solutions are placed in a thermostated chamber, which is then evacuated and left for several days. Solvent evaporates from some solutions and condenses in others until all of the solutions are in equilibrium. At this point, the vapor pressure of the solvent is the same above all the solutions.[hh]

---

[gg] In the next section we will see how to calculate $a_2$, the activity of the solute from $a_1$.

[hh] Care must be taken to ensure equilibrium. Several days are usually required to be sure that this happens. Stirring or rocking the solutions often helps.

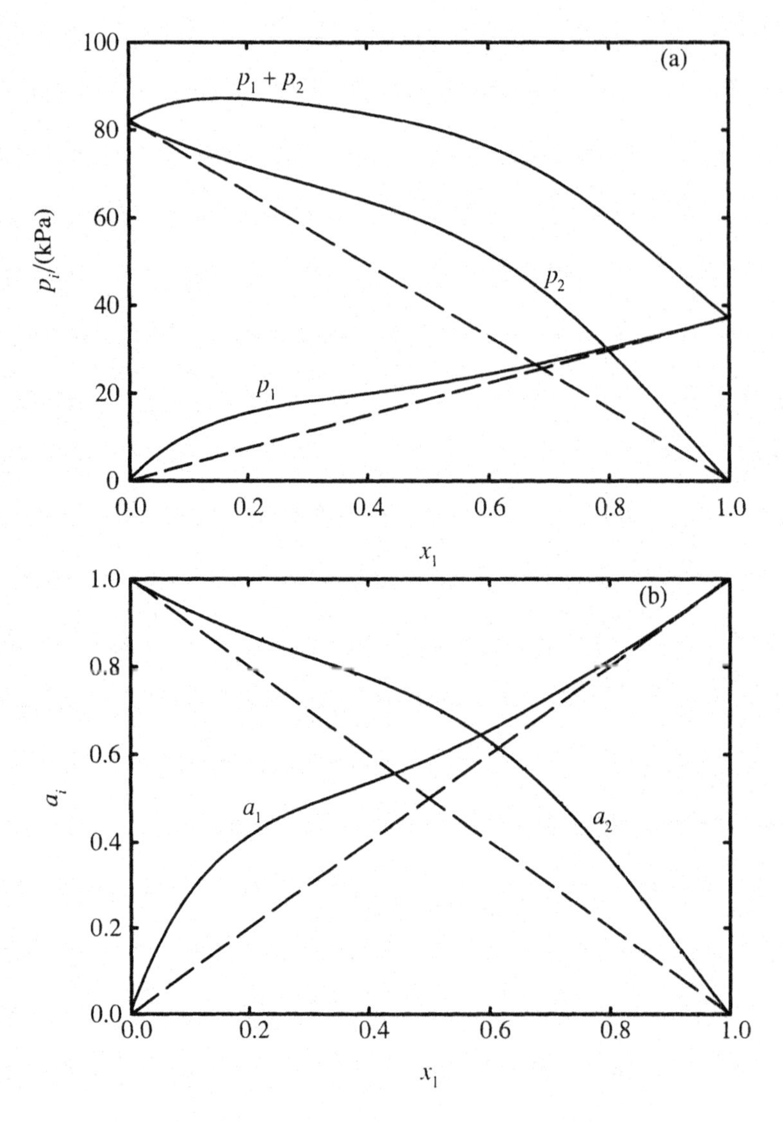


**Figure 6.16** (a) Vapor pressures and (b), activities for $(x_1C_2H_5OH + x_2CHCl_3)$ at $T = 328.15$ K. The solid lines represent the experimental results; the dashed lines are those predicted by Raoult's Law.

The apparatus is then disassembled, and the composition of each solution is determined. From the final concentration in A, the solvent activity can be determined in that mixture. Since the solvent vapor pressures above all the solutions are the same, so are the solvent activities.

The isopiestic method is often applied to electrolyte solutions, since volatility of the solute is not a problem. Sulfuric acid is often used as the

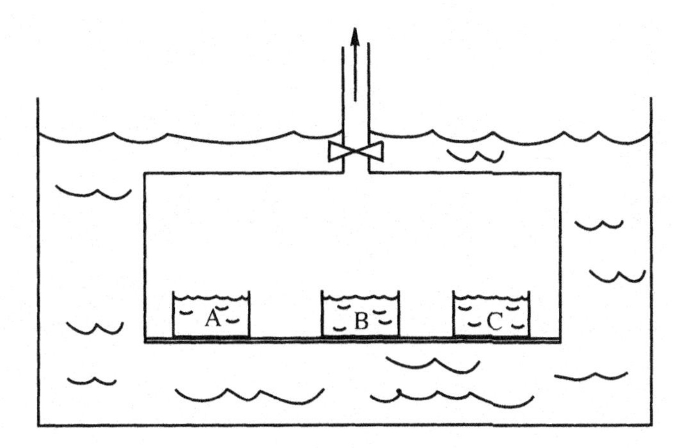


**Figure 6.17** The isopiestic apparatus.

reference solution (A), since its composition is easily determined, and the activity of the solvent (water) is accurately known as a function of concentration.

### 6.5d Solute Activities From Measurement of Partition Coefficients

The isopiestic method is based upon the equality of the solvent chemical potentials and fugacities when solutions of different solutes, but the same solvent, are allowed to come to equilibrium together. A method in which a solute is allowed to establish an equilibrium distribution between two solvents has also been developed to determine activities of the solute, usually based on the Henry's law standard state. In this case, one brings together two immiscible solvents, A and B, adds a solute, and shakes the mixture to obtain two phases that are in equilibrium, a solution of the solute in A with composition $x_2^{\mathrm{A}}$, and a solution of the solute in B with composition, $x_2^{\mathrm{B}}$.

From the equilibrium condition, we know that the chemical potentials $\mu_2^{\mathrm{A}}$ and $\mu_2^{\mathrm{B}}$ of the solute in the two solutions are equal. Since the chemical potentials are equal, the fugacities are also equal,

$$f_2^{\mathrm{A}} = f_2^{\mathrm{B}}, \tag{6.167}$$

but the activities are not because the Henry's law constants of the solute in the two solvents have different values. That is: $k_{\mathrm{H},x}^{\mathrm{A}} \neq k_{\mathrm{H},x}^{\mathrm{B}}$. However, there is a relationship between the two activities based on the ratio of the two constants.

If one multiplies and divides each side of equation (6.167) by the respective Henry's law constant, one can obtain, after some rearrangement

$$\frac{f_2^{\mathrm{A}}}{k_{\mathrm{H},x}^{\mathrm{A}}} = \frac{f_2^{\mathrm{B}}}{k_{\mathrm{H},x}^{\mathrm{B}}} \cdot \frac{k_{\mathrm{H},x}^{\mathrm{B}}}{k_{\mathrm{H},x}^{\mathrm{A}}}. \tag{6.168}$$

The activity of the solute in the two solvents is given by the respective fugacity over the Henry's law constant ratios given above, so that equation (6.168) becomes

$$a_2^{\mathrm{A}} = a_2^{\mathrm{B}} \cdot \frac{k_{\mathrm{H},x}^{\mathrm{B}}}{k_{\mathrm{H},x}^{\mathrm{A}}}$$

or

$$a_2^{\mathrm{A}} = k a_2^{\mathrm{B}} \tag{6.169}$$

where $k$ is the ratio of the two Henry's law constants. This ratio is often referred to as the distribution or partition coefficient.

In solutions that are dilute enough to follow Henry's law, $a_2^{\mathrm{A}} = x_2^{\mathrm{A}}$ and $a_2^{\mathrm{B}} = x_2^{\mathrm{B}}$ so that equation (6.169) becomes

$$x_2^{\mathrm{A}} = k \cdot x_2^{\mathrm{B}}. \tag{6.170}$$

That is, at low solute concentrations, the ratio of the concentrations of the solutes in the two solutions is constant.[ii] Thus, measurement of the solution compositions in equilibrium with one another at low concentrations where both solutions obey Henry's law yields a value of $k$. With the value of $k$ determined, one can obtain the activities of the solute in solvent A, if the activities of the solute in solvent B are known.[jj]

An example of this process is the study performed by Lewis and Storch[20] in 1917 to determine the activity of $Br_2(2)$ in aqueous solutions. First, they

---

[ii] Concentrations can be expressed in molalities (or molarities) to obtain equivalent expressions for equation (6.170). The result is expressions such as $m^{\mathrm{A}} = km^{\mathrm{B}}$ or $x_2^{\mathrm{A}} = km^{\mathrm{B}}$.

[jj] Strictly speaking, equation (6.170) applies to saturated solutions of A in B and B in A (rather than a solute distributed between pure A and pure B). Usually the difference is not important if the mutual solubilities are small, but occasionally a small amount of impurity can significantly affect the vapor fugacity of the solute. The result would be a difference between the activity of the solute in pure solvent and in the saturated mixture.

measured the vapor pressure of $Br_2$ in dilute solutions ($0.0025 \leqslant x_2 \leqslant 0.004$) of tetrachloromethane (1). They found that the ratio, $p_2/x_2$, was constant (within experimental error) at 54.7 kPa over this entire composition range. The constancy of the ratio indicates that $Br_2$ obeys Henry's law in this region, and $k^{l}_{H,x} = 54.7$ kPa. Lewis and Storch then equilibrated small amounts of $Br_2$ between tetrachloromethane and water (slightly acidified to prevent hydrolysis of the $Br_2$). These experiments showed that $m$, the molality of $Br_2$ in the water phase, divided by $x_2$, the mole fraction of $Br_2$ in tetrachloromethane was constant at 0.371. This value was taken as the distribution coefficient. Therefore,

$$a_2^{w} = 0.371 a_2^{l}. \tag{6.171}$$

where $a_2^{w}$ is the activity of $Br_2$ in the water phase and $a_2^{l}$ is the activity of $Br_2$ in the tetrachloromethane phase.[kk] In fact, the concentrations of $Br_2$ found in the tetrachloromethane solutions equilibrated with water were low enough to follow Henry's law, so that $a_2^{l} = x_2^{l}$. From measurements of the concentrations of $Br_2$ in the two solutions, Lewis and Storch were able to obtain the activities of the $Br_2$ in the water solutions, employing equation (6.171).

### 6.5e Calculation of the Activity of One Component From That of the Other

Most of the methods we have described so far give the activity of the solvent. Often the activity of the solute is of equal or greater importance. This is especially true of electrolyte solutions where the activity of the ionic solute is of primary interest, and in Chapter 9, we will describe methods that employ electrochemical cells to obtain ionic activities directly. We will conclude this chapter with a discussion of methods based on the Gibbs–Duhem equation that allow one to calculate activities of one component if the activities of the other are known as a function of composition.

In equation (5.27), we used the Gibbs–Duhem equation to relate changes in the chemical potentials of the two components in a binary system as the composition is changed at constant temperature and pressure. The relationship is

$$n_1\, d\mu_1 + n_2\, d\mu_2 = 0. \tag{5.27}$$

---

[kk] Note that the activity in the water phase is expressed in terms of a Henry's law molality standard state while the activity in the tetrachloromethane phase is expressed in terms of a Henry's law mole fraction standard state.

Dividing by the total number of moles, $(n_1 + n_2)$, expresses the Gibbs–Duhem equation in terms of mole fractions:

$$x_1 \, d\mu_1 + x_2 \, d\mu_2 = 0. \tag{6.172}$$

At constant $T$ and $p$, $d\mu_i = RT \, d \ln a_i$ since the chemical potential of the standard state is a constant, and equation (6.172) can be written as

$$x_1 \, d \ln a_1 + x_2 \, d \ln a_2 = 0. \tag{6.173}$$

A similar relationship relating the two activity coefficients can also be derived. We defined activity coefficients such that $a_1 = \gamma_1 x_1$ and $a_2 = \gamma_2 x_2$ where the activities and activity coefficients are established for the standard state that corresponds to $\mu_1^\circ$ and $\mu_2^\circ$, respectively. For both components, changes in the activity at constant temperature and pressure are given by

$$\begin{aligned} d \ln a_i &= d \ln \gamma_i x_i \\ &= d \ln \gamma_i + d \ln x_i. \end{aligned}$$

Substitution into equation (6.173) gives

$$x_1(d \ln x_1 + d \ln \gamma_1) + x_2(d \ln x_2 + d \ln \gamma_2) = 0. \tag{6.174}$$

But

$$\begin{aligned} x_1 \, d \ln x_1 + x_2 \, d \ln x_2 &= x_1 \frac{1}{x_1} dx_1 + x_2 \frac{1}{x_2} dx_2 \\ &= dx_1 + dx_2 \\ &= 0, \end{aligned}$$

since $x_1 = 1 - x_2$ and $dx_1 = -dx_2$. As a result, equation (6.174) becomes

$$x_1 \, d \ln \gamma_1 + x_2 \, d \ln \gamma_2 = 0, \tag{6.175}$$

which is of the same form as equation (6.173). It can be used to calculate the activity coefficient of one component if the activity coefficient of the other is known as a function of composition.

To make the calculation, we write

$$d \ln \gamma_2 = -\frac{x_1}{x_2} d \ln \gamma_1. \tag{6.176}$$

Integration of equation (6.176) gives

$$\int_{\ln \gamma_2}^{\ln \gamma_2'} d \ln \gamma_2 = -\int_{\ln \gamma_1}^{\ln \gamma_1'} \frac{x_1}{x_2} d \ln \gamma_1, \tag{6.177}$$

where $\gamma_1$ is the activity coefficient at the mole fraction ratio $x_1/x_2$, while $\gamma_1'$ is the activity coefficient at the mole fraction ratio $x_1'/x_2'$. Application of this equation requires a careful consideration of the limiting behavior of the activity coefficients associated with the standard states chosen for each component. When a Raoult's law standard state is chosen for both components, $\gamma_2 \rightarrow 1$ and $\ln \gamma_2 \rightarrow 0$ as $x_2 \rightarrow 1$. At this limit, $x_1 \rightarrow 0$, $x_1/x_2 = 0$, and $\ln \gamma_1$ is finite. Then, the integration corresponds to a change between two states, $x_2 = 1$ and a value of $x_2$, and the result can be represented as

$$\ln \gamma_{R,2} = -\int_{\ln \gamma_{R,1}}^{\ln \gamma_{R,1}'} \frac{x_1}{x_2} d \ln \gamma_1. \tag{6.178}$$

The value of the integral in equation (6.178) can be obtained by a graphical integration of a plot of $x_1/x_2$ against $\ln \gamma_{R,1}$. Figure 6.18(a) shows such a plot for the $\{x_1H_2O + x_2C_2H_5OH\}$ system at $T = 303.15$. Figure 6.18(b) shows this same plot with an expanded ordinate scale. The area marked "Area (1)" in Figure 6.18(b) gives the value of the integral in equation (6.178) with the integral taken from an initial point ($x_1/x_2 = 0$ and $\ln \gamma_{R,1} = 0.890$) to the final point ($x_1/x_2 = 1$ and $\ln \gamma_{R,1} = 0.382$). Its magnitude yields the value of $\ln \gamma_{R,2}$ for a solution with $x_2 = 0.5$. That is, "Area (1)" gives the activity coefficient for component 2 assuming a Raoult's law standard state.

Note, however, that as $\ln \gamma_{R,1} \rightarrow 0$, the value of $x_1/x_2 \rightarrow \infty$ since $x_2 \rightarrow 0$ at the same time. This integration method works well for determining the activity coefficient of species 2 with a Raoult's law standard state, but it poses difficulties when the integration must be extended to very dilute solutions of component 2 in component 1, as must be done when a Henry's law standard state is chosen for component 2.

If one wishes to determine activities for component 2 with a Henry's law standard state, the lower limit of the right-hand integral of equation (6.177) goes to 0 as $x_2 \rightarrow 0$. Thus, it would appear that a graphical integration could be performed by obtaining the area starting from an abscissa of $\ln \gamma_{R,1} = 0$ to the

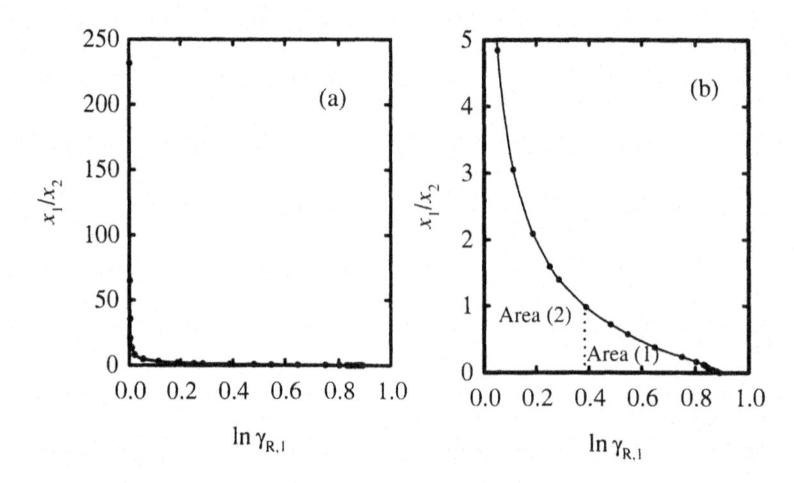


**Figure 6.18** Graphs of $\ln \gamma_{R,1}$ against $x_1/x_2$ for $\{x_1H_2O + x_2C_2H_5OH\}$ at $T = 303.15$ K. Graph (b) shows an expansion of the ordinate in graph (a) to better show the area under the curve, which can be used to calculate $\ln \gamma_2$ with either a Raoult's law standard state [Area (1)] or a Henry's law standard state [Area (2)].

value at the desired solution concentration. This corresponds to the area marked "Area (2)" in Figure 6.18(b). However, as we have noted above, the ordinate $x_1/x_2$ goes to infinity as $\ln \gamma_{R,1} \rightarrow 0$; that is as $x_2 \rightarrow 0$. This area can be shown to be finite, but it is difficult to measure accurately since $x_1/x_2 \rightarrow \infty$.

This difficulty can be overcome by writing equivalent expressions whose variables do not go to infinity at the limit of $x_2 \rightarrow 0$. A function known as the practical osmotic coefficient is one that can be used in a graphical method to obtain $\ln \gamma_{H,2}$. The practical osmotic coefficient expressed in terms of mole fraction is defined as

$$\phi = -\ln \frac{a_1}{r} \tag{6.179}$$

where $r = x_2/x_1$. This equation is differentiated to obtain

$$d\phi = -\frac{1}{r} \, d \ln a_1 + \frac{1}{r^2} \ln a_1 \, dr. \tag{6.180}$$

Solving for $d \ln a_1$ and substituting into equation (6.173) yields, after some rearrangement,

$$d \ln a_2 = d\phi + \phi \, d \ln r.$$

We then subtract $d \ln r$ from each side and integrate from $r = 0$ to some final composition $r$ to get

$$\ln \frac{a_2}{r} = \phi(r) - \phi(0) + \int_0^r \frac{\phi - 1}{r} \, dr. \tag{6.181}$$

When values of $\phi$ are known accurately enough in the dilute region to allow $\phi(0)$ to be obtained by extrapolation, a graphical evaluation of the integral gives $\ln a_2/r$, and thus $a_2$.

A form of equation (6.181) using the molality $m$ instead of $x_2$ is often used. The osmotic coefficient expressed in terms of molality is given by

$$\phi = -\frac{1}{M_1 \nu m} \ln a_1, \tag{6.182}$$

where $M_1$ is the molecular weight of the solvent in $kg{\cdot}mol^{-1}$ and $\nu$ is the total number of ions if the solute is a strong electrolyte. ($\nu = 1$ for a nonelectrolyte.) Starting with this equation and using a similar derivation, one obtains the relationship

$$\begin{aligned} \ln \gamma_{H,m} &= \ln \frac{a_2}{m} \\ &= \phi - 1 + \int_0^m \frac{\phi - 1}{m} \, dm. \end{aligned} \tag{6.183}$$

This expression is often used to determine activity coefficients ($\gamma_\pm$) of the solute in electrolyte solutions.

In a different method that applies equation (6.177), the integrations can be made between finite limits where the activity coefficients are $\gamma^{i}_{H,x_2}$ and $\gamma^{f}_{H,x_2}$. The result is

$$\ln \frac{\gamma^{f}_{H,x_2}}{\gamma^{i}_{H,x_2}} = -\int_{\ln \gamma^{i}_{R,1}}^{\ln \gamma^{f}_{R,1}} \frac{x_1}{x_2} \, d \ln \gamma_{R,1}. \tag{6.184}$$

If the same initial solution with an activity coefficient $\gamma^{i}_{H,x_2}$ is used, and the final (upper) limit varied, values for the activity coefficient ratio on the left can be obtained as a function of $x^{f}_{2}$. Extrapolation of this quantity to $x_2 = 0$ will allow the determination of $\gamma^{\mathrm{i}}_{\mathrm{H},x_2}$ (as $1/\gamma^{\mathrm{i}}_{\mathrm{H},x_2}$) at the intercept.[||] Once this is done $\gamma^{\mathrm{f}}_{\mathrm{H},x^2}$, in the various solutions can be calculated.

## Exercises

E6.1 Assume $O_2$ gas obeys the virial equation

$$pV_{\mathrm{m}} = RT + Bp$$

with $B = -12.5\ \mathrm{cm^3{\cdot}mol^{-1}}$ at 298.15 K, and calculate the fugacity of $O_2$ gas at $p = 10.0$ MPa pressure and $T = 298.15$ K.

E6.2 The fugacity of liquid water at 298.15 K is approximately 3.17 kPa. Take the ideal enthalpy of vaporization of water as 43,720 $\mathrm{J{\cdot}mol^{-1}}$, and calculate the fugacity of liquid water at 300 K.

E6.3 The second virial coefficient of methane is given by

$$B/(\mathrm{cm^3{\cdot}mol^{-1}}) = 460 - 88{,}000/T.$$

(a) Obtain an expression for the fugacity of methane as a function of temperature and pressure.
(b) Calculate $\Delta G_{\mathrm{m}}$ for the expansion of methane at $T = 175$ K from $p = 4.0$ MPa to $p = 0.2$ MPa.
(c) Determine $H^{\ddagger}_{\mathrm{m}} - H_{\mathrm{m}}$ for methane gas at $T = 175$ K and $p = 4.0$ MPa.

E6.4 If the difference between the free energies of $H_2O$ at $p = 100$ MPa and $T = 1073.15$ K, and at $p = 0.001$ MPa and $T = 1073.15$ K is 101.186 $\mathrm{kJ{\cdot}mol^{-1}}$, calculate the fugacity and fugacity coefficient of $H_2O$ at $p = 100$ MPa and $T = 1073.15$ K.

E6.5 At $T = 298.15$ K, the activities of mercury (1) and tin (2) in amalgams are given as a function of mole fraction $x$ by the van Laar expressions

$$\ln \frac{a_1}{x_1} = \frac{0.51}{\left(1 + 0.26\,\dfrac{x_1^2}{x_2}\right)}$$

[||] This method again requires accurate values of $\gamma_{\mathrm{R},1}$ in dilute solution so that reliable extrapolation can be obtained.

$$\ln \frac{a_2}{x_2} = \frac{0.13}{\left(\frac{x_2}{x_1} + 0.26\right)^2}$$

(a) Prove that the standard states for both tin and mercury are the pure metals.

(b) Calculate the pressure of mercury gas above a solution made by mixing one mole of Hg with one mole of Sn at $T = 298.15$ K. At this temperature, the vapor pressure of pure Hg is $2.25 \times 10^{-5}$ kPa.

E6.6 The partial pressure of $Br_2$ above a $(x_1CCl_4 + x_2Br_2)$ solution is 1.369 kPa. The composition of the solution is $x_2 = 0.0250$. The vapor pressure of pure bromine at the same temperature is 28.4 kPa. Assume a Raoult's law standard state for bromine and calculate the activity coefficient of $Br_2$ in the solution.

E6.7 Brønsted measured the solubility of solid orthorhombic and monoclinic sulfur at 298.15 K in various solvents (benzene, diethyl ether and ethanol). He found that the solubility of monoclinic sulfur is always 1.28 times that of orthorhombic sulfur. That is, the concentration of sulfur in the solution in equilibrium with monoclinic sulfur is always 1.28 times the concentration of sulfur in the solution in equilibrium with orthorhombic sulfur. Sulfur exists in these solvents exclusively in the form of $S_8$ molecules.

(a) Find $\Delta G_m$ for the transformation of monoclinic sulfur to orthorhombic sulfur at 298.15 K. (You may assume that the sulfur obeys Henry's law in the solutions.)

(b) Which of the two forms of sulfur is stable at 298.15 K? Why?

E6.8 A semi-empirical equation proposed by Margules expresses the vapor pressure $p_1$ of one of the components in a solution by

$$p_1 = p_1^* x_1 e^{\alpha x_2^2}$$

where $p_1^*$ is the vapor pressure of pure component 1, $x_1$, and $x_2$ are mole fractions, and $\alpha$ is a constant. If component 1 obeys the above equation, use the Gibbs–Duhem equation to prove that component 2 obeys the equation

$$p_2 = p_2^* x_2 e^{\alpha x_1^2}.$$

E6.9 At 373.15 K, the vapor pressure of normal water ($H_2O$) is 0.101 MPa (1.00 atm) while the vapor pressure of heavy water, ($D_2O$) is 0.0963 MPa

(D is deuterium, at. wt. = 2.01). A solution contains 0.10 moles of $D_2O$ and 0.90 moles of $H_2O$ at 373.15 K. Make reasonable assumptions to calculate
(a) the vapor pressure of the solution;
(b) the mole fraction of $D_2O$ in the vapor phase in equilibrium with this liquid solution.

E6.10 A 0.010 molal solution of cerous sulfate, $Ce_2(SO_4)_3$, has a mean ionic activity coefficient of 0.171 at 298.15 K. What are the values of $m_+$, $m_-$, $m_\pm$, $a_\pm$, and $a_2$?

E6.11 For a 1 : 1 electrolyte solute such as HCl, Henry's law is replaced the equation

$$f_2 = k_2 x_2^2$$

where $f_2$ and $x_2$ are the vapor fugacity and mole fraction of HCl. Use the Gibbs–Duhem equation to derive an equation to show how $f_1$, the vapor fugacity of the solvent, varies with $x_1$ when the solute obeys the above equation.

E6.12 The HCl pressure in equilibrium with a 1.20 molal solution is $5.15 \times 10^{-8}$ MPa and the mean ionic activity coefficient is known from emf measurements to be 0.842 at $T = 298.15$ K. Calculate the mean ionic activity coefficients of HCl in the following solutions from the given HCl pressures

| $m/(\text{mol·kg}^{-1})$ | 2.0 | 3.0 | 4.0 | 5.0 | 10.0 |
|---|---|---|---|---|---|
| $p/(\text{MPa})$ | $2.09 \times 10^{-7}$ | $7.88 \times 10^{-7}$ | $2.49 \times 10^{-6}$ | $7.10 \times 10^{-6}$ | $5.59 \times 10^{-4}$ |

E6.13 At $T = 298.15$ K, $\gamma_\pm = 0.025$ in a saturated solution of $CdCl_2$ where the molality is 6.62 mol·kg$^{-1}$. Calculate the minimum work necessary at $T = 298.15$ K to remove 1 mole of solid $CdCl_2$ from a large amount of solution where $a_2 = 0.0010$.

## Problems

P6.1 (a) Calculate the fugacity of $C_2H_4(g)$ at $T = 400$ K and each of the pressures given in the following table of densities that are measured at this temperature. A useful approach involves fitting the data to the virial equation

$$pV_m = RT + Bp + Cp^2 + Dp^3 + \cdots$$

and integrating using equation (6.19).

| $p$/(MPa) | $\rho$/(kg·m$^{-3}$) | $p$/(MPa) | $\rho$/(kg·mg$^{-3}$) |
|---|---|---|---|
| 0.10 | 0.8453 | 4.50 | 42.00 |
| 0.20 | 1.694 | 5.00 | 47.23 |
| 0.40 | 3.402 | 5.50 | 52.57 |
| 0.60 | 5.124 | 6.00 | 58.04 |
| 0.80 | 6.861 | 7.00 | 69.31 |
| 1.00 | 8.613 | 8.00 | 81.03 |
| 1.50 | 13.06 | 9.00 | 93.13 |
| 2.00 | 17.61 | 10.00 | 105.6 |
| 2.50 | 22.26 | 15.00 | 169.1 |
| 3.00 | 27.02 | 20.00 | 224.7 |
| 3.50 | 31.90 | 30.00 | 298.2 |
| 4.00 | 36.89 | 40.00 | 342.8 |

(b) Make a graph of $f$ against $p$ to show the pressures at which significant deviations from ideal behavior begin to occur.

(c) Make a graph of $\alpha = V_m^{\text{ideal}} - V_m$ using data in the table and use the intercept of this graph at $p = 0$ to obtain a value of the second virial coefficient $B$. Compare this value with $B$ obtained from the fit to the virial equation given in (a).

P6.2 The following are the vapor pressures at $T = 298.15$ K for $(x_1H_2O + x_2C_2H_5OH)$

| $x_2$ | $p_2$/(kPa) | $p_1$/(kPa) | $x_2$ | $p_2$/(kPa) | $p_1$/(kPa) |
|---|---|---|---|---|---|
| 0.00 | 0 | 3.169 | 0.60 | 5.363 | 2.070 |
| 0.02 | 0.571 | 3.108 | 0.70 | 5.858 | 1.755 |
| 0.05 | 1.328 | 3.022 | 0.80 | 6.432 | 1.318 |
| 0.08 | 1.978 | 2.942 | 0.90 | 7.126 | 0.717 |
| 0.10 | 2.353 | 2.893 | 0.93 | 7.352 | 0.510 |
| 0.20 | 3.603 | 2.700 | 0.96 | 7.582 | 0.297 |
| 0.30 | 4.164 | 2.578 | 0.98 | 7.736 | 0.151 |
| 0.40 | 4.524 | 2.466 | 1.00 | 7.893 | 0 |
| 0.50 | 4.914 | 2.305 | | | |

(a) Calculate the Henry's law constant for $C_2H_5OH$ as a solute in $H_2O$ as the solvent.

(b) Calculate the activity of ethanol in a solution with $x_2 = 0.25$, assuming a Henry's law standard state for the ethanol.

(c) Calculate the activity of water in a solution with $x_2 = 0.25$ assuming a Raoult's law standard state for the water.

P6.3 A young man, Horatio Le Chatelier Smith was involved in a traffic accident in which his automobile, traveling at 50 mph (80 $km \cdot h^{-1}$), became intimately associated with the pole which supported a mercury vapor street lamp. The wavelength of the brightest spectral line from the lamp was 435.835 nm. {Horatio's 3000 lb (1400 kg) car was red in color and was indeed a pitiful sight bent around the pole in the blue-violet light of the lamp.} A blood analysis indicated that there was more ethanol in Horatio's blood than would result from consumption of even very large quantities of bread made from stone ground flour. The total quantity of alcohol was estimated at 49 g. (Horatio was not of voting age but was able to name 800% more members of the Senate from the minority party than from the other party, which fact led some people to give special credence to the story he told.) We now quote Horatio: "I was asked — since I am a teetotaler — to act as a bartender at a party being held by some friends. I spent 3 hours breathing the fumes of the liquor and apparently inhaled enough alcohol to become intoxicated." (Many comments regarding this story were made by the policemen who were present.) The question is if ethanol ($C_2H_5OH$) and water formed an ideal solution (which they do not), if all the solutions opened during the 3-hour period had been 100 proof alcohol (50% water, 50% ethanol by volume; or 42.5% ethanol by weight), and if Horatio breathed 250 $cm^3$ of air every 4 s, could the fine lad have breathed enough ethanol in the time involved to have ingested the quantity estimated to be in his blood? The vapor pressure and density of pure ethanol are 8.0 kPa and 0.789 $g \cdot cm^{-3}$, respectively, at 293 K. The molecular weight of $C_2H_5OH$ is 0.046 $kg \cdot mol^{-1}$.

## References

1. Fugacities and fugacity coefficients in Figures 6.1 and 6.2 were taken from W. M. Haynes and R. D. Goodwin, "Thermophysical Properties of Normal Butane from 135 to 700 K at Pressure to 70 MPa", National Bureau of Standards Monograph 169, U.S. Government Printing Office, Washington, D.C., 1982.
2. See B. A. Younglove and J. F. Ely, "Thermophysical Properties of Fluids. II. Methane, Ethane, Propane, Isobutane, and Normal Butane", *J. Phys. Chem. Ref. Data*, **16**. No. 4, 577–798 (1987).
3. I. M. Klotz and R. M. Rosenburg, *Chemical Thermodynamics: Basic Theory and Methods*, Third Edition, W. A. Benjamin, Inc., Menlo Park, California, 1972, p. 247.
4. I. M. Klotz and R. M. Rosenburg, *Chemical Thermodynamics: Basic Theory and Methods*, Third Edition, W. A. Benjamin, Inc., Menlo Park, California, 1972, pp. 245–246.
5. G.-J. Su, "Modified Law of Corresponding States for Real Gases", *Ind. Eng. Chem.*, **38**, 803–806 (1946).

6. For more comprehensive calculations of the fugacity coefficients in mixtures, see J. M. Prausnitz, R. N. Lichtenthaler, and E. G. de Azevedo, *Molcular Thermodynamics of Fluid Phase Equilibria*, Prentice Hall, Englewood Cliffs, N.J., 1986, Chapter 5.
7. J. B. Ott, K. N. Marsh and R. H. Stokes, "Excess Enthalpies, Excess Gibbs Free Energies, and Excess Volumes for (Cyclohexane + *n*-Hexane), and Excess Gibbs Free Energies and Excess Volumes for (Cyclohexane + Methylcyclohexane) at 298.15 and 308.15 K", *J.Chem. Thermodyn.*, **12**, 1139–1148 (1980).
8. I. Brown and F. Smith, "Liquid–Vapor Equilibriums. VI. The Systems Acetonitrile–Benzene at 45° and Acetonitrile–Nitromethane at 60°", *Aust. J. Chem.*, 8, 62–67 (1955).
9. R. C. Pemberton and C. J. Mash, "Thermodynamic Properties of Aqueous Non-Electrolyte Mixtures II. Vapour Pressures and Excess Gibbs Energies for Water + Ethanol at 303.15 to 363.15 K Determined by an Accurate Static Method", *J. Chem. Thermodyn.*, **10**, 867–888 (1978).
10. Y. P. Handa, D. V. Fenby, and D. E. Jones, "Vapour Pressures of Triethylamine + Chloroform and of Triethylamine + Dichloromethane", *J. Chem. Thermodyn.*, **7**, 337–343 (1975).
11. J. R. Goates, J. B. Ott, and J. F. Moellmer, "Determination of Excess Volumes in Cyclohexane + Benzene and + *n*-Hexane with a Vibrating-Tube Densimeter", *J. Chem. Thermodyn.*, **9**, 249–257 (1977).
12. J. B. Ott, K. N. Marsh, and A. E. Richards, "Excess Enthalpies, Excess Gibbs Free Energies, and Excess Volumes for (di-*n*-Butyl Ether + Benzene) and Excess Gibbs Free Energies and Excess Volumes for (di-*n*-Butyl Ether + Tetrachloromethane) at 298.15 and 308.15 K", *J. Chem. Thermodyn.*, **13**, 447–455 (1981).
13. See O. Redlich and G. C. Hood, "Ionic Interaction, Dissociation, and Molecular Structure", *Discuss. Faraday Soc.*, **24**, 87–93 (1957).
14. The vapor pressures (fugacities) shown in Figure 6.14 were reported by J. J. Fritz and C. R. Fuget, "Vapor Pressure of Aqueous Hydrogen Chloride Solutions", *Chem. Eng. Data Ser.*, **1**, 10–12 (1956). The vapor pressures are too small to measure directly. The values reported were calculated from the results of emf measurements made on an electrochemical cell. In Chapter 9, we will describe this and other cells in detail.
15. For a detailed discussion of the calculation of activities (and excess Gibbs free energies) from freezing point measurements, see R. L. Snow, J. B. Ott, J. R. Goates, K. N. Marsh, S. O'Shea, and R. N. Stokes, "(Solid + Liquid) and (Vapor + Liquid) Phase Equilibria and Excess Enthalpies for (Benzene + *n*-Tetradecane), (Benzene + *n*-Hexadecane), (Cyclohexane + *n*-Tetradecane), and (Cyclohexane + *n*-Hexadecane) at 293.15, 298.15, and 308.15 K. Comparison of $G_m^E$ calculated from (Vapor + Liquid) and (Solid + Liquid) Equilibria", *J. Chem. Thermodyn.*, **18**, 107–130 (1986).
16. See G. N. Lewis, M. Randall, K. S. Pitzer, and L. Brewer, *Thermodynamics*, Second Edition, McGraw Hill Book Company, New York, 1961 pp. 404–416.
17. G. Scatchard, C. L. Raymond, and H. H. Gilmann, "Vapor–Liquid Equilibrium. I. Apparatus for the Study of Systems with Volatile Components", *J. Am. Chem. Soc.*, **60**, 1275–1278 (1938).
18. G. Scatchard and C. L. Raymond, "Vapor–Liquid Equilibrium. II. Chloroform–Ethanol Mixtures at 35, 45, and 55°", *J. Am. Chem. Soc.*, **60**, 1278–1287 (1938).
19. The isopiestic method was first introduced by W. R. Bousfield, "Isopiestic Solutions," *Trans. Faraday Soc.*, **13**, 401–410 (1918) and refined by D. A. Sinclair, "A Simple Method for Accurate Determinations of Vapor Pressure of Solutions", *J. Phys. Chem.*, **37**, 495–504 (1933).
20. G. N. Lewis and H. Storch, "The Potential of the Bromine Electrode; the Free Energy of Dilution of Hydrogen Bromide; the Distribution of Bromine Between Several Phases", *J. Am. Chem. Soc.*, **39**, 2544–2554 (1917).

## Chapter 7

# The Thermodynamic Properties of Solutions

With the definitions of activities, standard states, ideal solutions, and partial molar properties given in Chapters 5 and 6, we are ready to describe the properties of solutions. There are many topics of interest, but we will limit our discussion in this chapter to four — the change in the thermodynamic properties of nonelectrolyte solutions resulting from the mixing process, the calculation of the thermodynamic properties of solutions containing strong electrolyte solutes using the Debye–Hückel theory, the calculation of the change in the thermal properties of solutions due to mixing, and the effect of concentration on the osmotic pressure.

## 7.1 Change in the Thermodynamic Properties of Nonelectrolyte Solutions Due to the Mixing Process

We are interested in describing and calculating $\Delta_{\text{mix}}Z$, the change in the thermodynamic variable $Z$, when liquids (or solids) are mixed to form a solution. We will begin by deriving the relationship for calculating $\Delta_{\text{mix}}G$. Changes in the other thermodynamic properties can then be obtained.

For the process

$$n_1\text{A} + n_2\text{B} = \text{Solution}$$

the change in Gibbs free energy is given by

$$\Delta_{\text{mix}}G = G_{\text{solution}} - G_1^* - G_2^*$$

so that

$$\Delta_{\text{mix}}G = n_1\mu_1 + n_2\mu_2 - n_1\mu_1^* - n_2\mu_2^*, \tag{7.1}$$

where $\mu_1$ and $\mu_2$ are the chemical potentials of components A and B in the solution and $\mu_1^*$ and $\mu_2^*$ are the chemical potentials of the pure components.

Rearranging equation (7.1) gives

$$\Delta_{\text{mix}}G = n_1(\mu_1 - \mu_1^*) + n_2(\mu_2 - \mu_2^*). \tag{7.2}$$

If we assume a Raoult's law standard state for both components, we have

$$\mu_1 - \mu_1^* = RT \ln a_1$$

$$\mu_2 - \mu_2^* = RT \ln a_2$$

so that equation (7.2) can be written as

$$\Delta_{\text{mix}}G = RT(n_1 \ln a_1 + n_2 \ln a_2). \tag{7.3}$$

The general expression for mixing A, B, C, ... components is

$$\Delta_{\text{mix}}G = RT \sum_i n_i \ln a_i. \tag{7.4}$$

Each activity can be eliminated from equation (7.4) by substituting

$$a_i = \gamma_{\text{R},i} x_i$$

which upon rearrangement gives,

$$\Delta_{\text{mix}}G = RT \sum_i (n_i \ln x_i + n_i \ln \gamma_{\text{R},i}). \tag{7.5}$$

To obtain $\Delta_{\text{mix}}G_{\text{m}}$, the molar Gibbs free energy of mixing, we divide equation (7.5) by $n = \sum n_i$ to obtain

$$\Delta_{\text{mix}}G_{\text{m}} = RT \sum_i (x_i \ln x_i + x_i \ln \gamma_{\text{R},i}). \tag{7.6}$$

## 7.1a Change in Thermodynamic Properties Resulting from the Formation of Ideal Solutions

Equation (7.6) is the starting point for deriving equations for $\Delta_{\text{mix}}Z_{\text{m}}^{\text{id}}$, the change in $Z_{\text{m}}$ for forming an ideal mixture. For the ideal solution, $\gamma_{\text{R},i} = 1$ and equation (7.6) becomes

$$\Delta_{\text{mix}}G_{\text{m}}^{\text{id}} = RT \sum_i x_i \ln x_i. \tag{7.7}$$

The entropy change to form an ideal mixture from the pure components is obtained by differentiating equation (7.7) with respect to $T$. Since $x_i$ is independent of $T$, the result is

$$\left(\frac{\partial \Delta_{\text{mix}} G_{\text{m}}^{\text{id}}}{\partial T}\right)_{p,n} = -\Delta_{\text{mix}} S_{\text{m}}^{\text{id}} = R \sum_i x_i \ln x_i$$

or

$$\Delta_{\text{mix}} S_{\text{m}}^{\text{id}} = -R \sum_i x_i \ln x_i . \tag{7.8}$$ [a]

The ideal enthalpy of mixing is easily obtained from equations (7.7) and (7.8) and the relationship

$$\Delta_{\text{mix}} H_{\text{m}}^{\text{id}} = \Delta_{\text{mix}} G_{\text{m}}^{\text{id}} + T\Delta_{\text{mix}} S_{\text{m}}^{\text{id}}$$

$$= RT \sum_i x_i \ln x_i + T\left(-R \sum_i x_i \ln x_i\right).$$

$$\Delta_{\text{mix}} H_{\text{m}}^{\text{id}} = 0. \tag{7.9}$$

The volume change for forming an ideal mixture can be obtained by differentiating equation (7.7) with respect to $p$. Since $x_i$ is independent of $p$,

$$\left(\frac{\partial \Delta_{\text{mix}} G_{\text{m}}^{\text{id}}}{\partial p}\right)_{T,n} = \Delta_{\text{mix}} V_{\text{m}}^{\text{id}} = 0. \tag{7.10}$$

The change in internal energy on mixing can also be obtained from

$$\Delta_{\text{mix}} U_{\text{m}}^{\text{id}} = \Delta_{\text{mix}} H_{\text{m}}^{\text{id}} - p\Delta_{\text{mix}} V_{\text{m}}^{\text{id}} = 0. \tag{7.11}$$

Hence, $\Delta_{\text{mix}} H_{\text{m}}^{\text{id}}$, $\Delta_{\text{mix}} V_{\text{m}}^{\text{id}}$, and $\Delta_{\text{mix}} U_{\text{m}}^{\text{id}}$ are zero, while $\Delta_{\text{mix}} G_{\text{m}}^{\text{id}} = -T\Delta_{\text{mix}} S_{\text{m}}^{\text{id}} = RT \sum_i x_i \ln x_i$.

[a] Note that the equation for calculating $\Delta_{\text{mix}} S$ for an ideal liquid (or solid) mixture is the same as was derived in Chapter 2 for calculating $\Delta_{\text{mix}} S$ for ideal gases.

Figure 7.1 compares graphs of $\Delta_{\text{mix}}G_{\text{m}}^{\text{id}}$, $\Delta_{\text{mix}}S_{\text{m}}^{\text{id}}$ and $\Delta_{\text{mix}}H_{\text{m}}^{\text{id}}$ as a function of mole fraction, using equations (7.7), (7.8), and (7.9). The positive entropy change results from the increased disorder associated with mixing two fluids to form a solution. The negative Gibbs free energy change is a measure of the reversible work that can be obtained from the mixing process. The negative of $\Delta_{\text{mix}}G_{\text{m}}^{\text{id}}$ is the minimum work required to separate the two components.

### 7.1b Excess Thermodynamic Functions

The excess molar thermodynamic function $Z_{\text{m}}^{\text{E}}$ is defined as the difference in the property $Z_{\text{m}}$ for a real mixture and that for an ideal solution. That is,

$$Z_{\text{m}}^{\text{E}} = \Delta_{\text{mix}}Z_{\text{m}} - \Delta_{\text{mix}}Z_{\text{m}}^{\text{id}}. \tag{7.12}$$

For the enthalpy, internal energy, and volume changes, $\Delta_{\text{mix}}Z_{\text{m}}^{\text{id}} = 0$. Hence,

$$H_{\text{m}}^{\text{E}} = \Delta_{\text{mix}}H_{\text{m}} \tag{7.13}$$

$$U_{\text{m}}^{\text{E}} = \Delta_{\text{mix}}U_{\text{m}} \tag{7.14}$$

$$V_{\text{m}}^{\text{E}} = \Delta_{\text{mix}}V_{\text{m}}, \tag{7.15}$$

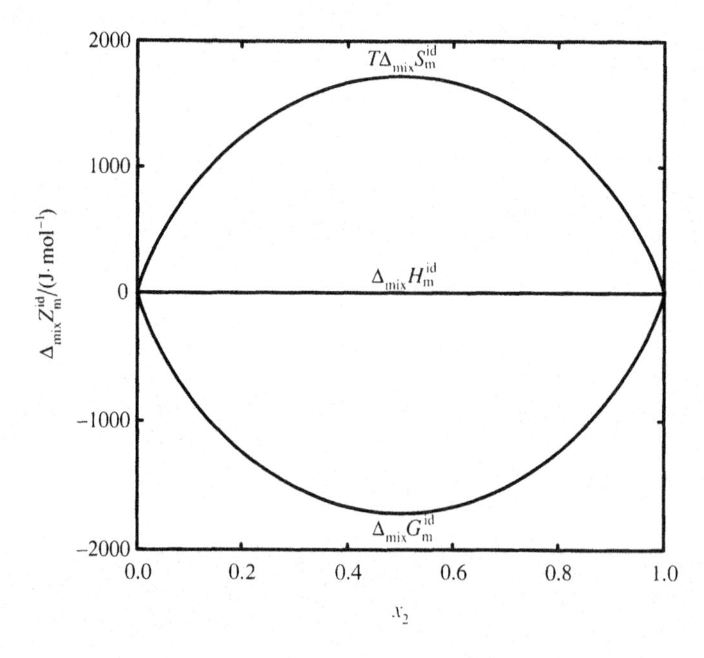


**Figure 7.1** Entropy, enthalpy, and Gibbs free energy changes at $T = 298.15$ K for forming one mole of an ideal mixture from the components.

and the value of the excess function is the total enthalpy, internal energy, and volume change upon mixing. We have shown earlier [equation (7.6)] that the molar Gibbs free energy change upon mixing is given by

$$\Delta_{\text{mix}} G_{\text{m}} = RT \sum_i (x_i \ln x_i + x_i \ln \gamma_{\text{R},i}) \tag{7.6}$$

where $\gamma_{\text{R},i}$ is the activity coefficient (Raoult's law standard state). Since

$$G_{\text{m}}^{\text{E}} = \Delta_{\text{mix}} G_{\text{m}} - \Delta_{\text{mix}} G_{\text{m}}^{\text{id}}$$

and

$$\Delta_{\text{mix}} G_{\text{m}}^{\text{id}} = RT \sum_i x_i \ln x_i,$$

then

$$G_{\text{m}}^{\text{E}} = RT \sum_i x_i \ln \gamma_{\text{R},i}. \tag{7.16}$$

The excess molar entropy is obtained from

$$G_{\text{m}}^{\text{E}} = H_{\text{m}}^{\text{E}} - TS_{\text{m}}^{\text{E}}$$

or

$$S_{\text{m}}^{\text{E}} = \frac{H_{\text{m}}^{\text{E}} - G_{\text{m}}^{\text{E}}}{T}. \tag{7.17}$$

Since these mixing processes occur at constant pressure, $H_{\text{m}}^{\text{E}}$ is the heat evolved or absorbed upon mixing. It is usually measured in a mixing calorimeter. The excess Gibbs free energy, $G_{\text{m}}^{\text{E}}$, is usually obtained from phase equilibria measurements that yield the activity of each component in the mixture[b] and $S_{\text{m}}^{\text{E}}$ is then obtained from equation (7.17). The excess volumes are usually obtained

[b] See section 6.5 of Chapter 6 for examples of methods for measuring the activity.

with a dilatometer, or from density measurements obtained from a pycnometer or a vibrating tube densimeter.

The extent of deviation from ideal solution behavior and hence, the magnitude and arithmetic sign of the excess function, depend upon the nature of the interactions in the mixture. We will now give some representative examples.

**Nonpolar + Nonpolar Mixtures:** Figure 7.2 summarizes the behavior of $H_m^E$, $G_m^E$, and $TS_m^E$ for mixtures of decane with hexane[1] at 298.15 K. The small $Z_m^E$ demonstrates the near ideal behavior of this system. The negative $G_m^E$ results from a somewhat greater lowering of the vapor fugacity of each component than is expected for an ideal mixture (small negative deviations from Raoult's law). Since $H_m^E$ is positive while $G_m^E$ is negative, $S_m^E$ must be positive, indicating an increase in disorder during the mixing process over that predicted for the ideal mixture.

**Polar + Nonpolar Mixtures:** Figure 7.3 summarizes $H_m^E$, $G_m^E$ and $TS_m^E$ for mixtures of acetonitrile with benzene at 318.15 K.[2] The large positive $H_m^E$ results principally from the energy that must be added to separate the highly polar acetonitrile molecules, {which are held together by strong (dipole + dipole)

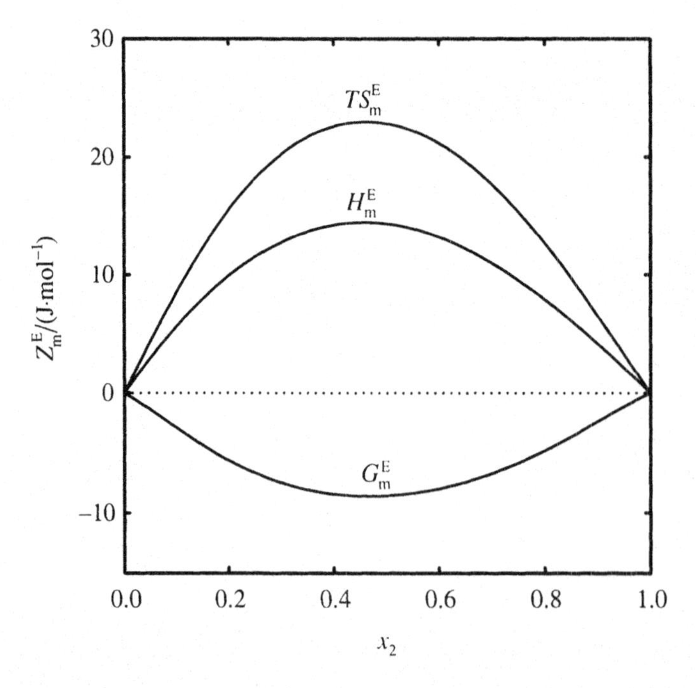


**Figure 7.2** Excess thermodynamic functions at $T = 298.15$ K for $\{x_1C_{10}H_{22} + x_2C_6H_{14}\}$, an example of a system where nonpolar chain-like molecules are mixed.

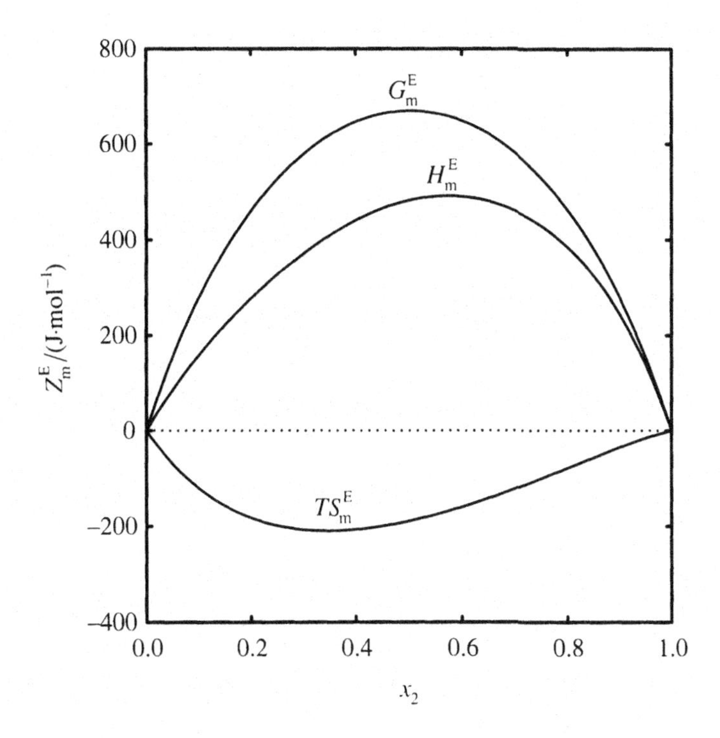


**Figure 7.3** Excess thermodynamic functions at $T = 318.15$ K for $\{x_1C_6H_6 + x_2CH_3CN\}$, an example of a system in which polar molecules are mixed with nonpolar molecules.

interactions}, as the nonpolar benzene molecules are added. The large positive $G^E_m$ reflects the increased vapor fugacities (positive deviations from Raoult's law) resulting from the decreased attractive forces in the mixtures.[c] Since $G^E_m$ is larger than $H^E_m$, $TS^E_m$ is less than zero, and increased ordering occurs in the mixture over that expected for the ideal mixture.

**Mixtures with Hydrogen Bonding:** Figure 7.4 summarizes $H^E_m$, $G^E_m$ and $TS^E_m$ for (ethanol + water)[3] at 303.15 K and 363.15 K. The negative $H^E_m$ and $TS^E_m$ result principally from the formation of hydrogen-bonded complexes in the liquid mixture. The complexes break down as the temperature increases, and the negative contributions decrease so that $H^E_m$ and $TS^E_m$ become less negative. At $T = 363.15$ K, $H^E_m$ has become s-shaped, with negative $H^E_m$ at low mole

[c] (Liquid + liquid) phase separation occurs with sufficiently large positive deviation from ideal behavior. The deviations are not large enough in (acetonitrile + benzene) to cause (liquid + liquid) equilibrium to occur. If benzene is replaced by an aliphatic hydrocarbon such as heptane, separation does occurs. In a later chapter, we will discuss (liquid + liquid) equilibrium in detail.

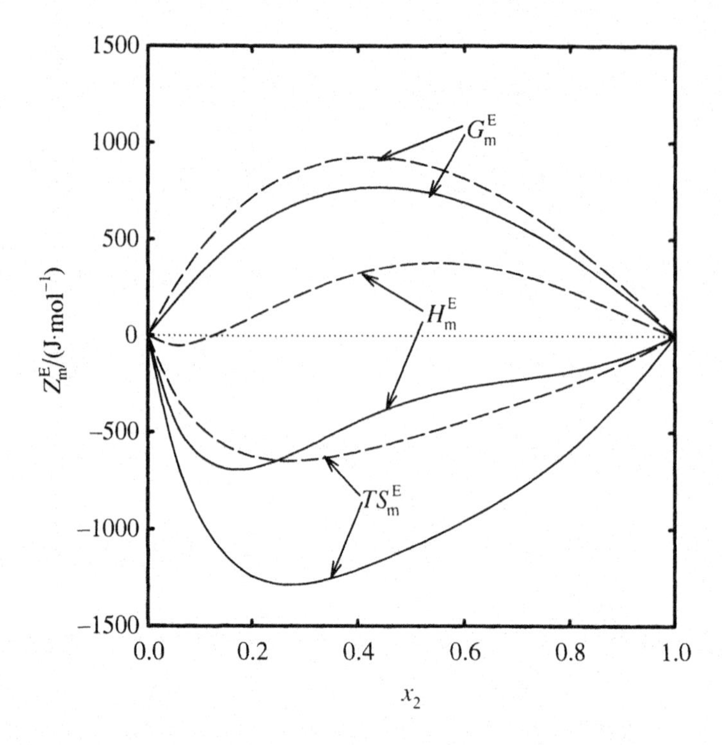


**Figure 7.4** Excess molar functions for $\{x_1H_2O + x_2C_2H_5OH\}$. The solid lines represent results at $T = 303.15$ K and the dashed lines are for results at $T = 363.15$ K.

fractions of ethanol,[d] changing to positive $H^{E}_{m}$ at high mole fractions. With increasing temperature, $H^{E}_{m}$ becomes more positive, until at high temperatures it has a large positive value over the entire mole fraction range.[4] The ordering effect in the mixture due to complex formation is larger than the energy effect so that $TS^{E}_{m}$ is larger in magnitude than $H^{E}_{m}$. The result is a positive $G^{E}_{m}$ and an increased escaping tendency.

**Excess Volume Comparison:** Figure 7.5 compares $V^{E}_{m}$ for the three systems for which we have compared $H^{E}_{m}$, $G^{E}_{m}$, and $S^{E}_{m}$, plus the (cyclohexane + decane) system.[5] The comparatively large negative $V^{E}_{m}$ for the (ethanol + water) system {curve (4)} can be attributed to the decrease in volume resulting from the formation of hydrogen-bonded complexes in those mixtures. The negative $V^{E}_{m}$ for the (hexane + decane) system {curve (3)} reflects an increased packing

[d] The stoichiometry of the complex is such that it contains considerably more water than ethanol. Since the decrease in energy due to the formation of the complex would be largest at the stoichiometric composition, the minimum in $H^{E}_{m}$ occurs at low mole fractions of ethanol.

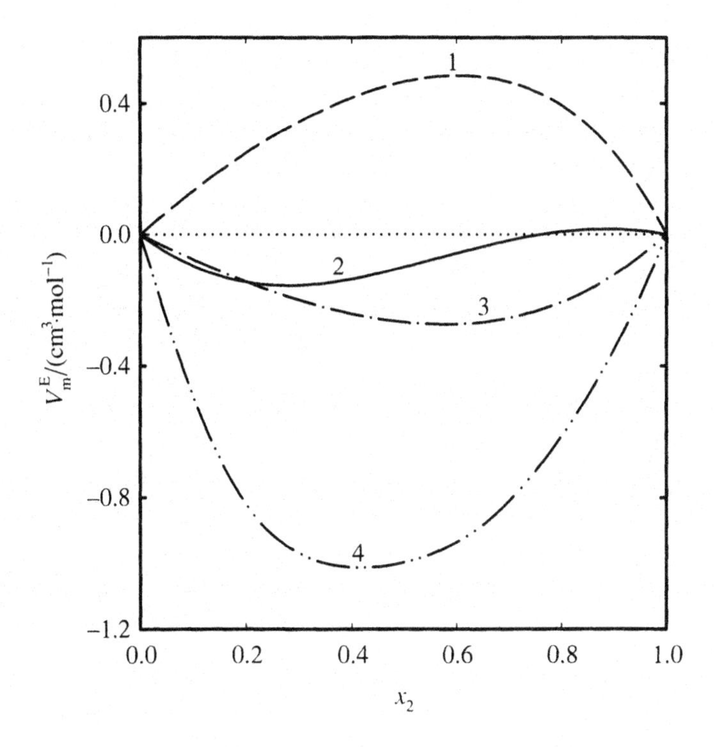


**Figure 7.5** Comparison of excess molar volumes for four mixtures as follows:
Curve 1: $\{x_1C_{10}H_{22} + x_2c\text{-}C_6H_{12}\}$ at $T = 313.15$ K.
Curve 2: $\{x_1C_6H_6 + x_2CH_3CN\}$ at $T = 318.15$ K.
Curve 3: $\{x_1C_{10}H_{22} + x_2C_6H_{14}\}$ at $T = 308.15$ K.
Curve 4: $\{x_1H_2O + x_2C_2H_5OH\}$ at $T = 318.15$ K.

efficiency when chain-like molecules of different lengths are mixed. The positive $V_m^E$ for the (cyclohexane + decane) system {curve (1)} indicates that mixing chain-like n-alkane molecules with globular cycloalkane molecules results in less efficient packing. The smaller and s-shaped $V_m^E$ curve for the (acetonitrile + benzene) system {curve (2)} has no simple explanation. Excess volumes for mixtures of molecules with different polarities and shapes vary from large positive $V_m^E$ to large negative $V_m^E$, with values in between also represented.

## 7.2 Calculation of the Thermodynamic Properties of Strong Electrolyte Solutes: The Debye–Hückel Theory

Solutions containing strong electrolyte solutes differ from those containing nonelectrolyte solutes in that deviations from Henry's law become important at much lower concentrations for the electrolyte solute than for the nonelectrolyte solute. For example, Figure 7.6 compares the activity coefficients (solute standard state) for aqueous solutions of ethanol, a nonelectrolyte, and of hydrochloric acid, an electrolyte. The activity coefficient for ethanol changes slowly with $m$ and is greater than 0.99 for $m$ less than 0.1. On the other hand, the activity coefficient ($\gamma_{\pm}$) for hydrochloric acid changes rapidly with $m$ and becomes less than 0.90 for $m$ greater than 0.01.

The large non-ideality in electrolyte solutions can be explained by the distribution of charged ions in the solution as shown schematically in Figure 7.7.[e] The ions are often spherically symmetrical (or at least near-spherically symmetrical) particles, with positive or negative charges that are large in comparison to the dipolar charges present in polar nonelectrolyte solutes. Electrostatic attractions and repulsions are much stronger than other types of interactions and dominate in the solution. Repulsions between ions of the same charge and attractions between ions of a different charge cause the ions to arrange themselves in solution so that positively charged ions are preferentially surrounded by negatively charged ions, while negatively charged ions are preferentially surrounded by positively charged ions. The result is a nonrandom

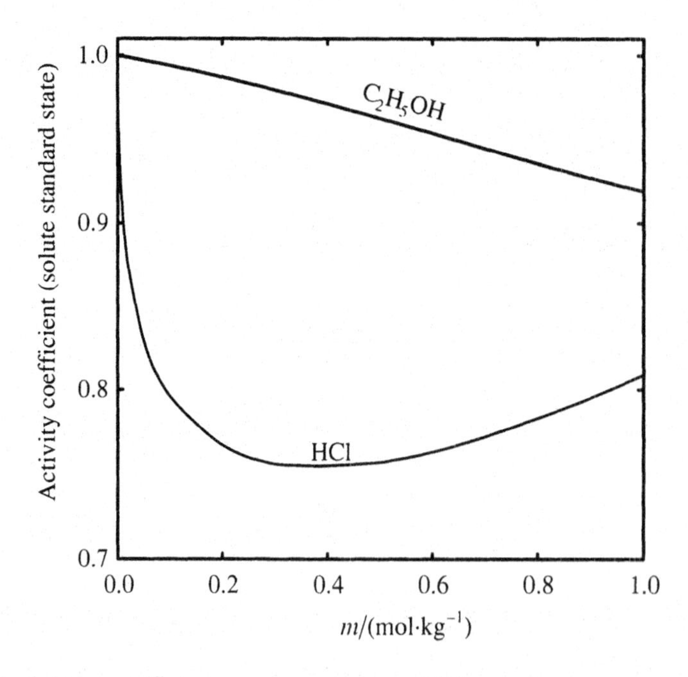


**Figure 7.6** Activity coefficients of aqueous HCl and aqueous $C_2H_5OH$ at $T = 303.15$ K.

[e] The ions are in thermal motion. Figure 7.7 represents the average distribution of charge.

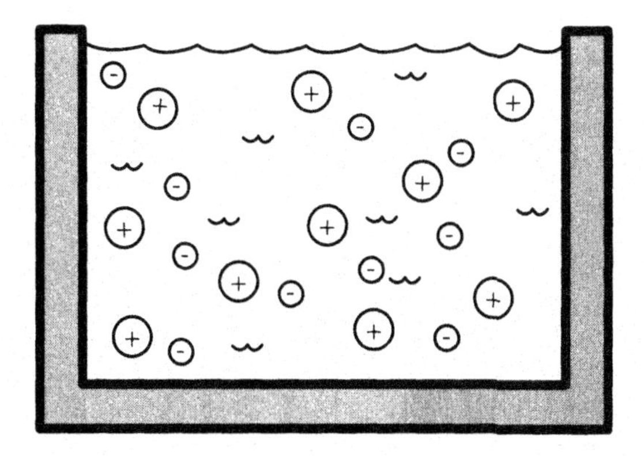


**Figure 7.7** In a strong electrolyte solution, negative ions are preferentially surrounded by positive ions while the positive ions are surrounded by negative ions.

distribution of ions in the solution, even at low concentrations. This causes deviations from ideal behavior, since in the ideal solution, the molecules are randomly distributed.

## 7.2a Derivation of the Activity Coefficient Equations

If the electrical potential in an ionic solution that results from the nonrandom distribution of charge could be calculated as a function of position of any given ion, then the activity coefficients of the ions could be calculated. Imagine a process in which the ions in their equilibrium arrangement are held in position while reducing their charges to zero. The ionic charges are then reversibly increased at constant temperature and pressure from zero to the value they have in solution, by bringing in electrical charge from infinite distance. If the electrical potential, $\psi$ is known, the electrical work $w_{\text{el}}$ added to the system as a result of this charging process can be calculated, since

$$\delta w_{\text{el}} = \psi \, \mathrm{d}q$$

where $q$ is the charge. Integration over the charge distribution gives the work. This electrical work, in turn, equals the free energy change for the process,[f] which can be used to calculate the excess chemical potential and hence, the activity coefficient.[g]

[f] We remember that $\Delta G$ for a constant $T$ and $p$ process equals the non (pressure–volume) work, that occurs in the process.

[g] An alternate approach is to add reversibly, an ion with no charge to the solution, and calculate the work needed in bringing the ion to full charge. Both methods lead to the same result for the activity coefficient.

The electrical potential is obtained by solving the Poisson equation

$$\nabla^2 \psi = -\frac{4\pi}{\varepsilon} \rho \tag{7.18}$$

where $\psi$ is the potential, $\rho$ is the charge density, $\varepsilon$ is the relative permittivity,[h] and $\nabla^2$ is the Laplacian operator given by

$$\nabla^2 = \frac{\partial^2}{\partial x^2} + \frac{\partial^2}{\partial y^2} + \frac{\partial^2}{\partial z^2}. \tag{7.19}$$

To solve the Poisson equation, we must express as a $\rho$ function of the coordinates of the system and solve the resulting second-order differential equation to obtain $\psi(x, y, z)$, from which $w_{el}$ and hence, $\Delta G$, $\mu - \mu^0$, and $\gamma$ can be calculated.

Solving the Poisson equation is not an easy thing to do. Assumptions must be made to relate, $\rho$ to $\psi$ so that $\rho$ can be eliminated from the equation. After this is done, approximations are required to simplify the resulting equation so that it can be solved. The assumptions used in the derivation limit the applicability of the resulting equations to low concentrations.

The method attributed to Debye and Hückel has been almost universally adopted by scientists. We will review the steps of their method and give the equations that are the end product of the derivation, but will leave it to others to present the mathematical details.[6]

The first assumption of the Debye–Hückel theory is that $\psi$ is spherically symmetric. With the elimination of any angular dependence, the Poisson equation (expressed in spherical-polar coordinates) reduces to

$$\frac{1}{r^2} \frac{\mathrm{d}}{\mathrm{d}r} \left( r^2 \frac{\mathrm{d}\psi}{\mathrm{d}r} \right) = -\frac{4\pi}{\varepsilon} \rho. \tag{7.20}$$

The second assumption is that the concentration $c_j$ (particles per unit volume) of type $j$ ions in the electrical field is related to $c_j^0$, the concentration at zero field, by the Maxwell–Boltzmann distribution function.[i]

$$c_j = c_j^0 \exp\left( -\frac{\epsilon_j}{kT} \right). \tag{7.21}$$

---

[h] The relative permittivity was formerly called the dielectric constant.

[i] In Chapter 10, we will derive the Maxwell-Boltzmann distribution function and describe its properties and applications.

In equation (7.21), $k$ is the Boltzmann constant, $T$ is the temperature, and $\epsilon_j$ is the potential energy associated with the field. It is given by

$$\epsilon_j = z_j e \psi_j \tag{7.22}$$

where e is the electronic charge ($1.6022 \times 10^{-19}$ coulombs), $z_j$ is the number of charges,[j] and $\psi_j$ is the electrical potential associated with the ion. Substituting equation (7.22) into equation (7.21) gives

$$c_j = c_j^0 \exp\left(-\frac{z_j e \psi_j}{kT}\right). \tag{7.23}$$

The charge density in solution is obtained by multiplying the concentration of particles by the charge on each particle and summing over the number of types of particles in the solution. That is,

$$\rho = \sum_j z_j e c_j. \tag{7.24}$$

Equation (7.23) is substituted into equation (7.24) to give the equation that relates $\rho$ to $\psi$

$$\rho = \sum_j z_j e c_j^0 \exp\left(-\frac{z_j e \psi_j}{kT}\right). \tag{7.25}$$

Equation (7.25) can be substituted into equation (7.20) to give a second order differential equation in $\psi$. In theory, the resulting equation can be solved to give $\psi$ as a function of $r$. However, it has an exponential term in $\psi$, that makes it impossible to solve analytically. In the Debye–Hückel approximation, the exponential is expanded in a power series to give

$$\rho = \sum_j z_j e c_j^0 - \sum_j \frac{z_j^2 e^2 c_j^0 \psi_j}{kT} + \sum_j \frac{z_j^3 e^3 c_j^0 \psi_j^2}{2k^2T^2} + \cdots \tag{7.26}$$

---

[j] For example, $z = +1$ for $H^+$, $Na^+$, $Ag^+$, ...; $z = +2$ for $Ca^{2+}$, $Mg^{2+}$, $Fe^{2+}$, ...; $z = -1$ for $Cl^-$, $NO_3^-$, ...; $z = -2$ for $SO_4^{2-}$, $CO_3^{2-}$, ...

The first term on the right-hand side of this equation is zero, since it is simply the sum of the electrical charge in solution, which must be zero for a neutral electrolyte solution. The third term is also zero for electrolytes with equal numbers of positive and negative ions, such as NaCl and $MgSO_4$. It would not be zero for asymmetric electrolytes such as $CaCl_2$. However, in the Debye–Hückel approach, all terms except the second are ignored for all ionic solutions. Substitution of the resulting expression into equation (7.20) gives the linear second-order differential equation

$$\frac{1}{r^2}\frac{\mathrm{d}}{\mathrm{d}r}\left(r^2\frac{\mathrm{d}\psi}{\mathrm{d}r}\right) = \kappa^2\psi, \tag{7.27}$$

where

$$\kappa^2 = \frac{4\pi e^2}{\varepsilon kT}\sum_j z_j^2 c_j^0. \tag{7.28}$$

The general solution to equation (7.27) has the form

$$\psi = A\frac{\exp(\kappa r)}{r} + B\frac{\exp(-\kappa r)}{r} \tag{7.29}$$

where $A$ and $B$ are constants that can be determined from the boundary conditions of the problem. The first condition is that $\psi$ must remain finite at large values of $r$. This requires that $A$ must be zero, in which case equation (7.29) becomes

$$\psi = B\frac{\exp(-\kappa r)}{r}. \tag{7.30}$$

The constant $B$ is evaluated by substituting equation (7.30) into equation (7.25) to obtain an equation relating the charge density $\rho$ to $r$. The result is applied to a central ion of charge $z_j e$ to give the charge density around this ion. That is,

$$\rho_j = -B\frac{\kappa^2 e}{4\pi}\frac{\exp(-\kappa r)}{r}. \tag{7.31}$$

Electrical neutrality in the solution requires that the total charge around the ion be balanced by the charge on the ion. This condition is expressed by the relationship

$$\int_a^\infty \rho_j 4\pi r_j^2 \, dr_j = -z_j e. \tag{7.32}$$

The lower limit of the integration assumes that the central ion has a finite size with $a$, the radius of the ion, representing a distance of closest approach of other ions to the central ion.

Substituting equation (7.31) into equation (7.32) and integrating results in a value for $B$ which, when substituted into equation (7.30), gives

$$\psi_j = \left(\frac{z_j e \exp(\kappa a)}{\varepsilon(1+\kappa a)}\right) \cdot \left(\frac{\exp(-\kappa r)}{r}\right). \tag{7.33}$$

Equation (7.33) can be used to express $w_{el}$, and hence the activity coefficient, as a function of $\sum_j z_j^2 c_j^0$. The concentration $c_j$ is usually converted to a quantity known as the molality-scale ionic strength $I_m$,[k] where

$$I_m = \frac{1}{2} \sum_j z_j^2 m_j. \tag{7.34}$$

In equation (7.34), $m_j$ is the molality of $j$ type of ions.[l]

---

[k] Ionic strength can be expressed in terms of the concentration $c$ (mol·dm$^{-3}$), in which case

$$I_c = \frac{1}{2} \sum_i z_i^2 c_i.$$

Mole fraction can also be used, in which case

$$I_x = \frac{1}{2} \sum_i z_i^2 x_i.$$

The form using molality [equation (7.34)] as the unit of concentration is the one most commonly encountered. It is often written as $I$ (instead of $I_m$).

[l] For a simple 1 : 1 electrolyte (e.g. HCl, NaCl, $AgNO_3$, etc.), $I_m = m$. This can be seen as follows:

$$I_m = \tfrac{1}{2}[z_+^2 m_+ + z_-^2 m_-] = \tfrac{1}{2}[(1)^2 m + (-1)^2 (m)] = m.$$

The Debye–Hückel final result is

$$\ln \gamma_+ = -\frac{z_+^2 C_\gamma I_m^{1/2}}{1 + B_\gamma a I_m^{1/2}} \tag{7.35}$$

$$\ln \gamma_- = -\frac{z_-^2 C_\gamma I_m^{1/2}}{1 + B_\gamma a I_m^{1/2}} \tag{7.36}$$

where

$$C_\gamma = (2\pi N_A \rho_A)^{1/2} \left( \frac{e^2}{4\pi\varepsilon_0\varepsilon_A kT} \right)^{3/2} \tag{7.37}$$

and

$$B_\gamma = e \left( \frac{2N_A \rho_A}{\varepsilon_0 \varepsilon_A kT} \right)^{1/2}. \tag{7.38}$$

In equations (7.37) and (7.38), $N_A$ is Avogadro's number, $\rho_A$ is the density of the solvent (in $kg{\cdot}m^{-3}$), $\varepsilon_0$ is the permittivity of vacuum, and $\varepsilon_A$ is the permittivity (dielectric constant) of the solvent. The other constants in equations (7.37) and (7.38) are as previously defined.

Equations (7.35) and (7.36) can be used to calculate the activity coefficients of individual ions. However, as we discussed in Chapter 6, $\gamma_+$ and $\gamma_-$ cannot be measured individually. Instead, $\gamma_\pm$, the mean ionic activity coefficient for the electrolyte, $M_{\nu+}X_{\nu-}$, given by

$$\gamma_\pm = (\gamma_+^{\nu+} \gamma_-^{\nu-})^{1/\nu} \tag{7.39}$$

is calculated, where $\nu = \nu_+ + \nu_-$. An expression for $\gamma_\pm$ can be obtained from equations (7.35) and (7.36). Taking the logarithm of both sides of equation (7.39) gives

$$\ln \gamma_\pm = \frac{\nu_+ \ln \gamma_+ + \nu_- \ln \gamma_-}{\nu}. \tag{7.40}$$

Since the electrolyte $M_{\nu+}X_{\nu-}$ is electrically neutral, we have

$$\nu_+ z_+ + \nu_- z_- = 0. \tag{7.41}$$

Equation (7.41) is multiplied by $z_+$ and added to a similar equation obtained by multiplying equation (7.41) by $z_-$. The result is

$$\nu_+ z_+^2 + \nu_- z_-^2 = -z_+ z_- \nu = | z_+ z_- | \nu \tag{7.42}$$

since $z_-$ is negative. Combining equations (7.35), (7.36), (7.40) and (7.42) gives

$$\ln \gamma_\pm = -| z_+ z_- | \frac{C_\gamma I_m^{1/2}}{1 + B_\gamma a I_m^{1/2}}. \tag{7.43}$$

SI values for all the constants can be substituted into equations (7.37) and (7.38), along with values for the density and permittivity of the solvent at the temperatures of interest, to give values for $C_\gamma$ and $B_\gamma$. For water solutions at $T = 298.15$ K, the values are

$$C_\gamma = 1.1745 \text{ kg}^{1/2}\cdot\text{mol}^{-1/2}$$

$$B_\gamma = 3.32384 \times 10^9 \text{ kg}^{1/2}\cdot\text{mol}^{-1/2}\cdot\text{m}^{-1}.$$

The coefficients vary slowly with temperature. In water at 273.15 K, $C_\gamma = 1.1301$ kg$^{1/2}\cdot$mol$^{-1/2}$, while at 373.15 K, it has a value of 1.3718 kg$^{1/2}\cdot$mol$^{-1/2}$. Values for $C_\gamma$ as a function of temperature for aqueous solutions are given in Table 7.1.[m]

Measurements of $a$ give values that typically range from $(3 \times 10^{-10}$ to $8 \times 10^{-10})$ m, with most ions around $3 \times 10^{-10}$ m. The product $aB_\gamma$ has a value of $(3 \times 10^{-10})(3.3 \times 10^9) \approx 1$. Equation (7.43) is often simplified by letting $aB_\gamma = 1$ so that

$$\ln \gamma_\pm = -\frac{C_\gamma | z_+ z_- | I_m^{1/2}}{1 + I_m^{1/2}}. \tag{7.44}$$

A further simplification is made by noting that at low $m$, $I_m^{1/2} \ll 1$ and can be neglected in the denominator of equation (7.44). The resulting equation, valid

[m] Other Debye–Hückel coefficients, which we will soon describe are also given in Table 7.1.

**Table 7.1** Debye–Hückel parameters for the activity coefficient, volume, enthalpy, and heat capacity

| $(T-273.15)$/(K) | $C_\gamma = 3C_\phi$/ $(\text{kg·mol})^{1/2}$ | $C_V$/ $(\text{cm}^3\cdot\text{kg}^{1/2}\cdot\text{mol}^{-3/2})$ | $C_H/RT$/ $(\text{kg·mol})^{1/2}$ | $C_J/R$/ $(\text{kg·mol})^{3/2}$ |
|---|---|---|---|---|
| 0.0 | 1.1301 | 2.256 | 0.834 | 4.43 |
| 10.0 | 1.1463 | 2.465 | 0.974 | 5.09 |
| 20.0 | 1.1646 | 2.690 | 1.124 | 5.64 |
| 25.0 | 1.1745 | 2.813 | 1.202 | 5.91 |
| 30.0 | 1.1847 | 2.943 | 1.281 | 6.20 |
| 40.0 | 1.2069 | 3.230 | 1.448 | 6.77 |
| 50.0 | 1.2309 | 3.558 | 1.622 | 7.38 |
| 60.0 | 1.2570 | 3.933 | 1.805 | 8.06 |
| 70.0 | 1.2849 | 4.364 | 1.997 | 8.79 |
| 80.0 | 1.3152 | 4.857 | 2.201 | 9.60 |
| 90.0 | 1.3473 | 5.423 | 2.417 | 10.50 |
| 100.0 | 1.3818 | 6.075 | 2.646 | 11.49 |
| 110.0 | 1.4181 | 6.825 | 2.891 | 12.60 |
| 120.0 | 1.4571 | 7.691 | 3.153 | 13.86 |
| 130.0 | 1.4982 | 8.693 | 3.435 | 15.26 |
| 140.0 | 1.5420 | 9.858 | 3.738 | 16.85 |
| 150.0 | 1.5885 | 11.216 | 4.068 | 18.68 |
| 160.0 | 1.6380 | 12.804 | 4.427 | 20.76 |
| 170.0 | 1.6902 | 14.669 | 4.820 | 23.21 |
| 180.0 | 1.7460 | 16.88 | 5.250 | 26.07 |
| 190.0 | 1.8051 | 19.49 | 5.729 | 29.48 |
| 200.0 | 1.8684 | 22.61 | 6.263 | 33.57 |
| 210.0 | 1.9359 | 26.4 | 6.864 | 38.58 |
| 220.0 | 2.0082 | 30.9 | 7.548 | 44.78 |
| 230.0 | 2.0859 | 36.5 | 8.334 | 52.58 |
| 240.0 | 2.1696 | 43.2 | 9.248 | 62.60 |
| 250.0 | 2.2605 | 51.6 | 10.328 | 75.69 |
| 260.0 | 2.3595 | 62.3 | 11.624 | 93.23 |
| 270.0 | 2.469 | 75.8 | 13.209 | 117.27 |
| 280.0 | 2.589 | 93.5 | 15.20 | 151.2 |
| 290.0 | 2.724 | 116.7 | 17.73 | 200.6 |
| 300.0 | 2.88 | 148.1 | 21.08 | 275.1 |
| 310.0 | 3.06 | 191 | 25.7 | 392 |
| 320.0 | 3.27 | 254 | 32.1 | 587 |
| 330.0 | 3.54 | 347 | 42.0 | 933 |
| 340.0 | 3.87 | 495 | 57.9 | 1590 |
| 350.0 | 4.32 | 740 | 86.0 | 2880 |

only at very low $m$, is given by

$$\ln \gamma_{\pm} = -C_{\gamma} \mid z_+ z_- \mid I_m^{1/2}. \qquad (7.45)$$

It is known as the limiting law expression. Note that for a simple 1 : 1 electrolyte with $\mid z_+ z_- \mid = 1$, $I_m = m$ and equation (7.45) becomes

$$\ln \gamma_{\pm} = -C_{\gamma} m^{1/2}.$$

This equation, which predicts that for a solution of a 1 : 1 electrolyte, $\ln \gamma_{\pm}$ varies as $m^{1/2}$, will be especially useful in extrapolating measurements to zero concentration.[n]

## 7.2b Comparison of the Debye–Hückel Prediction with Experimental Values

With all the approximations involved in its derivation, one well might wonder how well the Debye–Hückel theory works in predicting $\gamma_{\pm}$. Figure 7.8 compares experimental $\ln \gamma_{\pm}$ results against the Debye–Hückel predictions. We see from Figure 7.8(a) that the limiting law begins to work reasonably well below $I_m^{1/2} = 0.15$ for HCl and for $SrCl_2$, when $I_m \approx 0.02$. For HCl, (1 : 1 electrolyte), this corresponds to $m = 0.02$ and for $SrCl_2$ (2 : 1 electrolyte), $m = 0.0067$. We note, however, that even at this low ionic strength, $\gamma_{\pm}$ for $ZnSO_4$ deviates significantly from the Debye–Hückel limiting law.

It can be seen from Figure 7.8(b) that the curved lines predicted by the extended form of the Debye–Hückel equation follow the experimental results to higher ionic strengths than do the limiting law expressions for the (1 : 1) and (2 : 1) electrolytes. However, for the (2 : 2) electrolyte, the prediction is still not very good even at the lowest measured molality.[o]

Experience shows that solutions of other electrolytes behave in a manner similar to the examples we have used. The conclusion we reach is that the Debye–Hückel equation, even in the extended form, can be applied only at very low concentrations, especially for multivalent electrolytes. However, the behavior of the Debye–Hückel equation as we approach the limit of zero ionic strength appears to give the correct limiting law behavior. As we have said earlier, one of the most useful applications of Debye–Hückel theory is to

---

[n] The ionic strength $I_m = \frac{1}{2} \sum m_i z_i^2$ must include the concentration of all sources of ions. Thus, $I_m$ is not equal to $m$ for a 1 : 1 electrolyte if other sources of ions are present.

[o] Electrolytes of the 2 : 2 type, such as $ZnSO_4$, apparently have significant ion pairing, even in dilute solution. This ion pairing makes an important contribution to the deviation from the limiting law for $ZnSO_4$.

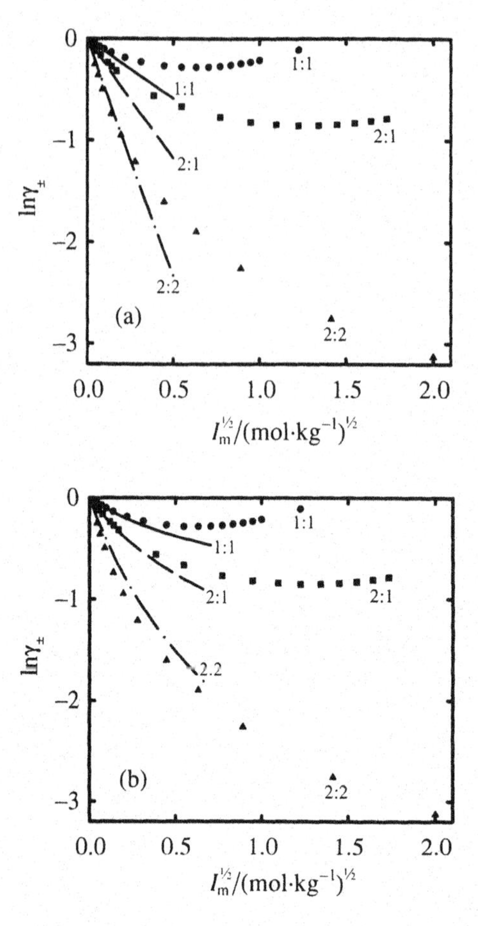


**Figure 7.8** Comparison of experimental $\ln \gamma_{\pm}$ for 1 : 1, 2 : 1, and 2 : 2 electrolytes. The symbols indicate the experimental results, with ● representing HCl ($z_{+} = 1$, $z_{-} = -1$); ■ representing $SrCl_2$ ($z_{+} = 2$, $z_{-} = -1$); and ▲ representing $ZnSO_4$ ($z_{+} = 2$, $z_{-} = -2$). The lines are the Debye–Hückel predictions, with the solid line giving the prediction for ($z_{+} = 1$, $z_{-} = -1$); the dashed line for ($z_{+} = 2$, $z_{-} = -1$); and the dashed-dotted line for ($z_{+} = 2$, $z_{-} = -2$). In (a), $\ln \gamma_{\pm}$ calculated from the limiting law [equation (7.45)] is shown graphed against $I_{\mathrm{m}}^{1/2}$. In (b), $\ln \gamma_{\pm}$ calculated from the extended form [equation (7.43)] is shown graphed against $I_{\mathrm{m}}^{1/2}$.

provide a method for extrapolating measurements to zero ionic strength. We will see examples later in this and other chapters.[p]

---

[p] $\gamma_{\pm}$ has not been measured to low enough concentrations for any electrolyte solution to definitively prove the square root dependence of $I_{\mathrm{m}}$ on $\ln\gamma_{\pm}$. Some other theories, in fact, predict a cube root dependence. With the absence of evidence that would contradict equations (7.44) and (7.45), the Debye–Hückel limiting law is usually used to extrapolate results to zero ionic strength.

### 7.2c The Debye–Hückel Prediction of the Osmotic Coefficient

Equation (7.45) is a limiting law expression for $\gamma_{\pm}$, the activity coefficient of the solute. Debye–Hückel theory can also be used to obtain limiting-law expressions for the activity $a_1$ of the solvent. This is usually done by expressing $a_1$ in terms of the practical osmotic coefficient $\phi$ that we described briefly in Chapter 6. For an electrolyte solute, it is defined in a general way as

$$\phi = -\frac{\ln a_1}{M_1 \sum_k \nu_k m_k} \tag{7.46}$$

where $\nu_k$ is the number of moles of ions produced by a solute whose molality is $m_k$, and $M_1$ is the molecular weight of the solvent in kg·mol$^{-1}$. The sum goes over all electrolytes present in solution. For a single electrolyte, equation (7.46) reduces to

$$\phi = -\frac{\ln a_1}{M_1 \nu m}. \tag{7.47}$$

For a nonelectrolyte solute, it simplifies even further to the expression

$$\phi = -\frac{\ln a_1}{M_1 m} \tag{7.48}$$

since $\nu = 1$.

The osmotic coefficient is often used as a measure of the activity of the solvent instead of $a_1$ because $a_1$ is nearly unity over the concentration range where $\gamma_{\pm}$ is changing, and many significant figures are required to show the effect of solute concentration on $a_1$. The osmotic coefficient also becomes one at infinite dilution, but deviates more rapidly with concentration of solute than does $a_1$.

The osmotic coefficient $\phi$ and activity coefficient $\gamma_{\pm}$ are related in a simple manner through the Gibbs–Duhem equation. We can find the relationship by writing this equation in a form that relates $a_1$ and $a_2$.

$$n_1 \,\mathrm{d}\ln a_1 + n_2 \,\mathrm{d}\ln a_2 = 0.$$

For a binary solution containing $n_2 = m$ moles of solute and $n_1 = 1/M_1$ moles of solvent (with $M_1$ in kg·mol$^{-1}$), the Gibbs–Duhem equation becomes

$$\frac{1}{M_1}\,\mathrm{d}\ln a_1 + m\,\mathrm{d}\ln a_2 = 0. \tag{7.49}$$

Substitutions are made for $a_1$ and $a_2$ in equation (7.49). For the solvent

$$\ln a_1 = -M_1 \nu m \phi$$

so that

$$\mathrm{d} \ln a_1 = -M_1 \nu (m \, \mathrm{d}\phi + \phi \, \mathrm{d}m). \tag{7.50}$$

For the solute

$$\begin{aligned} a_2 &= a_\pm^\nu \\ &= \gamma_\pm^\nu m^\nu (\nu_+^{\nu_+} \nu_-^{\nu_-}). \end{aligned}$$

Taking logarithms gives

$$\ln a_2 = \nu \ln \gamma_\pm + \nu \ln m + \ln (\nu_+^{\nu_+} \nu_-^{\nu_-}).$$

Differentiating gives

$$d \ln a_2 = \nu \, \mathrm{d} \ln \gamma_\pm + \nu \, \mathrm{d} \ln m. \tag{7.51}$$

Combining equations (7.49), (7.50), and (7.51) and rearranging gives

$$\mathrm{d} \ln \gamma_\pm = \mathrm{d}\phi + \frac{(\phi - 1)}{m} \mathrm{d} m. \tag{7.52}$$

Integration of equation (7.52) between $m = 0$ where $\phi = 1$ and $\gamma_\pm = 1$, and an upper limit $m$, $\phi$, and $\gamma_\pm$, gives an equation explicit in $\ln \gamma_\pm$

$$\ln \gamma_\pm = \phi - 1 + \int_0^m \frac{(\phi - 1)}{m} \mathrm{d} m. \tag{7.53}$$

A similar (reverse) manipulation gives equation (7.54) that is explicit in $\phi$

$$1 - \phi = -\frac{1}{m} \int_1^{\ln \gamma_\pm} m \, \mathrm{d} \ln \gamma_\pm. \tag{7.54}$$

Equation (7.53) can be used to obtain $\gamma_\pm$ from $\phi$, while equation (7.54) gives $\phi$ from $\gamma_\pm$. What must be known is the relationship between $\phi$ and $m$ [for equation (7.53)] or between $\gamma_\pm$ and $m$ [for equation (7.54)]. At higher concentrations, graphical integrations are usually used; at low $m$ where the

Debye–Hückel limiting law for $\gamma_\pm$ applies [equation (7.45)], equation (7.54) can be used to derive a limiting law for $\phi$. We start with

$$\ln \gamma_\pm = -C_\gamma \mid z_+z_- \mid I_\mathrm{m}^{1/2}. \tag{7.45}$$

Substituting

$$I_\mathrm{m} = \tfrac{1}{2}(\nu_+z_+^2 + \nu_-z_-^2)m \tag{7.55}$$

into equation (7.45) and differentiating gives

$$\mathrm{d}\ln \gamma_\pm = -\frac{A_\gamma}{2^{3/2}} \mid z_+z_- \mid (\nu_+z_+^2 + \nu_-z_-^2)^{1/2} m^{-1/2}\, \mathrm{d}m. \tag{7.56}$$

Substituting equation (7.56) into equation (7.54) gives

$$1 - \phi = \frac{C_\gamma}{2^{3/2}m} \mid z_+z_- \mid (\nu_+z_+^2 + \nu_-z_-^2)^{1/2} \int_0^m m^{1/2}\, \mathrm{d}m.$$

Integration gives

$$1 - \phi = \frac{C_\gamma}{2^{3/2}m} \mid z_+z_- \mid (\nu_+z_+^2 + \nu_-z_-^2)^{1/2} \frac{2}{3} m^{3/2}.$$

which simplifies to

$$1 - \phi = \frac{C_\gamma}{3} \mid z_+z_- \mid I_\mathrm{m}^{1/2}. \tag{7.57}$$

Thus, $\phi$ and $\gamma$ have a similar relationship to $I_\mathrm{m}$ in the limiting law. We usually rewrite equation (7.57) as

$$1 - \phi = C_\phi \mid z_+z_- \mid I_\mathrm{m}^{1/2} \tag{7.58}$$

with

$$C_\phi = \frac{C_\gamma}{3}. \tag{7.59}$$

As with $\gamma_\pm$, an extended form of equation (7.59) can be written

$$1 - \phi = C_\phi \mid z_+ z_- \mid \frac{I_m^{1/2}}{1 + I_m^{1/2}}, \tag{7.60}$$

with the extended form applying over a larger molality range than the limiting law expression.

### 7.2d The Debye–Hückel Prediction of Thermal and Volumetric Properties of the Solute

From the Debye–Hückel expressions for $\ln \gamma_\pm$, one can derive equations to calculate other thermodynamic properties. For example $\overline{L}_2$, the relative partial molar enthalpy,[q] and $\overline{V}_2$, the partial molar volume are related to $\gamma_\pm$ by the equations

$$\overline{L}_2 = \overline{H}_2 - \overline{H}_2^\circ = -\nu RT^2 \left( \frac{\partial \ln \gamma_\pm}{\partial T} \right)_p \tag{7.61}$$

$$\overline{V}_2 - \overline{V}_2^\circ = \nu RT \left( \frac{\partial \ln \gamma_\pm}{\partial p} \right)_T \tag{7.62}$$

where $\overline{H}_2^\circ$ and $\overline{V}_2^\circ$ are the partial molar enthalpy and the partial molar volume in the standard state. Starting with the limiting law expression equation (7.45) and differentiating gives

$$\overline{L}_2 = \frac{\nu}{2} \mid z_+ z_- \mid I_m^{1/2} \left[ 2RT^2 \left( \frac{\partial C_\gamma}{\partial T} \right)_p \right] \tag{7.63}$$

$$\overline{V}_2 - \overline{V}_2^\circ = \frac{\nu}{2} \mid z_+ z_- \mid I_m^{1/2} \left[ 2RT \left( \frac{\partial C_\gamma}{\partial p} \right)_T \right]. \tag{7.64}$$

---

[q] In Section 5.3 of Chapter 5, we introduced the concept of relative partial molar properties. In the next section of this chapter we will describe these properties in detail.

Differentiating the equation for $C_\gamma$ and substituting into equations (7.63) and (7.64) gives

$$\bar{L}_2 = C_H\left(\frac{\nu}{2}\right) | z_+ z_- | I_m^{1/2} \tag{7.65}$$

and

$$\bar{V}_2 - \bar{V}_2^\circ = C_v\left(\frac{\nu}{2}\right) | z_+ z_- | I_m^{1/2} \tag{7.66}$$

where

$$C_H = -3RT^2 C_\gamma\left[\frac{1}{T} + \left(\frac{\partial \ln \varepsilon}{\partial T}\right)_p + \frac{\alpha}{3}\right] \tag{7.67}$$

and

$$C_v = RTC_\gamma\left[3\left(\frac{\partial \ln \varepsilon}{\partial p}\right)_T - \left(\frac{\partial \ln V_m}{\partial p}\right)_T\right], \tag{7.68}$$

with $\alpha$ and $V_m$ as the coefficient of expansion and molar volume of the solvent.

Differentiation of equation (7.65) with respect to temperature gives an equation for $\bar{J}_2$, the relative partial molar heat capacity, given by

$$\bar{J}_2 = \bar{C}_{p,2} - \bar{C}_{p,2}^\circ = \left(\frac{\partial \bar{L}_2}{\partial T}\right)_p.$$

The result is

$$\bar{J}_2 = C_J\left(\frac{\nu}{2}\right) | z_+ z_- | I_m^{1/2} \tag{7.69}$$

where $C_J$ is an expression involving multiple derivatives of $\varepsilon$ and $V_m$ with respect to temperature. Values for the coefficients $C_H$, $C_v$, and $C_J$ as a function of temperature for aqueous solutions can be found in Table 7.1 from which the thermal and volumetric properties can be calculated.

## 7.3 Relative Partial Molar and Apparent Relative Partial Molar Thermal Properties

In Chapter 5, we defined the partial molar property $\overline{Z}_i$ and described how it could be used to determine the total thermodynamic property through the equation

$$Z = \sum_i n_i \overline{Z}_i.$$

We described methods for obtaining values for $\overline{V}_i$, $\overline{C}_{p,i}$, and $\overline{S}_i$, but did not apply the methods to $\overline{H}_i$ and $\overline{G}_i$ (or $\mu_i$), since absolute values of $G_m$ and $H_m$ cannot be obtained. We did describe a procedure for obtaining the volume difference $\overline{V}_i - \overline{V}_i^*$ using equations (5.40), (5.41) and (5.42),[r] where $\overline{V}_i^*$ is the volume of the pure substance, and indicated that equations of the same form can be used to obtain $\overline{H}_i - \overline{H}_i^*$. We will return to this method later in this chapter as we describe ways for measuring relative partial molar enthalpies.

### 7.3a Relative Partial Molar Enthalpies

The quantity $\overline{H}_i - \overline{H}_i^\circ$ is called the relative partial molar enthalpy and given the symbol $\overline{L}_i$. It is the difference between the partial molar enthalpy in the solution and the partial molar enthalpy in the standard state. That is,

$$\overline{L}_i = \overline{H}_i - \overline{H}_i^\circ. \tag{7.70}$$

For a Raoult's law standard state, $\overline{H}_i^\circ = H_i^*$ and $\overline{L}_i = \overline{H}_i - H_i^*$. These are the differences described in Chapter 5. For a Henry's law standard state, $\overline{H}_i^\circ$ is the enthalpy in a hypothetical $m = 1$ (or $x_2 = 1$ or $c = 1$) solution that obeys Henry's law. To help in understanding the nature of these standard state enthalpies, we will show that

$$\overline{H}_i(\text{hyp } m = 1) = \overline{H}_i(m = 0).$$

[r] The difference $\overline{G}_i - \overline{G}_i^*$ (or $\mu_i - \mu_i^*$) is obtained from the activity through the relationship

$$\mu_i - \mu_i^* = RT \ln a_i$$

where $a_i$ is the activity with a Raoult's law standard state.

This is done by starting with equation (6.84), which relates the chemical potential of a solute in solution with activity $a_i$, to the standard state chemical potential

$$\mu_i = \mu_i^\circ + RT\ln a_i. \qquad (6.84)$$

If the solute obeys Henry's law, $a_i = m_i$ and the above equation becomes

$$\mu_i - \mu_i^\circ = RT\ln m_i.$$

Dividing by $T$ and differentiating gives

$$\left(\frac{\partial(\mu_i/T)}{\partial T}\right)_p - \left(\frac{\partial(\mu_i^\circ/T)}{\partial T}\right)_p = R\left(\frac{\partial \ln m_i}{\partial T}\right)_p = 0$$

since $m_i$ cannot change with $T$. Recalling that

$$\left(\frac{\partial(\mu_i/T)}{\partial T}\right)_p = -\frac{\overline{H}_i}{T^2} \quad \text{and} \quad \left(\frac{\partial(\mu_i^\circ/T)}{\partial T}\right)_p = -\frac{\overline{H}_i^\circ}{T^2}$$

we get $\overline{H}_i = \overline{H}_i^\circ$ when Henry's law is valid. In the infinitely dilute solution, Henry's law is obeyed. Hence,

$$\overline{H}_i(m = 0) = \overline{H}_i^\circ$$

and $\Delta_{\text{dil}}H = 0$ for the change in concentration from the standard state to the infinitely dilute solution.

We can generalize this result by stating that extrapolating enthalpy changes in solution to infinite dilution gives the enthalpy change $\Delta H^\circ$ for the process[s]

$$\text{A(s, l, or solute)} = \text{A (hyp } m = 1).$$

Thus, with a Henry's law standard state, $\overline{H}_i^\circ$ is the enthalpy in an infinitely dilute solution. For mixtures, in which we choose a Raoult's law standard state for the solvent and a Henry's law standard state for the solute, we can

[s]This statement also works for $\Delta C_p$ but does not work for $\Delta G$. It works for $\Delta H$ because $H$ is independent of concentration in a solution which obeys Raoult's law or Henry's law, but this is not true for $G$, which does depend upon concentration, even in ideal solutions.

write

$$\overline{L}_1 = \overline{H}_1 - H_1^*$$

$$\overline{L}_2 = \overline{H}_2 - \overline{H}_2(m = 0),$$

and in the infinitely dilute solution, $\overline{L}_1 = \overline{L}_2 = 0$.

### 7.3b Calculation of $\Delta H$ from Relative Partial Molar Enthalpies

Relative partial molar enthalpies can be used to calculate $\Delta H$ for various processes involving the mixing of solute, solvent, and solution. For example, Table 7.2 gives values for $\overline{L}_1$ and $\overline{L}_2$ for aqueous sulfuric acid solutions[7] as a function of molality at 298.15 K. Also tabulated is $A$, the ratio of moles $H_2O$ to moles $H_2SO_4$.[t] We note from the table that $\overline{L}_1 = \overline{L}_2 = 0$ in the infinitely dilute solution. Thus, a Raoult's law standard state has been chosen for $H_2O$ and a Henry's law standard state is used for $H_2SO_4$. The value $\overline{L}_2^* = 95{,}281\ \text{J·mol}^{-1}$ is the extrapolated relative partial molar enthalpy of pure $H_2SO_4$. It is the value for $\overline{H}_2^* - \overline{H}_2^\circ$.

The results given in Table 11.1 can be used to calculate $\Delta H$ for various mixing processes.

**Example 7.1:** Calculate $\Delta H$ at 298.15 K for the process

$$H_2SO_4 + 25H_2O = \text{Solution}(H_2SO_4 + 25H_2O).$$

The enthalpy change for this process, in which we mix pure liquids, is known as an **integral enthalpy of solution.**

**Solution:** $\Delta H = n_1\overline{H}_1 + n_2\overline{H}_2 - n_1\overline{H}_1^* - n_2\overline{H}_2^*$
with $n_1 = 25$ and $n_2 = 1$.
Rearranging gives

$$\Delta H = n_1(\overline{H}_1 - \overline{H}_1^*) + n_2(\overline{H}_2 - \overline{H}_2^*).$$

With the choice of standard states used in Table 7.2, $\overline{H}_1^* = \overline{H}_1^\circ$. Making this substitution and adding and subtracting $n_2\overline{H}_2^\circ$ gives

$$\Delta H = n_1(\overline{H}_1 - \overline{H}_1^\circ) + n_2(\overline{H}_2 - \overline{H}_2^* + \overline{H}_2^\circ - \overline{H}_2^\circ)$$

[t] Also tabulated is $\phi L$, the apparent partial molar enthalpy. We will define this quantity and describe its application later.

**Table 7.2** Partial molar thermal properties for aqueous sulfuric acid solutions at $T = 298.15$ K. $m$ is the molality of the $H_2SO_4$ and $A$ is the ratio (moles $H_2O$/moles $H_2SO_4$)*

| $m$/ (moles·kg$^{-1}$) | $A$ | $\overline{L}_1$/ (J·mol$^{-1}$) | $\overline{L}_2$/ (J·mol$^{-1}$) | $\phi L$/ (J·mol$^{-1}$) |
|---|---|---|---|---|
| 0 | ∞ | 0 | 0 | 0 |
| 1.1101 | 50 | −25.5 | 23785 | 22510 |
| 2.2202 | 25 | −99.2 | 26166 | 23686 |
| 3.7004 | 15 | −373 | 31245 | 25650 |
| 5.551 | 10 | −978 | 38525 | 28745 |
| 6.938 | 8 | −1517 | 43328 | 31192 |
| 9.251 | 6 | −2416 | 49542 | 35046 |
| 13.876 | 4 | −4166 | 58056 | 41392 |
| 18.502 | 3 | −5937 | 64169 | 46358 |
| 27.75 | 2 | −9485 | 72800 | 53830 |
| 55.51 | 1 | −20,163 | 87641 | 67478 |
| 111.01 | 0.5 | −28,213 | 93837 | 79731 |
| ∞ | 0 | | 95281 | 95281 |

*Values taken from W. F. Giauque, E. W. Horning, J. E. Kunzler, and T. R. Rubin, "The Thermodynamic Properties of Aqueous Sulfuric Acid Solutions and Hydrates from 15 to 300°C," *J. Am. Chem. Soc.*, **82**, 62–70 (1960). Values of $\overline{L}_2$ were changed to the infinitely dilute solution standard state (instead of pure $H_2SO_4$) using $\Delta_f H^\circ$ (aq) data from D. D. Wagner, W. H. Evans, V. B. Parker, R. H. Schunm, I. Halow, S. M. Bailey, K. L. Churney, and R. L. Nuttall, "The NBS Tables of Chemical Thermodynamic Properties. Selected Values for Inorganic $C_1$ and $C_2$ Organic Substances in SI units," *J. Phys. Chem. Ref. Data*, **11** (1982). Supplement No. 2.

or

$$\Delta H = n_1(\overline{H}_1 - \overline{H}_1^\circ) + n_2(\overline{H}_2 - \overline{H}_2^\circ) - n_2(\overline{H}_2^* - \overline{H}_2^\circ).$$

Substitution of relative partial molar enthalpies gives

$$\Delta H = n_1\overline{L}_1 + n_2\overline{L}_2 - n_2\overline{L}_2^*. \tag{7.71}$$

From Table 7.2

$$\Delta H = (25)(-99.2) + (1)(26166) - (1)(95281)$$

$$= -71591 \text{ J}.$$

An easier way to solve this problem is to define a quantity $L$ that we call the **total relative enthalpy**. It is given by

$$L = H - H^\circ = n_1\overline{H}_1 + n_2\overline{H}_2 - n_1\overline{H}_1^\circ - n_2\overline{H}_2^\circ.$$

Collecting terms gives

$$L = n_1\overline{L}_1 + n_2\overline{L}_2. \tag{7.72}$$

A useful application of $L$ comes from the fact that for any solution process,

$$\Delta H = \sum_{\text{prod}} n_i\overline{H}_i - \sum_{\text{react}} n_i\overline{H}_i$$

$$= \sum_{\text{prod}} n_i(\overline{H}_i - \overline{H}_i^\circ) - \sum_{\text{react}} n_i(\overline{H}_i - \overline{H}_i^\circ)$$

or

$$\Delta H = \Delta L. \tag{7.73}$$

In our example

$$\Delta L = L(\text{solution}) - L(H_2SO_4) - L(H_2O)$$

$$L(\text{solution}) = n_1\overline{L}_1 + n_2\overline{L}_2$$

$$L(H_2SO_4) = n_2\overline{L}_2^*$$

$$L(H_2O) = n_1\overline{L}_1^* = n_1(\overline{H}_1^* - \overline{H}_1^\circ) = 0$$

$$\Delta H = n_1\overline{L}_1 + n_2\overline{L}_2 - n_2\overline{L}_2^* - n_1\overline{L}_1^*.$$

Substituting values from Table 7.2 gives the same answer as before.

**Example 7.2:** Calculate $\Delta H$ for the mixing process

$$25H_2O + \text{solution}(1H_2SO_4 + 25H_2O) = \text{solution}'(1H_2SO_4 + 50H_2O).$$

In this process, the original solution is diluted by the addition of pure solvent, and hence, the enthalpy change is called an **integral enthalpy of dilution**.

**Solution:** $n_1 = 25$, $n_2 = 1$, $n'_1 = 50$, $n'_2 = 1$
From Table 7.2:

$$A = 25: \quad \bar{L}_1 = -99.2 \quad \bar{L}_2 = 26{,}166$$

$$A = 50: \quad \bar{L}'_1 = -25.5 \quad \bar{L}'_2 = 23{,}785$$

$$\Delta H = \Delta L$$
$$= (1)(23{,}785) + (50)(-25.5) - (1)(26{,}166) - (25)(-99.2) - (25)(0)$$
$$= -1176 \text{ J}.$$

The last term in the calculation is $\bar{L}_1^*$, the relative partial molar enthalpy for pure water, which is zero.

**Example 7.3:** Calculate $\Delta H$ when one mole of $H_2SO_4$ is added to a large volume of the solution ($H_2SO_4 + 10H_2O$).

**Solution:** In this example, it is assumed that we add a solute to a large enough volume of solution so that the composition of the mixture does not change. The enthalpy change for this process is referred to as a **differential enthalpy of solution**. We can represent this process by

$$n'_2H_2SO_4 + \text{solution } (n_1H_2O + n_2H_2SO_4) =$$
$$\text{solution } [n_1H_2O + (n_2 + n'_2)H_2SO_4]$$

$$\Delta H = n_1\bar{L}_1 + (n_2 + n'_2)\bar{L}_2 - n_1\bar{L}_1 - n_2\bar{L}_2 - n'_2\bar{L}_2^*.$$

Since the concentrations are assumed to stay the same, $\bar{L}_1$ and $\bar{L}_2$ have the same values in both terms of the expression, they cancel, and we get

$$\Delta H = n'_2\bar{L}_2 - n'_2\bar{L}_2^*.$$

From Table 7.2

$$A = 10: \quad \bar{L}_2 = 38{,}525.$$

Since $\bar{L}_2^* = 95{,}281$ and $n'_2 = 1$

$$\Delta H = 38{,}525 - 95{,}281 = -56{,}756 \text{ J}.$$

A similar calculation involves the addition of a mole of $H_2O$ (instead of $H_2SO_4$) to a large amount of solution. When we add a mole of $H_2O$ ($n'_1 = 1$) to a solution with $A = 10$, $\overline{L}_1 = -978$ J. Since $L_1^* = 0$ and

$$\Delta H = n_1'\overline{L}_1 - n_1'\overline{L}_1^*,$$

then

$$\Delta H = (1)(-978) = -978 \text{ J}.$$

## 7.3c Relative Apparent Molar Enthalpy

In Chapter 5, with equation (5.43), we defined the apparent molar property $\phi Z$ as

$$\phi Z = \frac{Z - n_1 Z_1^*}{n_2}. \tag{5.43}$$

Applying this definition, the apparent molar enthalpy would be given by

$$\phi H = \frac{H - n_1 H_1^*}{n_2} \tag{7.74}$$

where $H_1^*$ is the molar enthalpy of the solvent and $H$ is the total enthalpy of a mixture containing $n_1$ moles of solvent and $n_2$ moles of solute.

We cannot obtain values for $\phi H$ since we cannot obtain absolute values for $H$ or $H_1^*$. To overcome this problem, we define a quantity $\phi L$, which we call the **relative apparent molar enthalpy**, by the equation

$$\phi L = \phi H - \phi H^\circ. \tag{7.75}$$

Thus, the relative apparent molar enthalpy is the difference between the apparent molar enthalpy in the mixture and the apparent molar enthalpy in the standard state.

We can relate $\phi L =$ to $\overline{L}_1$ and $\overline{L}_2$. To do so, we start with

$$H = n_1\overline{H}_1 + n_2\overline{H}_2. \tag{7.76}$$

Substitution of equation (7.76) into equation (7.74) and the resulting relationship into equation (7.75) gives

$$n_2\phi L = n_1\overline{L}_1 + n_2\overline{L}_2 \tag{7.77}$$

or

$$n_2\phi L = L. \tag{7.78}$$

Equation (7.78) allows us to use $\phi L$ to calculate $\Delta H$ for a solution process since

$$\Delta H = \Delta L = n_2 \Delta\phi L. \tag{7.79}$$

Values of $\phi L$ are tabulated and can be used to calculate $\Delta H$ for mixing processes. For example, for the integral enthalpy of solution process given in Example 7.1, we used $\bar{L}_1$ and $\bar{L}_2$ values to show that for the process

$$H_2SO_4 + 25H_2O = \text{solution}(H_2SO_4 + 25H_2O)$$

$$\Delta H = -71{,}595 \text{ J}.$$

We could get this same value by writing

$$\Delta H = n_2\phi L - n_2\phi L^*$$

where $\phi L = 23{,}686$ J is the relative apparent molar enthalpy for the solution (see Table 7.2) and $\phi L^* = 95{,}281$ is the relative apparent molar enthalpy of pure $H_2SO_4$. Thus

$$\Delta H = (1)(23{,}686) - (1)(95{,}281)$$

$$= -71{,}595 \text{ J}.$$

From the nature of apparent properties, we note that the apparent molar enthalpy assigns all of the enthalpy change in forming a mixture to the solute. The result, as shown in equation (7.79), is that all we need to do to calculate $\Delta H$ for a solution process is find the difference in $\phi L$ between the products and reactants. Thus, to solve Example 7.2 using apparent molar enthalpies, we would write

$$25H_2O + \text{solution}(1H_2SO_4 + 25H_2O) = \text{solution}'(1H_2SO_4 + 50H_2O)$$

$$\Delta \text{H} = n_2(\phi L' - \phi L).$$

From Table 7.2

$$\Delta H = (1)(22{,}510 - 23{,}686) = -1176 \text{ J}$$

which agrees with the value obtained from $\bar{L}_1$ and $\bar{L}_2$.

### 7.3d Determination of Relative Apparent Molar Enthalpies

The relative apparent molar enthalpy, $\phi L$, is usually obtained from enthalpy of dilution measurements in which the moles of solute are held constant and additional solvent is added to dilute the starting solution. The process can be represented as

$$n_1 \text{ solvent} + \text{solution}\{n'_1 \text{ solvent} + n_2 \text{ solute}\}$$
$$= \text{solution}\{(n'_1 + n_1)\text{solvent} + n_2 \text{ solute}\}.$$

The enthalpy change for this process is an integral enthalpy of dilution for which we saw earlier that

$$\Delta_{\text{dil}} H_{\text{m}} = n_2\phi L' - n_2\phi L. \tag{7.80}$$

If the resulting solution is infinitely dilute, then $\phi L' = 0$ and $\Delta_{\text{dil}} H_{\text{m}} = \Delta_{\text{dil}}^{\infty} H_{\text{m}} = -n_2\phi L$, where $\Delta_{\text{dil}}^{\infty} H_{\text{m}}$ is the notation we will use to represent the enthalpy of a dilution process that takes a starting solution to the infinitely dilute state.

To show how we can calculate relative apparent molar enthalpies from enthalpies of dilution, consider as an example, a process in which we start with a HCl solution of molality $m$ = 18.50 mol·kg$^{-1}$ and dilute it to a concentration of $m$ = 11.10 mol·kg$^{-1}$. The initial solution contains 3 moles of $H_2O$ per mole of HCl ($A$ = 3) while the final solution has $A$ = 5. The enthalpy change for that process is measured. Then the $m$ = 11.10 mol·kg$^{-1}$ solution is diluted to one with $m$ = 4.63 mol·kg$^{-1}$ and its enthalpy of dilution measured. The series continues as illustrated below.

As indicated by the final equation, the dilution steps are continued until the infinitely dilute solution is approached. The sum of all of the steps represents the change to infinite dilution from the given starting solution. Thus, the sum of all

| Process | Starting $m$/Ending $m$ | $\Delta_{\text{dil}} H_{\text{m}}$ | $\sum \Delta_{\text{dil}} H_{\text{m}}$/(kJ·mol$^{-1}$) |
|---|---|---|---|
| $2H_2O + \{1HCl + 3H_2O\}$ $= \{1HCl + 5H_2O\}$ | 18.50/11.1 | −7.197 | −7.197 |
| $5H_2O + \{1HCl + 5H_2O\}$ $= \{1HCl + 10H_2O\}$ | 11.10/4.63 | −6.318 | −13.515 |
| $15H_2O + \{1HCl + 10H_2O\}$ $= \{1HCl + 25H_2O\}$ | 4.63/2.22 | −2.176 | −15.691 |
| ... | ... | ... | ... |
| $\{\infty - 3\}H_2O + \{1HCl + 3H_2O\}$ $= \{1HCl + \infty H_2O\}$ | 0 | | −18.74 |

the dilution enthalpies yields the desired $\Delta_{\text{dil}}^{\infty}H_{\text{m}}$. As can be seen above, the enthalpy of dilution increments get smaller as the solution becomes more dilute. An extrapolation of this sum can be made to $m = 0$ to obtain a value for $\Delta_{\text{dil}}^{\infty}H_{\text{m}}$, and thus $-\phi L$ for the starting solution. A plot to obtain the extrapolated $\Delta_{\text{dil}}^{\infty}H_{\text{m}}$ for the solution of $\{1 \text{ mole HCl} + 3 \text{ mole H}_2\text{O}\}$ ($m = 18.50$ or $m^{1/2} = 4.30$) is shown in Figure 7.9a. This graph is made in terms of $\sum \Delta_{\text{dil}}H_{\text{m}}$ vs $m^{1/2}$, where $m$ is the molality of the solution on the left side of the cycle. This function of molality is chosen because HCl is a strong electrolyte and Debye–Hückel theory predicts that such a graph should become linear with a known limiting slope as

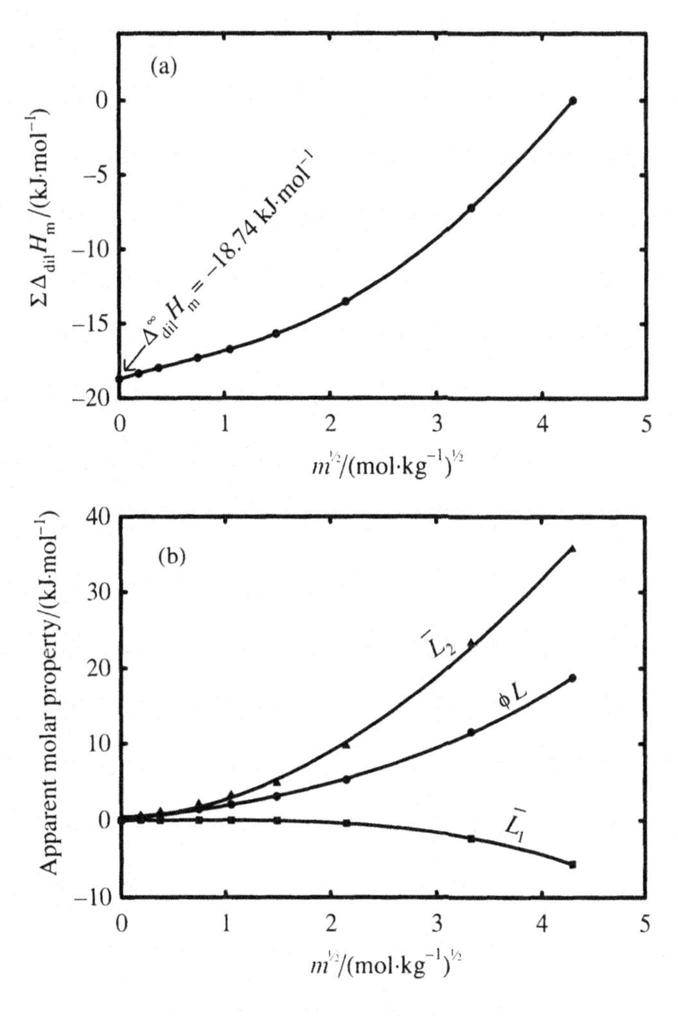


**Figure 7.9** (a) Enthalpies of dilution starting with a solution containing 3 moles $H_2O$ per 1 mole HCl ($m = 18.50$ or $m^{1/2} = 4.30$); and (b), relative apparent molar enthalpies ($\phi L$) and relative partial molar enthalpies ($\bar{L}_1$ and $\bar{L}_2$) for the resulting mixtures.

$m$ goes to 0. The extrapolated value ($\Delta_{\text{dil}}^{\infty} H_{\text{m}}$) for this solution of HCl is found to be $-18.74\ \text{kJ·mol}^{-1}$.

Note that from this extrapolation, one can also obtain $\phi L$ for all of the subsequent solutions, since

$$\phi L(\text{solution } j) = \phi L(\text{solution } i) + \Delta_{\text{dil}} H \text{ (process to dilute solution } i \text{ to solution } j). \tag{7.81}$$

Figure 7.9(b) shows a graph of values of $\phi L$ obtained from the series of dilutions of the starting solution as calculated from equation (7.81). Also shown in Figure 7.9(b) are values for $\overline{L}_1$ and $\overline{L}_2$. These relative partial molar enthalpies can be obtained from $\phi L$. To find the relationship, we start with the equation

$$L = n_2 \phi L. \tag{7.82}$$

Differentiating with respect to $n_2$ at constant $n_1$ gives

$$\left(\frac{\partial L}{\partial n_2}\right)_{n_1} = n_2 \left(\frac{\partial \phi L}{\partial n_2}\right)_{n_1} + \phi L. \tag{7.83}$$

But

$$\left(\frac{\partial L}{\partial n_2}\right)_{n_1} = \overline{L}_2. \tag{7.84}$$

Hence

$$\overline{L}_2 = \phi L + n_2 \left(\frac{\partial \phi L}{\partial n_2}\right)_{n_1}. \tag{7.85}$$

Equivalent forms of equation (7.85) are

$$\overline{L}_2 = \phi L + m \left(\frac{\partial \phi L}{\partial m}\right)_{n_1} \tag{7.86}$$

or

$$\overline{L}_2 = \phi L + \frac{1}{2} m^{1/2} \left(\frac{\partial \phi L}{\partial m^{1/2}}\right)_{n_1} \tag{7.87}$$

since $m$ is moles of solute for a constant number of moles of solvent and

$$m\left(\frac{\partial \phi L}{\partial m}\right)_{n_1} = m\left(\frac{\partial \phi L}{\partial m^{1/2}}\right)_{n_1}\left(\frac{\partial m^{1/2}}{\partial m}\right)_{n_1}$$

$$= \frac{1}{2}m^{1/2}\left(\frac{\partial \phi L}{\partial m^{1/2}}\right)_{n_1}.$$

Equation (7.86) is usually used for obtaining $\bar{L}_2$ from $\phi L$ for a nonelectrolyte mixture. A graph of $\phi L$ against $m$ allows one to obtain the slope as $(\partial \phi L/\partial m)_{n_1}$ from which $\bar{L}_2$ can be calculated. For an electrolyte, equation (7.87) is usually used instead since it is easier to find the slope of $\phi L$ against $m^{1/2}$ [as seen in Figure 7.9(b)].

Substitution of equation (7.85) into equation (7.77) gives an equation for calculating $\bar{L}_1$. The result is[u]

$$\bar{L}_1 = -\frac{n_2^2}{n_1}\left(\frac{\partial \phi L}{\partial n_2}\right)_{n_1}. \tag{7.88}$$

Equivalent expressions involving the molality $m$ are

$$\bar{L}_1 = -m^2 M_1\left(\frac{\partial \phi L}{\partial m}\right)_{n_1} \tag{7.89}$$

or

$$\bar{L}_1 = -\frac{m^{3/2} M_1}{2}\left(\frac{\partial \phi L}{\partial m^{1/2}}\right)_{n_1} \tag{7.90}$$

where $M_1$ is the molecular weight of the solvent (kg·mol$^{-1}$).

---

[u] $\bar{L}_1$ can also be obtained from $\bar{L}_2$ by integration of the Gibbs–Duhem equation

$$\bar{L}_1 = -\int_0^{\bar{L}_2} \frac{n_2}{n_1}\,\mathrm{d}\bar{L}_2.$$

Other methods can be used to obtain $\overline{L}_1$ and $\overline{L}_2$ from calorimetric measurements. For example, the derivation that lead to equations (5.33) and (5.34) in Chapter 5 for calculating $\overline{V}_1$ and $\overline{V}_2$ can be applied to give

$$\overline{H}_1 - H^*_{\mathrm{m},1} = \frac{H^{\mathrm{E}}_{\mathrm{m}}}{x_1} + x_1 x_2 \left[\frac{\partial(H^{\mathrm{E}}_{\mathrm{m}}/x_1)}{\partial x_1}\right]_{p,T} \tag{7.91}$$

$$\overline{H}_2 - \overline{H}^*_{\mathrm{m},2} = \frac{\mathrm{H}^{\mathrm{E}}_{\mathrm{m}}}{x_2} + x_1 x_2 \left[\frac{\partial(H^{\mathrm{E}}_{\mathrm{m}}/x_2)}{\partial x_2}\right]_{p,T}. \tag{7.92}$$

where $H^{\mathrm{E}}_{\mathrm{m}}$ is the excess molar enthalpy. If we choose a Raoult's law standard state for both components, $\overline{H}^{\circ}_1 = H^*_{\mathrm{m},1}$ and $\overline{H}^{\circ}_2 = H^*_{\mathrm{m},2}$ and equations (7.91) and (7.92) become

$$\overline{L}_1 = \frac{H^{\mathrm{E}}_{\mathrm{m}}}{x_1} + x_1 x_2 \left[\frac{\partial(H^{\mathrm{E}}_{\mathrm{m}}/x_1)}{\partial x_1}\right]_{p,T} \tag{7.93}$$

$$\overline{L}_2 = \frac{H^{\mathrm{E}}_{\mathrm{m}}}{x_2} + x_1 x_2 \left[\frac{\partial(H^{\mathrm{E}}_{\mathrm{m}}/x_2)}{\partial x_2}\right]_{p,T}. \tag{7.94}$$

Equations (7.93) and (7.94) are usually applied to mixtures of non-electrolytes where Raoult's law standard states are chosen for both components. For these mixtures, $H^{\mathrm{E}}_{\mathrm{m}}$ is often expressed as a function of mole fraction by the Redlich–Kister equation given by equation (5.40). That is

$$H^{\mathrm{E}}_{\mathrm{m}} = x_1(1 - x_1) \sum_{j=0}^{n} a_j(2x_1 - 1)^j. \tag{7.95}$$

Differentiation and substitution into equations (7.93) and (7.94) in a manner similar to that described in the derivation of equations (5.41) and (5.42) gives

$$\overline{L}_1 = \frac{H^{\mathrm{E}}_{\mathrm{m}}}{x_1} - x_1(1 - x_1)\left\{a_0 + \sum_{j=1}^{n} a_j(2x_1 - 1)^{j-1}[2(j+1)x_1 - (2j+1)]\right\} \tag{7.96}$$

$$\bar{L}_2 = \frac{H_m^E}{x_2} - x_2(1-x_2)\left\{a_0 + \sum_{j=1}^{n} a_j(1-2x_2)^{j-1}[2(j+1) - 2(j+1)x_2]\right\}. \tag{7.97}$$

## 7.3e Relative Partial Molar Heat Capacities

In equation (7.70), we defined the relative partial molar enthalpy $\bar{L}_i$ as

$$\bar{L}_i = \bar{H}_i - \bar{H}_i^\circ. \tag{7.70}$$

Differentiation gives

$$\left(\frac{\partial \bar{L}_i}{\partial T}\right)_{p,n} = \left(\frac{\partial \bar{H}_i}{\partial T}\right)_{p,n} - \left(\frac{\partial \bar{H}_i^\circ}{\partial T}\right)_{p,n}. \tag{7.98}$$

Substitution into equation (7.98) the relationships

$$\left(\frac{\partial \bar{H}_i}{\partial T}\right)_{p,n} = \bar{C}_{p,i}$$

and

$$\left(\frac{\partial \bar{H}_i^\circ}{\partial T}\right)_{p,n} = \bar{C}_{p,i}^\circ$$

gives

$$\left(\frac{\partial \bar{L}_i}{\partial T}\right)_p = \bar{C}_{p,i} - \bar{C}_{p,i}^\circ.$$

The difference $\bar{C}_{p,i} - \bar{C}_{p,i}^\circ$ is the **relative partial molar heat capacity** $\bar{J}_i$. Thus

$$\bar{J}_i = \bar{C}_{p,i} - \bar{C}_{p,i}^\circ \tag{7.99}$$

and

$$\left(\frac{\partial \bar{L}_i}{\partial T}\right)_p = \bar{J}_i. \tag{7.100}$$

The quantity $\bar{J}_i$ can be used to calculate the effect of temperature on $\bar{L}_i$ through equation (7.100).

**Example 7.4:** At $T = 298.15$ K, for a 0.100 M aqueous NaCl solution, $\gamma_\pm = 0.778$, $\bar{L}_2 = 431$ J·mol$^{-1}$ and $\bar{J}_2 = 20.7$ J·K$^{-1}$·mol$^{-1}$. Assume $\bar{J}_2$ is constant with temperature and calculate $\bar{L}_2$ and $\gamma_\pm$ for 0.100 M NaCl at 373.15 K.

**Solution:** First, we obtain an expression for $\bar{L}_2$ as a function of temperature

$$\left(\frac{\partial \bar{L}_2}{\partial T}\right)_p = \bar{J}_2 = 20.7.$$

Separating variables and doing an indefinite integration gives

$$\int \mathrm{d}\bar{L}_2 = \int 20.7\ \mathrm{d}T$$

$$\bar{L}_2 = I + 20.7T$$

where $I$ is a constant of integration. Its value can be obtained from the $T = 298.15$ K data

$$I = 431 - (20.7)(298.15) = -5741$$

so that

$$\bar{L}_2 = -5741 + 20.7T. \tag{7.101}$$

When $T = 373.15$ K

$$\bar{L}_2 = -(5741) + (20.7)(373.15) = 1983\ \mathrm{J{\cdot}mol^{-1}}.$$

To get $\gamma_\pm$ at 373.15 K, we start with equation (6.90)

$$\left(\frac{\partial \ln a_2}{\partial T}\right)_p = -\frac{\bar{L}_2}{RT^2}.$$

With $a_2 = \gamma_\pm^2 m^2$

$$\left(\frac{\partial \ln a_2}{\partial T}\right)_p = 2\left(\frac{\partial \ln \gamma_\pm}{\partial T}\right)_p$$

or

$$\left(\frac{\partial \ln \gamma_\pm}{\partial T}\right)_p = -\frac{\bar{L}_2}{2RT^2}. \tag{7.102}$$

Substituting equation (7.101) into equation (7.102) and integrating gives

$$\int_{\ln\gamma_\pm}^{\ln\gamma'_\pm} \mathrm{d}\ln\gamma_\pm = -\int_T^{T'} \frac{(-5742 + 20.7T)}{2RT^2}\,\mathrm{d}T$$

$$\ln\frac{\gamma'_\pm}{\gamma_\pm} = -\frac{2871}{R}\left(\frac{1}{T'} - \frac{1}{T}\right) - \frac{10.35}{R}\ln\frac{T'}{T}.$$

Substituting $\gamma_\pm = 0.778$, $T = 298.15$ K, and $T' = 373.15$ K gives $\gamma_\pm = 0.743$ at the higher temperature.

## 7.3f Relative Apparent Molar Heat Capacity:

The **relative apparent molar heat capacity** $\phi J$ is defined as

$$\phi J = \frac{J - n_1 J_1^*}{n_2} \tag{7.103}$$

where $J$ is the total relative heat capacity given by

$$J = n_1\bar{J}_1 + n_2\bar{J}_2. \tag{7.104}$$

Starting with $\bar{J}_i = \bar{C}_{p,i} - \bar{C}^{\circ}_{p,i}$ and equations (7.103) and (7.104), it is easy to find the relationship between $\phi J$ and $\phi C_p$, the apparent molar heat capacity defined as[v]

$$\phi C_p = \frac{C_p - n_1 C^*_{p,1}}{n_2}. \tag{7.105}$$

The relationship is

$$\phi J = \phi C_p - \bar{C}^{\circ}_{p,2} \tag{7.106}$$

where $\bar{C}^{\circ}_{p,2}$ is the partial molar heat capacity of the solute in the standard state. That is, it is the partial molar heat capacity for the solute extrapolated to infinite dilution.

In Chapter 5 we saw how to determine $\bar{C}_{p,2}$ from heat capacity measurements. $\phi J$ is easily obtained from these same measurements using equation (7.106), after the value of $\bar{C}^{\circ}_{p,2}$ is obtained by extrapolation of $\bar{C}_{p,2}$ to infinite dilution.

$\phi J$ can be used to calculate the effect of temperature on $\phi L$ through the relationship

$$\left(\frac{\partial \phi L}{\partial T}\right)_p = \phi J. \tag{7.107}$$

Of more importance, just as $\phi L$ provides a way for obtaining $\bar{L}_1$ and $\bar{L}_2$, $\phi J$ provides a method for obtaining $\bar{J}_1$ and $\bar{J}_2$. The derivation is similar to that used to obtain $\bar{L}_1$ and $\bar{L}_2$ from $\phi L$. One starts with equation (7.103) written as

$$J = n_2 \phi J + n_1 J^*_1$$

and differentiates with respect to $n_2$ to get

$$\bar{J}_2 = \phi J + n_2 \left(\frac{\partial \phi J}{\partial n_2}\right)_{n_1}. \tag{7.108}$$

---

[v] We are assuming a Raoult's law standard state for the solvent and a Henry's law standard state for the solute.

Differentiating with respect to $n_1$ gives

$$\bar{J}_1 = n_2\left(\frac{\partial \phi J}{\partial n_1}\right) + J_1^*. \tag{7.109}$$

Conversions similar to those used to obtain equations (7.86), (7.87), (7.89), and (7.90) can be used to obtain equations involving $m$ (or $m^{1/2}$) instead of $n_2$.

## 7.4 The Osmotic Pressure

The osmotic pressure is a property that has proven to be especially valuable in the study of solutions of macromolecules, including those of biologic and polymeric interest. The apparatus for measuring this quantity is shown schematically in Figure 7.10. Two compartments are separated by a membrane that will allow the flow of liquid solvent between the two chambers. If solvent is added, flow will occur until the liquid level on the two sides of the membrane is the same.

We now add a solute to the compartment on the left. The solute is confined to this compartment, since we have chosen a membrane that will allow the flow of solvent, but not the flow of solute. Such a device is called a **semipermeable membrane**.

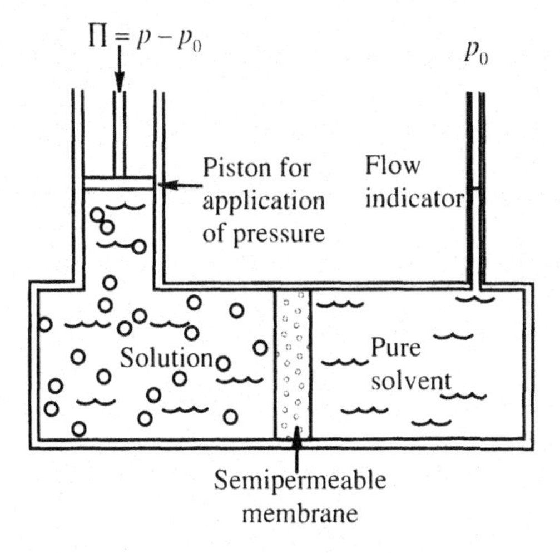


**Figure 7.10** Schematic representation of the apparatus for measuring osmotic pressure. The flow of solvent through the semipermeable membrane is followed by observing the movement of the meniscus of the flow indicator. The osmotic pressure Π is the pressure that must be applied to the solution to prevent the flow.

Before adding the solute, equilibrium has been established between the two compartments so that the chemical potentials, which are a function of the temperature $T$ and the pressure $p^\circ$ of the atmosphere above the solvent, are the same in the two sides. Adding a solute to the left side decreases the chemical potential, and unless we provide some mechanism for preventing it, solvent will flow from the right side to the left side, raising the liquid level on the left and lowering it on the right. The result is a pressure difference between the two sides that will continue to increase due to the flow until the chemical potentials are again equal on the two sides, at which point equilibrium is again established and the flow stops.

We can prevent the flow of solvent by placing a piston on top of the chamber on the left and applying a pressure $p$ to maintain equilibrium so that the liquid levels on the two sides of the membrane stay equal. The difference in pressure $p - p^\circ$ is known as the osmotic pressure $\Pi$ so that

$$\Pi = p - p^\circ. \tag{7.110}$$

By applying the pressure $p$ to the mixture, we keep the chemical potential of the solvent in the solution on the left equal to the chemical potential of the pure solvent on the right where the pressure is $p^\circ$. We can write this equality as

$$\mu_1(p, T, x_1) = \mu_{1,0}(p^\circ, T)$$

where $\mu_1$ is the chemical potential of the solvent at pressure $p$ and temperature $T$ in the mixture in which $x_1$ is the mole fraction of solvent, and $\mu_{1,0}$ is the chemical potential of the pure solvent at pressure $p_1^\circ$ and temperature $T$.

If we change the $(p, T, x_1)$ conditions of the mixture such that the chemical potential changes by $\mathrm{d}\mu_1$, then $\mu_{1,0}$ would need to change by an equivalent amount to maintain equilibrium. That is

$$\mathrm{d}\mu_1(p, T, x_1) = \mathrm{d}\mu_{1,0}(p^\circ, T).$$

But

$$\mu_{1,0} = f(p^\circ, T)$$

so that

$$\mathrm{d}\mu_{1,0} = \left(\frac{\partial \mu_{1,0}}{\partial p^\circ}\right)_T \mathrm{d}p^\circ + \left(\frac{\partial \mu_{1,0}}{\partial T}\right)_{p^\circ} \mathrm{d}T,$$

and if we keep $p^\circ$ and $T$ constant then $d\mu_{1,0} = 0$. This requires that $d\mu_1 = 0$ to maintain equilibrium.

Since $\mu_1 = f(p, T, x_1)$ we can write

$$d\mu_1 = \left(\frac{\partial\mu_1}{\partial T}\right)_{p,x_1} dT + \left(\frac{\partial\mu_1}{\partial p}\right)_{T,x_1} dp + \left(\frac{\partial\mu_1}{\partial x_1}\right)_{T,p} dx_1 = 0. \tag{7.111}$$

Since $T$ is constant, $dT = 0$, and equation (7.111) reduces to

$$\left(\frac{\partial\mu_1}{\partial p}\right)_{T,x_1} dp + \left(\frac{\partial\mu_1}{\partial x_1}\right)_{T,p} dx_1 = 0 \tag{7.112}$$

where $p$ is the equilibrium pressure above the mixture.

We can evaluate the derivatives in equation (7.112) by recognizing that

$$\left(\frac{\partial\mu_1}{\partial p}\right)_{T,x_1} = \bar{V}_1 \tag{7.113}$$

where $\bar{V}_1$ is the partial molar volume of the solvent in the mixture. Also, assuming a Raoult's law standard state for the solvent, we can write

$$\mu_1 = \mu_1^\circ + RT \ln \gamma_{R,1} x_1 \tag{7.114}$$

where $\gamma_{R,1} x_1 = a_1$, with $a_1$ as the activity of the solvent. Differentiating equation (7.114) gives

$$\left(\frac{\partial\mu_1}{\partial x_1}\right)_{p,T} = \left(\frac{\partial\mu_1^\circ}{\partial x_1}\right)_{p,T} + RT\left[\left(\frac{\partial \ln \gamma_{R,1}}{\partial x_1}\right)_{p,T} + \left(\frac{\partial \ln x_1}{\partial x_1}\right)_{p,T}\right]. \tag{7.115}$$

Equation (7.115) can be simplified by recognizing that

$$\left(\frac{\partial\mu_1^\circ}{\partial x_1}\right)_{p,T} = 0$$

since the standard state chemical potential does not depend upon composition, and

$$\left(\frac{\partial \ln x_1}{\partial x_1}\right)_{p,T} = \frac{1}{x_1}.$$

Hence

$$\left(\frac{\partial \mu_1}{\partial x_1}\right)_{p,T} = RT\left[\left(\frac{\partial \ln \gamma_{R,1}}{\partial x_1}\right)_{p,T} + \frac{1}{x_1}\right] \tag{7.116}$$

Substitution of equations (7.113) and (7.116) into equation (7.112) gives an equation that relates $p$ to $x_1$. The result is

$$\bar{V}_1 \, \mathrm{d}p + RT\left(\frac{\partial \ln \gamma_{R,1}}{\partial x_1}\right)_{p,T} \mathrm{d}x_1 + RT\frac{\mathrm{d}x_1}{x_1} = 0. \tag{7.117}$$

Equation (7.117) can be integrated to relate $p$ (and hence $T$) to $x_1$ when it is known how $\bar{V}_1$ and $\gamma_{R,1}$ are related to $p$ and $x_1$. We will use it as the starting point to derive equations for $\Pi$ that apply in dilute solution.

For the dilute solution, we assume that Raoult's law applies for the solvent, so that $\bar{V}_1 = V_1^*$, the volume of the pure solvent $\gamma_{R,1} = 1$, and $(\partial \ln \gamma_{R,1}/\partial x_1)_{p,T} = 0$. Under these conditions

$$V_1^* \, \mathrm{d}p = -RT \, \mathrm{d} \ln x_1.$$

Integration between $\ln x_1$ where the pressure is $p$, and $\ln(1) = 0$ where the pressure is $p°$ gives

$$\int_{p°}^{p} V_1^* \, \mathrm{d}p = -RT \int_0^{\ln x_1} \mathrm{d} \ln x_1. \tag{7.118}$$

The integration of the left side of equation (7.118) requires that we know how $V_1^*$ changes with $p$. Compressibility data would provide this information. However, $p$ is not a large pressure and liquid solvents are not usually very compressible, so that it is a reasonable approximation to assume $V_1^*$ is independent of pressure, in which case, integration of equation (7.118) gives

$$V_1^*(p - p^\circ) = -RT \ln x_1 \tag{7.119}$$

or, since $p - p^\circ = \Pi$, we get

$$\Pi = -\frac{RT}{V_1^*} \ln x_1. \tag{7.120}$$

Equation (7.120) can be used to calculate $\Pi$ in a dilute solution where Raoult's law is a good approximation of the behavior of the solvent. This equation is often put in an alternate form by making further approximations that apply to the dilute solution. First, we write

$$\ln x_1 = \ln(1 - x_2)$$

and expand $\ln(1 - x_2)$ in a series expansion

$$\ln(1 - x_2) = -x_2 - \frac{x_2^2}{2} - \frac{x_2^3}{3} \cdots$$

which can be truncated in dilute solution to

$$\ln(1 - x_2) = -x_2.$$

Substitution into equation (7.120) gives

$$\Pi = x_2 \frac{RT}{V_1^*}. \tag{7.121}$$

Still further approximation can be made by writing

$$x_2 = \frac{n_2}{n_1 + n_2} \cong \frac{n_2}{n_1}$$

since $n_2 \ll n_1$. Thus

$$\Pi = \frac{n_2 RT}{n_1 V_1^*}. \tag{7.122}$$

In dilute solutions, $n_1 V_1^* = V$, the volume of the solution, and $n_2/V = c$, the molar concentration of the solute. Substitution into equation (7.122) gives

$$\Pi = cRT. \tag{7.123}$$

Equation (7.123)[w] is often referred to as the law of Van't Hoff, since it was originally proposed by J. H. Van't Hoff. It is interesting to note that equation (7.123) is of the same form as the ideal gas equation, if one takes $c$ as $n/V$.[x]

Table 7.3 compares experimental values of $\Pi$ for aqueous sucrose solutions with those calculated using equations (7.120) and (7.123). We see that neither of the equations predicts $\Pi$ with high accuracy. However, the superiority of equation (7.120), especially at higher concentrations, is apparent.

It is also interesting to note from Table 7.3 the large values for $\Pi$, even in dilute solution. For example, we see that with $c = 0.098\ \text{mol·dm}^{-3}$, $\Pi = 262$ kPa. Assuming this solution has the same density as pure water, this pressure corresponds to a column of solution approximately 26 m high. In comparison, assuming that this solution is dilute enough so that Raoult's law applies, the vapor pressure is 99.82% of that of pure water (since $x_1 = 0.9982$).

**Table 7.3** Osmotic pressures of aqueous sucrose solutions at $T = 293.15$ K

| | | | $\Pi$/(kPa) | | |
|---|---|---|---|---|---|
| $m$/(mol·kg$^{-1}$) | $x_1$ | $c$/(mol·dm$^{-3}$) | Observed | From $\Pi = cRT$ | From $\Pi = -(RT \ln x_1)/V_1^*$ |
| 0.1 | 0.9982 | 0.098 | 262 | 238 | 244 |
| 0.2 | 0.9964 | 0.192 | 513 | 468 | 488 |
| 0.3 | 0.9946 | 0.282 | 771 | 687 | 733 |
| 0.4 | 0.9928 | 0.370 | 1027 | 902 | 978 |
| 0.5 | 0.9911 | 0.453 | 1292 | 1104 | 1209 |
| 0.6 | 0.9893 | 0.533 | 1559 | 1299 | 1455 |
| 0.7 | 0.9875 | 0.610 | 1837 | 1487 | 1701 |
| 0.8 | 0.9858 | 0.685 | 2119 | 1669 | 1935 |
| 0.9 | 0.9840 | 0.757 | 2403 | 1845 | 2182 |
| 1.0 | 0.9823 | 0.825 | 2699 | 2011 | 2416 |

[w] With $c$ expressed in mol·dm$^{-3}$, $\Pi$ will have the units of kPa.

[x] One should not take the correspondence between the equation for $\Pi$ and the equation for the pressure of the ideal gas too far. The ideal gas, composed of molecules that do not interact, is certainly very different than a liquid solvent containing a solute.

Also, the freezing point of this solution would be lowered 0.2 K below that of pure water.

The large $\Pi$ makes osmotic pressure measurements useful for determining molecular weights of high molecular weight solutes, such as polymers, since a very dilute solution still gives a substantial effect. The method is often used in biochemistry to determine the molecular weight of solutes such as polypeptides. For example, for a polypeptide with a molecular weight of 10 $kg{\cdot}mol^{-1}$, a solution obtained by dissolving 10 g of this substance in 1 kg of water gives an osmotic pressure of 2.5 kPa, a value that is easily measured. In contrast, a freezing point lowering of 0.002 K would need to be accurately measured if one tried to determine the molecular weight of the polypeptide from the freezing point lowering.

As a caution, however, one must carefully purify the sample since small amounts of low molecular weight impurities can contribute significantly to $\Pi$, even though their mass percentage in the sample is small.

### 7.4a Osmosis

If the pressure on the piston in the left cell in Figure 7.10 is less than $\Pi$, solvent will flow from the pure solvent on the right to the solution on the left. This process is called **osmosis**. The process can be reversed by increasing the pressure on the left to a value greater than $\Pi$. The result will be a flow from left to right. This process is known as **reverse osmosis** and is used to purify water in desalination plants. The key to making this process work is to find membranes that will allow the passage of water at a reasonable rate, that are strong enough to withstand the pressure difference, and are substantially impermeable to salt that is present as a solute in the impure water solute. Membranes of cellulose acetate or hollow fibers with very small pores are often used.

Osmosis across cell membranes is very important in biology, but the process is more complicated than we have pictured it in Figure 7.10. Cell membranes are permeable to water, and also to other substances such as $CO_2$, $O_2$, $N_2$, and low molecular weight organic compounds such as amino acids and simple sugars. Large polymer molecules such as proteins and polysaccharides, however, will not pass through the membrane. Intermediate to these two extremes are inorganic ions (such as NaCl) and disaccharides (such as sucrose) that pass slowly through the membrane. In a living organism, the cells are bathed in body fluids that contain various solutes while the fluid within the cell contains other solutes. Thus solutes are present on both sides of the membrane, and a flow of fluid in or out of the cell depends upon the difference in chemical potential across the cell membrane, which in turn depends upon the relative concentration of solutes on the two sides. If the activity (approximately concentration) of solutes in the fluid surrounding a cell is higher than in the cell fluid, then the cell will lose water by osmosis. When this occurs, the surrounding fluid is said to be **hypertonic**. For the reverse process when fluid flows into the cell, the surrounding fluid is said to be **hypotonic**. When there is no net flow, the two fluids are said to be **isotonic**.

Blood and lymph are approximately isotonic to a cell so that cells do not gain or lose liquid when bathed in these fluids. Pure water is hypotonic and may cause cells to swell and burst. During intravenous feeding, injections, and storage of cell tissue, a salt (saline) solution is used with a concentration of solutes that is essentially isotonic with blood (and hence, with the cell) to prevent cell damage.

## Exercises

E7.1 For mixtures of 1-hexene (component 1) + n-hexane (component 2), the total vapor pressure $p$ above the mixture is related to $x_1$ and $y_1$, the mole fraction in the liquid and vapor phases as follows:

$T/(\mathrm{K}) = 333.15$

| $x_1$ | $y_1$ | $p/(\mathrm{kPa})$ | $x_1$ | $y_1$ | $p/(\mathrm{kPa})$ |
|---|---|---|---|---|---|
| 0.00 | 0.000 | 76.45 | 0.55 | 0.589 | 84.79 |
| 0.05 | 0.060 | 77.27 | 0.60 | 0.637 | 85.47 |
| 0.10 | 0.118 | 78.08 | 0.65 | 0.684 | 86.15 |
| 0.15 | 0.175 | 78.87 | 0.70 | 0.731 | 86.82 |
| 0.20 | 0.230 | 79.65 | 0.75 | 0.777 | 87.48 |
| 0.25 | 0.284 | 80.42 | 0.80 | 0.822 | 88.12 |
| 0.30 | 0.338 | 81.18 | 0.85 | 0.867 | 88.76 |
| 0.35 | 0.390 | 81.92 | 0.90 | 0.912 | 89.39 |
| 0.40 | 0.441 | 82.66 | 0.95 | 0.956 | 90.01 |
| 0.45 | 0.491 | 83.38 | 1.00 | 1.000 | 90.62 |
| 0.50 | 0.541 | 84.09 | | | |

(a) Construct the (vapor + liquid) phase diagram by graphing $p$ against $x_1$ and $y_1$. Label the regions.

(b) Calculate the excess molar Gibbs free energy $G_\mathrm{m}^\mathrm{E}$ at each mole fraction and make a graph of $G_\mathrm{m}^\mathrm{E}$ against mole fraction.

E7.2 The following table gives the partial molal volumes at $T = 298.15$ K of ethyl acetate (1) and carbon tetrachloride (2) in solutions of the two.

(a) What is the volume of the solution when 3 moles of ethyl acetate are mixed with 7 moles of carbon tetrachloride?

(b) Calculate the change in volume when 0.6 moles of ethyl acetate are mixed with 0.4 moles of carbon tetrachloride. This quantity is the excess molar volume $V_\mathrm{m}^\mathrm{E}$.

| $x_1$ | $\bar{V}_1/(\mathrm{cm^3 \cdot mol^{-1}})$ | $\bar{V}_2/(\mathrm{cm^3 \cdot mol^{-1}})$ |
|---|---|---|
| 1.0 | 97.81 | 96.74 |
| 0.9 | 97.81 | 96.68 |
| 0.8 | 97.82 | 96.63 |
| 0.7 | 97.83 | 96.59 |
| 0.6 | 97.87 | 96.55 |
| 0.5 | 97.87 | 96.52 |
| 0.4 | 97.91 | 96.49 |
| 0.3 | 97.96 | 96.47 |
| 0.2 | 98.03 | 96.45 |
| 0.1 | 98.13 | 96.44 |
| 0.0 | 98.25 | 96.43 |

E7.3 Given the following partial molar volumes and vapor pressures for ($x_1$ n-$C_6H_{14}$ + $x_2$ c-$C_6H_{12}$) at $T = 298.15$ K.

| $x_2$ | $\bar{V}_1/(\mathrm{cm^3 \cdot mol^{-1}})$ | $\bar{V}_2/(\mathrm{cm^3 \cdot mol^{-1}})$ | $p_1/(\mathrm{kPa})$ | $p_2/(\mathrm{kPa})$ |
|---|---|---|---|---|
| 0.0 | 131.596 | 109.021 | 2.689 | 0.000 |
| 0.1 | 131.596 | 109.016 | 2.421 | 0.188 |
| 0.2 | 131.596 | 109.018 | 2.157 | 0.372 |
| 0.3 | 131.598 | 109.012 | 1.891 | 0.551 |
| 0.4 | 131.608 | 108.993 | 1.635 | 0.725 |
| 0.5 | 131.634 | 108.963 | 1.373 | 0.897 |
| 0.6 | 131.685 | 108.922 | 1.112 | 1.067 |
| 0.7 | 131.777 | 108.873 | 0.848 | 1.233 |
| 0.8 | 131.941 | 108.819 | 0.579 | 1.399 |
| 0.9 | 132.224 | 108.771 | 0.300 | 1.564 |
| 1.0 | 132.706 | 108.747 | 0.000 | 1.735 |

For the process at $T = 298.15$ K:

0.3 moles n-hexane + 0.7 moles cyclohexane = solution

(a) What is the excess volume $V_m^E$?

(b) Assuming a Raoult's law standard state for both components, calculate $G_m^E$,the excess Gibbs free energy for the process.

E7.4 At $T = 298.15$ K, the excess molar enthalpies $H_m^E$, and the excess molar Gibbs free energies $G_m^E$ for ($x_1$ c-$C_6H_{12}$ + $x_2$ c-$C_6H_{11}CH_3$) are given by

$$H_m^E/(\mathrm{J{\cdot}mol^{-1}}) = x_1x_2[73.79 + 56.05(1 - 2x_1) - 30.72(1 - 2x_1)^2 + 2.93(1 - 2x_1)^3 + 8.28(1 - 2x_1)^4]$$

$$G_m^E/(\mathrm{J{\cdot}mol^{-1}}) = x_1x_2[34.9 - 1.2(1 - 2x_1) - 0.8(1 - 2x_1)^2].$$

Make a graph of $G_m^E$, $H_m^E$, and $TS_m^E$ against $x_1$ and compare the values.

E7.5 Use the Debye–Hückel theory to calculate the activity at 298.15 K of $CaCl_2$ in the following aqueous solutions:
(a) A solution 0.0050 $m$ in $CaCl_2$
(b) A solution 0.0025 $m$ in $CaCl_2$ and 0.0050 $m$ in KCl.

E7.6 Given the following permittivities (dielectric constants) $\varepsilon_A$ for several liquid solvents at 293.15 K.

| Solvent | $\varepsilon_A$ |
|---|---|
| $CH_3CN$ | 37.5 |
| $CH_3OH$ | 33.62 |
| $(CH_3)_2SO$ | 48.9 |
| HCN | 116 |
| $H_2O$ | 80.10 |

Use the Debye–Hückel theory to calculate $\gamma_\pm$ for 0.0050 $m$ HCl dissolved in each of these solvents and make a graph of $\ln \gamma_\pm$ against $1/\varepsilon_A^{3/2}$.

E7.7 Use the Debye–Hückel theory to calculate at $T = 298.15$ K, $[\mu_2(m) - \mu_2(m = 0.010)]$, the change in Gibbs free energy for the solute when a 0.010 $m$ NaOH aqueous solution is diluted to 0.0050 $m$. Repeat the calculation for successive dilutions to 0.0025 $m$, 0.0010 $m$, and 0.00050 $m$. Graph the results and decide what happens to $[\mu_2(m) - \mu_2(m = 0.010)]$ as $m \longrightarrow 0$.

E7.8 Use the Debye–Hückel theory to calculate at $T = 298.15$ K, $[\overline{H}_2(m) - \overline{H}_2\ (m = 0.010)]$, the change in enthalpy for the solute when a 0.010 $m$ NaOH aqueous solution is diluted to 0.0050 $m$. Repeat the calculation for successive dilutions to 0.0025 $m$, 0.0010 $m$, and 0.00050$m$. Graph the results and decide what happens to $[\overline{H}_2\ (m) - \overline{H}_2\ (m = 0.010)]$ as $m \longrightarrow 0$.

E7.9 The following thermal properties have been measured for sodium hydroxide solutions at 298.15 K.

| $m/(\text{mol}\cdot\text{kg}^{-1})$ | $\bar{L}_1/(\text{J}\cdot\text{mol}^{-1})$ | $\bar{L}_2/(\text{J}\cdot\text{mol}^{-1})$ | $\bar{C}_{p,1}/(\text{J}\cdot\text{K}^{-1}\cdot\text{mol}^{-1})$ | $\bar{C}_{p,2}/(\text{J}\cdot\text{K}^{-1}\cdot\text{mol}^{-1})$ |
|---|---|---|---|---|
| 0.5 | 2.5 | −58.6 | 75.14 | −47.3 |
| 1.0 | 10.0 | −678 | 74.52 | −6.7 |
| 1.5 | 18.4 | −1046 | 73.81 | 18.4 |
| 2.0 | 22.2 | −1151 | 73.05 | 32.6 |
| 3.0 | 1.3 | −711 | 71.80 | 40.6 |
| 4.0 | −67.4 | 368 | 70.88 | 32.2 |

$\bar{L}_2^* = -41{,}840\ \text{J}\cdot\text{mol}^{-1}$ for solid NaOH.

Starting at an initial temperature of 298.15 K, assume $C_{p,\text{m}}$ is independent of temperature and determine the final temperature when:
(a) 1.00 mole of solid NaOH is dissolved in 1.000 kg $H_2O$.
(b) A solution containing 1.000 mole of NaOH and 0.500 kg of $H_2O$ is mixed with one containing 0.500 mole of NaOH and 1.000 kg of $H_2O$.

E7.10 It has been shown that the equation

$$\phi L = 128.9m$$

expresses the relative apparent molar enthalpy of aqueous sucrose solutions at 298.15 K. Find expressions for $\bar{L}_2$ and $\bar{L}_1$ as a function of the molality $m$.

E7.11 Given the following data for {Na(1) + K(2)} liquid mixtures at $T = 384$ K,

| $x_2$ | $\gamma_1$ | $\gamma_2$ | $\bar{L}_1/(\text{J}\cdot\text{mol}^{-1})$ | $\bar{L}_2/(\text{J}\cdot\text{mol}^{-1})$ |
|---|---|---|---|---|
| 1.0 | 2.438 | 1.000 | 2510 | 0 |
| 0.9 | 2.107 | 1.008 | 2096 | 21 |
| 0.8 | 1.835 | 1.033 | 1707 | 92 |
| 0.7 | 1.615 | 1.078 | 1435 | 180 |
| 0.6 | 1.437 | 1.148 | 1180 | 318 |
| 0.5 | 1.295 | 1.250 | 929 | 523 |
| 0.4 | 1.186 | 1.391 | 640 | 883 |
| 0.3 | 1.103 | 1.592 | 393 | 1343 |
| 0.2 | 1.046 | 1.870 | 184 | 1979 |
| 0.1 | 1.012 | 2.260 | 46 | 2749 |
| 0.0 | 1.000 | 2.816 | 0 | 3648 |

(a) What are the standard states chosen for K and Na in generating the data? Justify your answer.
(b) Calculate $\Delta H$ and $\Delta G$ (in Joules) at 384 K for the process:

$$\text{K(4 moles)} + \text{Na(1 mole)} = \text{solution.}$$

E7.12 The following table gives the osmotic pressures as a function of concentration for the polymer polyisobutylene dissolved in benzene at 298.15 K.

| $c/(\text{g}\cdot\text{dm}^{-3})$ | $\Pi/(\text{Pa})$ |
|---|---|
| 20.0 | 210.3 |
| 15.0 | 150.4 |
| 10.0 | 100.5 |
| 5.0 | 49.5 |

Extrapolate the data to infinite dilution to obtain a value for the molecular weight of the polymer. (Note that an average molecular weight is obtained since the polymer consists of a mixture of molecules of different chain lengths.)

E7.13 Twenty mg of a protein are dissolved in 10 g of water. The osmotic pressure at 298.15 K is determined to be 0.040 kPa. Estimate the molecular weight of the protein.

E7.14 Estimate the vapor pressure lowering and the osmotic pressure at 293.15 K for an aqueous solution containing 50.0 g of sucrose ($M_2 = 0.3423\ \text{kg}\cdot\text{mol}^{-1}$) in 1 kg of water. At this temperature, the density of pure water is $0.99729\ \text{g}\cdot\text{cm}^{-3}$ and the vapor pressure is 2.33474 kPa. Compare your results with those given in Table 7.3.

## Problems

P7.1 Vapor pressure data for ethanol (1) + 1,4-dioxane (2) at $T = 323.15$ K is given in the following table, where $x_1$ is the mole fraction in the liquid phase and $y_1$ is the mole fraction in the vapor phase.
(a) Graph the data to construct the (vapor + liquid) phase diagram and determine the composition of the minimum boiling azeotrope.
(b) Use the vapor pressure data to calculate $G_m^E$ at $T = 323.15$ K at each of the mole fractions.

$T/(\mathrm{K}) = 323.15$

| $x_1$ | $y_1$ | $p/(\mathrm{kPa})$ | $x_1$ | $y_1$ | $p/(\mathrm{kPa})$ |
|---|---|---|---|---|---|
| 0.00 | 0.0000 | 15.697 | 0.50 | 0.6625 | 28.666 |
| 0.05 | 0.1763 | 18.161 | 0.55 | 0.6870 | 29.154 |
| 0.10 | 0.2943 | 20.220 | 0.60 | 0.7106 | 29.566 |
| 0.15 | 0.3797 | 21.953 | 0.65 | 0.7341 | 29.908 |
| 0.20 | 0.4449 | 23.420 | 0.70 | 0.7583 | 30.183 |
| 0.25 | 0.4969 | 24.668 | 0.75 | 0.7841 | 30.388 |
| 0.30 | 0.5398 | 25.733 | 0.80 | 0.8126 | 30.511 |
| 0.35 | 0.5762 | 26.644 | 0.85 | 0.8454 | 30.534 |
| 0.40 | 0.6080 | 27.425 | 0.90 | 0.8849 | 30.419 |
| 0.45 | 0.6364 | 28.094 | 0.95 | 0.9345 | 30.109 |
| 0.50 | 0.6625 | 28.666 | 1.00 | 1.0000 | 29.510 |

(c) The excess enthalpy $H_m^E$ at $T = 298.15$ K for this mixture is given by the equation

$$H_m^E/(\mathrm{J{\cdot}mol^{-1}}) = x_1(1 - x_1)[4805.1 + 251.2(1 - 2x_1) + 421.2(1 - 2x_1)^2 + 1069.2(1 - 2x_1)^3].$$

Assume $H_m^E$ is constant with temperature and correct the $G_m^E$ in part (b) to $T = 298.15$ K.

(d) Make a graph of $G_m^E$, $H_m^E$, and $TS_m^E$ at each of the mole fractions, at $T = 298.15$ K.

P7.2 In a regular solution, the activity coefficients are given by the equations

$$\ln \gamma_1 = x_2^2 \frac{\omega}{RT}$$

$$\ln \gamma_2 = x_1^2 \frac{\omega}{RT}$$

where $\omega$ is a constant.

(a) Start with the equation relating $\gamma_1$, $\gamma_2$, $x_1$ and $x_2$ to $G_m^E$ and appropriate thermodynamic relationships to show that for a regular solution

$$G_m^E = x_1 x_2\, \omega$$

$$S_m^E = -x_1 x_2 \left(\frac{\partial \omega}{\partial T}\right)_{p,n},$$

and

$$H_{\rm m}^{\rm E} = x_1 x_2 \left[ \omega - T \left( \frac{\partial \omega}{\partial T} \right)_{p,n} \right].$$

(b) For mixtures of (benzene + cyclohexane) at $T = 293.15$ K, $\omega = 1280$ J·mol$^{-1}$ and $(\partial\omega/\partial T)_{p,n} = -7.0$ J·K$^{-1}$·mol$^{-1}$. Make a graph that compares $RT\ln\gamma_1$, $RT\ln\gamma_2$, $G_{\rm m}^{\rm E}$, $H_{\rm m}^{\rm E}$, and $TS_{\rm m}^{\rm E}$ as a function of $x_2$ for this mixture at this temperature.

P7.3 (a) Use the Debye–Hückel theory to calculate $\ln\gamma_\pm$ for a 0.050 M aqueous NaCl solution at 10 K intervals from 273.15 K to 373.15 K. Compare the results by making a graph of $\ln\gamma_\pm$ against $1/T$.

(b) With the aid of equation (7.65), calculate $\bar{L}_2$ at $T = 298.15$ K for this solution and compare it to a value for $\bar{L}_2$ obtained from the slope of the line at $T = 298.15$ K in the figure of part (a).

P7.4 The following table summarizes some of the properties of water (component 1) and 1,4-$C_4H_8O_2$ (component 2).

| Property | $H_2O$ | 1,4-$C_4H_8O_2$ |
|---|---|---|
| $\Delta_{\rm fus}H_{\rm m}/$J·mol$^{-1}$ | 6009.5 | 12840 |
| $\Delta_{\rm fus}C_{p,\rm m}/$J·K$^{-1}$·mol$^{-1}$ | 37.28 | −11.05 |
| $T_0/$(K) (normal melting temperature) | 273.15 | 284.93 |
| $p^*$ kPa (vapor pressure at 298.15 K) | 3.17 | 5.28 |

Water and dioxane are miscible. The following table lists thermodynamic properties at $T = 298.15$ K of the mixtures $\{x_1H_2O + x_2(1,4\text{-}C_4H_8O_2)\}$.

(a) Calculate $H_{\rm m}^{\rm E}$, $G_{\rm m}^{\rm E}$, and $S_{\rm m}^{\rm E}$ at 298.15 K for the process

$$0.4 \text{ mole } H_2O + 0.6 \text{ mole } 1,4\text{-}C_4H_8O_2 = \text{solution}.$$

(b) Calculate the total vapor pressure above the solution in part (a).

(c) The (solid + liquid) phase diagram is a simple eutectic system with no solid compound formation and no solid solutions. Assume $\bar{L}_1$ and $\bar{L}_2$ are independent of temperature and use the data at the beginning of the problem to calculate the best possible value for the melting temperature of a solution with $x_2 = 0.250$.

(d) Make a graph of vapor pressure against mole fraction at 298.15 K for each of the components to show the deviation in the solution from Raoult's law.

Thermodynamic Properties of $x_1 H_2O + x_2(1,4\text{-}C_4H_8O_2)$

| $x_2$ | $\bar{L}_1/(J \cdot mol^{-1})$ | $\bar{L}_2/(J \cdot mol^{-1})$ | $a_1$ | $a_2$ |
|---|---|---|---|---|
| 0.000 | 0 | ... | 1.0000 | 0 |
| 0.1000 | −314 | −2469 | 0.9151 | 0.4002 |
| 0.1500 | −690 | 21 | 0.8770 | 0.5450 |
| 0.2000 | −858 | 774 | 0.8476 | 0.6368 |
| 0.2500 | −941 | 1088 | 0.8236 | 0.6966 |
| 0.3000 | −962 | 1130 | 0.8068 | 0.7375 |
| 0.3500 | −962 | 1130 | 0.7963 | 0.7585 |
| 0.4000 | −962 | 1130 | 0.7843 | 0.7778 |
| 0.4500 | −962 | 1130 | 0.7723 | 0.7946 |
| 0.5000 | −962 | 1130 | 0.7638 | 0.8044 |
| 0.5500 | −962 | 1130 | 0.7544 | 0.8134 |
| 0.6000 | −900 | 1088 | 0.7408 | 0.8249 |
| 0.6500 | −586 | 920 | 0.7242 | 0.8366 |
| 0.7000 | −126 | 690 | 0.7004 | 0.8502 |
| 0.7500 | 167 | 565 | 0.6711 | 0.8647 |
| 0.8000 | 649 | 439 | 0.6264 | 0.8826 |
| 0.8500 | 1234 | 314 | 0.5631 | 0.9031 |
| 0.9000 | 2092 | 209 | 0.4764 | 0.9261 |
| 1.0000 | ... | 0 | 0 | 1.0000 |

P7.5 The osmotic coefficients of aqueous $CaCl_2$ solutions at 298.15 K are as follows:

| $m/(mol \cdot kg^{-1})$ | $\phi$ | $m/(mol \cdot kg^{-1})$ | $\phi$ |
|---|---|---|---|
| 0.0001 | 0.9869 | 0.2 | 0.862 |
| 0.0005 | 0.9719 | 0.3 | 0.876 |
| 0.001 | 0.9615 | 0.4 | 0.894 |
| 0.005 | 0.9250 | 0.5 | 0.917 |
| 0.01 | 0.9048 | 0.6 | 0.940 |
| 0.02 | 0.884 | 0.7 | 0.963 |
| 0.04 | 0.867 | 0.8 | 0.988 |
| 0.05 | 0.861 | 0.9 | 1.017 |
| 0.1 | 0.854 | 1.0 | 1.046 |

(a) Make a graph of $\ln \phi$ against $m$ for the experimental results given in the table. Also show on the graph the predictions of $\ln \phi$ against $m$ as predicted by the Debye–Hückel limiting equation and the extended form of the Debye–Hückel equation.

(b) Use the osmotic coefficients in the table to calculate $\gamma_{\pm}$ of a 0.001, 0.01, 0.1, and 1.0 molal $CaCl_2$ solution and compare your answers with those obtained from both the limiting and extended forms of the Debye–Hückel equation.

## References

1. K. N. Marsh, J. B. Ott, and M. J. Costigan, "Excess Enthalpies, Excess Volumes, and Excess Gibbs Free Energies for (*n*-Hexane + *n*-Decane) at 298.15 K and 308.15 K", *J. Chem. Thermodyn.*, **12**, 343–348 (1980).
2. D. A. Palmer and B. D. Smith, "Thermodynamic Excess Property Measurements for Acetonitrile–Benzene–*n*-Heptane System at 45 °C", *J. Chem. Eng. Data*, **17**, 71–76 (1972).
3. R. C. Pemberton and C. J. Mash, "Thermodynamic Properties of Aqueous Non-Electrolyte Mixture II. Vapour Pressures and Excess Gibbs Energies for Water + Ethanol at 303.15 to 363.15 K Determined by an Accurate Static Method", *J. Chem. Thermodyn.*, **10**, 867–888 (1978).
4. J. B. Ott, C. E. Stouffer, G. V. Cornett, B. F. Woodfield, C. Guanquan and J. J. Christensen, "Excess Enthalpies for (Ethanol + Water) at 398.15 K, 423.15, 448.15, and 473.15 K and at Pressures of 5 and 15 MPa. Recommendations for Choosing (Ethanol + Water) as an $H_m^E$ Reference Mixture", *J. Chem. Thermodyn.*, **19**, 337–348 (1987).
5. The $V_m^E$ results for (cyclohexane + decane) are obtained from J. R. Goates, J. R. Ott, and R. B. Grigg, "Excess Volumes of Cyclohexane + *n*-Hexane, +*n*-Heptane, +*n*-Octane, +*n*-Nonane, and +*n*-Decane", *J. Chem. Thermodyn.*, **11**, 497–506 (1979). Excess volumes for the (ethanol + water) system were obtained from K. N. Marsh and A. E. Richards, "Excess Volumes for Ethanol + Water Mixtures at 10-K intervals from 278.15 to 338.15 K", *Aust. J. Chem.*, **33**, 2121–2132 (1980). Excess volumes for the (acetonitrile + benzene) and the (hexane + decane) systems were obtained from the same source as the $H_m^E$, $G_m^E$ and $S_m^E$ results referenced earlier.
6. We refer those who are interested in the details of the Debye–Hückel derivation to the following sources: R. A. Robinson and R. H. Stokes, "*Electrolyte Solutions*", Academic Press, Inc., New York (1955). The Robinson/Stokes reference does an especially good job of summarizing and evaluating the assumptions made in the derivation; H. S. Harned and B. B. Owen, "*The Physical Chemistry of Electrolytic Solutions*," Reinhold Publishing Corporation, New York (1958); K. S. Pitzer, "*Thermodynamics*," Third Edition, McGraw Hill, Inc., New York (1995).
7. Values taken from S. Glasstone, *Thermodynamics for Chemists*, D. Van Nostrand Company Inc., Toronto, p. 443 (1947). The values tabulated in this reference were taken from D. N. Craig and G. W. Vinal, *J. Res. Natl. Bur. Stand.*, "Thermodynamic Properties of Sulfuric Acid Solutions and Their Relation to the Electromotive Force and Heat of Reaction of the Lead Storage Battery", **24**, 475–490 (1940). More recent values at the higher molality can be found in W. F. Giauque, E. W. Hornung, J. E. Kunzler and T. R. Rubin, "The Thermodynamic Properties of Aqueous Sulfuric Acid Solutions and Hydrates from 15 to 300° K", *J. Am. Chem. Soc.*, **82**, 62–70 (1960).

## Chapter 8

# The Equilibrium Condition Applied to Phase Equilibria

We now have the foundation for applying thermodynamics to chemical processes. We have defined the potential that moves mass in a chemical process and have developed the criteria for spontaneity and for equilibrium in terms of this chemical potential. We have defined fugacity and activity in terms of the chemical potential and have derived the equations for determining the effect of pressure and temperature on the fugacity and activity. Finally, we have introduced the concept of a standard state, have described the usual choices of standard states for pure substances (solids, liquids, or gases) and for components in solution, and have seen how these choices of standard states reduce the activity to pressure in gaseous systems in the limits of low pressure, to concentration (mole fraction or molality) in solutions in the limit of low concentration of solute, and to a value near unity for pure solids or pure liquids at pressures near ambient.

Chemical processes are central to the study of chemistry. The thermodynamic principles and relationships we have developed provide powerful tools for describing these processes, especially in predicting the spontaneity of the process and the equilibrium conditions that apply.

The chemical processes we will describe are generally classified as phase changes, which involve the change in state between solids, liquids, or gases; and chemical changes, which involve changes in chemical bonding in chemical reactions. Phase equilibrium is the equilibrium process that can occur in a phase change, while chemical equilibrium is of interest in the chemical change. We will describe phase equilibria in this chapter, where we will find that the Clapeyron equation and the Gibbs phase rule are especially useful in helping to understand the process. We will then describe chemical equilibrium in Chapter 9 where the equilibrium constant that we will define, derive, and describe will be of primary interest.

As we proceed, we should keep in mind that the classification of a chemical process as a phase change or as a chemical reaction is somewhat arbitrary, and that, in fact, a phase change can be thought of as a special case of a chemical reaction. For example the vaporization process

$$H_2O(l) = H_2O(g)$$

would almost always be thought of as a phase change, although bonding, principally in the form of the hydrogen bonds, definitely changes. The process

$$C(graphite) = C(diamond)$$

is also usually thought of as a phase change, although changes in chemical bonds and structure occur. Finally, the change

$$C(graphite) = C(fullerine)$$

would probably be thought of as a chemical reaction since forming the $C_{60}$ "Bucky balls" in fullerine from the flat plates in graphite requires a major rearrangement of chemical bonds and the formation of distinct molecular species. The formation of silicon carbide

$$Si(s) + C(graphite) = SiC(s)$$

would usually be treated as a chemical reaction (except perhaps in the high temperature melt), while the peritectic reaction (incongruent melting)

$$Na_2SO_4 \cdot 10H_2O(s) = Na_2SO_4(s) + \text{saturated solution(l)}$$

is often treated as a phase change in binary (solid + liquid) phase equilibria, even though a chemical reaction must be written to describe the process.

In Chapter 1, we described a phase as a homogeneous region. Two or more homogeneous regions or phases can be brought together to form a heterogeneous mixture. Phase equilibrium occurs when the chemical potential for each component is the same in each phase in the mixture. Thus, for the equilibrium process involving phases A, B, C, ..., the equilibrium condition is given by

$$A = B = C = \cdots,$$

and the chemical potentials of each ($i$th) component in the various phases are related by[a]

$$\mu_{i,\mathrm{A}} = \mu_{i,\mathrm{B}} = \mu_{i,\mathrm{C}} = \cdots$$

---

[a] See Chapter 5 for a derivation of the equilibrium condition in terms of the chemical potential.

Thus, for each component, its chemical potential is the same in all phases that are in equilibrium. We will see below that the relationships involving the pressure and temperature variations of the chemical potential that we have developed earlier will be helpful in explaining the effect of these variables on phase equilibria.

## 8.1 Phase Equilibria For Pure Substances

In this section we limit our discussion to the phase equilibria involved with pure substances. In this case, the condition for equilibrium between phases A, B, C, ..., becomes

$$\mu_A = \mu_B = \mu_C = \cdots,$$

and the Gibbs phase rule derived in Chapter 5 becomes

$$f = C - P + 2 = 3 - P,$$

where $f$ is the number of degrees of freedom with $P$ phases in equilibrium for $C$ components.

### 8.1a The Phase Diagram and the Gibbs Phase Rule

Phase properties are most easily described and summarized by a phase diagram. Figure 8.1 shows such a diagram for carbon dioxide.[b] The lines represent the conditions of temperature and pressure where two phases of $CO_2$ are in equilibrium. They are sometimes called two-phase lines. These lines separate the regions of temperature and pressure where only one phase — solid, liquid or gas — is stable. For example, the line db separates the solid and liquid regions of $CO_2$. Points on this line give the $T$ and $p$ conditions under which the solid and liquid phases of $CO_2$ coexist in equilibrium. The line ab gives the (vapor + liquid) equilibrium conditions, while bc shows how the (vapor + solid) equilibrium conditions vary with $T$ and $p$. Where the two-phase lines involve a vapor, the variable $p$ represents the pressure of the gas. Thus, the pressure corresponding to the ab and bc lines is the equilibrium vapor pressure of the

[b] In Figure 8.1 pressure is graphed against temperature. (Physicists usually construct the graph backwards from the way chemists do it — that is, they graph temperature vertically against pressure horizontally.) These are the most common variables chosen for the phase diagrams of pure materials, although volume is sometimes used instead of pressure or temperature. Other variables can be used to emphasize the effect of magnetic field, surface area, etc. Composition is often chosen as a variable when mixtures are involved. Three-dimensional diagrams are sometimes used to simultaneously show the effect of $p$, $V$, and $T$; $p$, $T$, and $x$; $p$, $V$, and $x$; or other combinations.

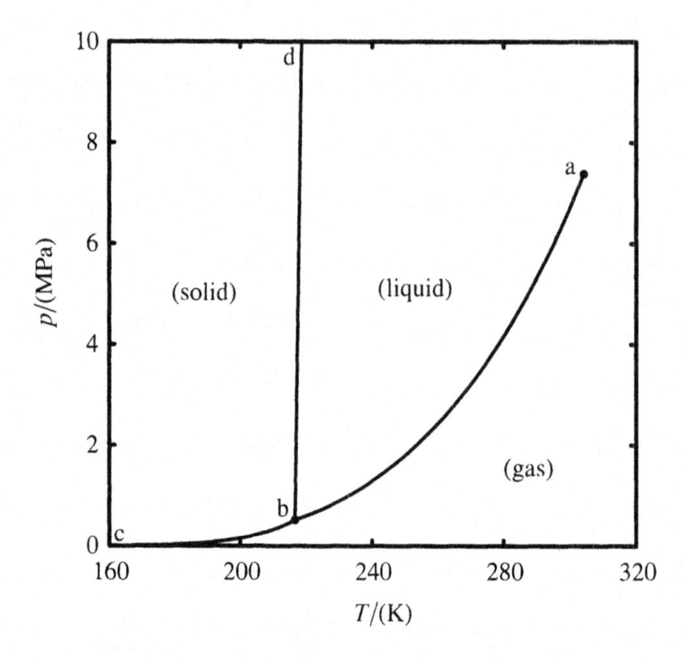


**Figure 8.1** Phase diagram for $CO_2$. Point (a) is the critical point and point (b) is the triple point. Line ab gives the vapor pressure of the liquid, line bc gives the vapor pressure of the solid, and line bd gives the melting temperature as a function of pressure.

condensed phase at a particular temperature. However, when only condensed phases are involved in the equilibrium, or only one phase is present so that equilibrium does not occur, the pressure is a mechanically-applied pressure.

The concept of chemical potentials, the equilibrium criterion involving chemical potentials, and the various relationships derived from it (including the Gibbs phase rule derived in Chapter 5) can be used to explain the effect of pressure and temperature on phase equilibria in both a qualitative and quantitive way.

The Gibbs phase rule, $f = C - P + 2$, derived in Chapter 5, can be used to qualitatively understand the relationship between different phases at equilibrium, and can be used to explain the existence of one-, two-, and three-phase regions in a one-component system. As we have seen, since $C = 1$ for a pure substance, $f = 3 - P$. For such a system where the composition is not changing, only two variables, usually temperature and pressure, can be changed. In one-phase regions, $P = 1$, and $f = 2$. This implies that both $T$ and $p$ can be varied independently within a one-phase region. Along a two-phase equilibrium line, $P = 2$ and $f = 1$. Thus, only one of the two variables, $p$ or $T$, can be set independently. If $T$ is specified, then the requirement of two-phase equilibria means that $p$ is determined for the system. For example, for any $T$ on the (vapor + liquid) line in Figure 8.1 a single value exists for the equilibrium vapor pressure of the liquid. Alternately, if one wishes to establish a system with a given vapor pressure of the solid, the temperature required to achieve that pressure would be set.

When $C = 1$ and $P = 3, f = 0$. Thus, if three phases are in equilibrium, there are no degrees of freedom. The $T$ and $p$ conditions at which three phases will be in equilibrium are determined by the system. The equilibrium occurs at a single $(T, p)$ point, commonly referred to as a **triple point**. Such a point is also referred to as an **invariant point** because all properties are fixed for the system. For $CO_2$ in Figure 8.1, point b is a triple point, with solid, liquid, and gas in equilibrium.

From these kinds of considerations, it is easy to see why a one component system can have no more than three phases in equilibrium. Any number of phases greater than three would yield a negative number of degrees of freedom, a situation that makes no physical sense.

While the Gibbs phase rule provides for a qualitative explanation, we can apply the Clapeyron equation, derived earlier [equation (5.71)], in conjunction with studying the temperature and pressure dependences of the chemical potential, to explain quantitatively some of the features of the one-component phase diagram.

## 8.1b Solid + Liquid Equilibrium

Line db in Figure 8.1 represents the equilibrium melting line for $CO_2$. Note that the equilibrium pressure is very nearly a linear function of $T$ in the $(p, T)$ range shown in this portion of the graph, and that the slope of the line, $(\mathrm{d}p/\mathrm{d}T)_{\mathrm{s-l}}$, is positive and very steep, with a magnitude of approximately 5 $\mathrm{MPa \cdot K^{-1}}$. These observations can be explained using the Clapeyron equation. For the process

$$CO_2(\text{solid}) = CO_2(\text{liquid}),$$

the slope of the pressure against temperature line is given by

$$\left(\frac{\mathrm{d}p}{\mathrm{d}T}\right)_{\mathrm{s-l}} = \frac{\Delta_{\mathrm{fus}} S_{\mathrm{m}}}{\Delta_{\mathrm{fus}} V_{\mathrm{m}}}. \tag{8.1}$$

For this process, both $\Delta_{\mathrm{fus}} S_{\mathrm{m}}$ and $\Delta_{\mathrm{fus}} V_{\mathrm{m}}$ are greater than zero so that the slope is positive; the ratio is large and, hence, the slope is steep because the volume change upon melting is much smaller than the entropy change. Equation (8.1) can be integrated to derive an equation that relates melting temperature to pressure. After separation of variables for the integration we get

$$\int \mathrm{d}T = \int \frac{\Delta_{\mathrm{fus}} V_{\mathrm{m}}}{\Delta_{\mathrm{fus}} S_{\mathrm{m}}} \mathrm{d}p. \tag{8.2}$$

The volume change and the entropy change between two condensed phases both show only a very small dependence on pressure so that the ratio can be treated as approximately constant for the integration over pressure if the pressure change is not too large. Assuming that $\Delta_{\text{fus}}V_{\text{m}}/\Delta_{\text{fus}}S_{\text{m}}$ is constant, and integrating with indefinite limits gives, after some rearrangement,

$$p = \left(\frac{\Delta_{\text{fus}}S_{\text{m}}}{\Delta_{\text{fus}}V_{\text{m}}}\right)T + \text{constant} \tag{8.3}$$

The straight line behavior shown in Figure 8.1 is in accord with equation (8.3). When a larger pressure range is considered, the (solid + liquid) line is shown to have curvature. Figure 8.2 shows the (solid + liquid) equilibrium portion of the $CO_2$ phase diagram[1] extended to a pressure of 1200 MPa, and the curvature is clearly evident. When the integration of equation (8.2) is performed over such a large interval, the assumption that the $\Delta_{\text{fus}}V_{\text{m}}/\Delta_{\text{fus}}S_{\text{m}}$ ratio is constant is no longer valid, and a nonlinear curve is expected. Keep in mind, however, that at every point on this line, the slope is still given by equation (8.1), where the volume and entropy changes are those for that value of $p$ and $T$.

The sign of the slope of the melting line is governed by the signs of the two quantities, $\Delta_{\text{fus}}S_{\text{m}}$ and $\Delta_{\text{fus}}V_{\text{m}}$. Since for most substances $\Delta_{\text{fus}}S_{\text{m}} > 0$ for solid

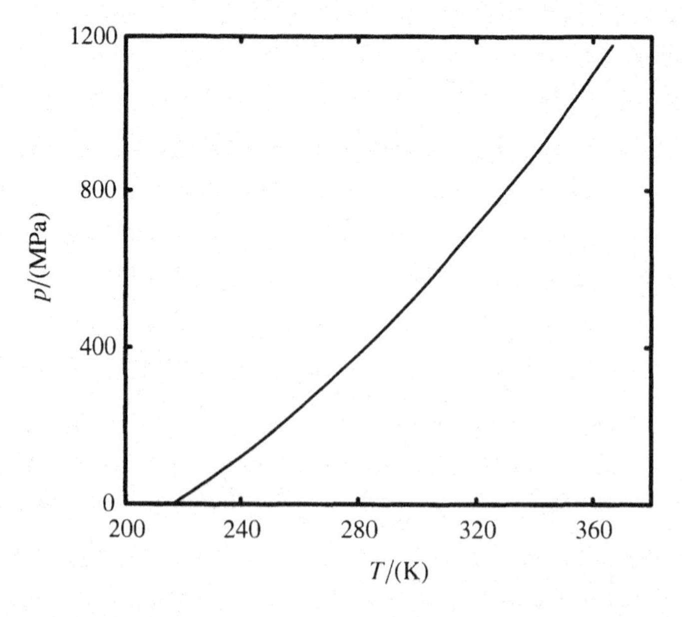


**Figure 8.2** Melting temperature of $CO_2$ as a function of pressure.

changing to liquid, the direction of the slope is usually governed by the sign of the volume change. For $CO_2$ and most other substances, $V_{m,l} > V_{m,s}$ so that $\Delta_{fus}V_m > 0$, and $(dp/dT)_{s-l} > 0$ as shown in Figures 7.1 and 7.2. For the few substances[c] where $V_{m,l} < V_{m,s}$, $\Delta_{fus}V_m < 0$ and $(dp/dT)_{s-l} < 0$. Qualitatively speaking, the (solid + liquid) line can be said to lean toward the less dense phase, since this is the one with the largest molar volume.

## 8.1c Equilibria Involving a Condensed Phase and the Vapor Phase

It is apparent from Figure 8.1 that the lines ab and bc, which give the (vapor + liquid) and (vapor + solid) equilibrium lines, respectively, differ from line db in two important ways. Their slopes are much smaller, and they exhibit much more curvature. An analysis of equation (8.1) with the quantities appropriate for these two phase lines, $(dp/dT)_{v-l}$ with $\Delta_{vap}S_m$ and $\Delta_{vap}V_m$, and $(dp/dT)_{v-s}$ with $\Delta_{sub}S_m$ and $\Delta_{sub}V_m$, provides the explanation for this behavior. The magnitude of the slopes are smaller because the volume change for both vaporization (liquid → vapor) and sublimation (solid → vapor) are considerably larger than that for melting. The curvature arises because both the volume and entropy differences associated with vaporization and sublimation change markedly with pressure. Since the volume and entropy changes upon vaporization and sublimation are always positive, the slopes of the (vapor + liquid) and (solid + liquid) lines are always positive.

**The Clausius–Clapeyron equation** The Clapeyron equation can be used to derive an approximate equation that relates the vapor pressure of a liquid or solid to temperature. For the vaporization process

liquid = gas

the Clapeyron equation becomes

$$\left(\frac{dp}{dT}\right)_{v-l} = \frac{\Delta_{vap}S_m}{\Delta_{vap}V_m} = \frac{\Delta_{vap}H_m}{T\Delta_{vap}V_m}. \tag{8.4}$$

Because the molar volume of the gas, $V_m(g)$, is much greater than that of the liquid, $V_m(l)$, $\Delta V_m = V_m(g) - V_m(l) \approx V_m(g)$. If we further approximate the gas

[c] Water is the pre-eminent example of a system of this type.

as being ideal then

$$\Delta_{\text{vap}} V_{\text{m}} \approx \frac{RT}{p} \tag{8.5}$$

and equation (8.4) can be rewritten, after the separation of variables, as

$$\frac{\mathrm{d}p}{p} = \frac{\Delta_{\text{vap}} H_{\text{m}}}{RT^2}\, \mathrm{d}T. \tag{8.6}$$

If the enthalpy of vaporization has only a small dependence on $T$, or if the integral is evaluated over only a small change, $\mathrm{d}T$, then $\Delta_{\text{vap}}H_{\text{m}}$ can be assumed to be constant, and integration between pressure limits, $p_1$ and $p_2$, and temperature limits, $T_1$ and $T_2$, yields

$$\ln \frac{p_2}{p_1} = -\frac{\Delta_{\text{vap}} H_{\text{m}}}{R} \left( \frac{1}{T_2} - \frac{1}{T_1} \right). \tag{8.7}$$

Indefinite integration yields

$$\ln p = -\frac{\Delta_{\text{vap}} H_{\text{m}}}{R} \left( \frac{1}{T} \right) + \text{const} \tag{8.8}$$

Equivalent expressions can be obtained for the (solid + vapor) equilibrium to give the vapor pressure of a solid, by substituting the sublimation enthalpy, $\Delta_{\text{sub}}H_{\text{m}}$, for $\Delta_{\text{vap}}H_{\text{m}}$.

Equations (8.7) and (8.8) are two forms of the **Clausius–Clapeyron equation**. Under the conditions for which these equations are valid, a graph of $\ln p$ against $1/T$ should be a straight line with slope $(-\Delta_{\text{vap}}H_{\text{m}})/R$ or $(-\Delta_{\text{sub}}H_{\text{m}})/R$. Figure 8.3 shows such graphs for $CO_2$, using temperatures and vapor pressures for the sublimation of solid $CO_2$ below the triple point, and the vaporization of the liquid between the triple point and the critical point.[d] In both graphs, the approximations inherent in the derivations of equations (8.7) and (8.8) appear to be valid because both plots are very nearly linear.

---

[d] The critical point is the $(p, T)$ condition above which liquid cannot exist. We will describe this condition in the next section in more detail.

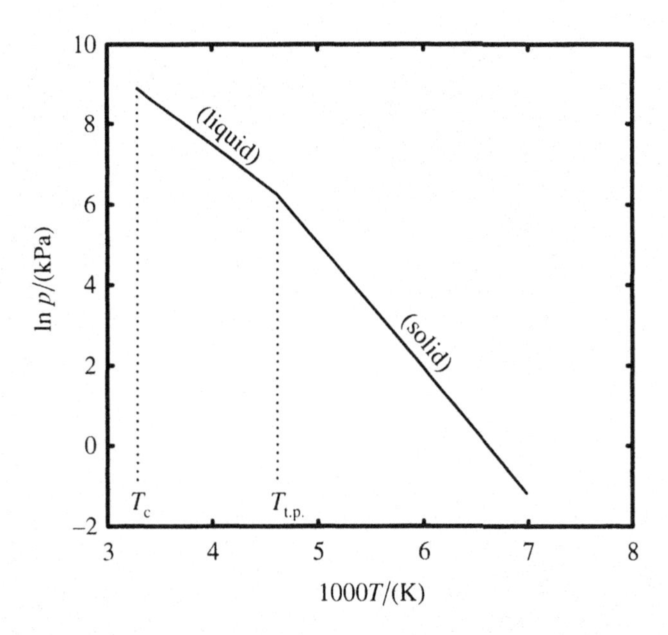


**Figure 8.3** Vapor pressure (in kPa) of solid and liquid $CO_2$ as a function of temperature. (Graphed as $\ln p$ against $1/T$.) $T_c$ is the critical temperature and $T_{t.p.}$ is the triple point temperature.

When vapor pressure data are collected over a large temperature range, or if the temperature variation of the enthalpy change is great, curvature of the $\ln p$ vs. $1/T$ plot may be evident. The approach then is to express the enthalpy change as an explicit function of temperature before integrating equation (8.6). A common form[e] expresses the enthalpy change as

$$\Delta H_m - B + CT + DT^2 + \cdots$$

Substitution of this expression into equation (8.6) gives, after integration,

$$\ln p = A - \frac{B}{RT} + \frac{C}{R}\ln T + \frac{D}{R}T + \cdots \tag{8.9}$$

where $A$ is the constant of integration. Experimental vapor pressure results can be fit successfully to equation (8.9) with high precision.

---

[e] Equation (6.165) is an even more extensive relationship between $\Delta_{fus}H_m$ and temperature. Similar equations can be written that relate $\Delta_{vap}H_m$ to temperature.

Another expression that does a reasonably good job of representing vapor pressures of liquids, if the temperature range is not too great, is the **Antoine equation**:

$$\log p = A - \frac{B}{t + C} \tag{8.10}$$

where $p$ is usually expressed in torr, $t$ is the Celsius temperature and $A$, $B$ and $C$ are constants.[f]

## 8.1d Vapor + Liquid Equilibrium: The Critical Point

The (vapor + liquid) equilibrium line for a substance ends abruptly at a point called the **critical point**. The critical point is a unique feature of (vapor + liquid) equilibrium where a number of interesting phenomena occur, and it deserves a detailed description. The temperature, pressure, and volume at this point are referred to as the critical temperature, $T_c$, critical pressure, $p_c$, and critical volume, $V_c$, respectively. For $CO_2$, the critical point is point a in Figure 8.1. As we will see shortly, properties of the critical state make it difficult to study experimentally.

In addition to phase diagrams such as that shown in Figure 8.1 where $p$ is plotted against $T$, other variables can be plotted in the phase diagram that are particularly informative with regard to the critical point. Common examples are a graph of $T$ against the molar volume, $V_m$, or $T$ against the density, $\rho$. Examples of such plots for $CO_2$ are shown in Figure 8.4a and 8.4b, respectively.

The curves shown in Figure 8.4a and 8.4b give the molar volumes (Figure 8.4a) or densities (Figure 8.4b) of the two phases in equilibrium with one another at a temperature $T$. At a temperature below $T_c$, where the gas and liquid are distinguishable, there are two values of $V_m$ or $\rho$, one for the liquid and one for the gas. They are given by the end points of the horizontal segment intersecting the two portions of the curve at each temperature. This segment is often referred to as a **tie line** because it ties together the two phases in equilibrium with one another. As $T \rightarrow T_c$, the liquid and gaseous volumes molar approach each other and the densities behave similarly. Above $T_c$, only a single phase exists; it has properties of both a liquid and a gas, and it is often referred to as a **fluid** or more specifically as a **supercritical fluid**.

---

[f] Note that $A$, $B$, and $C$ in equation (8.10) are not the same as the $A$, $B$, and $C$ in equation (8.9) The Antoine equation reduces to the Clausius–Clapeyron equation if $B = \Delta_{vap}H_m/2.303R$ and $C = 273.15$. For most substances, the Antoine constants are close to these values, but not exactly equal to them. Care must be taken in applying the Antoine equation. It works well for interpolation within the temperature range used in establishing $A$, $B$, and $C$, but it does not work well in extrapolating to temperatures outside this range.

Figure 8.4b is a particularly useful way of representing the equilibrium phase information. The plot of $T$ against $\rho$ is relatively symmetrical, and it lends itself to a method for determining the actual critical point by making use of the **Law of Rectilinear Diameters**, which states that the average density of the two phases in equilibrium is a linear function of temperature. Thus, a line which plots $T$ against the average density, $\frac{1}{2}(\rho_l + \rho_g)$, should intersect the equilibrium curve at $T_c$, where $\rho_l = \rho_g$. The nearly-vertical dashed line in Figure 8.4b is a graph of $T$ against the average density of $CO_2$, and is a representation of this law.

A condition of phase equilibrium is the equality of the chemical potentials in the two phases. Therefore, at all points along the two-phase line, $\mu(g) = \mu(l)$. But, as we have noted above, the approach to the critical point brings the liquid and gas closer and closer together in density until they become indistinguishable. At the critical point, all of the thermodynamic properties of the liquid become equal to those of the gas. That is, $H_m(g) = H_m(l)$, $U_m(g) = U_m(l)$, $S_m(g) = S_m(l)$, and $V_m(g) = V_m(l)$. Thus, quantities such as $\Delta_{vap}H_m = H_m(g) - H_m(l)$ and $\Delta_{vap}V_m = V_m(g) - V_m(l)$ go to zero as the critical point is reached. Figures 8.5 and 8.6 show this behavior for $CO_2$.

In the region of the critical point, the decline of $\Delta_{vap}V_m = V_m(g) - V_m(l)$, shown in Figure 8.6, can be represented by an equation of the form:

$$\{V_m(g) - V_m(l)\} = k'(T_c - T)^{\beta}. \tag{8.11}$$

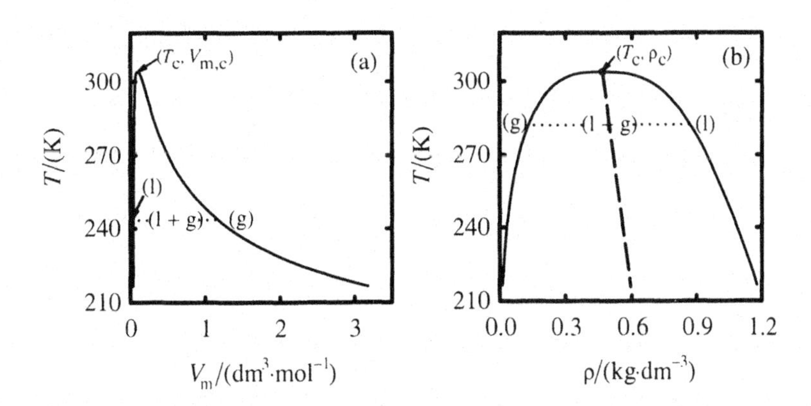


**Figure 8.4** Graph of temperature against molar volume (a), and density (b), for $CO_2$ (gas) and $CO_2$ (liquid) in the temperature range from the triple point to the critical point. The dashed line in (b) is the average density. The area enclosed within the curves is a two-phase region, with the molar volume or the density of the gas and liquid at a particular temperature given by the horizontal (dotted) tie-lines connecting the gas and liquid sides of the curve.

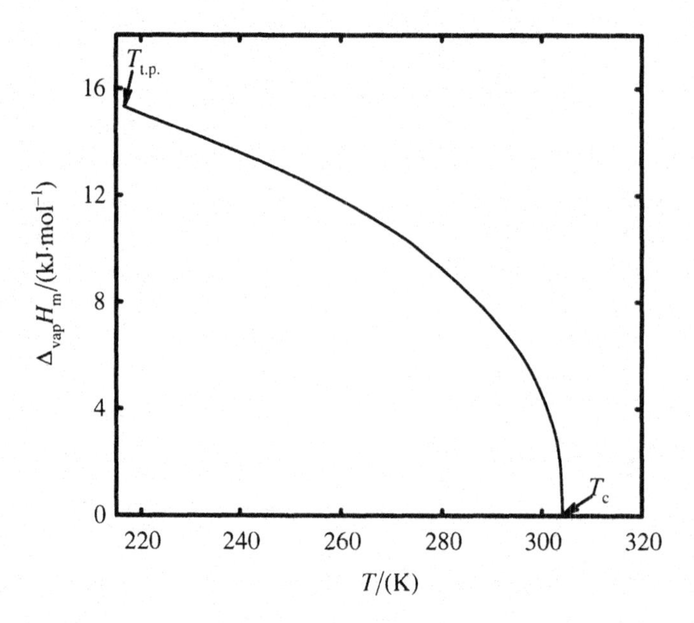


**Figure 8.5** Enthalpy of vaporization of $CO_2$ in the temperature range from the triple point temperature ($T_{t.p.}$) to the critical temperature ($T_c$).

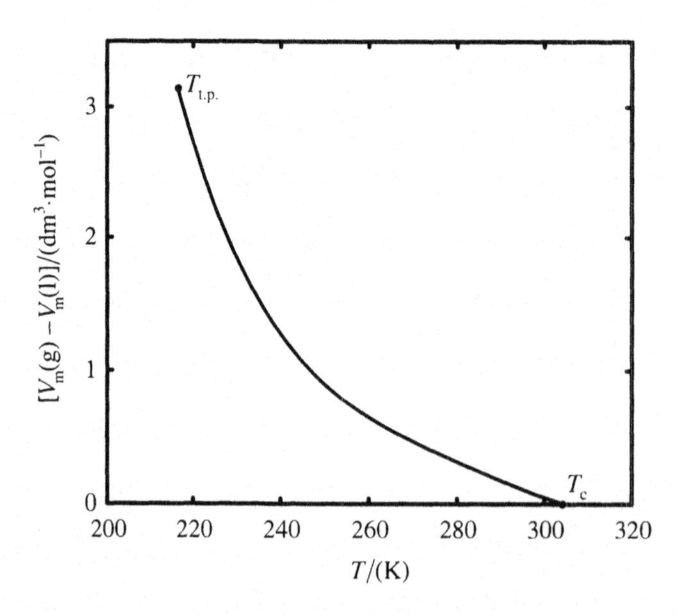


**Figure 8.6** Difference in molar volume between $CO_2(g)$ and $CO_2(l)$ in the temperature range from the triple point temperature ($T_{t.p.}$) to the critical temperature ($T_c$).

A similar form can be used to represent the difference of densities as well

$$\{\rho(\mathrm{l}) - \rho(\mathrm{g})\} = k(T_\mathrm{c} - T)^\beta. \tag{8.12}$$

In both equations, $k$ and $k'$ are proportionality constants and $\beta$ is a constant known as the **critical exponent**. Experimental measurements have shown that $\beta$ has the same value for both equations and for all gases. Analytic[g] equations of state, such as the Van der Waals equation, predict that $\beta$ should have a value of $\frac{1}{2}$. Careful experimental measurement, however, gives a value of $\beta = 0.32 \pm 0.01$.[h] Thus, near the critical point, $\rho$ or $V_\mathrm{m}$ varies more nearly as the cube root of temperature than as the square root predicted from classical equations of state.

The form of equations (8.11) and (8.12) turns out to be general for properties near a critical point. In the vicinity of this point, the value of many thermodynamic properties at $T$ becomes proportional to some power of $(T_\mathrm{c} - T)$. The exponents which appear in equations such as (8.11) and (8.12) are referred to as **critical exponents**. The exponent $\beta = 0.32 \pm 0.01$ describes the temperature behavior of molar volume and density as well as other properties,[i] while other properties such as heat capacity and isothermal compressibility are described by other critical exponents. A significant scientific achievement of the 20th century was the observation of the nonanalytic behavior of thermodynamic properties near the critical point and the recognition that the various critical exponents are related to one another.[j]

The implications of the approach to indistinguishability of the two phases at the critical point are far-reaching. As we will show next, quantities such as the compressibility and coefficient of expansion become infinite.

Consider Figure 8.7, which is a graph of $p$ against $V_\mathrm{m}$ at different temperatures (isotherms) for $CO_2$. The dotted line encompasses the two-phase

---

[g] An equation is said to show analytic behavior if a Taylor's series expansion about a point in the solution set of the equation converges in the neighborhood of the point.

[h] Historically, the derivation giving $\beta = \frac{1}{2}$ has been obtained with equations of state such as the Van der Waals equation. In fact, it can be shown that the classical behavior results from the general assumption that the free energy behaves analytically around the critical point. This derivation is left to more advanced books.

[i] For example, $\beta$ describes the temperature dependence of composition near the upper critical solution temperature for binary (liquid + liquid) equilibrium, of the susceptibility in some magnetic phase transitions, and of the order parameter in (order + disorder) phase transitions.

[j] For a detailed discussion of critical exponents, see J.S. Rowlinson and F.L. Swinton, *Liquids and Liquid Mixtures*; Third Edition, Chapter 3, "The Critical State," Butterworth Scientific, London, 1982.

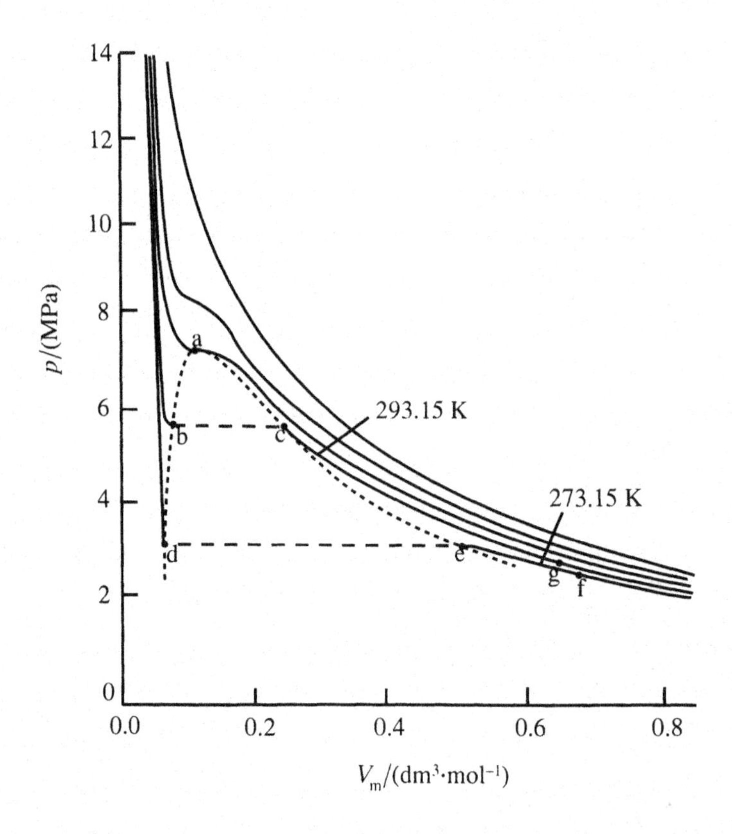


**Figure 8.7** Experimental $p-V$ isotherms for $CO_2$. Point a is the critical point where the critical isotherm ($T = 304.19$ K) has zero slope. The dashed line encloses the (liquid + vapor) two-phase region. At $T = 293.15$ K, points b and c give the molar volumes of the liquid and vapor in equilibrium, while points d and e give the same information at 273.15 K.

region, and the dashed lines are the equivalent of the two-phase tie lines in Figure 8.4. If one starts at a low pressure and at a temperature well below $T_c$, such as point f in the figure where $T = 273.15$ K, $CO_2$ is a gas. Compression along this isotherm causes the pressure to increase until point e is reached. Liquid begins to condense at this point. Further decreases in volume occur along the dashed line ed, but the pressure stays fixed, so that in this region, $(\partial p/\partial V_m)_T = 0$. We remember that the compressibility, $\kappa$, is defined as

$$\kappa = -\frac{1}{V_m}\left(\frac{\partial V_m}{\partial p}\right)_T.$$

Since

$$\left(\frac{\partial p}{\partial V_{\mathrm{m}}}\right)_T = \left(\frac{\partial V_{\mathrm{m}}}{\partial p}\right)_T^{-1},$$

$\kappa$ can be seen to be infinite.

Repeating the compression at higher temperatures gives the same result, except that the onset of condensation of the liquid occurs at a higher pressure and smaller volume and the coexistence region gets shorter. For example at $T = 293.15$ K, the co-existence line bc occurs at $p \approx 5.6$ MPa, while the line at 273.15 K was at $p \approx 3.1$ MPa. Along bc, $\kappa$ is again infinite.

At the temperature of the critical isotherm ($T_{\mathrm{c}} = 304.19$ K for $CO_2$), the coexistence region has collapsed to a single point and represents a point of inflection in the isotherm. From calculus we know that at an inflection point, the first and second derivatives are equal to zero so that

$$\left(\frac{\partial p}{\partial V_{\mathrm{m}}}\right)_T = \left(\frac{\partial^2 p}{\partial V_{\mathrm{m}}^2}\right)_T = 0$$

and $\kappa$ will again be infinite. At temperatures above $T_{\mathrm{c}}$, the isotherms no longer have a point at which $(\partial p/\partial V_{\mathrm{m}})_T = 0$ and $\kappa$ remains finite. At temperatures well above $T_{\mathrm{c}}$, the fluid behaves much like an ideal gas, where the $p-V$ isotherm is hyperbolic and $\kappa = 1/p$.

The coefficient of expansion, $\alpha$, can also be related to $(\partial p/\partial V_{\mathrm{m}})_T$. From the definition of $\alpha$ and the properties of the exact differential, we can write

$$\alpha = \frac{1}{V_{\mathrm{m}}}\left(\frac{\partial V_{\mathrm{m}}}{\partial T}\right)_p = -\frac{1}{V_{\mathrm{m}}}\frac{\left(\frac{\partial p}{\partial T}\right)_{V_{\mathrm{m}}}}{\left(\frac{\partial p}{\partial V_{\mathrm{m}}}\right)_T}.$$

Since $(\partial p/\partial T)_{V_{\mathrm{m}}} \neq 0$, $\alpha$ also becomes infinite at the critical point.

The heat capacity ($C_p$ and $C_V$) also becomes infinite at the critical point.[k] For example, Figure 8.8 shows a graph of the heat capacity of $CO_2$ (expressed

---

[k] This can be inferred from the observation that the slope of the curve of $\Delta_{\mathrm{vap}}H_{\mathrm{m}}$ against $T$ becomes infinite at $T_{\mathrm{c}}$ in Figure 8.5. Since $(\partial\Delta_{\mathrm{vap}}H_{\mathrm{m}}/\partial T)_p = \Delta C_{p,\mathrm{m}}$, and this derivative goes to infinity, $\Delta C_{p,\mathrm{m}}$ must go to infinity.

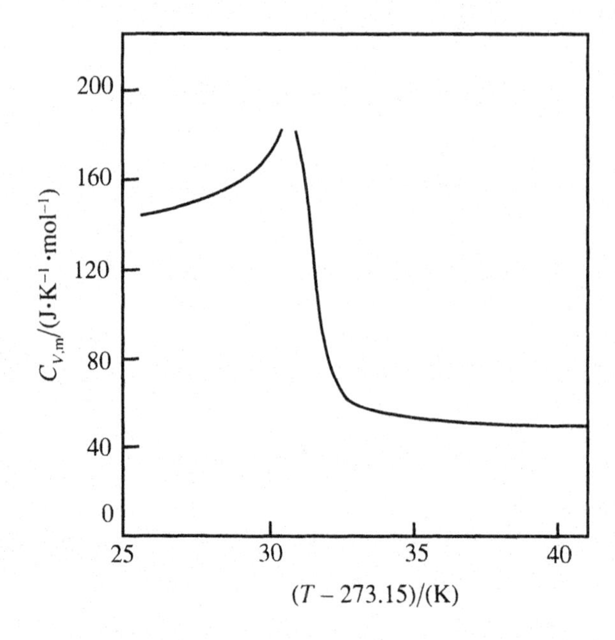


**Figure 8.8** Heat capacity of $CO_2$ at temperatures near $T_c$ and densities spanning the critical pressure region. The line is at a density of 451.2 kg·m$^{-3}$, which is very nearly the critical density. (The pressure is close to the critical pressure.)

as $C_{V,m}$) against $T$ at temperatures near $T_c$ and at densities spanning the ($p$, $V_m$, $T$) region near the critical point.[2] Note that $C_{V,m}$ increases rapidly at temperatures just below $T_c$. It becomes infinitely large at $T_c$, then decreases rapidly at temperatures above $T_c$. The shape of the heat capacity curve in this region gave rise to the name "lambda" transition because of the resemblance to the Greek letter $\lambda$. In the next section we will describe other types of critical phenomena and we will see that the heat capacities associated with these also result in a lambda peak.

The infinite values for $\kappa$, $\alpha$, $C_{V,m}$ and $C_{p,m}$ at the critical point give rise to unusual behavior for $CO_2$ (or any other substance) in this ($p$, $T$) region. Normally, the change of gas pressure with height in a vertical tube, due to the effect of gravity, is too small to be observed, except in very tall columns of gas, such as in the atmosphere. In gases near the critical point, however, the abnormally large value of $\kappa$ means that even small gravitationally-induced pressure changes can lead to large density fluctuations with height in the container. Because of the large $\alpha$, small temperature gradients can also cause large density gradients in the fluid. The temperature gradients also set up convection currents which, because of the large heat capacities, allow the fluid to transport large amounts of heat. Therefore, the thermal conductivity of the material also becomes very large.

The density gradients give rise to a phenomenon known as **critical opalescence**. Because the variations in density throughout the fluid near the critical point cause variations in the refractive index, an anomalous light scattering occurs that causes the fluid to appear cloudy. Consider an experiment in which a fluid at a temperature well above its $T_c$ is cooled. Initially, at high $T$, the fluid will be transparent. With continued cooling, the critical temperature will be approached, density fluctuations will become large, and critical opalescence will set in, causing the fluid to look cloudy. Continued cooling will decrease the temperature below $T_c$ into the two phase region, and droplets of liquid will begin to appear, but the droplets remained suspended as in a fog, because the two phases still have comparable densities. With more cooling, the density of the liquid increases faster than the density of the gas, the mixture separates into two distinct layers, and the fog disappears.[l]

## 8.1e Solid + Solid Phase Transitions

In the $CO_2$ phase diagram of Figure 8.1, we considered only (solid + liquid), (vapor + solid) and (vapor + liquid) equilibria. A (solid + solid) phase transition has not been observed in $CO_2$,[m] but many substances do have one or more. Equilibrium can exist between the different solid phases I, II, III, etc., so that

$$\mu_{\text{I}} = \mu_{\text{II}} = \mu_{\text{III}} = \cdots$$

The Clapeyron equation applies to these transitions so that

$$\frac{\mathrm{d}p}{\mathrm{d}T} = \frac{\Delta_{\text{trans}} S_{\text{m}}}{\Delta_{\text{trans}} V_{\text{m}}}, \tag{8.13}$$

where $\Delta_{\text{trans}} S_{\text{m}}$ and $\Delta_{\text{trans}} V_{\text{m}}$ are the entropy and volume changes for the transition, respectively.

---

[l] The critical point is unique for (vapor + liquid) equilibrium. That is, no equivalent point has been found for (vapor + solid) or (liquid + solid) equilibria. There is no reason to suspect that any amount of pressure would eventually cause a solid and liquid (or a solid and gas) to have the same $V_{\text{m}}$, $H_{\text{m}}$, $S_{\text{m}}$, and $U_{\text{m}}$, with an infinite $\alpha$ and $\kappa$ at that point.

[m] $CO_2$ was chosen for Figure 8.1 because of the very high vapor pressure at the (vapor + liquid + solid) triple point. In fact, it probably has the highest triple point pressure of any known substance. As a result, one can show on an undistorted graph both the triple point and the critical point. For most substances, the triple point is at so low a pressure that it becomes buried in the temperature axis on a graph with a pressure axis scaled to include the critical point.

Figure 8.9 is the phase diagram for Sn, a system that shows (solid + solid) phase transitions.[n] Solid II is the form of tin stable at ambient conditions, and it is the shiny, metallic element that we are used to observing. Line ab is the melting line for solid II. Points on this line represent the values of $p$ and $T$ for which

Solid II = Liquid

is an equilibrium process. Note that line ab leans toward the liquid, which means that the liquid is of lower density than the solid.

At atmospheric pressure, solid II converts to solid I at a temperature of 291 K. Solid I is a grey semiconductor[o] with a face-centered cubic crystal structure similar to that of C(diamond), Si, and Ge at atmospheric pressure. It is

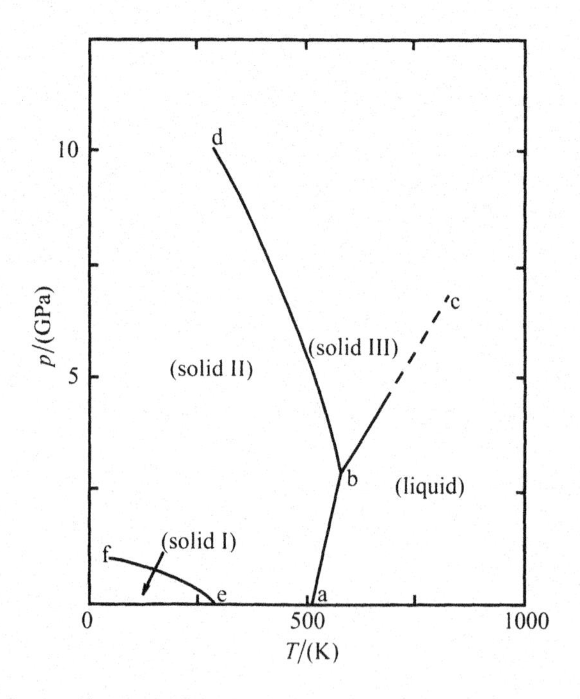


**Figure 8.9** The phase diagram for Sn. The pressure is expressed in gigapascals (1 GPa = $10^9$ Pa). The dashed line represents an extrapolation of the melting temperature to higher pressures.

---

[n] Tin also has a vapor phase, but its vapor pressures are so low that the vapor phase does not show on this diagram.

[o] In cold climates, metallic tin (solid II) slowly changes to solid I. (Solid phase transitions are often slow.) The change from a shiny metallic surface to a brittle and flaky grey surface is known as tin disease.

apparent that solid I is of lower density than solid II. Line ef shows how this transition temperature changes with pressure.

At high pressures, solid II can be converted (slowly) to solid III. Solid III has a body-centered cubic crystal structure. Line bd is the equilibrium line between solid II and solid III, while line bc is the melting line for solid III.[p] A triple point is present between solid II, solid III, and liquid at point b. Two other triple points are present in this system, but they are at too low a pressure to show on the phase diagram. One involves solid II, liquid, and vapor; while the other has solid I, solid II, and vapor in equilibrium.

Phase diagrams can be used to predict the changes in phase that occur when the $(p, T)$ conditions are changed.

**Example 8.1:** Use Figure 8.1 to predict the phase changes that would occur when solid $CO_2$ at 160 K is heated isobarically to 320 K at (a), $p = 0.101$ MPa, (b), $p = 5$ MPa, and (c), $p = 10$ MPa.

**Solution:** (a) The pressure $p = 0.101$ MPa is less than the vapor pressure of $CO_2$ at the triple point so that solid goes directly to gas. A pressure of 0.101 MPa is 1 atm. The temperature where the solid will sublime to form gas at $p = 1$ atm is the normal sublimation temperature of 194.70 K.[q]

(b) At $p = 5$ MPa, solid $CO_2$ melts to form liquid at approximately $T = 217.5$ K and then vaporizes to form gas at $T = 288$ K.

(c) A pressure of 10 MPa is above the critical pressure. At this pressure, solid $CO_2$ melts to form $CO_2$ (fluid) at approximately $T = 218.5$ K. No other phase changes occur as the $CO_2$ (fluid) is heated to 320 K.

**Example 8.2:** Use Figure 8.9 to predict the phase changes that would occur when solid Sn at $p = 0.1$ MPa is compressed isothermally to $p = 15$ GPa at (a) $T = 600$ K; (b) $T = 550$ K; and (c) $T = 250$ K. Assume that the equilibrium phase changes occur rapidly enough to keep up with the change in pressure.

**Solution:** (a) At $T = 600$ K, liquid Sn freezes at 3 GPa to form solid III. Apparently, no other phase changes occur with increasing pressure. (b) At 550 K, liquid Sn freezes to form solid II at 1.5 GPa, then changes to solid III at 3.5 MPa. (c) At 250 K, solid I converts to solid II at 0.3 GPa, which presumably would convert to solid III at approximately 11 MPa. (The equilibrium line stops at 10 GPa.)

---

[p] The phase lines bc, bd, and ef in Figure 8.9 do not terminate at points c, d and f. They merely end at the pressure and temperature conditions above which experimental measurements have not been made and data are not available.

[q] At ambient pressure, solid $CO_2$ is subliming rather than melting. Hence the name, "dry ice".

**First-Order Phase Transitions:** The phase transitions we have described are examples of first-order phase transitions. In a phase transition represented as

$$\mathrm{A} = \mathrm{B}$$

the chemical potentials are equal at the transition temperature, so that $\Delta G_{\mathrm{m}} = \Delta\mu = 0$, and the free energy is continuous through the phase transition.[r] The name "first-order phase transition", comes from the fact that, although $\Delta G$ is zero at the transition, it does change slope there, so that the **first derivatives** of $\Delta G$ are not continuous. Thus,

$$\left(\frac{\partial \Delta G_{\mathrm{m}}}{\partial T}\right)_p = -\Delta S_{\mathrm{m}} \neq 0$$

$$\left(\frac{\partial \Delta G_{\mathrm{m}}}{\partial p}\right)_T = \Delta V_{\mathrm{m}} \neq 0.$$

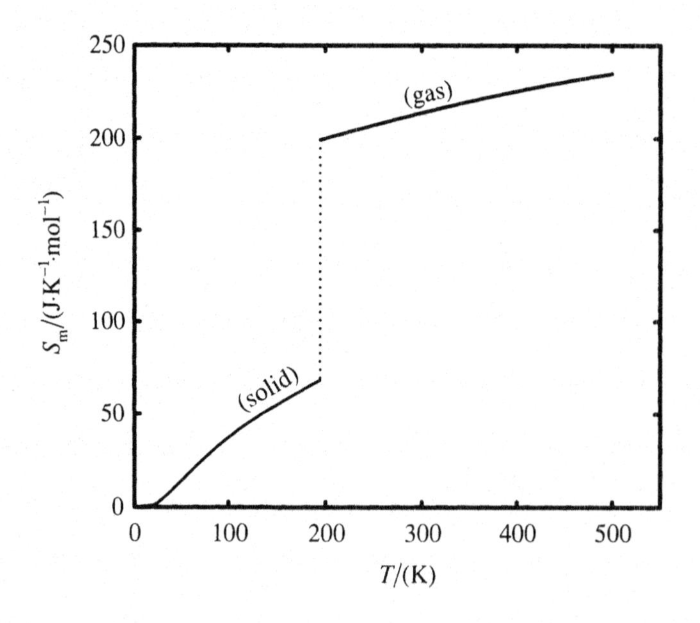


**Figure 8.10** Entropy of $CO_2$ at a pressure of 0.1 MPa. At this pressure, the solid changes to gas at $T = 194.52$ K.

---

[r] We established this equality in Chapter 5 where we derived the condition for equilibrium for a phase change.

As an example, Figures 8.10 and Figure 8.11 show how the entropy and volume of $CO_2$ change in going through the equilibrium phase change from solid to gas. Similar changes in $\Delta S_m$ and $\Delta V_m$ are obtained for the equilibrium processes: (solid = liquid), (liquid = gas), and (solid I = solid II), and they are also first-order phase transitions. An exception is the change from liquid to gas that occurs at the critical temperature. That change is an example of a continuous phase transition in which the liquid changes continuously to gas at the critical point, with $\Delta_{vap}S_m$ and $\Delta_{vap}V_m$ equal to zero.[s] The change from normal to

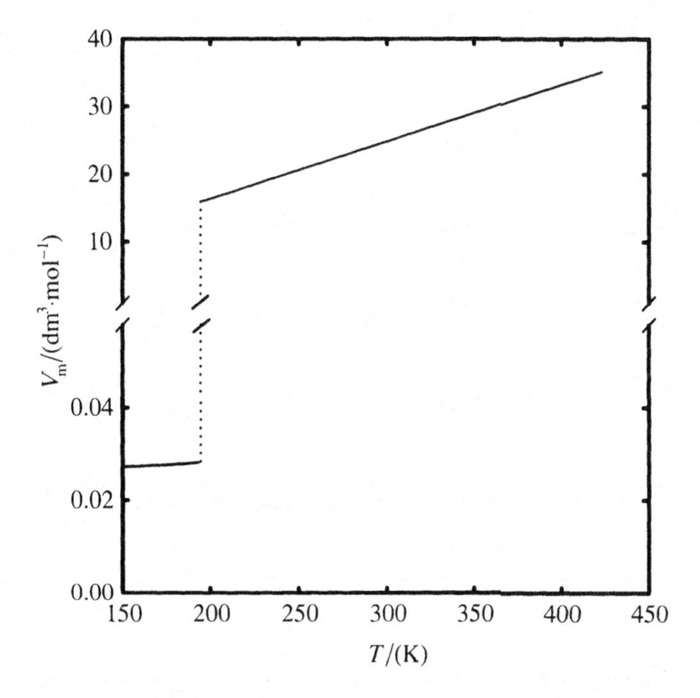


**Figure 8.11** Molar volumes of $CO_2(s)$ and $CO_2(g)$ at $p = 0.100$ MPa. At this pressure, the solid is in equilibrium with gas at $T = 194.52$ K. (Note the change in the volume scale.)

[s]The lambda transitions involving critical phenomena are examples of continuous phase transitions in which the first derivatives of $\Delta G$, $\Delta S$ and $\Delta V$ are zero, but the second derivatives (such as $C_p$ and $\kappa$) are not zero. Note that in a continuous phase transition, the Clapeyron equation does not apply since

$$\frac{dp}{dT} = \frac{\Delta S}{\Delta V} = \frac{0}{0}$$

and is indeterminate.

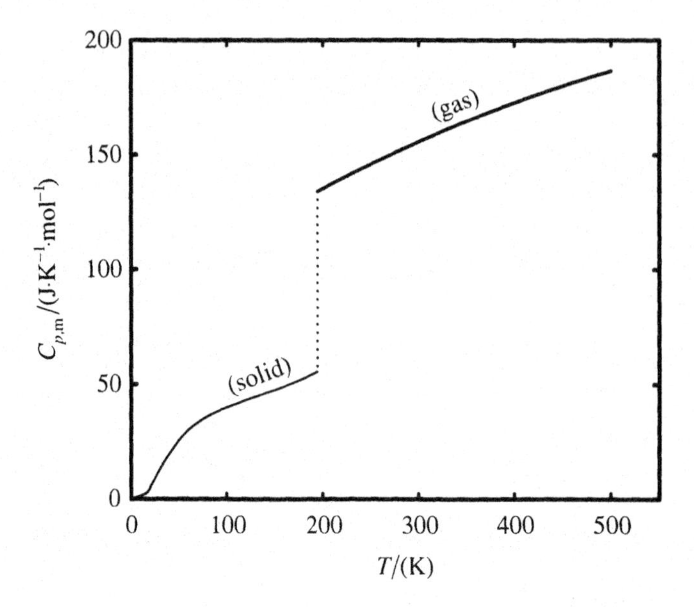


**Figure 8.12** Heat capacity of $CO_2$ at a pressure of 0.1 MPa. At this pressure, the solid is in equilibrium with gas at $T = 194.52$ K.

superfluid helium and the magnetic transitions described in Chapter 4 are also examples of continuous phase transitions. There are other examples. In this chapter, we will restrict our discussion to first-order phase transitions.

In a first-order phase transition, the second derivatives of $\Delta G$ are also not zero. That is

$$\left(\frac{\partial^2 \Delta G_m}{\partial T^2}\right)_p = -\left(\frac{\partial \Delta S_m}{\partial T}\right)_p = -\Delta C_{p,m} \neq 0$$

$$\left(\frac{\partial^2 \Delta G_m}{\partial p^2}\right)_T = \left(\frac{\partial \Delta V_m}{\partial p}\right)_T = V_m \Delta\kappa \neq 0.$$

Again using $CO_2$ as an example, Figure 8.12 shows the discontinuity in $C_{p,m}$ at the equilibrium transition temperature where $CO_2$ changes from solid to gas. A similar discontinuity is found for the compressibility $\kappa$.[t]

---

[t] Actually, the temperature does not change as heat is added to change the solid to gas at the equilibrium sublimation temperature. Hence, the heat capacity becomes infinite at this temperature, and the dotted line shown in Figure 8.12 should extend vertically to infinity. The compressibility and coefficient of expansion would show a similar behavior.

## 8.2 Phase Equilibria for Mixtures

So far, we have described the effect of pressure and temperature on the phase equilibria of a pure substance. We now want to describe phase equilibrium for mixtures. Composition, usually expressed as mole fraction $x$ or $y$, now becomes a variable, and the effect of composition on phase equilibrium in mixtures becomes of interest and importance.

In Chapter 5, we showed that the condition of phase equilibrium for multicomponent phases is that the chemical potential of each component must be the same in all the phases. That is

$$\mu_{A,1} = \mu_{B,1} = \mu_{C,1} = \cdots = \mu_{P,1}$$

$$\mu_{A,2} = \mu_{B,2} = \mu_{C,2} = \cdots = \mu_{P,2}$$

$$\vdots \quad \vdots \quad \vdots \quad \vdots$$

$$\mu_{A,C} = \mu_{B,C} = \mu_{C,C} = \cdots = \mu_{P,C}$$

where 1, 2 ... $C$ are the components and A, B, ... P are the phases in equilibrium.

We will be looking at first-order phase transitions in a mixture so that the Clapeyron equation, as well as the Gibbs phase rule, apply. We will describe mostly binary systems so that $C = 2$ and the phase rule becomes

$$f = C - P + 2 = 4 - P. \tag{8.14}$$

Pressure $p$, temperature $T$, and mole fraction $x$ (condensed phases) or $y$ (gas phase) are the variables we will usually employ in constructing phase diagrams. Rather than depicting phase diagrams in three dimensions, we will usually hold one variable constant and look at isotherms (constant $T$) with $p$ and $x$ (or $y$) as the variables; at isobars (constant $p$) with $T$ and $x$ (or $y$) as the variables; or at **isopleths** (constant $x$ or $y$) with $T$ and $p$ as the variables. The result is a two-dimensional phase diagram.[u] As with the single-component phase diagram, lines will represent equilibrium between two phases. Along these lines, one degree of freedom is present, since a degree of freedom has already been spent by specifying constant $T$, $p$, or $x$, and specifying one of the two remaining variables fixes the other. Lines come together at triple points where three phases are in equilibrium; such points are invariant. In areas of the phase diagram where a single phase is present, two degrees of freedom are allowed so that $p$ and $T$, $p$ and $x$, or $T$ and $x$ can be varied independently. In regions where two phases are

[u] Sometimes the third variable will be shown as contours on the two-dimensional phase diagrams. Thus, a series of isobars may be shown on a $(T, x)$ phase diagram.

present, one degree of freedom is allowed. The connection between the two phases in equilibrium is established through tie lines.

We will begin our discussion by describing (vapor + liquid) equilibrium, which we will extend into the supercritical fluid region as (fluid + fluid) equilibrium. (Liquid + liquid) equilibrium will then be described and combined with (vapor + liquid) equilibrium in the (fluid + fluid) equilibrium region. Finally, we will describe some examples of (solid + liquid) equilibrium.

### 8.2a Vapor + Liquid Equilibrium

In Chapter 6, we showed that the total vapor fugacity $f$ above an ideal solution was linearly related to composition by

$$f = f_1^* + (f_2^* - f_1^*)x_2, \tag{8.15}$$

where $x_2$ is the mole fraction of component 2 in the liquid mixture, and $f_1^*$ and $f_2^*$ are the vapor fugacities of pure components 1 and 2.

We used the system ($x_1$c-$C_6H_{11}CH_3$ + $x_2$c-$C_6H_{12}$) as an example of a system that closely approximates ideal behavior. Figure 6.5 showed the linear relationship between vapor pressure and mole fraction for this system. In this figure, vapor pressure could be substituted for vapor fugacity, since at the low pressure involved, the approximation of ideal gas behavior is a good one, and

$$p = p_1^* + (p_2^* - p_1^*)x_2. \tag{8.16}$$

The total vapor pressure line in Figure 6.5 for ($x_1$c-$C_6H_{11}CH_3$ + $x_2$c-$C_6H_{12}$) at $T = 308.15$ K is reproduced as the upper line in Figure 8.13. This line is often known as the **bubble-pressure curve**. We will refer to it as the liquid line. The straight line relationship for this line as predicted by equation (8.16) is evident.

Because c-$C_6H_{12}$ has a higher vapor pressure than c-$C_6H_{11}CH_3$, the vapor phase above the liquid mixture is richer in c-$C_6H_{12}$ than is the liquid phase. Thus, if $x_2$ is the mole fraction of c-$C_6H_{12}$ in the liquid phase and $y_2$ is the corresponding mole fraction in the vapor phase, then $y_2 > x_2$.

If we assume that Dalton's law of partial pressure holds in the vapor phase, it is easy to show that $y_2$ is related to the total pressure, $p$, by the equation

$$p = \frac{p_1^* p_2^*}{p_2^* - (p_2^* - p_1^*)y_2}. \tag{8.17}$$

Thus, $y_2$ is not linearly related to $p$, and a graph of $p$ against $y_2$ is not a straight line. The graph of the vapor line, which is also known as the **dew pressure curve**, is shown in Figure 8.13 for ($y_1$c-$C_6H_{11}CH_3$ + $y_2$c-$C_6H_{12}$) as the lower curve at $T = 308.15$ K.

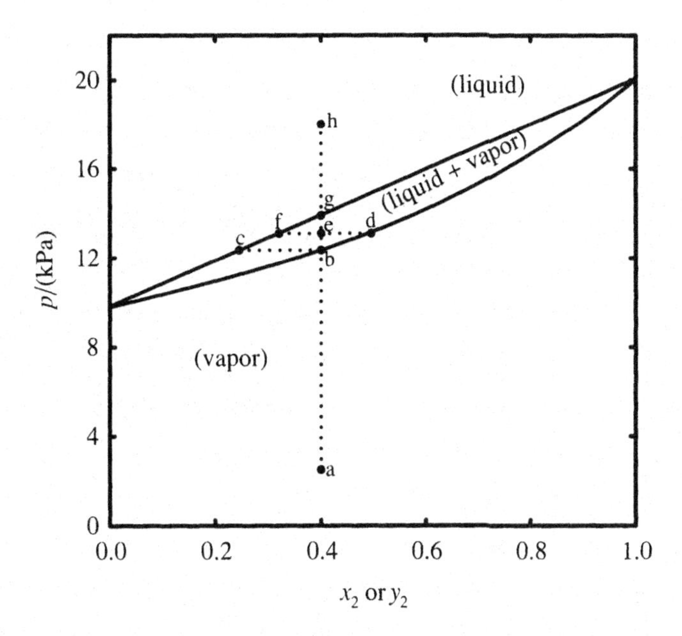


**Figure 8.13** Vapor pressures at $T = 308.15$ K for $\{(x_1 \text{ or } y_1)\ \text{c-C}_6\text{H}_{11}\text{CH}_3 + (x_2 \text{ or } y_2)\ \text{c-C}_6\text{H}_{12}\}$.

The liquid line and vapor line together constitute a binary (vapor + liquid) phase diagram, in which the equilibrium (vapor) pressure is expressed as a function of mole fraction at constant temperature. At pressures less than the vapor (lower) curve, the mixture is all vapor. Two degrees of freedom are present in that region so that $p$ and $y_2$ can be varied independently. At pressures above the liquid (upper) curve, the mixture is all liquid. Again, two degrees of freedom are present so that $p$ and $x_2$ can be varied independently.[v]

Two phases are present in the region between the two curves; the compositions of the two phases in equilibrium with each other are given by the intersection of a horizontal tie-line with the vapor and liquid curves. Lines cb and fd in Figure 8.13 are two examples. One degree of freedom is present in this region. Thus, specifying the pressure fixes the compositions of the phases in equilibrium; conversely, specifying the composition of one of the phases in equilibrium sets the pressure and the composition of the other phase.[w]

[v] Since vapor is not present above the liquid curve, this pressure is not a vapor pressure. Rather, it is a mechanical pressure exerted by a device such as a piston in a cylinder containing the liquid mixture.

[w] Setting the composition of one of the phases specifies the composition of the other since the two phases are related through the tie-line.

To better understand these relationships, consider a process in which a mixture containing 0.4 moles of c-$C_6H_{12}$ and 0.6 moles of c-$C_6H_{11}CH_3$ is placed in a cylinder at a pressure given by point a in Figure 8.13, and then compressed isothermally at $T = 308.15$ K. Initially, the mixture is all vapor. Compression along the vertical dotted line increases the pressure of the gas until point b on the vapor curve is reached. This pressure equals the equilibrium vapor pressure of the mixture. With further compression, droplets of liquid form, with the first drop having a composition given by point c. Thus, the liquid that forms is richer in c-$C_6H_{11}CH_3$ than the vapor. As a result, condensation of the liquid causes the remaining vapor to become richer in c-$C_6H_{12}$. As we compress, condensation continues and the compositions of the phases in equilibrium change. The composition of the vapor phase moves up along the vapor curve and the composition of the liquid phase in equilibrium with the vapor moves up the liquid curve. At a pressure corresponding to point e, for example, the overall composition of the mixture is still (0.4 moles of c-$C_6H_{12}$ + 0.6 moles of c-$C_6H_{11}CH_3$), but the actual compositions of the vapor and liquid phases are given by points d and f. Furthermore, the relative amounts of the two phases present are given by the **lever rule**, which states that the ratio of the number of moles present in the two phases is given by the ratio of the lengths de and ef. Thus

$$\frac{\text{moles vapor}}{\text{moles liquid}} = \frac{\text{ef}}{\text{de}}. \tag{8.18}$$

Since, in this two-phase region, the compositions of the phases in equilibrium are fixed by a horizontal tie-line (for example, line def), specifying the pressure, which specifies the position of the tie-line, sets the composition, as expected with one degree of freedom present.

Continued compression increases the pressure along the vertical dotted line. The compositions and amounts of the vapor and liquid phases continue to change along the liquid and vapor lines and the relative amounts change as required by the lever rule. When a pressure corresponding to point g is reached, the last drop of vapor condenses. Continued compression to a point such as h simply increases the total pressure exerted by the piston on the liquid.

Increasing the temperature increases the vapor pressures and moves the liquid and vapor curves to higher pressure. This effect can best be seen by referring to Figure 8.14, which is a schematic three-dimensional representation for a binary system that obeys Raoult's law, of the relationship between pressure, plotted as the ordinate, mole fraction plotted as abscissa, and temperature plotted as the third dimension perpendicular to the page. The liquid and vapor lines shown in Figure 8.13 in two dimensions (with $T$ constant)

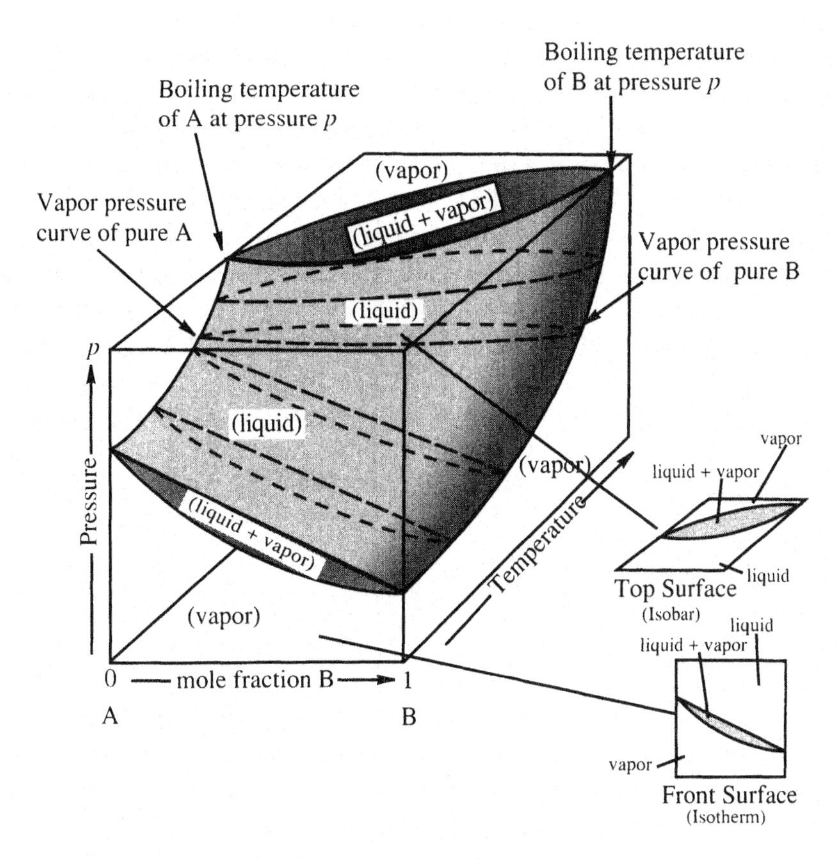


**Figure 8.14** Relationship between the $(p, x)$ and $(T, x)$ (vapor + liquid) phase diagrams for an ideal solution.

become surfaces as $T$ varies; the upper surface gives the composition of the liquid phase as a function of pressure and temperature, in equilibrium with the gas phase whose composition as a function of pressure and temperature is given by the lower surface. A single liquid phase is present at pressures above the upper surface, while at pressures below the lower surface, only vapor is present. Between the two surfaces is a volume where the two phases can coexist, with tie-lines again connecting the compositions of the phases at equilibrium.

Cutting through the space shown in Figure 8.14 with planes at constant temperature generates two-dimensional pressure versus mole fraction phase diagrams similar to the one shown as our example in Figure 8.13. Slices through the vapor and liquid surfaces yield the vapor and liquid lines in the resulting diagram. Planes cut at increasingly higher temperatures give liquid and vapor lines at increasingly higher pressures, since the vapor pressures of the pure components are increasing with increasing temperature as shown in Figure 8.14.

We can also cut through the volume with planes at constant pressure. These planes would be perpendicular to the plane of the paper. They cut through the liquid and vapor surfaces to give a temperature against mole fraction phase diagram such as that shown in Figure 8.14 as the top plane. Figure 8.15 gives an example for $\{(x_1 \text{ or } y_1)\ \text{c-C}_6\text{H}_{11}\text{CH}_3 + (x_2 \text{ or } y_2)\ \text{c-C}_6\text{H}_{12}\}$ at $p = 101.3$ kPa. Phase diagrams of this type are known as boiling point diagrams, since they give the boiling temperatures of mixtures at an external (atmospheric) pressure equal to the total vapor pressure. When the pressure is 101.3 kPa (or 1 atm) as in Figure 8.15, the temperatures are the normal boiling temperatures of the mixtures.

In Figure 8.15, a two-phase (liquid + vapor) region is again enclosed by a liquid line and a vapor line. However, the lines have inverted from those in the pressure against mole fraction phase diagram, with the vapor line now on top and the liquid line on the bottom. At temperatures below the liquid line, only liquid is present while above the vapor line, only vapor is present.

A boiling point phase diagram such as that shown in Figure 8.15 can be better understood by taking an example in which we start with a liquid mixture with temperature and mole fraction given by a point such as point a. Heating the liquid at a constant atmospheric pressure, in this case 101.3 kPa, causes the temperature to rise along the vertical dotted line until point b is reached. At this temperature, the vapor pressure of the mixture has become equal to

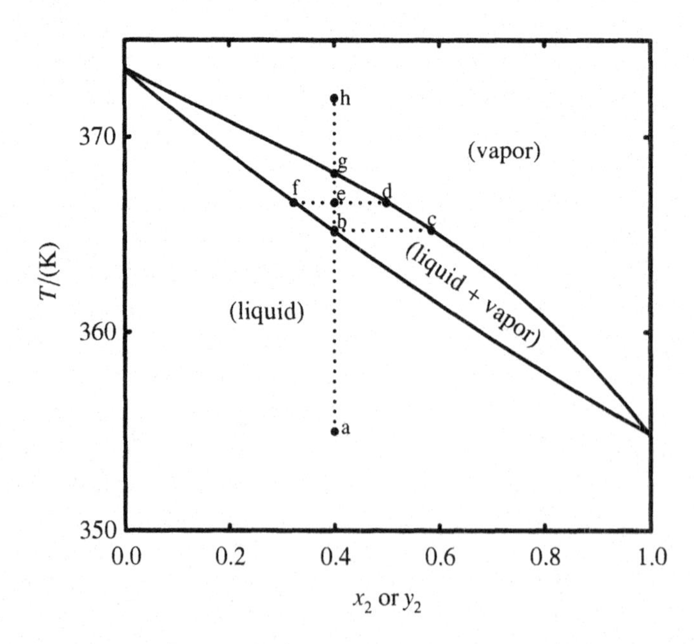


**Figure 8.15** Boiling temperatures at $p = 101.3$ kPa for $\{(x_1 \text{ or } y_1)\ \text{c-C}_6\text{H}_{11}\text{CH}_3 + (x_2 \text{ or } y_2)\ \text{c-C}_6\text{H}_{12}\}$.

atmospheric pressure, and the liquid starts to boil, with the first vapor (known as the distillate) that is produced having a composition given by the end of the tie-line at point c.

Since the vapor is richer in c-$C_6H_{12}$, continued boiling enriches the liquid (known as the residue) in c-$C_6H_{11}CH_3$. As the residue becomes richer in c-$C_6H_{11}CH_3$, the boiling temperature increases, following the liquid line. At the same time, the distillate follows the vapor line and becomes richer in c-$C_6H_{11}CH_3$. At a boiling temperature corresponding to point e, for example, the distillate will have a composition given by point d while the residue will have a composition given by point f. Again, the lever rule applies so that the ratio of distillate to residue is given by

$$\frac{\text{moles of distillate}}{\text{moles of residue}} = \frac{\text{ef}}{\text{de}}.$$

With continued boiling, the temperature increases. When point g is reached, the last drop of residue boils away, and the distillate has the composition given by point g, which is the same as the starting composition of the liquid at point a. Continued heating raises the temperature of the gaseous mixture to a value such as that given by point h.[x]

The relationship between (pressure against composition) and (temperature against composition) phase diagrams for near-ideal mixtures is summarized schematically in the (fluid + fluid) phase diagram shown in Figure 8.16. Lines ab and cd give the vapor pressures of the pure components 1 and 2. They end at the critical points b and d. Vertical sections at temperatures, $T_1$, $T_2$ ... $T_5$, give the isothermal ($p$, $x$ or $y$) phase diagrams, with the liquid and vapor lines enclosing the two phase (darker) shaded region. Horizontal sections at pressures, $p_1, p_2, \ldots p_4$, give the isobaric ($T$, $x$ or $y$) phase diagrams, with liquid and vapor lines enclosing the two-phase (lighter) shaded region. Also shown, is an example of a ($p$, $T$) isopleth (constant $x$) at composition, point e, with the vapor line given by ef and the liquid line by fg.

Of special interest are the isothermal sections at $T_4$ and $T_5$, which are at temperatures above the critical temperature of component 1, but below the

[x] Collecting the distillate during the distillation and then condensing it gives a liquid richer in c-$C_6H_{12}$. Repeating this process a number of times in a fractional distillation apparatus allows one to separate and purify the components in a mixture. We leave a discussion of such procedures to books describing purification techniques. See, for example, J. B. Ott and J. R. Goates, "Temperature Measurement with Application to Phase Equilibria Studies," Chapter 6, B. W. Rossiter and R. C. Baetzold, ed., "Physical Methods of Chemistry", Second Edition, John Wiley and Sons, New York, 1992.

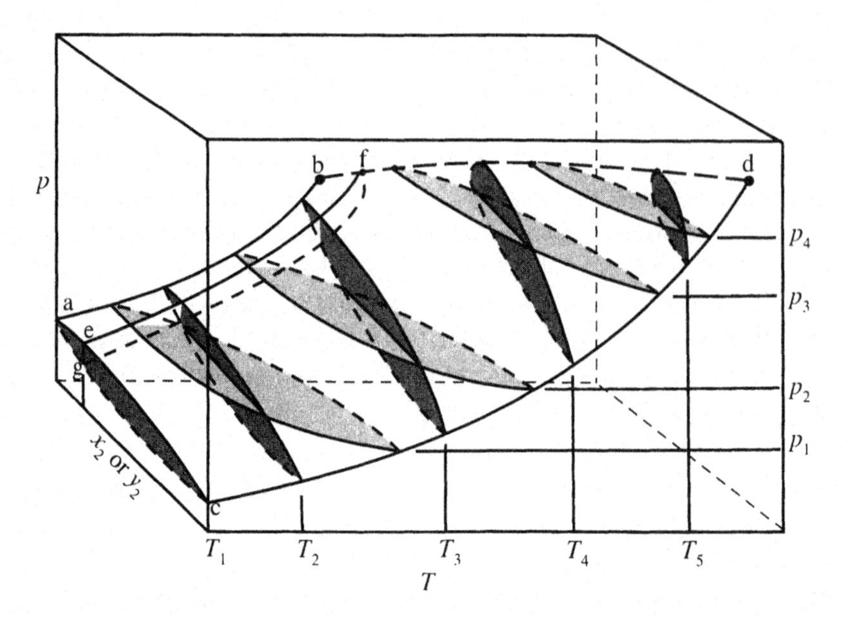


**Figure 8.16** (Fluid + fluid) phase diagram for a near-ideal system. Reproduced with permission from W. B. Streett, Chapter 1 in *Chemical Engineering at Supercritical Fluid Conditions*, M. E. Paulaitis, J. M. L. Penninger, R. D. Gray Jr., and P. Davidson, editors, Ann Arbor Science Press, Michigan, 1983.

critical temperature of component 2. In this temperature region, the liquid and vapor lines come together along line bd, instead of line ab. Line bd is known as the **critical locus**. It gives the critical point of the mixture as a function of composition. Since the critical locus extends from pure component 1 to pure component 2, the two-phase regions at temperatures between the two critical temperatures do not extend to pure component 1, and with increasing temperature extend over a smaller and smaller mole fraction region, until at the critical temperature of component 2, the two-phase region disappears. Above this temperature, there is no distinction between liquid and vapor, and only a fluid phase is present.

In our discussion of (vapor + liquid) phase equilibria to date, we have limited our description to near-ideal mixtures. As we saw in Chapter 6, positive and negative deviations from ideal solution behavior are common. Extreme deviations result in azeotropy, and sometimes to (liquid + liquid) phase equilibrium. A variety of critical loci can occur involving a combination of (vapor + liquid) and (liquid + liquid) phase equilibria, but we will limit further discussion in this chapter to an introduction to (liquid + liquid) phase equilibria and reserve more detailed discussion of what we designate as (fluid + fluid) equilibria to advanced texts.

## 8.2b Liquid + Liquid Equilibrium

The origins of (liquid + liquid) phase equilibrium can be understood by referring to a schematic (vapor + liquid) diagram such as that shown in Figure 8.17, which shows a graph of the vapor fugacity of component 2 in a binary mixture as a function of mole fraction, at a series of temperatures. At the highest temperature $T_1$, significant positive deviations from Raoult's law are apparent. As the temperature is lowered, we often find that the curvature of the vapor fugacity line becomes more pronounced and deviations from ideal behavior increase until a temperature is reached where the curve becomes

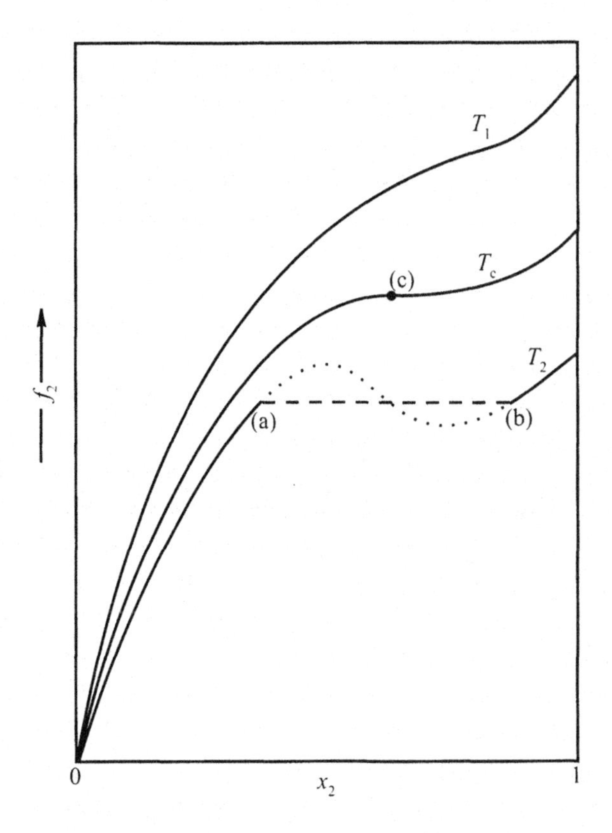


**Figure 8.17** Vapor fugacity for component 2 in a liquid mixture. At temperature $T_1$, large positive deviations from Raoult's law occur. At a lower temperature, the vapor fugacity curve goes through a point of inflection (point c), which becomes a critical point known as the upper critical end point (UCEP). The temperature $T_c$ at which this happens is known as the upper critical solution temperature (UCST). At temperatures less than $T_c$, the mixture separates into two phases with compositions given by points a and b. Component 1 would show similar behavior, with a point of inflection in the $f_1$ against $x_2$ curve at $T_c$, and a discontinuity at $T_2$.

horizontal at a point of inflection c. If the solution is further cooled, it separates into two phases, one richer in component 1 and the other richer in component 2.

Point c is a critical point known as the upper critical end point (UCEP).[y] The temperature, $T_c$, where this occurs is known as the upper critical solution temperature (UCST) and the composition as the critical solution mole fraction, $x_{2,c}$. The phenomenon that occurs at the UCEP is in many ways similar to that which happens at the (liquid + vapor) critical point of a pure substance. For example, at a temperature just above $T_c$, critical opalescence occurs, and at point c, the coefficient of expansion, compressibility, and heat capacity become infinite.

Although not shown in Figure 8.17, the vapor fugacity of component 1 also goes through a point of inflection at the same $T_c$ and $x_{2,c}$. This can be shown by noting that since the slope is horizontal at point c, then

$$\frac{\partial f_2}{\partial x_2} = 0.$$

From the Duhem–Margules equation [equation (6.77)], we know that

$$\frac{x_2}{f_2}\frac{\partial f_2}{\partial x_2} + \frac{x_1}{f_1}\frac{\partial f_1}{\partial x_2} = 0.$$

Hence, when the slope $\partial f_2/\partial x_2 = 0$, the slope $\partial f_1/\partial x_2 = 0$.

We should note that point c is a point of inflection rather than a maximum so that the second derivatives are also zero. That is[z]

$$\frac{\partial^2 f_2}{\partial x_2^2} = \frac{\partial^2 f_1}{\partial x_2^2} = 0. \tag{8.19}$$

---

[y] We will see later that in some circumstances, a solution breaks into two equilibrium phases as the temperature is increased. A comparable critical point is obtained in that circumstance that is known as a lower critical end point (LCEP), with the temperature known as the lower critical solution temperature (LCST).

[z] Equation (8.19) can be expressed in terms of chemical potential rather than fugacities. That is, at the critical point

$$\frac{\partial^2 \mu_2}{\partial x_2^2} = \frac{\partial^2 \mu_1}{\partial x_2^2} = 0. \tag{8.20}$$

If we could prevent the mixture from separating into two phases at temperatures below $T_c$, we would expect the point of inflection to develop into curves similar to those shown in Figure 8.17 as the dotted line for $f_2$, with a maximum and minimum in the fugacity curve. This behavior would require that the fugacity of a component decreases with increasing mole fraction. In reality, this does not happen, except for the possibility of a small amount of supersaturation that may persist briefly. Instead, the mixture separates into two phases. These phases are in equilibrium so that the chemical potential and vapor fugacity of each component is the same in both phases. That is, if we represent the phase equilibrium as

$$\mathrm{A} = \mathrm{B} \tag{8.21}$$

then

$$f_{\mathrm{A},1} = f_{\mathrm{B},1}, \text{ and } \mu_{\mathrm{A},1} = \mu_{\mathrm{B},1}$$

$$f_{\mathrm{A},2} = f_{\mathrm{B},2}, \text{ and } \mu_{\mathrm{A},2} = \mu_{\mathrm{B},2}.$$

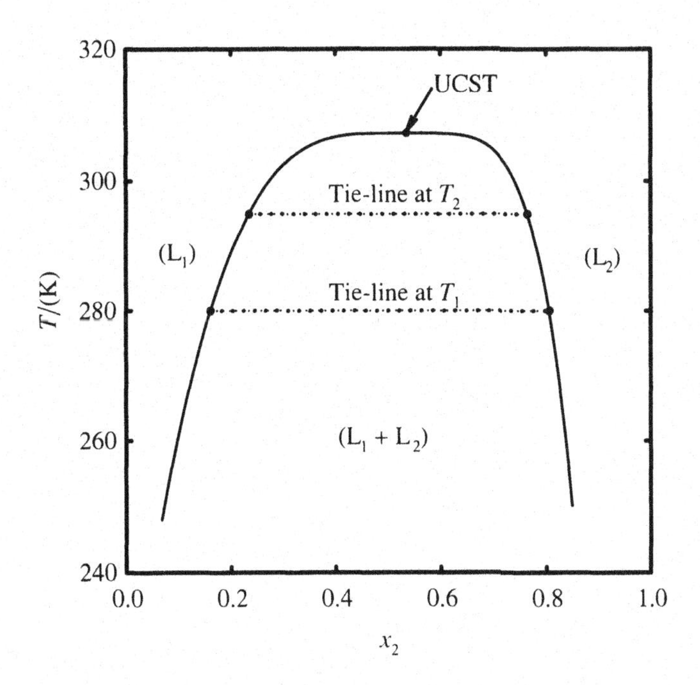


**Figure 8.18** (Liquid + liquid) equilibria at $p = 0.1$ MPa for ($x_1$n-$C_6H_{14}$ + $x_2CH_3OH$). The intersections of the tie-lines with the equilibrium curve give the compositions of the phases in equilibrium.

As an example of (liquid + liquid) phase equilibrium, Figure 8.18 is the (liquid + liquid) phase diagram for the system ($x_1$n-$C_6H_{14}$ + $x_2CH_3OH$) at a pressure of 0.1 MPa. The equilibrium curve is of the type that would result from the changes in vapor fugacity shown schematically in Figure 8.17 that results in a UCEP. The area under the curve is a two-phase region, with liquids $L_1$ and $L_2$ in equilibrium. The compositions of the two liquids are given by horizontal tie-lines that connect the two sides of the curve. We note that as the temperature increases, the two phases become closer and closer in composition until at the temperature at the maximum in the curve (the UCST), the two liquids have the same composition. We can see that the UCST is a critical point, above which only one liquid phase is present.

As with (vapor + liquid) equilibrium, only one degree of freedom is available when two phases are in equilibrium, while two degrees of freedom are allowed when only one phase is present.[aa] In Figure 8.18, $T$ and $x_2$ can be varied independently outside the two-phase region. However, under the curve in the two-phase region, only one of the remaining variables need be specified to fix the other. For example, if we specify the temperature, the compositions of the phases in equilibrium are fixed by the tie-lines. Conversely, specifying the composition of one of the phases fixes the temperature and the composition of the other phase along the tie-line.

The UCST behaves like other critical points in that critical exponents describe the properties of the mixture near the critical point. For example, $T_c$, $T_1$, $x_{2,c}$ and $x_2$ are related by the equation

$$T - T_c = k \mid x_2 - x_{2,c} \mid^{1/\beta}, \tag{8.22}$$

where $\beta = 0.32 \pm 0.02$ is the critical exponent and $k$ is a constant. Equation (8.22) applies only at values of $x_2$ very near the UCST. An extended semi-empirical form of this equation, which takes into account the asymmetry of the equilibrium curve, can be used to fit experimental (liquid + liquid) results over an extended range of temperature and mole fraction. It has the form

$$T = T_c + k \mid y_2 - y_{2,c} \mid^{1/\beta} \tag{8.23}$$

where

$$y_2 = \frac{\alpha x_2}{1 + x_2(\alpha - 1)} \tag{8.24}$$

[aa] Remember that we have already used one degree of freedom by holding the pressure constant.

and

$$y_{2,c} = \frac{\alpha x_{2,c}}{1 + x_{2,c}(\alpha - 1)}, \tag{8.25}$$

with $\alpha$ as a constant that skews the curve. [With $\alpha = 1$, equation (8.23) reduces to equation (8.22).] In equation (8.23), $\beta$ can be approximated as the critical exponent. Often, better fits to the experimental data can be obtained by letting $\beta$ vary in the fitting process, and a value somewhat different than $\beta = 0.32$ may result.

As mentioned earlier, the physical properties of a liquid mixture near a UCST have many similarities to those of a (liquid + gas) mixture at the critical point. For example, the coefficient of expansion and the compressibility of the mixture become infinite at the UCST. If one has a solution with a composition near that of the UCEP, at a temperature above the UCST, and cools it, critical opalescence occurs. This is followed, upon further cooling, by a cloudy mixture that does not settle into two phases because the densities of the two liquids are the same at the UCEP. Further cooling results in a density difference and separation into two phases occurs. Examples are known of systems in which the densities of the two phases change in such a way that at a temperature well below the UCST, the solutions connected by the tie-line again have the same density.[bb] When this occurs, one of the phases separates into a shapeless mass or "blob" that remains suspended in the second phase. The tie-lines connecting these phases have been called **isopycnics** (constant density). Isopycnics usually occur only at a specific temperature. Either heating or cooling the mixture results in density differences between the two equilibrium phases, and separation into layers occurs.

Increasing the pressure may either increase or decrease the UCST and, thus, raise or lower the (liquid + liquid) equilibrium line. Figure 8.19, for example, shows the effect of pressure on the (liquid + liquid) equilibrium curves for ($x_1$n-$C_6H_{14}$ + $x_2CH_3OH$).[3] In this example, we see that increasing $p$ increases the UCST.

So far, we have considered only one type of binary (liquid + liquid) equilibrium. Examples can be found in the literature where a lower critical end point (LCEP) is obtained instead of a UCEP, and where both a LCEP and a UCEP are obtained for the same system. Also, (liquid + liquid) equilibria can be combined with (vapor + liquid) equilibria to give interesting (fluid + fluid)

[bb] An example is the (water + aniline) system at T = 350 K. See A. W. Francis, *Liquid–Liquid Equilibrium*, Interscience Publishers, New York, 1963, p. 14.

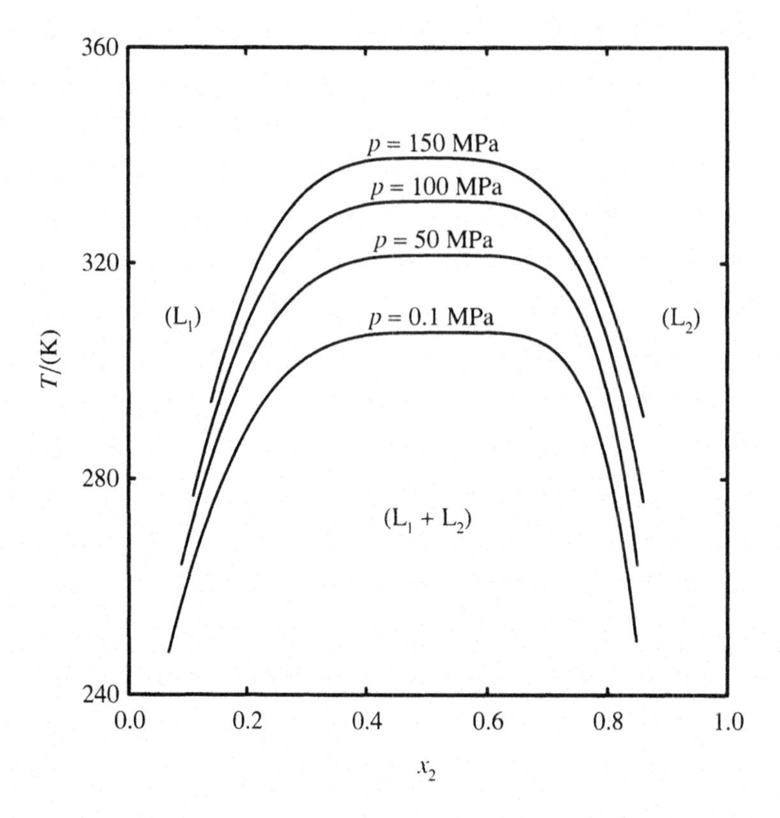


**Figure 8.19** Comparison of (liquid + liquid) equilibria for ($x_1$n-$C_6H_{14}$ + $x_2CH_3OH$) at $p = 0.1$ MPa, 50 MPa, 100 MPa, and 150 MPa.

phase diagrams, especially when the phase diagrams are extended to pressures and temperatures in the supercritical fluid region.

## 8.2c Solid + Liquid Equilibrium

In Chapter 6, we derived equation (6.161) shown below, which relates the activity, $a_i$, of a component in solution to the equilibrium melting temperature, $T$, of that substance.

$$\left(\frac{\partial \ln a_i}{\partial T}\right)_p = \frac{\Delta_{\text{fus}} H_{\text{m},i}}{RT^2}, \tag{6.161}$$

where $\Delta_{\text{fus}}H_{\text{m},i}$ is the enthalpy of fusion of pure component $i$. This equation was shown to be valid when component $i$ freezes out as a pure solid and a Raoult's law standard state is chosen for it so that the activity of the pure substance is one.

Since Raoult's law activities become mole fractions in ideal solutions, a simple substitution of $x_i = a_i$ into equation (6.161) yields an equation that can be applied to (solid + liquid) equilibrium where the liquid mixtures are ideal. The result is

$$\left(\frac{\partial \ln x_i}{\partial T}\right)_p = \frac{\Delta_{\text{fus}} H_{\text{m},i}}{RT^2}. \tag{8.26}$$

Equation (8.26) relates the melting temperature, $T$, of an ideal solution to the mole fraction, $x_i$, of the (pure) component that freezes from solution. It can be integrated by separating variables and setting the integration limits between $T$, the melting temperature where the mole fraction is $x_i$, and $T^*$, the melting temperature of the pure component, $i$, where $x_i = 1$. The result is

$$\int_{\ln x_i = 0}^{\ln x_i} \mathrm{d} \ln x_i = \int_{T_i^*}^{T} \frac{\Delta_{\text{fus}} H_{\text{m},i}}{RT^2} \mathrm{d}T \tag{8.27}$$

$$\ln x_i = \int_{T_i^*}^{T} \frac{\Delta_{\text{fus}} H_{\text{m},i}}{RT^2} \mathrm{d}T. \tag{8.28}$$

When the enthalpy of fusion can be assumed to be constant, equation (8.28) can be integrated to give

$$\ln x_i = -\frac{\Delta_{\text{fus}} H_{\text{m},i}}{R}\left(\frac{1}{T} - \frac{1}{T_i^*}\right). \tag{8.29}$$

In Chapter 6, we considered the more general behavior of the activity with temperature and showed with equations (6.163) to (6.166) how to integrate the right hand of equation (8.28) when the enthalpy of fusion is not constant over the temperature range of interest. Those same considerations apply directly here and they will not be repeated, except to give the final result:

$$R \ln x_i = \Delta H_{1,i}\left(\frac{1}{T_i^*} - \frac{1}{T}\right) + \Delta a_i \ln \frac{T}{T_i^*} + \frac{\Delta b_i}{2}(T - T_i^*)$$
$$+ \frac{\Delta c_i}{2}\left(\frac{1}{T^2} - \frac{1}{T_i^{*2}}\right) + \frac{\Delta d_i}{6}(T^2 - T_i^{*2}) \tag{8.30}$$

where the constants come from the differences in the heat capacities of pure solid and liquid component "$i$" expressed in the form of equation (6.164).

Figure 8.20 shows the (solid + liquid) equilibrium in the system $\{x_1C_6H_6 + x_21,4\text{-}C_6H_4(CH_3)_2\}$, a binary system that might be expected to approximate ideal solution behavior.[4] Line ac is the freezing line for benzene, while line bc is the freezing line for 1,4-dimethylbenzene. The dashed lines represent the ideal solution prediction using equation (8.30), with the appropriate set of heat capacity constants used to generate each side of the diagram. That is, the dashed line that is an approximation to line ac has been generated using the heat capacities of liquid and solid benzene as a function of temperature, the melting temperature of pure benzene, and its enthalpy of fusion, while the dashed line that is an approximation to line bc has been generated using similar data for 1,4-dimethylbenzene. The agreement between the experimental results and the ideal solution prediction shows that this system is very nearly ideal and has no solid-phase solubility.

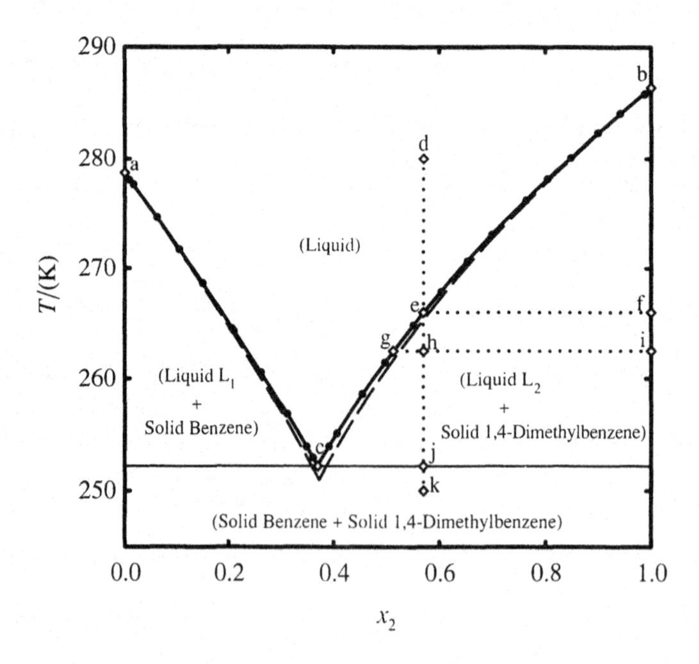


**Figure 8.20** (Solid + liquid) phase equilibria for $\{x_1C_6H_6 + x_21,4\text{-}C_6H_4(CH_3)_2\}$. The circles are the experimental results; the solid lines are the fit of the experimental results to equation (8.31); the dashed lines are the ideal solution predictions using equation (8.30); the solid horizontal line is at the eutectic temperature; and the diamonds are $(x, T)$ points referred to in the text.

The solid lines in Figure 8.20 were obtained by fitting the experimental results to the equation

$$T = T^* \left[ 1 + \sum_{j=1}^{n} a_j (x_2 - x_2^*)^j \right] \tag{8.31}$$

where $T^*$ is the melting temperature of the pure substance that freezes from solution and $x_2^*$ is the value of $x_2$ at $T^*$. In simple systems such as this, $x_2^* = 1$ for component 2 and $x_2^* = 0$ for component 1. It has been shown[5] that this equation works well for most binary (solid + liquid) systems, including those for which the phase diagram is much more complicated than the one shown in Figure 8.20.

Figure 8.20 demonstrates that adding a solute to a solvent to form an ideal (or near-ideal) mixture lowers the melting temperature of the solvent. Thus, adding benzene to 1,4-dimethylbenzene lowers the melting temperature of 1,4-dimethylbenzene along line bc, and adding 1,4-dimethylbenzene to benzene lowers the melting temperature of benzene along line ac. Point c represents the lowest melting mixture for this system. It is known as the **eutectic point** with composition, $x_{2,\mathrm{e}}$.

In Figure 8.20, a single-phase liquid mixture is present in the area above the (solid + liquid) equilibria lines. Since pressure is fixed, two degrees of freedom are present in this region so that temperature and mole fraction can be varied independently.

Below the equilibrium lines, but above the eutectic temperature, a liquid and solid are in equilibrium. Under line ac, solid benzene, and liquid $L_1$, whose composition is given by line ac, are present. Under line bc, the phases present are solid 1,4-dimethylbenzene and liquid $L_2$, whose composition is given by line bc. Below point c, solid benzene and solid 1,4-dimethylbenzene are present. In the two phase regions, one degree of freedom is present. Thus, specifying $T$ fixes the composition of the liquid, or specifying $x_2$ fixes the temperature.[cc] Finally, at point c (the eutectic) three phases (solid benzene, solid 1,4-dimethylbenzene, and liquid with $x_2 = x_{2,\mathrm{e}}$) are present. This is an invariant point, since no degrees of freedom are present.

To demonstrate the changes in phase predicted by Figure 8.20, let us start with a liquid mixture, with temperature and composition given by point d, and lower the temperature along the vertical dotted line. When a temperature

[cc] In a two-phase region, the composition of the two phases in equilibrium are given by the end points of tie-lines. Thus, under curve ac, the phases are pure solid benzene ($x_2 = 0$) and a liquid with composition given by line ac.

corresponding to point e is reached, a solid crystallizes from the liquid. The composition of this solid is that of pure 1,4-dimethylbenzene, given by point f along the tie-line ef.

As the mixture freezes, 1,4-dimethylbenzene is removed from solution, the liquid mixture becomes richer in benzene, and the melting temperature falls along line bc. For example, when the temperature given by point h is reached, solid 1,4-dimethylbenzene (point i) and a liquid solution with a composition given by point g are present. The lever rule gives the ratio of solid to liquid as

$$\frac{\text{moles solid}}{\text{moles liquid}} = \frac{\text{gh}}{\text{hi}}.$$

Continued cooling results in more solid 1,4-dimethylbenzene and a liquid enriched in benzene until a temperature corresponding to point j is reached. This is the eutectic temperature and the liquid has a $x_{2,\text{e}}$ composition given by point c. At this temperature solid benzene also begins to crystallize from solution. Thus, three phases are in equilibrium, resulting in an invariant point for which neither composition nor temperature can change. Solid 1,4-dimethylbenzene and solid benzene continue to crystallize from solution at the eutectic temperature in a ratio to hold $x_2$ fixed at $x_{2,\text{e}}$, until all of the liquid is gone. After this happens, we have two phases in equilibrium (solid benzene + solid 1,4-dimethylbenzene), and the temperature falls along line jk until point k is reached where we terminate the cooling process.[dd]

When a mixture with $x_2 < x_{2,\text{e}}$ is cooled, solid benzene crystallizes from solution at a temperature given by line ac. For such mixtures, the liquid changes composition along line ac until point c is again reached. Continued cooling causes 1,4-dimethylbenzene to crystallize from solution and the temperature stays constant until all liquid is gone. Thus, cooling any mixture eventually results in a liquid with composition given by $x_{2,\text{e}}$ that freezes at the eutectic temperature. It is interesting to cool a liquid mixture with the eutectic composition. At the eutectic temperature (point c), both solids crystallize from solution, and the composition and temperature remain constant until all the liquid is frozen. Thus, the freezing behavior is similar to that for a pure substance.

**Effect of Pressure on Solid + Liquid Equilibrium:** Equation (6.84) is the starting point for deriving an equation that gives the effect of pressure on (solid + liquid) phase equilibria for an ideal mixture in equilibrium with a pure

[dd] A tie-line at point k would have end points $x_2 = 0$ and $x_2 = 1$, the compositions of the pure solid benzene and pure solid 1,4-dimethylbenzene in equilibrium.

solid. For the ideal solution with a Raoult's law standard state, $a_i = x_i$ and $\mu_i^\circ(\mathrm{l}) = \mu_i^*(\mathrm{l})$. Furthermore, the equilibrium condition gives $\mu_i(\mathrm{l}) = \mu_i^*(\mathrm{s})$. When these relationships are substituted into equation (6.84), it becomes

$$RT \ln x_i = \mu_i^*(\mathrm{s}) - \mu_i^*(\mathrm{l}), \tag{8.32}$$

where $T$ is the melting temperature.

Differentiation of equation (8.32) with respect to pressure gives

$$RT\left(\frac{\partial \ln x_i}{\partial p}\right)_T = \left(\frac{\partial \mu_i^*(\mathrm{s})}{\partial p}\right)_T - \left(\frac{\partial \mu_i^*(\mathrm{l})}{\partial p}\right)_T. \tag{8.33}$$

Substituting $(\partial \mu_i^*/\partial p)_T = V_{\mathrm{m},i}^*$ (for both solid and liquid) into equation (8.33) and collecting terms gives

$$\left(\frac{\partial \ln x_i}{\partial p}\right)_T = -\frac{\Delta_{\mathrm{fus}} V_{\mathrm{m},i}^*}{RT} \tag{8.34}$$

where $\Delta_{\mathrm{fus}} V_{\mathrm{m},i} = V_{\mathrm{m},i}^*(\mathrm{l}) - V_{\mathrm{m},i}^*(\mathrm{s})$ is the change in volume when the pure solid melts to form liquid.

Equation (8.34) can be integrated to calculate the change in mole fraction with pressure at a melting temperature $T$. The result is

$$\int_{\ln x_i}^{\ln x_i'} \mathrm{d} \ln x_i = \int_p^{p'} -\frac{\Delta_{\mathrm{fus}} V_{\mathrm{m},i}^*}{RT}\, \mathrm{d}p. \tag{8.35}$$

If the pressure change is not too large, it is a reasonable approximation to assume that $\Delta_{\mathrm{fus}} V_{\mathrm{m},i}^*$ is independent of pressure,[ee] in which case, equation (8.35) becomes

$$\ln \frac{x_i'}{x_i} = -\frac{\Delta_{\mathrm{fus}} V_{\mathrm{m},i}}{RT}(p' - p). \tag{8.36}$$

Figure 8.21 gives the ideal solution prediction {equation (8.36)} of the effect of pressure on the (solid + liquid) phase diagram for $\{x_1 C_6H_6 + x_2 1,4\text{-}C_6H_4(CH_3)_2\}$. The curves for $p = 0.1$ MPa are the same as those shown in Figure 8.20. As

---

[ee] Before integrating equation (8.35), one can express $\Delta_{\mathrm{fus}} V_{\mathrm{m},i}^*$ as a function of pressure, usually in terms of $\kappa$, the compressibility, if the necessary information is available.

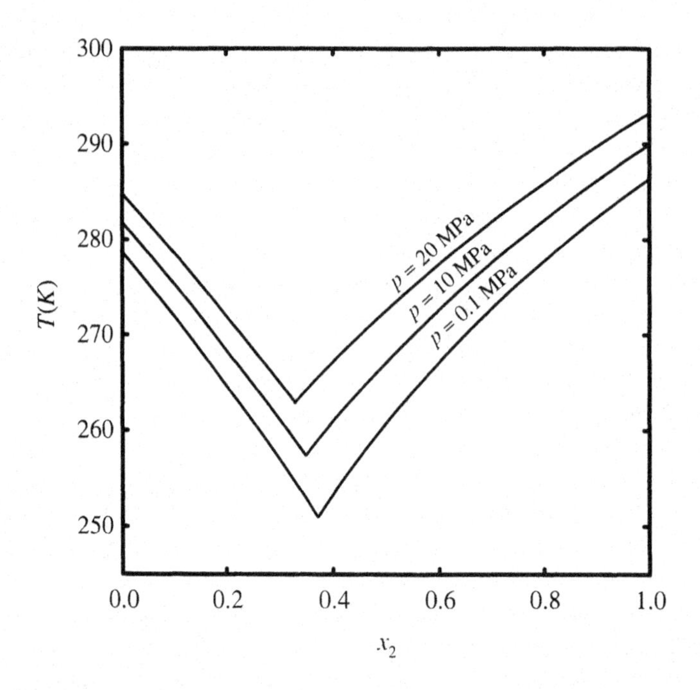


**Figure 8.21** Effect of pressure on the melting temperatures of $\{x_1C_6H_6 + x_2 1,4\text{-}C_6H_4(CH_3)_2\}$ assuming ideal solution behavior.

expected,[ff] increasing the pressure increases the melting curves, since for both $C_6H_6$ and $1,4\text{-}C_6H_4(CH_3)_2$, $V^*_m(l) > V^*_m(s)$ so that $\Delta_{fus}V^*_m$ is greater than zero.

**Solid + Liquid Equilibria in Less Ideal Mixtures:** We should not be surprised to find that the near-ideal (solid + liquid) phase equilibria behavior shown in Figures 8.20 and 8.21 for (benzene + 1,4-dimethylbenzene) is unusual. Most systems show considerably larger deviations. For example, Figure 8.22 shows the phase diagram for $\{x_1 n\text{-}C_{16}H_{34} + x_2C_6H_6\}$. The solid line is the fit of the experimental results to equation (8.31) and the dashed line is the ideal solution prediction.[gg] Thus, even in this mixture of two nonpolar hydrocarbons, significant deviations occur when one hydrocarbon is aliphatic and the other is aromatic.

---

[ff] LeChatelier's principle states that when a stress is applied to a system at equilibrium, the equilibrium shifts to relieve the stress. For this system, at a given temperature, the stress caused by increased pressure is relieved by freezing liquid to decrease the volume. Thus, the system is no longer at equilibrium at the higher pressure and must be heated to a higher temperature to reach the equilibrium line again.

[gg] Note that the ideal solution prediction is followed in dilute solution. We have shown earlier that in the limit of dilute solution, the solvent (the component that freezes from solution) obeys Raoult's law.

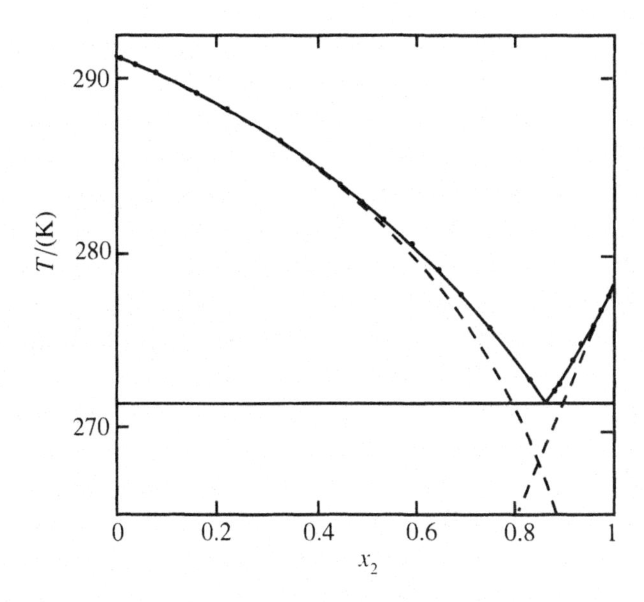


**Figure 8.22** (Solid + liquid) phase diagram for $\{x_1 \text{n-}C_{16}H_{34} + x_2C_6H_6\}$. The circles are the experimental melting temperatures and the lines are the fit of the experimental results to equation (8.31). The dashed lines are the ideal solution predictions from equation (8.30).

Polarity differences between the two components give rise to even larger deviations from ideal solution behavior in binary systems. Figure 8.23 shows the phase diagram for $\{x_1CCl_4 + x_2CH_3CN\}$. Again, the solid lines represent the experimental results and the dashed lines are the ideal solution predictions. In this system, where nonpolar molecules ($CCl_4$) are mixed with molecules with a large dipole moment ($CH_3CN$), large positive deviations from Raoult's law occur. The equilibrium line for $CCl_4$ almost has a point of inflection with zero slope around $x_2 = 0.4$. An inflection point with zero slope would be a critical point representing the beginnings of liquid phase separation and (liquid + liquid) equilibria. The situation would be similar to that described earlier for (vapor + liquid) equilibria in Figure 8.17. Also shown in Figure 8.23 are (solid + solid) phase transitions in both the $CCl_4$ and the $CH_3CN$, represented by the horizontal lines.[hh]

In addition to solid phase transitions and the combination of (solid + liquid) with (liquid + liquid) equilibria, solid solutions can form and a variety of

[hh] The solid-phase transition in $CH_3CN$ is at a temperature below the eutectic and is not shown in the "actual" (solid + liquid) diagram, although it is present as a solid line extending from pure $CH_3CN$ to pure $CCl_4$. It is involved in the (solid + liquid) equilibria for the ideal solution and is drawn with a dashed line to indicate its relationship to the ideal solution prediction of the phase diagram.

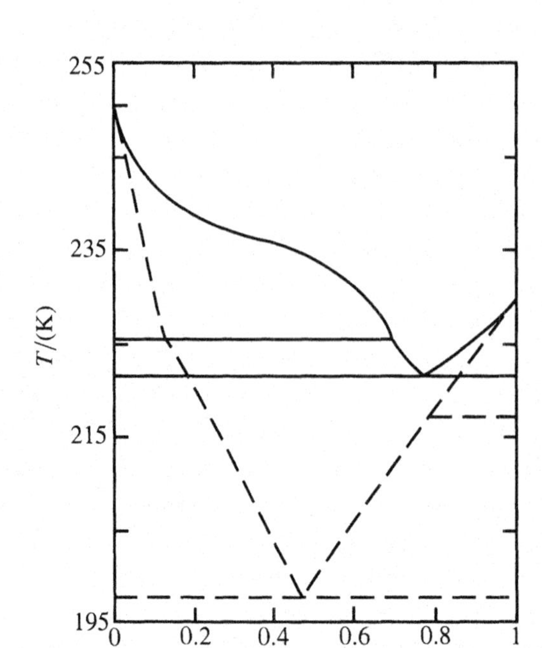


**Figure 8.23** (Solid + liquid) phase diagram for ($x_1CCl_4 + x_2CH_3CN$), an example of a system with large positive deviations from ideal solution behavior. The solid line represents the experimental results and the dashed line is the ideal solution prediction. Solid-phase transitions (represented by horizontal lines) are present in both $CCl_4$ and $CH_3CN$. The $CH_3CN$ transition occurs at a temperature lower than the eutectic temperature. It is shown as a dashed line that intersects the ideal $CH_3CN$ (solid + liquid) equilibrium line.

molecular addition compounds can occur. Examples can be found in the literature.[6]

## Exercises

E8.1 From the following data, sketch the phase diagram for nitrogen at low temperatures as accurately as possible:

- There are three crystal forms, $\alpha$, $\beta$, and $\gamma$ that coexist at $p = 471$ MPa and $T = 44.5$ K.
- At this triple point, the volume changes $\Delta V_m$ for the various transitions are 0.165 $cm^3 \cdot mol^{-1}$ for $\alpha \rightarrow \gamma$; 0.209 $cm^3 \cdot mol^{-1}$ for $\beta \rightarrow \gamma$; and 0.043 $cm^3 \cdot mol^{-1}$ for $\beta \rightarrow \alpha$.
- The $\Delta S_m$ values for the transitions cited are 1251, 5879, 4590, and 6519 $J \cdot K \cdot mol^{-1}$, respectively.
- At $p = 0.10$ MPa and $T = 36$ K, $\beta \rightarrow a$ with $\Delta V_m = 0.22$ $cm^3 \cdot mol^{-1}$.

E8.2 Solid monoclinic sulfur spontaneously converts to solid rhombic sulfur at 298.15 K and 0.101 MPa pressure. For the conversion process:

$$\text{Monoclinic Sulfur (298.15 K, 0.101 MPa)}$$
$$\rightarrow \text{Rhombic Sulfur (298.15 K, 0.101 MPa)}$$

Calculate (a) $\Delta_{\text{trans}}S_{\text{m}}$, (b) $\Delta_{\text{trans}}H_{\text{m}}$, (c) $\Delta_{\text{trans}}G_{\text{m}}$
Given:

$$C^{\circ}_{p,\text{m}} = 22.59 \text{ J·K}^{-1}\text{·mol}^{-1} \text{ for rhombic sulfur}$$
$$C^{\circ}_{p,\text{m}} = 23.64 \text{ J·K}^{-1}\text{·mol}^{-1} \text{ for monoclinic sulfur}$$
$$\Delta_{\text{trans}}H^{\circ}_{\text{m}} = 397.9 \text{ J·mol}^{-1}$$

for the transition of rhombic sulfur to monoclinic sulfur at the reversible transition temperature of 368.60 K and $p = 0.101$ MPa.

E8.3 The vapor pressures of solid and liquid hydrogen cyanide are given by

$$\text{solid:} \quad \ln p/(\text{kPa}) = 19.489 - \frac{4252.4}{T} \quad \text{(from 243.7 K to 258 K)}$$

$$\text{liquid:} \quad \ln p/(\text{kPa}) = 15.818 - \frac{3345.8}{T} \quad \text{(from 265 K to 300.4 K)}$$

Calculate (a) the molar enthalpy of sublimation, (b) the molar enthalpy of vaporization, (c) the molar enthalpy of fusion, (d) the triple point temperature and pressure, and (e) the normal boiling point for HCN.

E8.4 Solid $O_2$ has a solid phase transition at 23.66 K and 0.101 MPa pressure. The molar enthalpy of transition is 93.81 J· mol$^{-1}$. $\alpha O_2$ (stable below 23.66 K) has a density of 1.470 g·cm$^{-3}$ while $\beta O_2$ (stable above 23.66 K) has a density of 0.702 g·cm$^{-3}$. For the process at 23.66 K and 0.101 MPa pressure

$$\alpha O_2 = \beta O_2.$$

(a) Calculate $w$, $\Delta_{\text{trans}}U_{\text{m}}$, $\Delta_{\text{trans}}H_{\text{m}}$, $\Delta_{\text{trans}}S_{\text{m}}$, and $\Delta_{\text{trans}}G_{\text{m}}$.
(b) Calculate the transition temperature at 1 MPa pressure (assume $\Delta_{\text{trans}}S_{\text{m}}/\Delta_{\text{trans}}V_{\text{m}}$) is constant.

E8.5 The vapor pressure of liquid helium can be expressed by the equation

$$p = AT^{5/2}\exp\{-[(a/T) + bT^{11/2}]\}$$

in which $A$, $a$, and $b$ are constants. Derive an equation for $\Delta_{vap}H_m$ of He as a function of temperature.

E8.6 Use the following vapor pressure data for solid palladium metal as a function of temperature,[7] to calculate $\Delta_{sub}H_m$, the mean enthalpy of sublimation of palladium.

| $T/(K)$ | $p/(10^3\ Pa)$ | $T/(K)$ | $p/(10^3\ Pa)$ |
|---|---|---|---|
| 1294 | 0.289 | 1396 | 3.27 |
| 1308 | 0.425 | 1406 | 3.87 |
| 1322 | 0.697 | 1426 | 6.37 |
| 1333 | 0.801 | 1459 | 11.81 |
| 1350 | 1.096 | 1488 | 22.4 |

E8.7 The equilibrium vapor pressure of water at $T = 298.15$ K is 3.168 kPa.
(a) What is $\Delta G_m$ for the conversion of one mole of liquid water to water vapor at $p = 3.168$ kPa?
(b) Calculate $\Delta G_m$ for the process

$$H_2O(l, 100\ kPa, 298.15\ K) = H_2O(g, 100\ kPa, 298.15\ K).$$

Treat the vapor as an ideal gas, and assume that the liquid obeys the equation of state

$$V_m = V_{m,0}(1 - \alpha p)$$

where $V_{m,0} = 18.074\ cm^3{\cdot}mol^{-1}$ and $\alpha = 4.8 \times 10^{-4}\ MPa^{-1}$ at 298.15 K.

E8.8 (a) Construct the binary (vapor + liquid) phase diagram for an ideal mixture in which $p_1^* = p_2^*$.
(b) Construct the binary (vapor + liquid) phase diagram for an ideal mixture in which $p_1^* = 2p_2^*$.
(c) Compare the results of attempting to separate by fractional distillation, the components of the solutions in (a) and (b). Can you generalize to make a statement concerning the effect of the relative volatility of the components on the ease of separation of an ideal mixture by fractional distillation?

E8.9 Given below are total vapor pressures for $\{(x_1 \text{ or } y_1)\text{n-}C_6H_{14} + (x_2 \text{ or } y_2)\text{c-}C_6H_{12}\}$ at $T = 308.15$ K, where $x$ is the mole fraction in the liquid phase and $y$ is the mole fraction in the vapor phase.

| $x_2$ | $y_2$ | $p/(\text{kPa})$ | $x_2$ | $y_2$ | $p/(\text{kPa})$ |
|---|---|---|---|---|---|
| 0.0000 | 0.0000 | 30.567 | 0.5487 | 0.4462 | 25.421 |
| 0.0806 | 0.0588 | 29.850 | 0.6327 | 0.5287 | 24.543 |
| 0.1803 | 0.1339 | 28.954 | 0.7220 | 0.6231 | 23.585 |
| 0.2714 | 0.2053 | 28.118 | 0.7276 | 0.6295 | 23.511 |
| 0.3639 | 0.2811 | 27.243 | 0.8237 | 0.7426 | 22.401 |
| 0.4540 | 0.3591 | 26.367 | 0.8238 | 0.7429 | 22.396 |
| 0.4568 | 0.3616 | 26.341 | 0.9177 | 0.8694 | 21.210 |
| 0.5441 | 0.4418 | 25.468 | 1.0000 | 1.0000 | 20.050 |

(a) Make a graph of $x_2$ against $p$, $p_1$, and $p_2$, where $p_1$ and $p_2$ are the vapor pressures of the individual components. For comparison, include on the graph the ideal solution predictions for $p_1$, $p_2$, and $p$.

(b) Construct the (vapor + liquid) phase diagram at $T = 308.15$ K, by making a graph of $p$ against $x_2$, and against $y_2$. Label all regions of the phase diagram with the phases present.

E8.10 Given below are boiling temperatures for $\{(x_1 \text{ or } y_1)CHCl_3 + (x_2 \text{ or } y_2)(CH_3)_2CO\}$ at $p = 0.101$ MPa, where $x$ is the mole fraction in the liquid phase and $y$ is the mole fraction in the vapor phase.

| $x_2$ | $y_2$ | $T/(\text{K})$ | $x_2$ | $y_2$ | $T/(\text{K})$ |
|---|---|---|---|---|---|
| 0.0000 | 0.0000 | 333.35 | 0.5902 | 0.6683 | 334.65 |
| 0.0883 | 0.0655 | 334.55 | 0.7177 | 0.8088 | 332.75 |
| 0.2600 | 0.2302 | 336.15 | 0.8649 | 0.9209 | 330.55 |
| 0.3282 | 0.3196 | 336.35 | 1.0000 | 1.0000 | 328.45 |
| 0.4467 | 0.4859 | 336.05 | | | |

(a) Construct the (vapor + liquid) phase diagram at $p = 0.101$ MPa, by making a graph of $T$ against $x_2$, and against $y_2$. Label all regions in the diagram with the phases present. This system has a maximum boiling azeotrope (where the liquid and vapor phases have the same composition). Estimate from the phase diagram, the temperature and composition of this azeotrope.

(b) An $x_2 = 0.80$ liquid mixture is heated, with atmospheric pressure equal to 0.101 MPa. Use the phase diagram constructed in (a) to estimate:

(i) The temperature where the liquid will start to boil.

(ii) The temperature where the last drop of liquid evaporates.

(iii) The temperature where 50 mol% of the liquid is left in the pot.

E8.11 The three possible geometrical isomers of dinitrobenzene are 1,2-, 1,3-, and 1,4-dinitrobenzene. Given the following data at 0.101 MPa pressure

| Isomer | Melting temperature/(K) | $\Delta_{\text{fus}}H/(\text{J}\cdot\text{g}^{-1})$ |
|---|---|---|
| 1,2- | 390.08 | 135.1 |
| 1,3- | 363.23 | 103.3 |
| 1,4- | 446.65 | 167.4 |

Assume ideal solution behavior with no solid solutions and constant $\Delta_{\text{fus}}H_{\text{m}}$ and

(a) Calculate the melting temperature of a mixture of {1,2-dinitrobenzene(1) + 1,3-dinitrobenzene(2)} with $x_1 = 0.500$.

(b) Calculate the eutectic temperature and composition for the binary system of {1,2-dinitrobenzene(1) + 1,4-dinitrobenzene(2)}.

E8.12 The melting point of 1,4-dichlorobenzene is 326.4 K and that of naphthalene is 353.4 K. The eutectic point occurs at a temperature of 303.4 K and a mole fraction of naphthalene in the liquid phase of 0.394. Assume ideal liquid solutions, no solid solubility, and $\Delta_{\text{fus}}C_{p,\text{m}} = 0$ and calculate $\Delta_{\text{fus}}H_{\text{m}}$ for 1,4-dichlorobenzene.

## Problems

P8.1 The molar enthalpy of vaporization of liquid mercury is 59.229 $\text{kJ}\cdot\text{mol}^{-1}$ at its normal boiling point of 630.0 K. The heat capacities of the liquid and gaseous phases, valid over the temperature range from 250 to 630 K, are as follows:

$$C_{p,\text{m}}(\text{l})/(\text{J}\cdot\text{K}^{-1}\cdot\text{mol}^{-1}) = 31.097 - 1.425 \times 10^{-2}T + 1.275 \times 10^{-5}T^2$$

$$C_{p,\text{m}}(\text{g})/(\text{J}\cdot\text{K}^{-1}\cdot\text{mol}^{-1}) = 20.79$$

Calculate the vapor pressure of liquid mercury at 298.15 K.

P8.2 The vapor pressure of $AsH_3$ gas is given by the equation[8]

$$\ln p = -\frac{3231.26}{T} - 9.43935 \ln T + 0.018506T + 61.95534$$

where $p$ is the vapor pressure in bars at the Kelvin temperature $T$.

At moderate to low pressures, gaseous $AsH_3$ can be represented by the equation of state

$$pV_m = RT + \frac{bp}{T^2}$$

with $b = -18\ m^3 \cdot K^2 \cdot mol^{-1}$.
The density of liquid $AsH_3$ in $kg \cdot m^{-3}$ is given by the equation

$$\rho = 2074 - 1.241T - 4.321 \times 10^{-3}T^2.$$

Assume $AsH_3$ is incompressible and calculate

(a) The enthalpy of vaporization at the normal boiling point where the vapor pressure is 1 atm (101.325 kPa).

(b) The vapor pressure at $T = 200$ K and $p = 10.00$ bar total pressure.

P8.3 Given the following (solid + liquid) equilibrium data for $(x_1C_6H_6 + x_2SiCl_4)$.

| $x_2$ | $T/(K)$ | $x_2$ | $T/(K)$ | $x_2$ | $T/(K)$ |
|---|---|---|---|---|---|
| 0.0000 | 278.66 | 0.4646 | 254.54 | 0.8823 | 215.65 |
| 0.0622 | 274.83 | 0.5323 | 250.81 | 0.9042 | 210.48 |
| 0.1212 | 271.47 | 0.6268 | 244.97 | 0.9275 | 204.16 |
| 0.2060 | 267.15 | 0.6958 | 239.51 | 0.9428 | 202.00 |
| 0.2903 | 263.08 | 0.7542 | 234.31 | 0.9616 | 202.67 |
| 0.3762 | 258.94 | 0.8171 | 226.85 | 0.9801 | 203.51 |

(a) Construct the binary (solid + liquid) phase diagram and label all the regions.

(b) Use the $(T, x_2)$ data near $x_2 = 0$ and $x_2 = 1$ to calculate $\Delta_{fus}H_m$ for both components. Assume this $\Delta_{fus}H_m$ is constant with temperature and calculate the ideal freezing point at every 0.1 mole fraction unit. Plot these results on the diagram constructed in part (a) to compare the actual freezing points with the ideal solution prediction.

P8.4 The (solid + liquid) phase diagram for $(x_1\text{n-}C_6H_{14} + x_2\text{c-}C_6H_{12})$ has a eutectic at $T = 170.59$ K and $x_2 = 0.3317$. A solid phase transition occurs in c-$C_6H_{12}$ at $T = 186.12$ K, resulting in a second invariant point in the phase diagram at this temperature and $x_2 = 0.6115$, where liquid and the two solid forms of c-$C_6H_{12}$ are in equilibrium. A fit of the experimental data to equation (8.31)

$$T = T^* \left[ 1 + \sum_{j=1}^{n} a_j (x_2 - x_2^*)^j \right] \qquad (8.31)$$

results in the following parameters for this system:

| $x_{2,\,\mathrm{min}}$ | $x_{2,\,\mathrm{max}}$ | $x_2$ | $T^*/(\mathrm{K})$ | $a_1$ | $a_2$ |
|---|---|---|---|---|---|
| 0 | 0.332 | 0 | 177.79 | −0.10706 | −0.04256 |
| 0.332 | 0.612 | 1 | 201.93 | 0.15998 | −0.10822 |
| 0.612 | 1 | 1 | 279.73 | 0.85306 | −0.03873 |

In the table, $x_{2,\,\mathrm{min}}$ and $x_{2,\,\mathrm{max}}$ give the mole fraction range over which the parameters $a_1$ and $a_2$ apply.

(a) Use the parameters to construct the binary (solid + liquid) phase diagram ($T$ against $x_2$) for this system. Be sure to label the phases present in all the regions.

(b) Differentiate equation (8.31) applied to component 2 and combine with equation (8.26), again applied to component 2, that relates mole fraction and temperature for an ideal solution

$$\left( \frac{\partial \ln x_2}{\partial T} \right)_p = \frac{\Delta_{\mathrm{fus}} H_{\mathrm{m},\,2}}{RT^2}. \qquad (8.26)$$

The result is an equation relating $\Delta_{\mathrm{fus}} H_{\mathrm{m},\,2}$, $T$, $x_2$, and the $a_j$ parameters. Extrapolate the equation to $x_2 = 1$ {a condition where equation (8.26) applies, even if the solution is not ideal} to show that $\Delta_{\mathrm{fus}} H_{\mathrm{m},\,2}$ at the melting temperature of the pure substance is given by

$$\Delta_{\mathrm{fus}} H_{\mathrm{m},\,2} = \frac{RT^*}{a_1}$$

where $T^*$ is the melting temperature at $x_2 = 1$.

(c) Use the equation in (b) to calculate $\Delta_{\mathrm{fus}} H_{\mathrm{m},\,2}$ for the high temperature form of c-$C_6H_{12}$, and compare the result with the experimental value given in Table 2.2. Also, use the extrapolated $T^*$ for the low temperature form of solid c-$C_6H_{12}$ to calculate $\Delta_{fus}H_{m,2}$ for this substance, and then use the two enthalpies of fusion to estimate $\Delta_{trans}H_{m,2}$, the enthalpy of transition. Compare the $\Delta_{fus}H_{m,2}$ values with others in Table 2.2 for hydrocarbons (such as $C_6H_6$ and $C_6H_{14}$) that have similar molecular weights. (Molecular rotation in the solid c-$C_6H_{12}$ above the transition temperature is the cause of the exceptionally small $\Delta_{fus}H_{m,2}$ for this substance. The large $\Delta_{trans}H_{m,2}$ is due largely to this rotation.)

## References


1. Taken from *The Collected Paper of P.W. Bridgeman*, Vol. 2, Harvard University Press, Cambridge, Massachusetts, 1962, pp. 141–203.
2. A. Michels and J. Strijland, "The Specific Heat at Constant Volume of Compressed Carbon Dioxide", *Physica*, **18**, 613–628 (1952).
3. I. F. Hölscher, G. M. Schneider and J. B. Ott, "Liquid–Liquid Phase Equilibria of Binary Mixtures of Methanol with Hexane, Nonane, and Decane at Pressures up to 150 MPa", *Fluid Phase Equilib.*, **27**, 153–169 (1986).
4. J. R. Goates, J. B. Ott, J. F. Moellmer, and D. W. Farrell, "(Solid + Liquid) Phase Equilibria in *n*-Hexane + Cyclohexane and Benzene + *p*-Xylene", *J. Chem. Thermodyn.*, **11**, 709–711 (1979).
5. J. B. Ott and J. R. Goates, "(Solid + Liquid) Phase Equilibria in Binary Mixtures Containing Benzene, a Cycloalkane, an *n*-Alkane, or Tetrachloromethane. An equation for Representing (Solid + Liquid) Phase Equilibria", *J. Chem. Thermodyn.*, **15**, 267–278 (1983).
6. For a discussion of representative examples, see J. B. Ott and J. R. Goates, "Temperature Measurement With Application to Phase Equilibria Studies," Chapter 6 of *Physical Methods of Chemistry, Volume VI, Determination of Thermodynamic Properties*, B. W. Rossiter and R. C. Baetzold, Editors, John Wiley & Sons, New York, 1992.
7. R. F. Hampson and R. F. Walker, "The Vapor Pressure of Palladium", *J. Res. Natl. Bur. Std.*, **66A**, 177–178 (1962).
8. R. H. Sherman and W. F. Giauque, "Arsine. Vapor Pressure, Heat Capacity, Heats of Transition, Fusion, and Vaporization. The Entropy from Calorimetric and from Molecular Data", *J. Am. Chem. Soc.*, 77, 2154–2160 (1955).

## Chapter 9

# The Equilibrium Condition Applied to Chemical Processes

In Chapter 5, we wrote the generalized chemical reaction

$$\nu_1 A_1 + \nu_2 A_2 + \cdots = \nu_m A_m + \nu_{m+1} A_{m+1} + \cdots$$

as

$$\sum_i \nu_i A_i = 0, \tag{9.1}$$

where the $\nu_i$ are positive for products and negative for reactants.

We also showed that at equilibrium

$$\sum_i \nu_i \mu_i = 0. \tag{9.2}$$

We will start with these equations and the relationship between activity and chemical potential

$$\mu_i = \mu_i^\circ + RT \ln a_i \tag{9.3}$$

to derive an equation that relates the activities of the various species in a chemical reaction to $\Delta_r G$, the free energy change in the reaction. Of special interest is the relationship between the free energy change and the activities at equilibrium.

## 9.1 The Equilibrium Constant

Equation (9.3) can be substituted into equation (9.2) to give

$$\sum_i \nu_i \mu_i = \sum_i \nu_i \mu_i^\circ + RT \sum_i \nu_i \ln a_i. \tag{9.4}$$

But

$$\sum_i \nu_i \mu_i = \Delta_r G$$

and

$$\sum_i \nu_i \mu_i^\circ = \Delta_r G^\circ,$$

where $\Delta_r G^\circ$ is the free energy change for the chemical reaction with reactants and products in their standard states. Also, since the sum of logarithms equals the logarithms of the product, we can write the last term as

$$RT \sum_i \nu_i \ln a_i = RT \ln \prod_i a_i^{\nu_i}.$$

Combining terms gives

$$\Delta_r G = \Delta_r G^\circ + RT \ln \prod_i a_i^{\nu_i}. \tag{9.5}$$

Equation (9.5) enables us to calculate $\Delta_r G$ for a chemical reaction under a given set of activity conditions when we know the free energy change for the reaction under the standard state condition. Of special interest are the activities when reactants and products are at equilibrium. Under those conditions,

$$\Delta_r G = 0.$$

Substitution into equation (9.5) gives[a]

$$\Delta_r G^\circ = -RT \ln \prod_i a_i^{\nu_i}. \tag{9.6}$$

---

[a] We must remember in applying equation (9.6) that the activities are the values at equilibrium. Also, it is important to note that $\Delta_r G^\circ$ applies to the free energy change with all species in specified standard states. Different values for $\Delta_r G^\circ$ (and for activities) will be obtained for different choices of standard states.

The standard free energy change is a function only of temperature. That is, it is independent of pressure and concentration for a specified standard state. Thus, at a given temperature, the term $\prod_i a_i^{\nu_i}$ is constant and we can write

$$\prod_i a_i^{\nu_i} = K \tag{9.7}$$

and

$$\Delta_r G^\circ = -RT \ln K. \tag{9.8}$$

Equations (9.7) and (9.8) define $K$, the equilibrium constant for the reaction.[b] It is sometimes referred to as the **thermodynamic equilibrium constant**. As we shall see, this ratio of activities can be related to ratios of pressure or concentration which, themselves, are sometimes called equilibrium constants. But $K$, as defined in equations (9.7) and (9.8), is the fundamental form that is directly related to the free energy change of the reaction.

## 9.1a Alternate Forms of the Equilibrium Constant

Alternative forms of the equilibrium constant can be obtained as we express the relationship between activities, and pressures or concentrations. For example, for a gas phase reaction, the standard state we almost always choose is the ideal gas at a pressure of 1 bar (or $10^5$ Pa). Thus

$$a_i = \frac{f_i}{f_i^\circ} = \frac{f_i}{1\ \text{bar}}$$

and

$$K = \prod_i f_i^{\nu_i} \tag{9.9}$$

---

[b] IUPAC recommends* $K^\circ$ as the symbol for what is referred to as the "standard equilibrium constant." We think this terminology is confusing and misleading. The symbol $K^\circ$ implies that the activities are the standard state values, which they are not. Instead they are equilibrium activities, which may differ considerably from the standard state values. For example, in the reaction $H_2(g) + Cl_2(g) = 2HCl(g)$, $p^\circ_{H_2} = 1$ bar (or $10^5$ Pa), but at 298.15 K and a total pressure as 1 bar, the equilibrium pressure of $H_2$ is much less than 1 bar. IUPAC does allow for the symbol $K$ and the name "thermodynamic equilibrium constant." This is our choice.

* See I. Mills, T. Cuitaš, K. Homann, N. Kallay and K. Kuchitsu, *Quantities, Units and Symbols in Physical Chemistry*, Second Edition, Blackwell Science, London, 1988.

where the fugacity must be expressed in bars.[c] Substituting $f_i = \phi_i p_i$ into equation (9.9), where $\phi_i$ is the fugacity coefficient, gives

$$K = \prod_i \phi_i^{\nu_i} \cdot \prod_i p_i^{\nu_i}. \tag{9.10}$$

The ratio of pressures in equation (9.10) is known as $K_p$ and the ratio of fugacity coefficients is written as $J_\phi$. Hence,

$$K = J_\phi K_p. \tag{9.11}$$

At low pressures $J_\phi \cong 1$, and

$$K \cong K_p. \tag{9.12}$$

for ideal gases, or gases at low pressure, $p_i$ is related to $c_i$ the molar concentration in the gas phase by

$$p_i = c_i RT.$$

Substitution into equation (9.11) gives

$$K = J_\phi \cdot (RT)^{\sum \nu_i} \cdot \prod_i c_i^{\nu_i}. \tag{9.13}$$

The expression $\prod_i c_i^{\nu_i}$ is the concentration equilibrium constant $K_c$, and equation (9.13) becomes

$$K = J_\phi \cdot (RT)^{\sum \nu_i} \cdot K_c. \tag{9.14}$$

---

[c] An alternative for a reaction would be to write (with $f_i^\circ = p_i^\circ$)

$$k = \prod_i \left(\frac{f_i}{p_i^\circ}\right)^{\nu_i}$$

or

$$K = \prod_i \left(\frac{f_i}{10^5\ \text{Pa}}\right)^{\nu_i}$$

with the pressure expressed in Pa. We prefer the simplicity of the expression involving pressure in bars.

Finally, $p_i$ is related to $x_i$, the mole fraction in the gas phase[d] by Dalton's law

$$p_i = x_i p$$

where $p$ is the total pressure. Substitution into equation (9.11) gives

$$K = J_\phi \cdot p^{\sum \nu_i} \cdot K_x \tag{9.15}$$

where

$$K_x = \prod_i x_i^{\nu_i} \tag{9.16}$$

is the mole fraction equilibrium constant.

In summary

$$\frac{K}{J_\phi} = K_p = K_c \cdot (RT)^{\sum \nu_i} = K_x \cdot p^{\sum \nu_i} \tag{9.17}$$

and at low pressures when $J_\phi = 1$,

$$K = K_p = K_c \cdot (RT)^{\sum \nu_i} = K_x \cdot p^{\sum \nu_i}. \tag{9.18}$$

Alternate expressions can also be written for equilibrium constants in condensed phase reactions. For example, for the reaction

$$H_2O(l) = H^+(aq) + OH^-(aq) \tag{9.19}$$

$$K = \frac{a_{H^+} a_{OH^-}}{a_{H_2O}}. \tag{9.20}$$

The activity product $a_{H^+} a_{OH^-}$ is given the symbol $K_w$ so that

$$K_w = a_{H^+} a_{OH^-} \tag{9.21}$$

[d] In earlier chapters, we have used $y_i$ to represent mole fraction in the gas phase. Here, for chemical equilibrium calculations where we generally have no need to refer to vapor and liquid compositions simultaneously, we bow to tradition and use $x_i$ as mole fraction.

and

$$K = \frac{K_W}{a_{H_2O}}. \tag{9.22}$$

The standard state for liquid water is chosen as the pure substance at 1 bar total pressure. Thus, at this pressure condition, $a_{H_2O} = 1$ and

$$K = K_W. \tag{9.23}$$

If pressure differs considerably from one bar, equation (6.98) must be used to give

$$a_{H_2O} = \Gamma_{H_2O}(l)$$

and

$$K = \frac{K_W}{\Gamma_{H_2O}}. \tag{9.24}$$

In reaction (9.19), $H^+$ and $OH^-$ are solutes, and a Henry's law standard state is chosen so that

$$a_{H^+} = \gamma_+ m_{H^+}$$

$$a_{OH^-} = \gamma_- m_{OH^-}$$

and equation (9.21) becomes

$$K_W = \gamma_+ m_{H^+} \gamma_- m_{OH^-} \tag{9.25}$$

or[e]

$$K_W = \gamma_\pm^2 m_{H^+} m_{OH^-}. \tag{9.26}$$

Equation (9.24) then becomes

$$K = \frac{\gamma_\pm^2 m_{H^+} m_{OH^-}}{\Gamma_{H_2O}}. \tag{9.27}$$

---

[e] We recall from Chapter 6 that for a 1 : 1 electrolyte

$$\gamma_+ \gamma_- = \gamma_\pm^2.$$

For (acid + base) reactions, expressions can be written for the equilibrium constant that involve activity coefficients as well. For example, for the reactions

$$H_2CO_3(aq) = H^+(aq) + HCO_3^-(aq) \tag{9.28}$$

and

$$NH_4OH(aq) = NH_4^+(aq) + OH^-(aq) \tag{9.29}$$

the equilibrium constants are

$$K_a = \frac{a_{H^+} a_{HCO_3^-}}{a_{H_2CO_3}} \tag{9.30}$$

and

$$K_b = \frac{a_{NH_4^+} a_{OH^-}}{a_{NH_4OH}}. \tag{9.31}$$

A Henry's law standard state is chosen for the acid, base, and ions, so that

$$K_a = \frac{\gamma_\pm^2 m_{H^+} m_{HCO_3^-}}{\gamma_{H_2CO_3} m_{H_2CO_3}} \tag{9.32}$$

and

$$K_b = \frac{\gamma_\pm^2 m_{NH_4^+} m_{OH^-}}{\gamma_{NH_4OH} m_{NH_4OH}}. \tag{9.33}$$

Reactions (9.19), (9.28) and (9.29) are often written in the alternate forms

$$2H_2O(aq) = H_3O^+(aq) + OH^-(aq) \tag{9.34}$$

$$CO_2(aq) + 2H_2O = H_3O^+(aq) + HCO_3^-(aq) \tag{9.35}$$

$$NH_3(aq) + H_2O = NH_4^+(aq) + OH^-(aq). \tag{9.36}$$

Expressing the reactions in this manner gives

$$K_w = \frac{a_{H_3O^+} a_{OH^-}}{a^2_{H_2O}} \tag{9.37}$$

$$K_a = \frac{a_{H_3O^+} a_{HCO_3^-}}{a_{CO_2} a^2_{H_2O}} \tag{9.38}$$

$$K_b = \frac{a_{NH_4^+} a_{OH^-}}{a_{NH_3} a_{H_2O}}. \tag{9.39}$$

The reason for the ambiguity in the form of the equations we use for expressing $K_w$, $K_a$, and $K_b$ is that we do not know the nature of the species present in solution. We do not know the extent to which $H^+$ is hydrated to form $H_3O^+$ according to the reaction

$$H^+(aq) + H_2O = H_3O^+(aq), \tag{9.40}$$

and we do not know how much $CO_2$ or $NH_3$ are reacted to form $H_2CO_3$ or $NH_4OH$ according to the reactions

$$CO_2(aq) + H_2O = H_2CO_3(aq) \tag{9.41}$$

$$NH_3(aq) + H_2O = NH_4OH(aq). \tag{9.42}$$

Thermodynamics handles this problem by choosing the standard state so that the $K$ for reactions (9.40), (9.41) and (9.42) is 1. That is, we choose[f] the standard state so that $\mu^\circ_{H_3O^+} = \mu^\circ_{H^+} + \mu^\circ_{H_2O}$ and $\Delta_r G^\circ = 0$ for reaction (9.40). This makes $K = 1$ for this reaction and

$$a_{H_3O^+} = a_{H^+} a_{H_2O}.$$

In like manner, $\Delta_r G^\circ = 0$ and, hence, $K = 1$ for reactions (9.41) and (9.42) and

$$a_{H_2CO_3} = a_{CO_2} a_{H_2O}$$

$$a_{NH_4OH} = a_{NH_3} a_{H_2O}.$$

[f] This is the same choice for the standard state that we used (for the same reason) for a strong electrolyte. For $MX(aq) = M^+(aq) + X^-(aq)$, $a_{M^+} a_{X^-} / a_{MX} = 1$ or $a_{M^+} a_{X^-} = a_{MX}$, and $\Delta_r G^\circ = 0$ for the dissociation of the salt.

With these choices of standard states, our examples become (with $a_{H_2O} = 1$)

$$K_w = a_{H^+}a_{OH^-} = a_{H_3O^+}a_{OH^-}$$

$$K_a = \frac{a_{H^+}a_{HCO_3^-}}{a_{H_2CO_3}} = \frac{a_{H_3O^+}a_{HCO_3^-}}{a_{H_2CO_3}}$$

$$= \frac{a_{H^+}a_{HCO_3^-}}{a_{CO_2}} = \frac{a_{H_3O^+}a_{HCO_3^-}}{a_{CO_2}}$$

$$K_b = \frac{a_{NH_4^+}a_{OH^-}}{a_{NH_4OH}} = \frac{a_{NH_4^+}a_{OH^-}}{a_{NH_3}}.$$

A flexibility that is inherent in the choosing of standard states is that we can set them to simplify such problems.

## 9.1b Effect of Pressure and Temperature on the Equilibrium Constant

**The Effect of Pressure:** With the addition of one restricting condition, we like the simple and direct statement of Rossini[1] in describing the effect of pressure on the equilibrium constant:

> *It is important to note that, for any given temperature, the [thermodynamic] equilibrium constant is directly related to the standard change in free energy. Since, at any given temperature, the free energy in the standard state for each reactant and product,* $G_i^\circ$, *is independent of the pressure, it follows that the standard change in free energy for the reaction,* $\Delta_r G^\circ$, *is independent of the pressure.*[g] *Therefore, at constant temperature, the equilibrium constant K ... is also independent of the pressure. That is,*
>
> $$\left(\frac{\partial K}{\partial p}\right)_T = 0 = \left(\frac{\partial \ln k}{\partial p}\right)_T. \qquad (9.43)$$

Rossini is correct of course. But his analysis does not agree with that found in other textbooks. Many start with equation (9.8)

$$\Delta_r G^\circ = -RT \ln K$$

[g] The condition we add is that $G_i^\circ$ is independent of $p$ after the standard state is set. $G_i^\circ$ will, of course, change if we choose a different pressure for the standard state.

and differentiate to get

$$\left(\frac{\partial \Delta_r G^\circ}{\partial p}\right)_T = \Delta_r V^\circ = -RT\left(\frac{\partial \ln K}{\partial p}\right)_T \tag{9.44}$$

so that

$$\left(\frac{\partial \ln K}{\partial p}\right)_T = -\frac{\Delta_r V^\circ}{RT}. \tag{9.45}$$

The ambiguity results from the failure to recognize that the Rossini statement applies to the thermodynamic equilibrium constant. As we have already noted, the other forms that we have derived do depend upon the pressure. For example, for a gas phase reaction, $K_p$ is related to $K$ by equation (9.11)

$$K = \prod_i \phi_i^{\nu_i} \cdot \prod_i p_i^{\nu_i}. \tag{9.11}$$

Taking logarithms and differentiating gives

$$\left(\frac{\partial \ln K}{\partial p}\right)_T = 0 = \left(\frac{\partial \ln J_\phi}{\partial p}\right)_T + \left(\frac{\partial \ln K_p}{\partial p}\right)_T. \tag{9.46}$$

But

$$\ln J_\phi = \sum_i \ln \phi_i^{\nu_i} = \sum_i \nu_i \ln \phi_i \tag{9.47}$$

so that

$$\left(\frac{\partial \ln J_\phi}{\partial p}\right)_T = \sum_i \nu_i \left(\frac{\partial \ln \phi_i}{\partial p}\right)_T. \tag{9.48}$$

Equation (6.46) can be used to find the effect of $p$ on $\phi_\mathrm{i}$

$$\left(\frac{\partial \ln \phi_i}{\partial p}\right)_T = \left(\frac{\overline{V}_i}{RT} - \frac{1}{p}\right). \tag{9.49}$$

Substitution of equation (9.49) into (9.48), and then the result into equation (9.46) gives

$$\left(\frac{\partial \ln K_p}{\partial p}\right)_T = \frac{\Delta(V_i^{\text{ideal}} - \overline{V}_i)}{RT}. \tag{9.50}$$

As another example,

$$K_\text{w} = a_{\text{H}^+} a_{\text{OH}^-} = K a_{\text{H}_2\text{O}}.$$

Taking logarithms and differentiating gives

$$\ln K_\text{w} = \ln K + \ln a_{\text{H}_2\text{O}}$$

$$\left(\frac{\partial \ln K_\text{w}}{\partial p}\right)_T = 0 + \left(\frac{\partial \ln a_{\text{H}_2\text{O}}}{\partial p}\right)_T.$$

From equation (6.85), for pure water

$$\left(\frac{\partial \ln a_{\text{H}_2\text{O}}}{\partial p}\right)_T = \frac{V_\text{m}^*}{RT}$$

so that

$$\left(\frac{\partial \ln K_\text{w}}{\partial p}\right)_T = \frac{V_\text{m}^*}{RT}. \tag{9.51}$$

Equation (9.51) can be integrated to give $K_\text{w}$ as a function of pressure. At $T = 298.15$ K, $K_\text{w} = 1.00 \times 10^{-14}$ at 1 bar pressure, and

$$\int_{\ln 1.00 \times 10^{-14}}^{\ln K_\text{w}} \text{d} \ln K_\text{w} = \int_{1\,\text{bar}}^{p} \frac{V_\text{m}^*}{RT}\, \text{d}p.$$

Assuming water is incompressible (that is, $V_\text{m}^*$ is independent of $p$) gives

$$\ln \frac{K_\text{w}}{1.00 \times 10^{-14}} = \frac{V_\text{m}^*}{RT}(p - 1). \tag{9.52}$$

Figure 9.1 is a graph of equation (9.52) showing how $K_w$ varies with pressure at 298.15 K. We see that it increases by a factor of approximately 2 as the pressure increases by a factor of 1000. The increase is due to the change in the activity of the water rather than to a change in the thermodynamic equilibrium constant with pressure.

**The Effect of Temperature:** The effect of temperature on $K$ can be obtained by dividing equation (9.8) by $T$ and differentiating[h]

$$\frac{\Delta_r G^\circ}{T} = -R \ln K$$

$$\left(\frac{\partial \Delta_r G^\circ / T}{\partial T}\right)_p = -\frac{\Delta_r H^\circ}{T^2} = -R\left(\frac{\partial \ln K}{\partial T}\right)_p \tag{9.53}$$

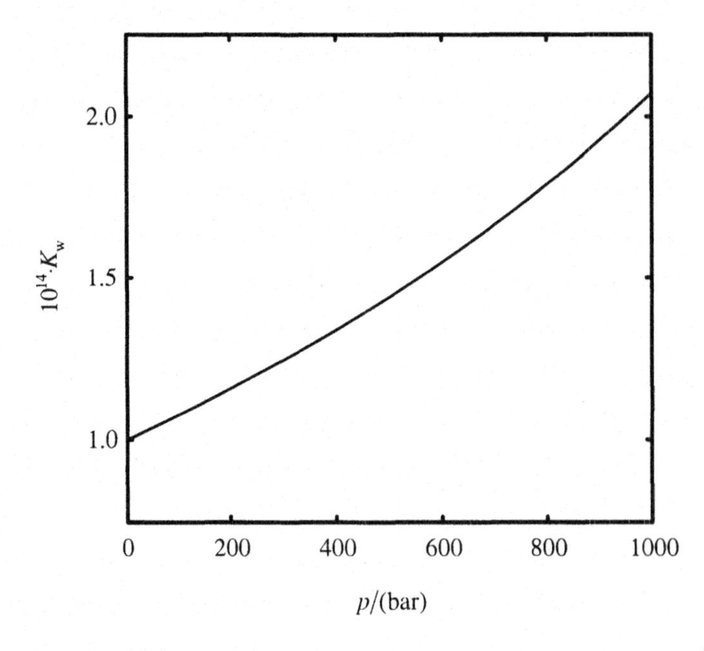


**Figure 9.1** Change in $K_w$ with pressure $p$. The temperature is 298.15 K.

[h] We remember that specification of temperature is not part of the standard state condition. Thus, $K$ can (and usually does) vary with $T$.

or

$$\left(\frac{\partial \ln K}{\partial T}\right)_p = \frac{\Delta_r H^\circ}{RT^2}. \tag{9.54}$$

Equation (9.54) can be integrated to give $K$ as a function of $T$

$$\int_{\ln K_1}^{\ln K_2} \mathrm{d}\ln K = \int_{T_1}^{T_2} \frac{\Delta_r H^\circ}{RT^2}\,\mathrm{d}T.$$

If the temperature range is small, it is a good approximation to assume that $\Delta_r H_m^\circ$ is independent of temperature, in which case

$$\ln\frac{K_2}{K_1} = -\frac{\Delta_r H^\circ}{R}\left(\frac{1}{T_2} - \frac{1}{T_1}\right). \tag{9.55}$$

For wider temperature intervals, we must express $\Delta_r H^\circ$ as a function of temperature before we integrate equation (9.55).

An equation for $\Delta_r H^\circ$ as a function of temperature can be obtained by following the procedure used to obtain equation (6.165), but applied to a chemical reaction instead of a phase change. The result is

$$\Delta_r H^\circ = \Delta H_1 + \Delta a T + \frac{\Delta b}{2}T^2 - \Delta c\left(\frac{1}{T}\right) + \frac{\Delta d}{3}T^3 + \cdots \tag{9.56}$$

where $\Delta a$, $\Delta b$, $\Delta c$, $\Delta d$ ... are the differences in constants in the heat capacity equations for products and reactants, and $\Delta H_1$ is a constant of integration. This equation can be substituted into equation (9.54) and integrated to get an expression for $\Delta_r G^\circ$ as a function of temperature. The result is

$$\frac{\Delta_r G^\circ}{T} = I + \frac{\Delta H_1}{T} - \left(\sum_i \nu_i a_i\right)\ln T - \frac{\left(\sum_i \nu_i b_i\right)}{2}T - \frac{\left(\sum_i \nu_i c_i\right)}{2T^2} - \frac{\left(\sum_i \nu_i d_i\right)}{6}T^2 + \cdots \tag{9.57}$$

or

$$R \ln K = -I - \frac{(\Delta H_1)}{T} + \left(\sum_i \nu_i a_i\right) \ln T + \frac{\left(\sum_i \nu_i b_i\right)}{2} T + \frac{\left(\sum_i \nu_i c_i\right)}{2T^2} + \frac{\left(\sum_i \nu_i d_i\right)}{6} T^2 + \cdots \tag{9.58}$$

where $I$ is the constant obtained for the second integration.

## 9.2 Enthalpies and Gibbs Free Energies of Formation

To find a numerical value for $\Delta H_1$, we need to know $\Delta_r H^\circ$ at one temperature, while evaluation of $I$ requires $\Delta_r G^\circ$ at one temperature. The usual choice is to obtain $\Delta_r H^\circ$ and $\Delta_r G^\circ$ at $T = 298.15$ K from **standard molar enthalpies of formation and standard molar Gibbs free energies of formation**. Earlier in this chapter we referred to examples of these quantities. It is now time to define $\Delta_f H^\circ$ and $\Delta_f G^\circ$ explicitly and describe methods for their measurement.

The formation reaction is the one in which the elements in their stable form, as they occur in nature at $T = 298.15$ K, react to form the chemical substance. The following are examples of formation reactions:

$$H_2(g) + \tfrac{1}{2}O_2(g) = H_2O(l)$$

$$H_2(g) + \tfrac{1}{8}S_8(\text{rhombic}) + 2O_2(g) = H_2SO_4(l)$$

$$Hg(l) + Cl_2(g) = HgCl_2(s)$$

$$2C(\text{graphite}) + 3H_2(g) = C_2H_6(g)$$

$$3Na(s) + \tfrac{1}{4}P_4(\text{white}) + 2O_2(g) = Na_3PO_4(s).$$ [i]

Standard molar enthalpies of formation, $\Delta_f H_m^\circ$, and standard molar Gibbs free energies of formation, $\Delta_f G_m^\circ$, are useful, since they can be used to calculate $\Delta_r H^\circ$ and $\Delta_r G^\circ$. The relationships are

$$\Delta_r H^\circ = \sum_i \nu_i \Delta_f H_{m,i}^\circ \tag{9.59}$$

[i] The use of white phosphorus ($P_4$) is an exception to the rule of using the most stable form, since red phosphorus is more stable (but its properties are less reproducible).

$$\Delta_r G^\circ = \sum_i \nu_i \Delta_f G^\circ_{m,i}. \tag{9.60}$$

The reason for these relationships can be seen by referring to Figure 9.2. For the reactants going to the elements, the $\nu_i$ are negative and the sums give $\Delta H^\circ$ and $\Delta G^\circ$ for separating the reactants into the elements. If we add to this the sums for the formation of the products from the elements, we get $\Delta_r H^\circ$ and $\Delta_r G^\circ$ for the reaction. In applying equations (9.59) and (9.60), we should keep in mind that $\Delta_f H^\circ_m$ and $\Delta_f G^\circ_m$ are equal to zero for the elements in their stable state at $T = 298.15$ K.

Extensive tables can be found in the literature[j] giving values at $T = 298.15$ K for $\Delta_f H^\circ_m$ and $\Delta_f G^\circ_m$ (usually along with $S^\circ_m$ and $C^\circ_{p,m}$) from which $\Delta_r H^\circ$ and $\Delta_r G^\circ$ can be obtained. We will now describe some of the methods used to determine $\Delta_f H^\circ_m$ and $\Delta_f G^\circ_m$.

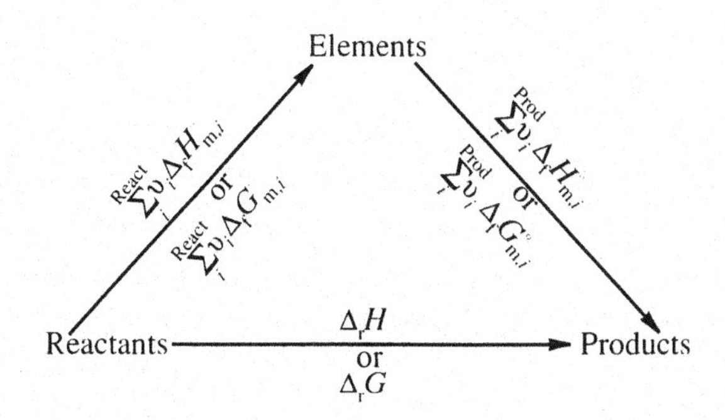


**Figure 9.2** The $\Delta_r H^\circ$ and $\Delta_r G^\circ$ for a reaction can be obtained by adding the $\nu_i \Delta_f H^\circ_{m,i}$ and $\nu_i \Delta_f G^\circ_{m,i}$ for separating the reactants into the elements, to the $\nu_i \Delta_f H^\circ_{m,i}$ and $\nu_i \Delta_f G^\circ_{m,i}$ for combining the elements to form the products. (Remember that the $\nu_i$ are negative for the reactants and positive for the products.)

[j] Two of the best sources are: D. D. Wagman, W. H. Evans, V. B. Parker, R. H. Schumm, I. Halow, S. M. Bailey, K. L. Churney, and R. L. Nuttall, "The NBS Tables of Chemical Thermodynamic Properties. Selected Values for Inorganic and $C_1$ and $C_2$ Organic Substance in SI units", *J. Phys. Chem. Ref. Data*, **11**, 1982, Supplement No. 2.

M. W. Chase, Jr., C. A. Davies, J. R. Downey, Jr., D. J. Fryrip, R. A. McDonald, and A. N. Syverud, "JANAF Thermochemical Tables, Third Edition. Part 1, Al–Co and Part 2 Cr–Zr", *J. Phys. Chem. Ref. Data*, **14**, 1985, Supplement No. 1. Other more specialized tables can be found, along with individual references, in the literature.

## 9.2a Determination of Standard Enthalpies and Gibbs Free Energies of Formation

**Enthalpies of Formation:** Because of the relationship of $\Delta H$ to $q$ in a constant pressure process, $\Delta_f H^\circ_m$ is often obtained from calorimetric measurements. The enthalpy of formation for substances that will react with oxygen is often determined in a combustion calorimeter, which measures **enthalpies of combustion, $\Delta_c H^\circ$**. In the procedure followed, the substance to be burned is loaded in a heavy-walled "bomb" calorimeter and brought in contact with a high-pressure oxygen atmosphere. Combustion is initiated; the exothermic reaction causes the temperature of the calorimetric apparatus to rise; and from the change in temperature, the heat of combustion, $q_c$, is obtained. Since the reaction occurs at constant volume, $q_c = \Delta_r U$. This $\Delta_r U$ can be corrected to constant pressure conditions at $T = 298.15$ K to give $\Delta_c H^\circ$. Examples are the burning of graphite and diamond in oxygen to give (at $T = 298.15$ K)

$$C(graphite) + O_2(g) = CO_2(g) \qquad \Delta_c H^\circ = -393.522 \text{ kJ} \tag{9.61}$$

$$C(diamond) + O_2(g) = CO_2(g) \qquad \Delta_c H^\circ = -395.417 \text{ kJ} \tag{9.62}$$

The first $\Delta_c H^\circ$ is $\Delta_f H^\circ_m$ for $CO_2$ at 298.15 K, since elements in their naturally occurring state are combining to give $CO_2(g)$. This combustion reaction is the standard state enthalpy of formation if we carry it out at $p = 1$ bar and make small corrections to change the $CO_2(g)$ to the ideal gas condition.

The second reaction is the enthalpy of combustion of diamond. It does not directly give $\Delta_f H^\circ_m$ of $CO_2$, since diamond is not the stable form of carbon. We can, however, get $\Delta_f H^\circ_m$ for diamond by subtracting the two reactions[k]

$$C(graphite) + O_2(g) = CO_2(g) \qquad \Delta_c H^\circ = -393.522 \text{ kJ} \tag{9.63}$$

$$CO_2(g) + O_2(g) = C(diamond) \qquad \Delta_c H^\circ = +395.417 \text{ kJ} \tag{9.64}$$

$$C(graphite) = C(diamond) \qquad \Delta_f H^\circ_m = 1.895 \text{ kJ·mol}^{-1} \tag{9.65}$$

As another example, the enthalpy of formation of water can be obtained by burning $H_2$ in oxygen

$$H_2(g) + \tfrac{1}{2} O_2(g) = H_2O(l) \qquad \Delta_c H^\circ = \Delta_f H^\circ_m = -285.830 \text{ kJ·mol}^{-1}.$$

---

[k] Note that the $\Delta_f H^\circ_m$ of diamond is not zero. Graphite is the stable form of carbon so that it is the form with $\Delta_f H^\circ_m = 0$.

One must be careful to indicate the physical state of the compound. For example, the enthalpy of formation of $H_2O(g)$ is given by

$$H_2(g) + \tfrac{1}{2}O_2(g) = H_2O(g) \qquad \Delta_f H^\circ_m = -241.826\ \text{kJ·mol}^{-1}. \tag{9.66}$$

The difference between the two enthalpies of formation is the enthalpy of vaporization of $H_2O$ at 298.15 K:

$$H_2O(l) = H_2O(g) \qquad \Delta_{vap} H^\circ_m = 44.00\ \text{kJ·mol}^{-1}. \tag{9.67}$$

Starting with the $\Delta_f H^\circ_m$ for $H_2O(l)$ and for $CO_2(g)$, $\Delta_f H^\circ_m$ for many organic compounds can be determined. For example, Bastos *et al.*[2] have measured the enthalpy of combustion of glycerol $C_3H_8O_3$. They obtain a value at $T = 298.15$ K of

$$C_3H_8O_3(l) + \tfrac{7}{2}O_2(g) = 3CO_2(g) + 4H_2O(l)$$

$$\Delta_c H^\circ = -1654.3\ \text{kJ}. \tag{9.68}$$

Combining with values for $\Delta_c H^\circ$ of other reactions as follows gives the desired $\Delta_f H^\circ_m$

$$\begin{array}{ll}
3CO_2(g) + 4H_2O(l) = C_3H_8O_3(l) + \tfrac{7}{2}O_2(g) & -\Delta_c H^\circ = 1654.3 \\
3C(\text{graphite}) + 3O_2(g) = 3CO_2(g) & \Delta_c H^\circ = (3)(-393.52) \\
4H_2(g) + 2O_2(g) = 4H_2O(l) & \Delta_c H^\circ = (4)(-285.83) \\
\hline
4H_2(g) + 3C(\text{graphite}) + \tfrac{3}{2}O_2(g) & \\
\quad = C_3H_8O_3(l) & \Delta_f H^\circ_m = -669.6\ \text{kJ·mol}^{-1}.
\end{array} \tag{9.69}$$

Bastos and co-workers also measured the enthalpy of vaporization of $C_3H_8O_3$ at $T = 298.15$ K and obtained a value of 91.7 kJ·mol$^{-1}$. By adding this to the above value, we get

$$3C(\text{graphite}) + \tfrac{3}{2}O_2(g) = C_3H_8O_3(g) \qquad \Delta_f H^\circ_m = -577.91\ \text{kJ·mol}^{-1}. \tag{9.70}$$

Finally, they measured the enthalpy of solution of $C_3H_8O_3$ in water as a function of concentration and extrapolated to infinite dilution to get a value of $-5.84\ \text{kJ·mol}^{-1}$ for the reaction

$$C_3H_8O_3(l) = C_3H_8O_3(aq, m = 0) \qquad \Delta H = 5.84\ \text{kJ} \tag{9.71}$$

where the (aq, $m = 0$) indicates an aqueous solution of infinite dilution. The standard state is not the infinitely dilute solution. Instead it is a hypothetical state with $m = 1$ that obeys Henry's law. To get $\Delta H^\circ$ for the standard state reaction, we must add to the above equation[1]

$$C_3H_8O_3(aq, m = 0) = C_3H_8O_3(\text{hyp}\ m = 1) \qquad \Delta_{\text{dil}}H = 0. \tag{9.72}$$

With this result, adding equations (9.71) and (9.72) to equation (9.69) gives

$$3C(\text{graphite}) + \tfrac{3}{2}O_2(g) = C_3H_8O_3(\text{hyp}\ m = 1)$$

$$\Delta_f H^\circ_m = -675.4\ \text{kJ·mol}^{-1}. \tag{9.73}$$

This value is the standard enthalpy of formation of glycerol dissolved in water in a hypothetical $m = 1$ solution that obeys Henry's law.

Different methods are required to get $\Delta_f H^\circ_m$ for substances that will not burn in oxygen. Sometimes the reaction can be made to proceed directly in a calorimeter, and a measurement of $q$ gives $\Delta_f H^\circ_m$. For example, at $T = 298.15$ K

$$\tfrac{1}{2}H_2(g) + \tfrac{1}{2}Cl_2(g) = HCl(g) \qquad \Delta_f H^\circ_m = -92.312\ \text{kJ·mol}^{-1} \tag{9.74}$$

$$Na(s) + \tfrac{1}{2}Cl_2(g) = NaCl(s) \qquad \Delta_f H^\circ_m = -411.120\ \text{kJ·mol}^{-1}. \tag{9.75}$$

In recent years, fluorine bomb calorimetry has been used effectively. A number of substances that will not burn in oxygen will burn in fluorine gas. The heat resulting from this fluorine reaction can be used to calculate $\Delta_f H^\circ_m$. For example, Murray and O'Hare[3] have reacted $GeS_2$ with fluorine and measured $\Delta_r H^\circ$. The reaction is

$$GeS_2(s) + 8F_2(g) = GeF_4(g) + 2SF_6(g) \qquad \Delta_r H^\circ = -3507.5\ \text{kJ}. \tag{9.76}$$

---

[1] We showed in Section 7.3a that $\Delta_{\text{dil}}H = 0$ for the change from the infinitely dilute solution to the standard state.

Combining this result with the standard enthalpies of formation of $SF_6$ and $GeF_4$

$$\frac{1}{8}S_8(\text{rhombic}) + 3F_2(g) = SF_6(g) \qquad \Delta_f H^\circ_m = -1220.8\ \text{kJ}\cdot\text{mol}^{-1} \tag{9.77}$$

$$Ge(s) + 2F_2(g) = GeF_4(g) \qquad \Delta_f H^\circ_m = -1190.6\ \text{kJ}\cdot\text{mol}^{-1} \tag{9.78}$$

gives the enthalpy of formation of $GeS_2$ as shown

$$Ge(s) + \frac{1}{4}S_8(\text{rhombic}) = GeS_2(s) \qquad \Delta_f H^\circ_m = -124.7\ \text{kJ}\cdot\text{mol}^{-1}. \tag{9.79}$$

A number of compounds react with $F_2(g)$ in this manner, and fluorine bomb calorimetry can be used to measure their $\Delta_f H^\circ_m$.

Finally, a number of solution calorimetric techniques can be used to measure $\Delta_f H^\circ_m$. For example, $q$ for the solution of a metal in acid can be measured. Additional thermochemical results are then used to complete a cycle that yields $\Delta_f H^\circ_m$ for the metallic salt. The process adds $\Delta_r H^\circ$ for the following reactions:

$$\begin{aligned} M(s) + HCl(aq) &= MCl(aq) + \tfrac{1}{2}H_2(g) \\ MCl(aq) &= MCl(s) \\ HCl(g) &= HCl(aq) \\ \tfrac{1}{2}H_2(g) + \tfrac{1}{2}Cl_2(g) &= HCl(g) \\ \hline M(s) + \tfrac{1}{2}Cl_2(g) &= MCl(s) \end{aligned}$$

In recent years, aqueous solutions of $XeO_3$ have been used to oxidize a species in solution, from which $\Delta_f H^\circ_m$ can be calculated when $\Delta H$ for the oxidation reaction is combined with $\Delta H$ for other reactions. The noble gas oxide $XeO_3$ is used as an oxidant because of its stability and the fact that the final reaction product is Xe(g), which has a zero enthalpy of formation and is easily removed from the reaction mixture. As an example, O'Hare[4] has reported $\Delta_f H^\circ_m$ for $UCl_4$. We will not go through the details of his procedure, but the critical step involved measuring $\Delta_r H$ for the reaction

$$UCl_4(s) + \frac{1}{3}XeO_3(aq) + H_2O(l) = UO_2Cl_2(aq) + 2HCl(aq) + \frac{1}{3}Xe(g).$$

Extensive tabulations of standard enthalpies of formation and related thermodynamic data can be found in the literature.[5] Table 9.1 summarizes selected values from these sources.

**Table 9.1** Standard heat capacities, entropies, enthalpies of formation, and Gibbs free energies of formation at $T = 298.15$ K*.

| Substance | $C^{\circ}_{p,m}$/ $(J \cdot K^{-1} \cdot mol^{-1})$ | $S^{\circ}_{m}$/ $(J \cdot K^{-1} \cdot mol^{-1})$ | $\Delta_f H^{\circ}_{m}$/ $(kJ \cdot mol^{-1})$ | $\Delta_f G^{\circ}_{m}$/ $(kJ \cdot mol^{-1})$ |
|---|---|---|---|---|
| AgCl(cr) | 50.79 | 96.2 | −127.068 | −109.789 |
| Al(cr) | 24.209 | 28.275 | 0 | 0 |
| Ar(g) | 20.786 | 154.845 | 0 | 0 |
| Br(g) | 20.786 | 175.017 | 111.860 | 82.369 |
| $Br_2$(g) | 30.048 | 245.394 | 30.910 | 3.126 |
| $Br_2$(l) | 75.674 | 152.206 | 0 | 0 |
| C(g) | 20.838 | 158.100 | 716.670 | 671.244 |
| C(graphite) | 8.517 | 5.740 | 0 | 0 |
| C(diamond)** | 6.113 | 2.377 | 1.895 | 2.900 |
| $CCl_4$(g) | 83.401 | 309.809 | −95.981 | −53.617 |
| CH(g) | 29.171 | 183.040 | 594.128 | 560.747 |
| $CH_2$(g) | 34.600 | 193.931 | 386.392 | 369.245 |
| $CH_3Cl$(g) | 40.731 | 234.367 | −83.680 | −60.146 |
| $CH_4$(g) | 35.639 | 186.251 | −74.873 | −50.768 |
| CO(g) | 29.142 | 197.653 | −110.527 | −137.163 |
| $CO_2$(g) | 37.129 | 213.795 | −393.522 | −394.389 |
| $CS_2$(g) | 45.664 | 237.977 | 116.943 | 66.816 |
| $C_2$(g) | 43.145 | 199.382 | 837.737 | 781.714 |
| $C_2H_2$(g) | 44.095 | 200.958 | 226.731 | 248.163 |
| $C_2H_4$(g) | 42.886 | 219.330 | 52.467 | 68.421 |
| $C_3O_2$(g) | 66.989 | 276.071 | −93.638 | −109.649 |
| Ca(cr) | 25.929 | 41.588 | 0 | 0 |
| CaO(cr) | 42.120 | 38.212 | −635.089 | −603.501 |
| Cl(g) | 21.838 | 165.189 | 121.302 | 105.306 |
| $Cl_2$(g) | 33.949 | 223.079 | 0 | 0 |
| Cr(cr) | 23.434 | 23.618 | 0 | 0 |
| CuO(cr) | 42.246 | 42.594 | −156.063 | −128.292 |
| $Cu_2O$(cr) | 62.543 | 92.360 | −170.707 | −147.886 |
| D(g) | 20.786 | 123.350 | 221.720 | 206.553 |
| $D_2$(g) | 29.194 | 144.960 | 0 | 0 |
| F(g) | 22.746 | 158.750 | 79.390 | 62.289 |
| $F_2$(g) | 31.302 | 202.789 | 0 | 0 |
| H(g) | 20.786 | 114.716 | 217.999 | 203.278 |
| HBr(g) | 29.141 | 198.699 | −36.443 | −53.513 |
| HCl(g) | 29.136 | 186.901 | −92.312 | −95.300 |
| HCN(g) | 35.857 | 201.828 | 135.143 | 124.725 |
| HD(g) | 29.200 | 143.803 | 0.320 | −1.463 |
| HDO(g) | 33.786 | 199.511 | −245.371 | −233.181 |
| HF(g) | 29.138 | 173.780 | −272.546 | −274.646 |

**Table 9.1** *Continued*

| Substance | $C^\circ_{p,m}$/ (J·K$^{-1}$·mol$^{-1}$) | $S^\circ_m$/ (J·K$^{-1}$·mol$^{-1}$) | $\Delta_f H^\circ_m$/ (kJ·mol$^{-1}$) | $\Delta_f G^\circ_m$/ (kJ·mol$^{-1}$) |
|---|---|---|---|---|
| $HNO_3$(g) | 53.336 | 266.400 | −134.306 | −73.941 |
| $H_2$(g) | 28.836 | 130.680 | 0 | 0 |
| $H_2O$(l) | 75.351 | 69.950 | −285.830 | −237.141 |
| $H_2O$(g) | 33.590 | 188.834 | −241.826 | −228.582 |
| $H_2S$(g) | 34.192 | 205.757 | −20.502 | −33.329 |
| $H_2SO_4$(g) | 83.761 | 298.796 | −735.129 | −653.366 |
| $H_2SO_4$(l) | 138.584 | 156.895 | −813.989 | −689.918 |
| He(g) | 20.786 | 126.152 | 0 | 0 |
| Hg(g) | 20.786 | 174.970 | 61.380 | 31.880 |
| HgO(cr) | 44.062 | 70.270 | −90.789 | −58.490 |
| I(g) | 20.786 | 180.786 | 106.762 | 70.174 |
| $I_2$(cr) | 54.436 | 116.142 | 0 | 0 |
| KCl(cr) | 51.287 | 82.554 | −436.684 | −408.761 |
| Kr(g) | 20.786 | 164.084 | 0 | 0 |
| MgO(cr) | 37.106 | 26.924 | −601.241 | −568.945 |
| N(g) | 20.786 | 153.300 | 472.683 | 455.540 |
| $NH_3$(g) | 35.652 | 192.774 | −45.898 | −16.367 |
| NO(g) | 29.845 | 210.758 | 90.291 | 86.600 |
| $NO_2$(g) | 36.974 | 240.034 | 33.095 | 51.258 |
| $N_2$(g) | 29.124 | 191.609 | 0 | 0 |
| $N_2O$(g) | 38.617 | 219.957 | 82.048 | 104.179 |
| $N_2O_3$(g) | 65.615 | 308.539 | 82.843 | 139.727 |
| $N_2O_4$(g) | 77.256 | 304.376 | 9.079 | 97.787 |
| $N_2O_5$(g) | 96.303 | 346.548 | 11.297 | 118.013 |
| NaCl(cr) | 50.509 | 72.115 | −411.120 | −384.024 |
| NaCl(g) | 35.786 | 229.793 | −181.418 | −201.334 |
| $NaHCO_3$(cr) | 87.61 | 101.7 | −950.81 | −851.0 |
| NaOH(cr) | 59.530 | 64.445 | −425.931 | −379.741 |
| $Na_2CO_3$(cr) | 111.003 | 138.797 | −1130.768 | −1048.009 |
| Ne(g) | 20.786 | 146.327 | 0 | 0 |
| O(g) | 21.911 | 161.058 | 249.173 | 231.736 |
| OH(g) | 29.986 | 183.708 | 38.987 | 34.277 |
| $O_2$(g) | 29.376 | 205.147 | 0 | 0 |
| $O_3$(g) | 39.238 | 238.932 | 142.674 | 163.184 |
| P(cr-black) | 21.547 | 22.586 | −12.851 | −7.338 |
| P(cr-white) | 23.825 | 41.077 | 0 | 0 |
| $PH_3$(g) | 37.102 | 210.243 | 22.886 | 30.893 |
| $P_4$(g) | 67.158 | 279.992 | 58.907 | 24.416 |

**Table 9.1** *Continued*

| Substance | $C^{\circ}_{p,\mathrm{m}}$/ (J·K$^{-1}$·mol$^{-1}$) | $S^{\circ}_{\mathrm{m}}$/ (J·K$^{-1}$·mol$^{-1}$) | $\Delta_f H^{\circ}_{\mathrm{m}}$/ (kJ·mol$^{-1}$) | $\Delta_f G^{\circ}_{\mathrm{m}}$/ (kJ·mol$^{-1}$) |
|---|---|---|---|---|
| S(cr-monoclinic) | 23.225 | 33.028 | 0.360 | 0.070 |
| S(cr-orthorhombic) | 22.698 | 32.056 | 0 | 0 |
| $SO_2$(g) | 39.878 | 248.212 | −296.842 | −300.125 |
| $SO_3$(g) | 50.661 | 256.769 | 395.765 | −371.016 |
| $S_2$(g) | 32.490 | 228.165 | 128.600 | 79.687 |
| Si(cr) | 20.000 | 18.820 | 0 | 0 |
| Xe(g) | 20.786 | 169.684 | 0 | 0 |

*Taken from M. W. Chase, C. A. Davies, J. R. Downey, Jr., D. J. Frurip, R. A. McDonald, and A. N. Syverud, "JANAF Thermochemical Tables, Third Edition", *J. Phys. Chem. Ref. Data*, **14**, Supplement No. 1, 1985.

**Taken from D. D. Wagman, W. H. Evans, V. B. Parker, R. H. Schumm, I. Halow, S. M. Bailey, K. L. Churney, and R. L. Nuttall, "The NBS Tables of Chemical Thermodynamic Properties. Selected Values for Inorganic and $C_1$ and $C_2$ Organic Substances in SI Units", *J. Phys. Chem. Ref. Data*, **11**, Supplement No. 2, 1982.

**Gibbs Free Energies of Formation:** A common method for obtaining $\Delta_f G^{\circ}_{\mathrm{m}}$ makes use of the equation

$$\Delta_f G^{\circ}_{\mathrm{m}} = \Delta_f H^{\circ}_{\mathrm{m}} - T\Delta_f S^{\circ}_{\mathrm{m}}. \tag{9.80}$$

The entropy of formation is calculated from $S^{\circ}_{\mathrm{m}}$ values obtained from Third Law measurements (Chapter 4) or calculated from statistical thermodynamics (Chapter 10). The combination of $\Delta_f S^{\circ}_{\mathrm{m}}$ with $\Delta_f H^{\circ}_{\mathrm{m}}$ gives $\Delta_f G^{\circ}_{\mathrm{m}}$. For example, for the reaction at 298.15 K

$$H_2(g) + \tfrac{1}{2}O_2(g) = H_2O(l)$$

$$\Delta_f S^{\circ}_{\mathrm{m}} = S^{\circ}_{\mathrm{m}}(H_2O) - S^{\circ}_{\mathrm{m}}(H_2) - \tfrac{1}{2}S^{\circ}_{\mathrm{m}}(O_2)$$

$$= 69.950 - 130.680 - (\tfrac{1}{2})(205.147)$$

$$= -163.31\ \mathrm{J{\cdot}K^{-1}{\cdot}mol^{-1}}.$$

Earlier we found that $\Delta_f H^{\circ}_{\mathrm{m}} = -285.830\ \mathrm{kJ{\cdot}mol^{-1}}$ for this reaction. Therefore

$$\Delta_f G^{\circ}_{\mathrm{m}} = -285.830 - (298.15)(-163.31)(10^{-3}\ \mathrm{kJ{\cdot}J^{-1}})$$

$$= -237.14\ \mathrm{kJ{\cdot}mol^{-1}}.$$

Values of the standard free energies of formation at $T = 298.15$ K for selected substances are summarized in Table 9.1.

Other measurements of $\Delta_f G_m^\circ$ involve measuring $\Delta G$ for equilibrium processes, such as the measurement of equilibrium constants, reversible voltages of electrochemical cells, and phase equilibrium measurements. These methods especially come into play in the measurement of $\Delta_f H_m^\circ$ and $\Delta_f G_m^\circ$ for ions in solution, which are processes that we will now consider.

## 9.2b Enthalpies of Formation and Gibbs Free Energies of Formation of Ions in Solution

Consider the reaction

$$H_2O(l) = H^+(aq) + OH^-(aq).$$

The calorimetric study[6] of this reaction at 298.15 K gives $\Delta_r H^\circ = 55.836\ \text{kJ·mol}^{-1}$. Electrochemical cell measurements yield $K = 1.008 \times 10^{-14}$ from which

$$\Delta_r G^\circ = -RT \ln K = -(8.314)(298.15)\ln 1.008 \times 10^{-14}$$

$$= 79.893\ \text{kJ}.$$

For this reaction

$$\Delta_r H^\circ = \Delta_f H_m^\circ(H^+) + \Delta_f H_m^\circ(OH^-) - \Delta_f H_m^\circ(H_2O) = 55.836\ \text{kJ}.$$

Earlier we showed that for $H_2O(l)$ at $T = 298.15$ K, $\Delta_f H_m^\circ = -285.830\ \text{kJ·mol}^{-1}$ and $\Delta_f G_m^\circ = -237.129\ \text{kJ·mol}^{-1}$. Substitution of the value for $\Delta_f H_m^\circ(H_2O)$ gives

$$\Delta_f H_m^\circ(H^+) + \Delta_f H_m^\circ(OH^-) = -229.994\ \text{kJ·mol}^{-1}. \tag{9.81}$$

In a similar manner

$$\Delta_f G_m^\circ(H^+) + \Delta_f G_m^\circ(OH^-) = -157.23\ \text{kJ·mol}^{-1}. \tag{9.82}$$

With our present techniques, it is not possible to measure $\Delta_f H_m^\circ$ or $\Delta_f G_m^\circ$ for individual ions. Instead we always end up with a combination of $\Delta_f H_m^\circ$ for two or more ions such as we found in equations (9.81) and (9.82). A convention has been established in which we assign $\Delta_f H^\circ_m(H^+) = \Delta_f G^\circ_m(H^+) = 0$. Equations (9.81) and (9.82) can then be used to obtain the thermochemical quantities for the $OH^-$ ion. From equations (9.81) and (9.82), we get

$$\Delta_f H^\circ_m(OH^-) = -229.994\ kJ{\cdot}mol^{-1}$$

$$\Delta_f G^\circ_m(OH^-) = -157.236\ kJ{\cdot}mol^{-1}.$$

With these values for $OH^-$, we have now begun the generation of a table of $\Delta_f H^\circ_m$ and $\Delta_f G^\circ_m$ for ions. For example, starting with the above $\Delta_f H^\circ_m$ and $\Delta_f G^\circ_m$, we can obtain the values for $Cl^-(aq)$. $\Delta_f H^\circ_m(Cl^-)$ is obtained from the sum of the following steps (all values are in kJ).

(1) $H_2(g, f_2 = p_2 = 1\ bar) + Cl_2(g, f_2 = p_2 = 1\ bar)$

$= HCl(g, f_2 = p_2 = 1\ bar)$ $\qquad \Delta_r H^\circ_1 = -92.307$

(2) $HCl(g, f_2 = p_2 = 1\ bar) = HCl(aq, m = 0)$ $\qquad \Delta_r H^\circ_2 = -74.852$

(3) $HCl(aq, m = 0)$

$= H^+(aq, m = 0) + Cl^-(aq, m = 0)$ $\qquad \Delta_r H^\circ_3 = 0$

(4) $H^+(aq, m = 0) + Cl^-(aq, m = 0)$

$= H^+(aq, \text{hyp } a_+ = m_+ = 1)$

$+ Cl^-(aq, \text{hyp } a_- = m_- = 1)$ $\qquad \Delta_r H^\circ_4 = 0$

---

(5) $H_2(g, f_2 = p_2 = 1\ bar) + Cl_2(g, f_2 = p_2 = 1\ bar)$

$= H^+(aq, \text{hyp } a_+ = m_+ = 1)$

$+ Cl^-(aq, \text{hyp } a_- = m_- = 1)$ $\qquad \Delta_r H^\circ_5 = -167.1595.$

The explanation of the $\Delta_r H^\circ$ for each of the steps is as follows:

(1) $\Delta_r H^\circ_1$ is obtained from the direct combination of the gaseous elements under the standard state condition.
(2) $\Delta_r H^\circ_2$ is obtained from the enthalpies of solution of HCl(g) in water, extrapolated to infinite dilution.
(3) $\Delta_r H^\circ_3$ recognizes that in the infinitely dilute solution HCl is already completely separated into ions so that no enthalpy change is involved in the ionization process.

(4) $\Delta_r H_4^\circ$ takes the infinitely dilute solution to the Henry's law standard state. We have shown earlier that $\Delta H = 0$ for this process.
(5) Reaction (5) is the sum of the four steps. Since all reactants and products are in their standard states,

$$\Delta_r H_5^\circ = \Delta_f H_m^\circ(H^+) + \Delta_f H_m^\circ(Cl^-).$$

But $\Delta_f H_m^\circ(H^+)$ is set at zero. Hence,

$$\Delta_f H_m^\circ(Cl^-) = -167.159\ \text{kJ}\cdot\text{mol}^{-1}.$$

The Gibbs free energy of formation of $Cl^-$(aq) is obtained from the following steps (again in kJ).

(1) $H_2(g, f_2 = p_2 = 1\ \text{bar}) + Cl_2(g, f_2 = p_2 = 1\ \text{bar})$

$= HCl(g, f_2 = p_2 = 1\ \text{bar})$ $\quad \Delta_r G_1^\circ = -95.299$

(2) $HCl(g, f_2 = p_2 = 1\ \text{bar})$

$= HCl(g, f_2 = 2.427 \times 10^{-5}\ \text{bar})$ $\quad \Delta_r G_2 = -26.342$

(3) $HCl(g, f_2 = 2.427 \times 10^{-5}\ \text{bar}) = HCl(aq, m = 4)$ $\quad \Delta_r G_3 = 0$

(4) $HCl(aq, m = 4) = HCl(aq, a_2 = 1)$ $\quad \Delta_r G_4 = -9.681$

(5) $HCl(aq, a_2 = 1)$

$= H^+(aq, a_+ = 1) + Cl^-(aq, a_- = 1)$ $\quad \Delta_r G_5^\circ = 0$

---

(6) $H_2(g, f_2 = p_2 = 1\ \text{bar}) + Cl_2(g, f_2 = p_2 = 1\ \text{bar})$

$= H^+(aq, a_+ = 1) + Cl^-(aq, a_- = 1)$ $\quad \Delta_r G_5^\circ = -131.326.$

The explanation of the $\Delta_r G$ for each of the steps is as follows:

(1) $\Delta_r G_1^\circ$ is $\Delta_f G_m^\circ$ for HCl(g), obtained from $\Delta_f G_m^\circ = \Delta_f H_m^\circ - T\Delta_f S_m^\circ$.
(2) $\Delta G_2$ is the free energy change for changing the gas from a fugacity of 1 bar to $f_2 = 2.427 \times 10^{-5}$ bar, the equilibrium vapor fugacity of HCl above an $m = 4$ aqueous solution of HCl. It is calculated from

$$\Delta_r G_2 = RT \ln \frac{2.427 \times 10^{-5}}{1} = -26.342\ \text{kJ}.$$

(3) Step (3) is an equilibrium process. Hence $\Delta_r G_3 = 0$.

(4) In an HCl solution with $m = 4$, $\gamma_\pm$ has a value of 1.762. This result is obtained from vapor pressure measurements as described earlier. Hence, $a_2 = \gamma_\pm^2 m^2 = (1.762)^2(4)^2 = 49.66$ and

$$\Delta_r G_4 = RT \ln \frac{1}{49.66} = -9681 \text{ J}.$$

(5) The definition of the strong electrolyte standard state gives $a_2 = a_+ a_-$ so that

$$\Delta G_5^\circ = -RT \ln \frac{a_+ a_-}{a_2} = -RT \ln 1 = 0.$$

(6) Equation (6) is the sum of the five steps

$$H_2(g, f_2 = p_2 = 1 \text{ bar}) + Cl_2(g, f_2 = p_2 = 1 \text{ bar})$$

$$= H^+(aq, a_+ = 1) + Cl^-(aq, a_- = 1). \qquad \Delta_r G_5^\circ = -131.32 \text{ kJ}.$$

Since all reactants and products are in their standard states,

$$\Delta_r G_6^\circ = \Delta_f G_m^\circ(H^+) + \Delta_f G_m^\circ(Cl^-) = -131.32 \text{ kJ}.$$

With $\Delta_f G_m^\circ(H^+) = 0$, we get

$$\Delta_r G_m^\circ(Cl^-) = -131.32 \text{ kJ·mol}^{-1}.$$ [m]

These types of processes can be repeated to generate values of $\Delta_f H_m^\circ$ and $\Delta_f G_m^\circ$ for other ions, all based upon $\Delta_f G_m^\circ(H^+) = \Delta_f H_m^\circ(H^+) = 0$. For example, addition of the following steps gives $\Delta_f G_m^\circ$ for $Na^+$ from the known

---

[m] This result differs slightly from the value of $-131.228$ kJ·mol$^{-1}$ given in Table 9.2, which is based on the weighted average of values obtained from several workers using different experimental techniques.

value of $\Delta_f G^\circ_m$ for $Cl^-$

$$Na(s) + \tfrac{1}{2}Cl_2(g) = NaCl(s) \qquad \Delta_f G^\circ_1 = \Delta_f H^\circ_m - T\Delta_f S^\circ_m$$

$$NaCl(s) = NaCl(aq,\ \text{saturated solution with } m = 6.12) \qquad \Delta_r G_2 = 0$$

$$NaCl(aq,\ m = 6.12,\ a_2 = 38.42) = NaCl(aq,\ a_2 = 1) \qquad \Delta_r G_3 = RT \ln \frac{1}{38.42}$$

$$NaCl(aq,\ a_2 = 1) = Na^+(aq,\ a_+ = 1) + Cl^-(aq,\ a_- = 1) \qquad \Delta_r G^\circ_4 = 0$$

$$Na(s) + \tfrac{1}{2}Cl_2(g) = Na^+(aq,\ a_+ = 1) + Cl^-(aq,\ a_- = 1) \qquad \Delta_r G^\circ_5 = \text{sum of above}$$

$$\Delta_r G^\circ_5 = \Delta_f G^\circ_m(Na^+) + \Delta_f G^\circ_m(Cl^-)$$

or

$$\Delta_f G^\circ_m(Na^+) = \Delta_r G^\circ_5 - \Delta_f G^\circ_m(Cl^-).$$

Since we already know $\Delta_f G^\circ_m(Cl^-)$, we can obtain a value for $\Delta_f G^\circ_m(Na^+)$. Table 9.2 summarizes $\Delta_f H^\circ_m$ and $\Delta_f G^\circ_m$ for selected ions. Other values can be found in the literature from sources such as those referred to earlier in the chapter.[n]

Before leaving our discussion of standard enthalpies and standard Gibbs free energies of formation, we should note that for the $H_3O^+(aq)$ ion, $\Delta_f H^\circ_m = -285.830\ \text{kJ·mol}^{-1}$ and $\Delta_f G^\circ_m = -237.129\ \text{kJ·mol}^{-1}$. These are the same values as $\Delta_f H^\circ_m$ and $\Delta_f G^\circ_m$ for $H_2O(l)$. The reasons for this can be seen as follows. For the reaction

$$H^+(aq) + H_2O(l) = H_3O^+(aq)$$

we want $K$ to equal one at all temperatures so that

$$a_{H^+}a_{H_2O} = a_{H_3O^+}.$$

[n] Remember that the results for ions found in tables are relative values. In our calculations, they work as well as absolute values, since the differences always subtract out.

**Table 9.2** Standard* heat capacities, entropies, enthalpies, and Gibbs free energies of formation** of some common ions in aqueous solution at $T = 298.15$ K

| Substance | $C^\circ_{p,m}$/ $(J \cdot K^{-1} \cdot mol^{-1})$ | $S^\circ_m$/ $(J \cdot K^{-1} \cdot mol^{-1})$ | $\Delta_f H^\circ_m$/ $(kJ \cdot mol^{-1})$ | $\Delta_f G^\circ_m$/ $(kJ \cdot mol^{-1})$ |
|---|---|---|---|---|
| $Ag^+$ | 21.8 | 72.68 | 105.579 | 77.107 |
| $Al^{3+}$ | – | −321.7 | −531 | −485 |
| $AsO_4^{3-}$ | – | −162.8 | −888.14 | −648.41 |
| $Ba^{2+}$ | – | 9.6 | −537.64 | −560.77 |
| $Be^{2+}$ | – | −129.7 | −382.8 | −379.73 |
| $CH_3COO^-$ | −6.3 | 86.6 | −486.01 | −369.31 |
| $CN^-$ | – | 94.1 | 150.6 | 172.4 |
| $CO_3^{2-}$ | – | −56.9 | −677.14 | −527.81 |
| $Ca^{2+}$ | – | −53.1 | −542.83 | −553.58 |
| $Ce^{3+}$ | – | −205 | −696.2 | −672.0 |
| $Ce^{4+}$ | – | −301 | −537.2 | −503.8 |
| $Cl^-$ | −136.4 | 56.5 | −167.159 | −131.228 |
| $ClO_3^-$ | – | 162.3 | −103.97 | −7.95 |
| $ClO_4^-$ | – | 182.0 | −129.33 | −8.52 |
| $CrO_4^{2-}$ | – | 50.21 | −881.15 | −727.75 |
| $Cr_2O_7^{2-}$ | – | 261.9 | −1490.3 | −1301.1 |
| $Cs^+$ | −10.5 | 133.05 | −258.28 | −292.02 |
| $Cu^+$ | – | 40.6 | 71.67 | 49.98 |
| $Cu^{2+}$ | – | −99.6 | 64.77 | 65.49 |
| $F^-$ | −106.7 | −13.8 | −332.63 | −278.79 |
| $Fe^{2+}$ | – | −137.7 | −89.1 | −78.90 |
| $Fe^{3+}$ | – | −315.9 | −48.5 | −4.7 |
| $H^+$ | 0 | 0 | 0 | 0 |
| $HCOO^-$ | −87.9 | 92 | −425.55 | −351.0 |
| $HCO_3^-$ | – | 91.2 | −691.99 | −586.77 |
| $Hg^{2+}$ | – | −32.2 | 171.1 | 164.40 |
| $Hg_2^{2+}$ | – | 84.5 | 172.4 | 153.52 |
| $I^-$ | −142.3 | 113.3 | −55.19 | −51.57 |
| $K^+$ | 21.8 | 102.5 | −252.38 | −283.27 |
| $Li^+$ | 68.6 | 13.4 | −278.49 | −293.31 |
| $Mg^{2+}$ | – | −138.1 | −466.85 | −454.8 |
| $Mn^{2+}$ | 50 | −73.6 | −220.75 | −228.1 |
| $MnO_4^-$ | −82.0 | 191.2 | −541.4 | −447.2 |
| $N_3^-$ | – | 107.9 | 275.14 | 348.2 |
| $NH_4^+$ | 79.9 | 113.4 | −132.51 | −79.31 |
| $NO_3^-$ | −86.6 | 146.4 | −205.0 | −108.74 |
| $NO_2^-$ | −97.5 | 123.0 | −104.6 | −32.2 |
| $Na^+$ | 46.4 | 59.0 | −240.12 | −261.905 |
| $OH^-$ | −148.5 | −10.75 | −229.994 | −157.244 |

**Table 9.2** *Continued*

| Substance | $C^\circ_{p,\mathrm{m}}$/ (J·K$^{-1}$·mol$^{-1}$) | $S^\circ_\mathrm{m}$/ (J·K$^{-1}$·mol$^{-1}$) | $\Delta_f H^\circ_\mathrm{m}$/ (kJ·mol$^{-1}$) | $\Delta_f G^\circ_\mathrm{m}$/ (kJ·mol$^{-1}$) |
|---|---|---|---|---|
| $PO_4^{3-}$ | – | −222 | −1277.4 | −1018.7 |
| $Rb^+$ | – | 121.50 | −251.17 | −283.98 |
| $S^{2-}$ | – | −14.6 | 33.1 | 85.8 |
| $SO_4^{2-}$ | −293 | 20.1 | −909.27 | −744.53 |
| $SO_3^{2-}$ | – | −29 | −635.5 | −486.5 |
| $Sr^{2+}$ | – | −32.6 | −545.80 | −559.48 |
| $Zn^{2+}$ | 46 | −112.1 | −153.89 | −147.06 |

*The standard state is the hypothetical $m = 1$ solution that obeys Henry's Law. The values are relative to those for $H^+$ being equal to zero.

**Taken from D. D. Wagman, W. H. Evans, V. B. Parker, R. H. Schumm, I. Halow, S. M. Bailey, K. L. Churney, and R. L. Nuttall, "The NBS Tables of Chemical Thermodynamic Properties. Selected Values for Inorganic and $C_1$ and $C_2$ Organic Substances in SI Units", *J. Phys. Chem. Ref. Data*, **11**, Supplement No. 2, 1982.

This requires that $\Delta_r H^\circ = \Delta_r G^\circ = 0$ for the above reaction. Thus

$$\Delta_f H^\circ_\mathrm{m}(H_3O^+) - \Delta_f H^\circ_\mathrm{m}(H^+) - \Delta_f H^\circ_\mathrm{m}(H_2O) = 0$$

$$\Delta_f G^\circ_\mathrm{m}(H_3O^+) - \Delta_f G^\circ_\mathrm{m}(H^+) - \Delta_f G^\circ_\mathrm{m}(H_2O) = 0.$$

Since

$$\Delta_f H^\circ_\mathrm{m}(H^+) = \Delta_f G^\circ_\mathrm{m}(H^+) = 0$$

then

$$\Delta_f H^\circ_\mathrm{m}(H_3O^+) = \Delta_f H^\circ_\mathrm{m}(H_2O)$$

and

$$\Delta_f G^\circ_\mathrm{m}(H_3O^+) = \Delta_f G^\circ_\mathrm{m}(H_2O).$$

The above analysis results from the choice of $\Delta_r H^\circ = \Delta_r G^\circ = 0$ for reaction (9.40). A similar treatment for reaction (9.36) is as follows. Measurements give $\Delta_f H^\circ_\mathrm{m} = -80.29$ kJ·mol$^{-1}$ and $\Delta_f G^\circ_\mathrm{m} = -26.50$ kJ·mol$^{-1}$ for $NH_3$(aq). This requires that $\Delta_f H^\circ_\mathrm{m} = -362.50$ kJ·mol$^{-1}$ and $\Delta_f G^\circ_\mathrm{m} = -263.65$ kJ·mol$^{-1}$ for $NH_4OH$(aq). The differences between the two sets of values are $\Delta_f H^\circ_\mathrm{m}$ and

$\Delta_f G_m^\circ$ of $H_2O(l)$. This difference is required so that for the reaction

$$NH_3(aq) + H_2O(l) = NH_4OH(aq) \qquad \Delta_r G^\circ = 0$$

and hence,

$$a_{NH_3} a_{H_2O} = a_{NH_4OH}.$$

Similar procedures can be used to relate $\Delta_f H_m^\circ$ and $\Delta_f G_m^\circ$ of aqueous species such as $CO_2(aq)$ and $H_2CO_3(aq)$ so that

$$a_{CO_2} a_{H_2O} = a_{H_2CO_3}.$$

## 9.3 Examples of Chemical Equilibrium Calculations

We are now prepared to use thermodynamics to make chemical equilibrium calculations. The following examples demonstrate some of the possibilities.

**Example 9.1:** Calculate the equilibrium sublimation pressure above diamond at 298.15 K.

**Solution:** The reaction is[o]

$$C(s) = C(g)$$

for which

$$K = \frac{a_g}{a_s}.$$

With our choice of standard states

$$a_g = \frac{f_g}{1 \text{ bar}} = \frac{\phi \cdot p_C}{1 \text{ bar}}$$

$$a_s = \Gamma(1)$$

so that

$$K = \frac{\phi \cdot p_C}{\Gamma}.$$

[o] This is usually thought of as a phase equilibrium problem. Earlier, we indicated that a phase equilibrium is nothing more than a simple chemical equilibrium. This problem is one such example.

The sublimation pressure will be small so that $\phi = 1$. Also, at the low pressure involved, we can let $\Gamma = 1$. Hence

$$K = p_{\mathrm{C}}\,.$$

From Table 9.1

$$\Delta_r G^\circ = \Delta_f G^\circ_m(\mathrm{g}) - \Delta_f G^\circ_m(\mathrm{s})$$

$$= 671.244\ \mathrm{kJ{\cdot}mol^{-1}} \times 1\ \mathrm{mol} - 2.900\ \mathrm{kJ{\cdot}mol^{-1}} \times 1\ \mathrm{mol}$$

$$= 668.344\ \mathrm{kJ}$$

$$\ln K = -\frac{\Delta_r G^\circ}{RT}$$

$$= -\frac{(668.344)(10^3\ \mathrm{J{\cdot}kJ^{-1}})}{(8.3145)(298.15)}$$

$$= -269.61$$

$$K = 8.126 \times 10^{-118}$$

so that

$$p_{\mathrm{C}} = 8.126 \times 10^{-118}\ \mathrm{bar}.$$

This extremely low pressure corresponds to much less than one gaseous carbon atom (on the average) in the universe at a time.

**Example 9.2:** $NaHCO_3$(s) dissociates according to the reaction

$$2\mathrm{NaHCO_3(s)} = \mathrm{Na_2CO_3(s)} + \mathrm{H_2O(g)} + \mathrm{CO_2(g)}.$$

Calculate the dissociation pressure above $NaHCO_3$(s) at (a) 298.15 K and (b) 400 K.

**Solution:**

$$K = \frac{a_{\mathrm{H_2O}}\,a_{\mathrm{CO_2}}\,a_{\mathrm{Na_2CO_3}}}{a^2_{\mathrm{NaHCO_3}}}\,.$$

With our choice of standard states and the low pressure involved, we can write

$$a_{H_2O} = \frac{p_{H_2O}}{1 \text{ bar}}$$

$$a_{CO_2} = \frac{p_{CO_2}}{1 \text{ bar}}$$

$$a_{Na_2CO_3} = a_{NaHCO_3} = 1$$

so that

$$K = p_{CO_2} p_{H_2O}.$$

Using Table 9.1, we find that at $T = 298.15$ K,

$$\begin{aligned}
\Delta_r G^\circ &= \Delta_f G^\circ_m(H_2O) + \Delta_f G^\circ_m(CO_2) + \Delta_f G^\circ_m(Na_2CO_3) \\
&\quad - 2\Delta_f G^\circ_m(NaHCO_3) \\
&= -228.582 - 394.389 - 1048.009 - (2)(-851.0) \\
&= 31.0 \text{ kJ}
\end{aligned}$$

$$\begin{aligned}
\Delta_r H^\circ &= \Delta_f H^\circ_m(H_2O) + \Delta_f H^\circ_m(CO_2) + \Delta_f H^\circ_m(Na_2CO_3) \\
&\quad - 2\Delta_f H^\circ_m(NaHCO_3) \\
&= -241.826 - 393.522 - 1130.768 - (2)(-950.81) \\
&= 135.50 \text{ kJ}.
\end{aligned}$$

At $T = 298.15$ K

$$\ln K = -\frac{\Delta_r G^\circ}{RT} = -\frac{(31.0)(10^3 \text{ J·kJ}^{-1})}{(8.314)(298.15)}$$

$$K = 3.70 \times 10^{-6}$$

$$p_{CO_2} = p_{H_2O}$$

$$p^2_{CO_2} = 3.70 \times 10^{-6}$$

$$p_{CO_2} = 1.92 \times 10^{-3} \text{ bar} = p_{H_2O}$$

$$p_{\text{dissoc}} = p_{CO_2} + p_{H_2O} = 3.85 \times 10^{-3} \text{ bar}.$$

To calculate $K$ at $T = 400$ K, we use equation (9.55) since the temperature interval is small. The result is

$$\ln \frac{K_2}{K_1} = -\frac{\Delta_r H^\circ}{R}\left(\frac{1}{T_2} - \frac{1}{T_1}\right)$$

$$\ln \frac{K_2}{3.70 \times 10^{-6}} = -\frac{(135.50 \text{ kJ})(10^3 \text{ J·kJ}^{-1})}{(8.314)}\left(\frac{1}{400.00} - \frac{1}{298.15}\right)$$

$$= 13.93$$

$$\frac{K_2}{3.70 \times 10^{-6}} = 1.12 \times 10^6$$

$$K_2 = 4.15 = p_{CO_2}^2$$

$$p_{CO_2} = 2.04 \text{ bar} = p_{H_2O}$$

$$p_{\text{dissoc}} = (2)(2.04) = 4.08 \text{ bar}.$$

At this temperature, the dissociation pressure is greater than atmospheric pressure, and the $CO_2$(g) and $H_2O$(g) will escape from the $NaHCO_3$(s), so that $Na_2CO_3$(s) remains.

**Example 9.3:** Calculate the pressure of atomic chlorine in $Cl_2$(g) at a total pressure of 1.00 bar and a temperature of 2000 K. Do the calculation (a) using the thermodynamic functions in Table 4.3, and (b) using equation (9.58), which requires $C_{p,\text{m}}$ expressed as a function of $T$, obtained from Table 2.1, and compare the results.

**Solution:** The reaction is

$$Cl_2(g) = 2Cl(g)$$

$$K = \frac{a_{Cl}^2}{a_{Cl_2}}.$$

With our choice of standard states and the low pressures involved,

$$a_{Cl} = \phi_{Cl} p_{Cl} = p_{Cl}$$

$$a_{Cl_2} = \phi_{Cl_2} p_{Cl_2} = p_{Cl_2}$$

$$K = K_p = \frac{p^2_{Cl}}{p_{Cl_2}}.$$

(a) Using the thermodynamic functions and equation (4.34), $\Delta_r G^\circ_T$ is obtained from

$$\Delta_r G^\circ_T = T \sum \nu_i \frac{(G^\circ_T - H^\circ_{298})_{i,\,T}}{T} + \Delta_r H^\circ_{298}. \tag{9.83}$$

For our reaction at $T = 2000$ K,

$$\sum_i \nu_i \frac{(G^\circ_T - H^\circ_{298})}{T} = (2)\frac{(G^\circ_T - H^\circ_{298})(\mathrm{Cl})}{T} - (1)\frac{(G^\circ_T - H^\circ_{298})(\mathrm{Cl_2})}{T}.$$

Using the values from Table 4.3 gives

$$\sum_i \nu_i \frac{(G^\circ_T - H^\circ_{298})}{T} = (2)(-188.749) - (-261.277)$$

$$= -116.221 \text{ J}.$$

From Table 9.1[p] (or Table 4.3)

$$\Delta_r H^\circ_{298} = 2\Delta_f H^\circ_{298}(\mathrm{Cl}) - \Delta_f H^\circ_{298}(\mathrm{Cl_2})$$

$$= (2 \text{ moles})(121.302 \text{ kJ·mol}^{-1})(10^3 \text{ J·kJ}^{-1}) - 0$$

$$= 242{,}604 \text{ J}.$$

---

[p] Note that $\Delta_f H^\circ_m$ and $\Delta_f G^\circ_m$ are zero for elements in their naturally occurring form. For chlorine, the elemental form is $Cl_2(g)$.

Hence

$$\Delta_r G^\circ_{2000} = (2000)(-116.221) + 242{,}604$$

$$= 10{,}162 \text{ J}$$

$$\ln K = -\frac{(10{,}162)}{(8.314)(2000)}$$

$$= -0.6111$$

$$K = 0.543 = \frac{p_{Cl}^2}{p_{Cl_2}}$$

$$p_{Cl_2} + p_{Cl} = 1 \text{ bar}$$

$$\frac{p_{Cl}^2}{1 - p_{Cl}} = 0.543$$

$$p_{Cl}^2 + 0.543 p_{Cl} - 0.543 = 0.$$

Solving with the quadratic formula gives

$$p_{Cl} = 0.514 \text{ bar}$$

$$p_{Cl_2} = 0.486 \text{ bar}.$$

(b) We now repeat the calculation using equation (9.58). From Table 2.1

$$C^\circ_{p,m}(\text{Cl}) = 23.033 - 0.749 \times 10^{-3} T - \frac{0.695 \times 10^5}{T^2}$$

$$C^\circ_{p,m}(\text{Cl}_2) = 36.90 + 0.25 \times 10^{-3} T - \frac{2.85 \times 10^5}{T^2}$$

$$\sum \nu_i C^\circ_{p,i} \doteq 9.17 - 1.74 \times 10^{-3} T + \frac{1.46 \times 10^5}{T^2}$$

so that $\sum \nu_i a_i = 9.17$, $\sum \nu_i b_i = -1.74 \times 10^{-3}$ and $\sum \nu_i c_i = 1.46 \times 10^5$.

From Table 9.1, at $T = 298.15$ K

$$\Delta_r H^\circ = 2\Delta_f H^\circ_m(\text{Cl}) - \Delta_f H^\circ_m(\text{Cl}_2)$$

$$= (2)(121.302) - 0$$

$$= (242.604 \text{ kJ})(10^3 \text{ J·kJ}^{-1}) = 2.4260 \times 10^5 \text{ J}$$

$$\Delta_r G^\circ = 2\Delta_f G^\circ_m(\text{Cl}) - \Delta_f G^\circ_m(\text{Cl}_2)$$

$$= (2)(105.306) - 0$$

$$= (210.610 \text{ kJ})(10^3 \text{ J·kJ}^{-1}) = 2.1061 \times 10^5 \text{ J.}$$

The enthalpy change $\Delta_r H^\circ$ for the reaction expressed as a function of temperature is given by equation (9.56)

$$\Delta_r H^\circ = \Delta H_1 + \Delta a T + \frac{\Delta b}{2} T^2 - \Delta c \left(\frac{1}{T}\right) + \frac{\Delta d}{3} T^3 + \cdots \tag{9.56}$$

Substituting the $T = 298.15$ value into equation (9.56) and solving for $\Delta_r H_1$ gives

$$\Delta H_1 = 2.4260 \times 10^5 - (9.17)(298.15)$$

$$+ \frac{(1.74 \times 10^{-3})}{(2)}(298.15)^2 + \frac{(1.46 \times 10^5)}{(298.15)}$$

$$= 2.404 \times 10^5 \text{ J.}$$

Solving for $I$ in equation (9.57) using $\Delta_r G^\circ$ at $T = 298.15$ K gives

$$I = \frac{(2.1061 \times 10^5)}{(298.15)} - \frac{(2.4043 \times 10^5)}{(298.15)} + (9.17) \ln 298.15$$

$$- \frac{(1.74 \times 10^{-3})}{(2)}(298.15) + \frac{(1.46 \times 10^5)}{(2)(298.15)^2}$$

$$= -47.25 \text{ J·K}^{-1}.$$

Substituting values into equation (9.58) gives an equation for $K$ as a function of $T$

$$R \ln K = 47.25 - \frac{2.404 \times 10^5}{T} + 9.17 \ln T - 0.87 \times 10^{-3} T + \frac{7.30 \times 10^4}{T^2}. \tag{9.84}$$

At $T = 2000$ K

$$R \ln K = -5.052$$

Hence,

$$K = 0.546.$$

This value can be compared with $K = 0.543$ obtained from the thermodynamic functions given in Table 4.3. The agreement is quite reasonable. Using this value of $K$ gives $p_{\text{Cl}} = 0.515$ bar instead of 0.514 bar as obtained using Table 4.1.

**Example 9.4:** Calculate the solubility (moles solute per kilogram of solvent) at $T = 298.15$ K of AgCl(s) in (a) pure water, (b) $0.010m$ $KNO_3$, and (c) $0.010m$ KCl.

**Solution:** The reaction is

$$\text{AgCl(s)} = \text{AgCl(aq)} \tag{9.85}$$

for which

$$K = \frac{a_2}{a_s}.$$

With the choice of a pure substance standard state for the solid, and a strong electrolyte standard state for the dissolved AgCl, we get

$$a_s = 1$$

$$a_2 = a_{Ag^+}a_{Cl^-} = m_+\gamma_+m_-\gamma_-$$

$$= \gamma_\pm^2 m_+m_-$$

$$K = \gamma_\pm^2 m_+m_- \,. \tag{9.86}$$

(a) For AgCl dissolved in water, $m_+ = m_- = m$, where $m$ is the solubility. In this case

$$K = K_{sp} = \gamma_\pm^2 m^2.$$

We obtain $K$ from the standard Gibbs free energies of formation. For reaction (9.85) we get

$$\Delta_r G^\circ = \Delta_f G_m^\circ(\text{AgCl, aq}) - \Delta_f G_m^\circ(\text{AgCl, s}).$$

Since AgCl is a strong electrolyte $\Delta_r G^\circ = 0$ for the reaction

$$\text{AgCl(aq)} = \text{Ag}^+\text{(aq)} + \text{Cl}^-\text{(aq)}.$$

Hence,

$$\Delta_f G_m^\circ(\text{AgCl, aq}) = \Delta_f G_m^\circ(\text{Ag}^+\text{, aq}) + \Delta_f G_m^\circ(\text{Cl}^-\text{, aq})$$

and

$$\Delta_r G^\circ = \Delta_f G_m^\circ(\text{Ag}^+) + \Delta_f G_m^\circ(\text{Cl}^-) - \Delta_f G_m^\circ(\text{AgCl, s}).$$

From Tables 9.1 and 9.2, we get

$$\Delta_f G_m^\circ(\text{Ag}^+) = 77.107\ \text{kJ·mol}^{-1}$$

$$\Delta_f G_m^\circ(\text{Cl}^-) = -131.228\ \text{kJ·mol}^{-1}$$

$$\Delta_f G_m^\circ(\text{AgCl, s}) = -109.789\ \text{kJ·mol}^{-1}$$

so that

$$\Delta_r G^\circ = 77.107 + (-131.228) - (-109.789)$$

$$= 55.668 \text{ kJ}$$

and

$$\ln K = -\frac{(55.668)(10^3 \text{ J}\cdot\text{kJ}^{-1})}{(8.3145)(298.15)} = -22.46$$

$$K = 1.77 \times 10^{-10}.$$

Thus

$$\gamma_\pm^2 m^2 = 1.77 \times 10^{-10}.$$

The solubility ($m$) is very small and it is reasonable to expect Henry's law to follow for the ions, so that $\gamma_\pm = 1$ and $m = 1.33 \times 10^{-5}$ moles $AgCl\cdot kg^{-1}$.

We have problems when we attempt to repeat this calculation in $0.010m$ KCl and $0.010m$ $KNO_3$, since $\gamma_\pm$ is a function of the total concentration of ions, and, as we saw in Chapter 7, at $m = 0.010$ mol·kg$^{-1}$, $\gamma_\pm$ differs significantly from one. Debye–Hückel theory provides a method for calculating $\gamma_\pm$ for dilute solutions.

With the Debye–Hückel equation (7.44), we can complete Example 9.4 and calculate the solubility of AgCl in (b) $0.010m$ $KNO_3$ and (c) $0.010m$ KCl.

**Example 9.4 (continued)** (a) $m = 1.33 \times 10^{-5}$ moles·kg$^{-1}$.

(b) In $0.010m$ $KNO_3$, equation (9.86) applies with $m_{Ag^+} = m_{Cl^-} = m$. To calculate $\gamma\pm$, we use equation (7.44)

$$\ln \gamma_\pm = -C_\gamma \mid z_+ z_- \mid \frac{I_m^{1/2}}{1 + I_m^{1/2}}, \tag{7.44}$$

with $C_\gamma = 1.1745$ kg$^{1/2}$·mol$^{1/2}$ obtained from Table 7.1. The ionic strength must include the concentrations of all ions in the solution. Thus

$$I_m = \frac{1}{2}(m_{K^+} + m_{NO_3^-} + m_{Ag^-} + m_{Cl^-}).$$

We will neglect $m_{Ag^+}$ and $m_{Cl^-}$, since they will be of the magnitude of the values calculated in (a). Hence,

$$I_m = \frac{1}{2}(0.010 + 0.010) = 0.010.$$

With $|z_+z_-| = 1$,

$$\ln \gamma_\pm = -\frac{(1.175)(0.010)^{1/2}}{[1 + (0.010)^{1/2}]}$$

$$\gamma_\pm = 0.899.$$

From equation (9.86) $m_{Ag^+} = m_{Cl^-} = m$, we get

$$m = \frac{(1.77 \times 10^{-10})^{1/2}}{\gamma_\pm}$$

$$= 1.48 \times 10^{-5} \text{ moles}\cdot\text{kg}^{-1}.$$

Thus, adding the $KNO_3$, increases the solubility from $1.33 \times 10^{-5}$ mol·kg$^{-1}$ to $1.48 \times 10^{-5}$ mol·kg$^{-1}$, or approximately 10%.

(c) In (0.010$m$ KCl), $m_{Cl^-} = 0.010 + m$ and $m_{Ag^+} = m$, so that

$$\gamma_\pm^2(m + 0.010)(m) = 1.77 \times 10^{-10}. \tag{9.87}$$

Again we neglect $m$ compared to 0.010 so that equation (9.87) becomes

$$m = \frac{1.77 \times 10^{-10}}{0.010\gamma_\pm^2}.$$

The ionic strength is again 0.010 so that $\gamma_\pm = 0.899$ and

$$m = \frac{1.77 \times 10^{-10}}{(0.010)(0.899)^2}$$

$$= 2.19 \times 10^{-8}.$$

Adding the common ion greatly decreases the silver chloride solubility.

## 9.4 Electrochemical Cells

In an electrochemical cell, electrical work is obtained from an oxidation–reduction reaction. For example, consider the process that occurs during the discharge of the lead storage battery (cell). Figure 9.3 shows a schematic drawing of this cell. One of the electrodes (anode)[q] is Pb metal and the other (cathode) is $PbO_2$ coated on a conducting metal (Pb is usually used). The two electrodes are immersed in an aqueous sulfuric acid solution.

The reaction in the lead storage cell is

$$Pb(s) + PbO_2(s) + 2H_2SO_4(aq) = 2H_2O + 2PbSO_4(s). \tag{9.88}$$

Under most conditions, the process is spontaneous.[r] A chemical potential difference drives the reaction and $\Delta G < 0$. When the reactants are separated as shown in Figure 9.3, the chemical potential difference can be converted to an electrical potential $E$. When the electrodes are connected through an external circuit, the electrical potential causes an electric current to flow. Because the electrical potential is the driving force for electrons to flow, it is sometimes

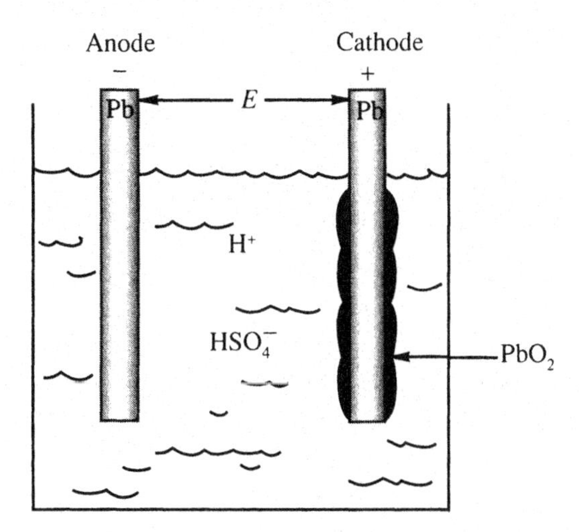


**Figure 9.3** The lead storage battery. The key to obtaining electrical energy from a redox chemical reaction is to physically separate the two half-cell reactions so that electrons are transferred from the anode through an external circuit to the cathode. In the process, electrical work is accomplished.

---

[q] The anode is the electrode where oxidation takes place. Reduction occurs at the cathode.

[r] The extent of the reaction depends mostly upon the activity (concentration) of $H_2SO_4$. For the usual range of sulfuric acid concentrations, $\Delta G < 0$.

called the electromotive force or emf. The electrons that flow in the circuit are produced by the chemical reaction at one electrode:

$$\text{Anode: } Pb(s) + HSO_4^-(aq) = PbSO_4(s) + H^+(aq) + 2e^-$$

and consumed by the chemical reaction at the other:

$$\text{Cathode: } PbO_2(s) + 3H^+(aq) + HSO_4^-(aq) + 2e^- = PbSO_4(s) + 2H_2O.$$

Adding these two half-cell reactions gives reaction (9.88).[s]

In theory, any oxidation–reduction reaction can be set up in a cell to do electrical work. The amount of reversible[t] work is easily calculated. If, during the discharge of a cell, a quantity of electricity $Q$ flows through the external circuit at a constant potential, the amount of electrical work, $w_e$, produced is given by

$$w_e = -EQ \tag{9.89}$$

where $E$ is the electrical potential. The negative sign is included to make the equation agree with our previous definition of positive and negative work.

The quantity $Q$ can be calculated from a knowledge of the chemical reaction. If $n$ moles of electrons are transferred from cathode to anode during the chemical reaction[u] and $F$ is the quantity of electrical change carried by a mole of electrons, then

$$Q = nF$$

and

$$w_e = -nFE. \tag{9.90}$$

$F$ is called the **Faraday constant**. It has a value of $9.6485309 \times 10^4$ C·mol$^{-1}$.

The electromotive force of a cell can be related to the Gibbs free energy change for the cell reaction by combining equations (9.5), (9.90), and (3.96). We recall that

$$dG = \delta w' \tag{3.96}$$

[s] Remember that for the strong electrolyte $H_2SO_4$, we can write $H_2SO_4$ (aq) as $H^+(aq) + HSO_4^-$ (aq).

[t] It is often difficult to discharge an electrochemical cell in a reversible manner. We will return to this problem later.

[u] $n = 2$ for the lead cell.

or

$$\Delta G = w',$$

where $w'$ is the work other than the pressure–volume work in a reversible constant temperature and pressure process. For the discharge of an electrochemical cell

$$w' = w_e$$

so that

$$\Delta G = -nFE. \tag{9.91}$$

For a chemical cell operated under standard state conditions, equation (9.91) becomes

$$\Delta G^\circ = -nFE^\circ \tag{9.92}$$

where $E^\circ$ is the voltage for the cell when its contents are in their standard state.

The voltages, $E$ and $E^\circ$, can be related to the activities of products and reactants in the chemical reaction by substituting equations (9.91) and (9.92) into equation (9.5)

$$\Delta_r G = \Delta_r G^\circ + RT \ln \prod_i a_i^{\nu_i}. \tag{9.5}$$

The result is

$$E = E^\circ - \frac{RT}{nF} \ln \prod_i a_i^{\nu_i}. \tag{9.93}$$

Equation (9.93) is known as the **Nernst equation**. It provides a way for calculating $E$ in a cell when $E^\circ$ is known and the activities of the products and reactants are specified.[v]

---

[v] Conversely, measuring $E$ allows one to calculate $a_i$. Thus, electrochemical cells provide a way for measuring activities.

Other thermodynamic relationships can be derived. Starting with equation (9.91) and differentiating gives

$$\left(\frac{\partial \Delta_r G}{\partial T}\right)_p = -\Delta S_r = -nF\left(\frac{\partial E}{\partial T}\right)_p$$

or

$$\Delta S_r = nF\left(\frac{\partial E}{\partial T}\right)_p. \tag{9.94}$$

Since

$$\Delta_r H = \Delta_r G + T\Delta_r S$$

$$\Delta_r H = -nF\left[E - T\left(\frac{\partial E}{\partial T}\right)_p\right]. \tag{9.95}$$

Other relationships are obtained from

$$\left(\frac{\partial \Delta_r S}{\partial T}\right)_p = \frac{\Delta_r C_p}{T} = nF\left(\frac{\partial^2 E}{\partial T^2}\right)_p$$

or

$$\Delta_r C_p = nFT\left(\frac{\partial^2 E}{\partial T^2}\right)_p. \tag{9.96}$$

Also,

$$\left(\frac{\partial \Delta_r G}{\partial p}\right)_T = \Delta_r V = -nF\left(\frac{\partial E}{\partial p}\right)_T$$

or

$$\Delta_r V = -nF\left(\frac{\partial E}{\partial p}\right)_T. \tag{9.97}$$

Finally, at equilibrium[w]

$$\Delta_r G = -nFE = 0$$

and

$$\prod_i a_i^{\nu_i} = K.$$

Substitution into equation (9.93) gives

$$RT \ln K = nFE^\circ. \tag{9.98}$$

## 9.4a Thermodynamic Applications of Electrochemical Cells

**Measurement of $E^\circ$ and Activities:** Electrochemical cells can be constructed to measure $E^\circ$ and thermodynamic properties such as $a_i$, $K$, $\Delta G$, $\Delta H$, $\Delta S$, $\Delta V$, and $\Delta C_p$ for a reaction. Consider as an example the cell shown schematically in Figure 9.4.[x] The cathode consists of an Ag metal rod coated with AgCl(s). The anode is a Pt metal rod around which $H_2$(g) is bubbled. The two electrodes are

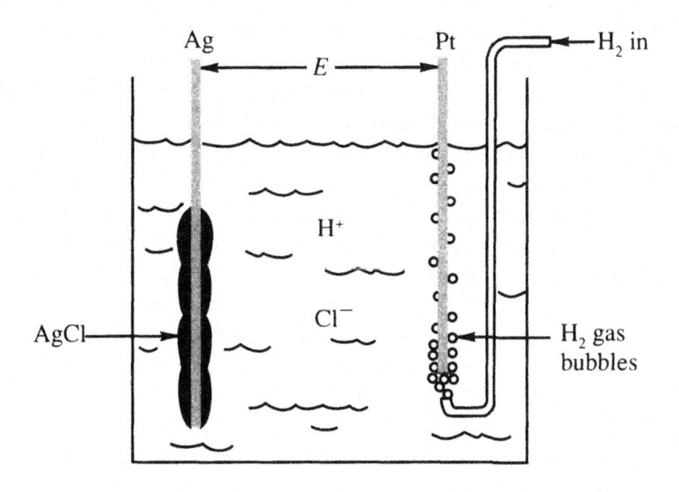


**Figure 9.4** The Pt(s), $H_2$(g) | HCl(aq) | Ag(s), AgCl(s) cell.

---

[w] As a cell discharges, $E$ decreases until at equilibrium, $E = 0$. No further reaction occurs and no more electrical work is obtained.

[x] In the cell designation shown in Figure 9.4, oxidation occurs at the half-cell given on the left, and reduction occurs at the half-cell given on the right.

immersed in an HCl(aq) solution. A voltage difference $E$ is present between the two electrodes. During discharge the half-cell reactions occur

$$\text{Cathode: } AgCl(s) + e^- = Ag(s) + Cl^-(aq) \tag{9.99}$$

$$\text{Anode: } \tfrac{1}{2}H_2(g) = H^+(aq) + e^-. \tag{9.100}$$

The cell reaction is the sum of the half cell reactions

$$AgCl(s) + \tfrac{1}{2}H_2(g) = Ag(s) + H^+(aq) + Cl^-(aq)$$

or

$$AgCl(s) + \tfrac{1}{2}H_2(g) = Ag(s) + HCl(aq).$$

The $E$ for the cell is given by equation (9.93) (with $n = 1$)

$$E = E^\circ - \frac{RT}{F}\ln\frac{a_{Ag}a_{HCl}}{a_{AgCl}a_{H_2}^{1/2}}. \tag{9.101}$$

With our choice of standard states

$$a_{Ag} = a_{AgCl} = 1$$

$$a_{H_2} = f_{H_2} = p_{H_2}$$

$$a_{HCl} = a_{H^+}a_{Cl^-} = \gamma_\pm^2 m^2$$

where $m$ is the molality of the acid. Substitution into equation (9.101) gives

$$E = E^\circ - \frac{RT}{F}\ln\frac{\gamma_\pm^2 m^2}{p_{H_2}^{1/2}}. \tag{9.102}$$

Equation (9.102) can be rearranged into the form

$$E + \frac{2RT}{F}\ln m - \frac{RT}{2F}\ln p_{H_2} = E^\circ - \frac{2RT}{F}\ln\gamma_\pm \tag{9.103}$$

or

$$\text{LHS} = E^\circ - \frac{2RT}{F}\ln\gamma_\pm \tag{9.104}$$

where[y]

$$\text{LHS} = E + \frac{2RT}{F}\ln m - \frac{RT}{2F}\ln p_{H_2}. \tag{9.105}$$

G. A. Linhart[7] made careful measurements of $E$ for this cell at a series of molalities, including dilute solutions. His work was reported in 1919 and is an example of excellent results reported many years ago by careful investigators. He took great pains to ensure reversible conditions in the cell. The (Ag, AgCl) electrode was made by mixing granular Ag, prepared by electrolysis, with solid AgCl prepared by adding an HCl solution to an $AgNO_3$ solution. This (Ag, AgCl) solid mixture was packed around a Pt wire that served to carry the electrical current from the cell.

Linhart reports that the most reversible ($H_2$, $H^+$) electrode was obtained by depositing, by electrolysis, finely divided Ir metal on a gold-plated Pt wire. Freshly generated $H_2$ gas was bubbled over this electrode to complete the half cell.

Measurements, made with a potentiometer to balance the cell voltage so that no current flowed, were continued until the readings were stable. It required times varying from 60 to 200 hours to achieve this stability.

Figure 9.5 is a graph of LHS from equation (9.105) obtained from Linhart's results, plotted against $m^{1/2}$. According to equation (9.104), an extrapolation to $m^{1/2} = 0$ gives LHS $= E^\circ$, since $\gamma_\pm$ would equal 1 and $\ln \gamma_\pm = 0$. The plot was made against $m^{1/2}$ instead of $m$ to take advantage of the Debye–Hückel limiting law in making the extrapolation. According to equation (7.46) obtained from the Debye–Hückel theory,

$$\ln \gamma_\pm = -C_\gamma m^{1/2}.$$

When the Debye–Hückel theory applies, equation (9.104) becomes

$$\text{LHS} = E^\circ + \frac{2RTC_\gamma}{F} m^{1/2}. \tag{9.106}$$

Equation (9.106) can be used to give the limiting slope of a graph of LHS against $m^{1/2}$, since it applies in the limit where $m^{1/2} = 0$. The dashed line shown in Figure 9.5 is drawn through the experimental results with the slope

[y] LHS is an abbreviation for the "left-hand side" of the equation.

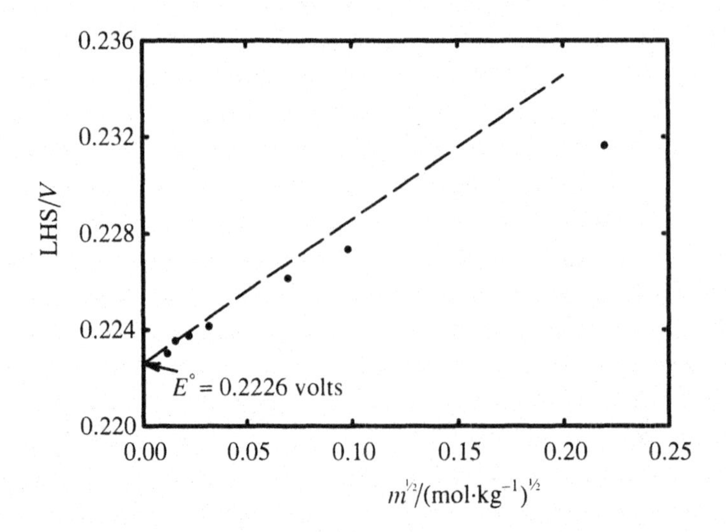


**Figure 9.5** Extrapolation of the cell emf data of G. A. Linhart [*J. Am. Chem. Soc.*, **41**, 1175–1180 (1919)] to obtain $E^\circ$ for the Ag/AgCl half-cell. The dashed line gives the limiting slope as predicted by Debye–Hückel theory. LHS = Left-hand side of equation; see text.

$2RTC_\gamma/F$. Extrapolation of the experimental results to $m^{1/2} = 0$ with the help of this line gives $E^\circ = 0.2226$ V.[z]

Once a value of $E^\circ$ is obtained by extrapolation, the $\gamma_\pm$ corresponding to each molality can be obtained from equation (9.104). Figure 9.6 is a graph of $\ln \gamma_\pm$ calculated from Linhart's results plotted against $m^{1/2}$. The Debye–Hückel limiting law value is shown as the dashed line. The agreement is excellent below $m^{1/2} = 0.10$ ($m = 0.03$), which attests to the reliability of Linhart's work and the validity of the Debye–Hückel limiting law.

$E^\circ$ values have been measured for many reactions and tabulated as standard half-cell potentials. Table 9.3 summarizes half-cell potentials as standard reduction potentials for a select set of reactions.[aa] In the tabulations, $E^\circ$ for

---

[z] Linhart's $E^\circ$ is based on a standard state for the $H_2$(g) of $p = 1$ atm (instead of $p = 1$ bar or $10^5$ Pa). Small corrections should be made to account for this change made in the standard state since 1912. Values of $E^\circ$ found in tables represent averages obtained from several investigators and may differ slightly from Linhart's value.

[aa] Table 9.3 is obtained from the extensive tabulation by W. M. Latimer, *The Oxidation States of the Elements and their potentials in Aqueous Solutions, Second Edition*, Prentice-Hall, Inc., Engelwood Cliffs, N.J. (1952). His tabulation was published at the time when $p = 1$ atm was the standard state for gases. His results should be corrected to the $p = 1$ bar standard state for comparison with other modern standard state thermodynamic data. The correction is given by $(RT/F) \ln (1.01325)$, which has a value of 0.0003 V at $T = 298.15$ K. The correction is small and probably negligible for all except the most precise work. ($E^\circ$ values measured to better than $10^{-3}$ V are unusual.)

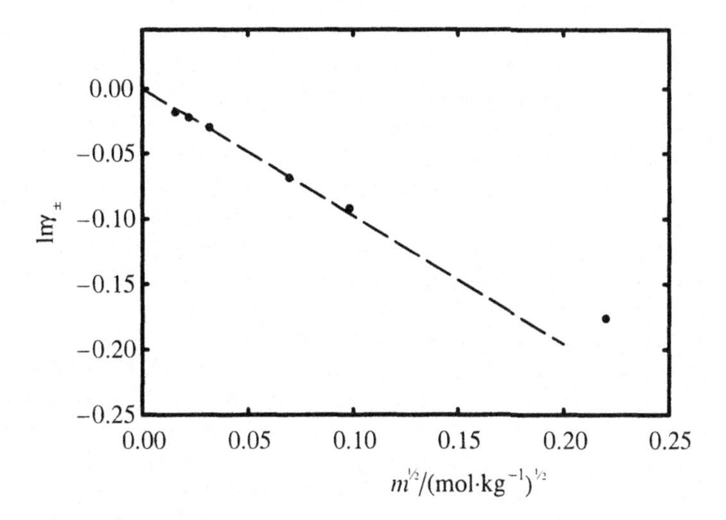


**Figure 9.6** Mean ionic activity coefficients for HCl(aq) at $T = 298.15$ K obtained from the emf results of G. A. Linhart, *J. Am. Chem. Soc.*, **41**, 1175–1180 (1919). The dashed line is the Debye–Hückel limiting law prediction.

the half-cell

$$\tfrac{1}{2}\,H_2(g) = H^+(aq) + e^- \qquad (9.107)$$

is set equal to zero. Thus, the tabulated $E^\circ$ are relative values that can be added together to give $E^\circ$ for a reaction. For example, $E^\circ$ values for the half-cells in the lead storage battery obtained from Table 9.3 are[bb]

$$Pb(s) + HSO_4^-(aq) = PbSO_4(s) + H^+(aq) + 2e^- \qquad E^\circ = 0.356\ V$$

$$PbO_2(s) + HSO_4^-(aq) + 3H^+(aq) + 2e^- = PbSO_4(s) + 2H_2O(l) \qquad E^\circ = 1.685\ V$$

$$PbO_2(s) + Pb(s) + 2H^+(aq) + 2HSO_4^-(aq) = 2PbSO_4(s) + 2H_2O(l) \qquad E^\circ = 2.041\ V.$$

[bb] $E^\circ$ values are intensive properties and care must be taken in adding them as we have done in our example. The correct procedure involves conversion to extensive $\Delta_r G^\circ$ values through the relationship $\Delta_r G^\circ = -nFE^\circ$, and adding the $\Delta_r G^\circ$. It is easy to show that $E^\circ$ values are additive when, as in our example, half-cells are added so that the electrons cancel. A more detailed discussion of calculations using $E^\circ$ values can be found in almost any general chemistry textbook.

**Table 9.3** Standard reduction potentials at 298.15 K*

| Half-reaction | $E^\circ/(V)$ | Half-reaction | $E^\circ/(V)$ |
|---|---|---|---|
| $Li^+(aq) + e^- = Li(s)$ | −3.045 | $H_2SO_3(aq) + 4H^+(aq) + 4e^- = S(s) + 3H_2O$ | 0.45 |
| $K^+(aq) + e^- = K(s)$ | −2.925 | $Cu^+(aq) + e^- = Cu(s)$ | 0.521 |
| $Ba^{2+}(aq) + 2e^- = Ba(s)$ | −2.90 | $I_2(s) + 2e^- = 2I^-(aq)$ | 0.535 |
| $Sr^{2+}(aq) + 2e^- = Sr(s)$ | −2.89 | $I_3^-(aq) + 2e^- = 3I^-(aq)$ | 0.536 |
| $Ca^{2+}(aq) + 2e^- = Ca(s)$ | −2.87 | $Cu^{2+}(aq) + Cl^-(aq) + e^- = CuCl(s)$ | 0.538 |
| $Na^+(aq) + e^- = Na(s)$ | −2.714 | $MnO_4^-(aq) + e^- = MnO_4^{2-}(aq)$ | 0.564 |
| $Mg^{2+}(aq) + 2e^- = Mg(s)$ | −2.37 | $Cu^{2+}(aq) + Br^-(aq) + e^- = CuBr(s)$ | 0.640 |
| $\frac{1}{2}H_2(g) + e^- = H^-(aq)$ | −2.25 | $O_2(g) + 2H^+(aq) + 2e^- = H_2O_2(aq)$ | 0.682 |
| $Be^{2+}(aq) + 2e^- = Be(s)$ | −1.85 | $Fe^{3+}(aq) + e^- = Fe^{2+}(aq)$ | 0.771 |
| $Al^{3+}(aq) + 3e^- = Al(s)$ | −1.66 | $Hg_2^{2+}(aq) + 2e^- = 2Hg(l)$ | 0.789 |
| $Mn^{2+}(aq) + 2e^- = Mn(s)$ | −1.18 | $Ag^+(aq) + e^- = Ag(s)$ | 0.799 |
| $V^{2+}(aq) + 2e^- = V(s)$ | −1.18 | $Cu^{2+}(aq) + I^-(aq) + e^- = CuI(s)$ | 0.86 |
| $TiO^{2+}(aq) + 2H^+(aq) + 4e^- = Ti(s) + H_2O$ | −0.89 | $2Hg^{2+}(aq) + 2e^- = Hg_2^{2+}(aq)$ | 0.920 |
| $H_3BO_3(aq) + 3H^+(aq) + 3e^- = B(s) + 3H_2O$ | −0.87 | $NO_3^-(aq) + 4H^+(aq) + 3e^- = NO(g) + 2H_2O$ | 0.96 |
| $SiO_2(s) + 4H^+(aq) + 4e^- = Si(s) + 2H_2O$ | −0.86 | $HNO_2(aq) + H^+(aq) + e^- = NO(g) + H_2O$ | 1.00 |
| $Zn^{2+}(aq) + 2e^- = Zn(s)$ | −0.763 | $Br_2(l) + 2e^- = 2Br^-$ | 1.065 |
| $Cr^{3+}(aq) + 3e^- = Cr(s)$ | −0.74 | $N_2O_4(g) + 2H^+(aq) + 2e^- = 2HNO_2(aq)$ | 1.07 |
| $H_3PO_2(aq) + H^+(aq) + e^- = P(s) + 2H_2O$ | −0.51 | $ClO_4^-(aq) + 2H^+(aq) + 2e^- = ClO_3^-(aq) + H_2O$ | 1.19 |
| $H_3PO_3(aq) + 2H^+(aq) + 2e^- = H_3PO_2(aq) + H_2O$ | −0.50 | $IO_3^-(aq) + 6H^+(aq) + 5e^- = \frac{1}{2}I_2(s) + 3H_2O$ | 1.195 |
| $Fe^{2+}(aq) + 2e^- = Fe(s)$ | −0.440 | $ClO_3^-(aq) + 3H^+(aq) + 2e^- = HClO_2(aq) + H_2O$ | 1.21 |
| $Cr^{3+}(aq) + e^- = Cr^{2+}C_r^{2+}(aq)$ | −0.41 | $O_2(g) + 4H^+(aq) + 4e^- = 2H_2O$ | 1.229 |
| $Cd^{2+}(aq) + 2e^- = Cd(s)$ | −0.403 | $MnO_2(s) + 4H^+(aq) + 2e^- = Mn^{2+}(aq) + 2H_2O$ | 1.23 |

| | $E^\circ/(V)$ | | $E^\circ/(V)$ |
|---|---|---|---|
| $PbI_2(s) + 2e^- = Pb(s) + 2I^-(aq)$ | −0.365 | $Cr_2O_7^{2-}(aq) + 14H^+(aq) + 6e^- = 2Cr^{3+}(aq) + 7H_2O$ | 1.33 |
| $PbSO_4(s) + H^+(aq) + 2e^- = Pb(s) + HSO_4^-(aq)$ | −0.356 | $Cl_2(g) + 2e^- = 2Cl^-(aq)$ | 1.3595 |
| $Co^{2+}(aq) + 2e^- = Co(s)$ | −0.277 | $HIO(aq) + H^+(aq) + e^- = \frac{1}{2}I_2(s) + H_2O$ | 1.45 |
| $H_3PO_4(aq) + 2H^+(aq) + 2e^- = H_3PO_3(aq) + H_2O$ | −0.276 | $PbO_2(s) + 4H^+(aq) + 2e^- = Pb^{2+}(aq) + 2H_2O$ | 1.455 |
| $Ni^{2+}(aq) + 2e^- = Ni(s)$ | −0.250 | $Au^{3+}(aq) + 3e^- = Au(s)$ | 1.50 |
| $CuI(s) + e^- = Cu(s) + I^-(aq)$ | −0.185 | $MnO_4^-(aq) + 8H^+(aq) + 5e^- = Mn^{2+}(aq) + 4H_2O$ | 1.51 |
| $AgI(s) + e^- = Ag(s) + I^-(aq)$ | −0.151 | $BrO_3^-(aq) + 6H^+(aq) + 5e^- = \frac{1}{2}Br_2(l) + H_2O$ | 1.52 |
| $Sn^{2+}(aq) + 2e^- = Sn(s)$ | −0.136 | $HBrO(aq) + H^+(aq) + e^- = \frac{1}{2}Br_2(l) + H_2O$ | 1.59 |
| $Pb^{2+}(aq) + 2e^- = Pb(s)$ | −0.126 | $H_5IO_6(aq) + H^+(aq) + 2e^- = IO_3^-(aq) + 3H_2O$ | 1.6 |
| $2H^+(aq) + 2e^- = H_2(g)$ | 0.000 | $Ce^{4+}(aq) + e^- = Ce^{3+}(aq)$ | 1.61 |
| $P(s) + 3H^+(aq) + 3e^- = PH_3(g)$ | 0.06 | $HClO(aq) + H^+(aq) + e^- = \frac{1}{2}Cl_2(g) + H_2O$ | 1.63 |
| $AgBr(s) + e^- = Ag(s) + Br^-(aq)$ | 0.095 | $HClO_2(aq) + 2H^+(aq) + 2e^- = HClO(aq) + H_2O$ | 1.64 |
| $CuCl(s) + e^- = Cu(s) + Cl^-(aq)$ | 0.137 | $Au^+(aq) + e^- = Au(s)$ | 1.68 |
| $S(s) + 2H^+(aq) + 2e^- = H_2S(g)$ | 0.141 | $NiO_2(s) + 4H^+(aq) + 2e^- = Ni^{2+}(aq) + 2H_2O$ | 1.68 |
| $Sn^{4+}(aq) + 2e^- = Sn^{2+}(aq)$ | 0.15 | $PbO_2(s) + HSO_4^-(aq) + 3H^+(aq) + 2e^- = PbSO_4(s) + 2H_2O$ | 1.685 |
| $Cu^{2+}(aq) + e^- = Cu^+(aq)$ | 0.153 | $MnO_4^-(aq) + 4H^+(aq) + 3e^- = MnO_2(s) + 2H_2O$ | 1.695 |
| $HSO_4^-(aq) + 3H^+(aq) + 2e^- = H_2SO_3(aq) + H_2O$ | 0.17 | $H_2O_2(aq) + 2H^+(aq) + 2e^- = 2H_2O$ | 1.77 |
| $AgCl(s) + e^- = Ag(s) + Cl^-(aq)$ | 0.222 | $Co^{3+}(aq) + e^- = Co^{2+}(aq)$ | 1.82 |
| $Cu^{2+}(aq) + 2e^- = Cu(s)$ | 0.337 | $O_3(g) + 2H^+(aq) + 2e^- = O_2(g) + H_2O$ | 2.07 |
| $Fe(CN)_6^{3-}(aq) + e^- = Fe(CN)_6^{4-}(aq)$ | 0.36 | $F_2(g) + 2e^- = 2F^-(aq)$ | 2.87 |
| | | $F_2(g) + 2H^+(aq) + 2e^- = 2HF(aq)$ | 3.06 |

The voltage of the lead storage cell is then given by

$$E = 2.041 + \frac{RT}{2F} \ln \frac{a^2_{H^+} a^2_{HSO_4^-}}{a^2_{H_2O}}.$$

At $T = 298.15$ K, this equation becomes

$$E = 2.041 + 0.0257 \ln \frac{(\gamma_\pm m)^2}{\gamma_{H_2O} x_{H_2O}}. \tag{9.108}$$

Table 9.3 can be used to calculate the point of equilibrium in an oxidation–reduction reaction.

**Example 9.5:** Use $E°$ values in Table 9.3 to show that $Cu^+$(aq) is unstable in solution at $T = 298.15$ K, and will disproportionate to Cu(s) and $Cu^{2+}$(aq).

**Solution:** The disproportionation reaction is obtained by adding the following half-cells whose $E°$ values are given in Table 9.3.

$$Cu^+(aq) + e^- = Cu(s) \qquad E° = 0.521 \text{ V}$$

$$Cu^+(aq) = Cu^{2+}(aq) + e^- \qquad E° = -0.153 \text{ V}$$

$$2Cu^+(aq) = Cu^{2+}(aq) + Cu(s) \qquad E° = 0.368 \text{ V}$$

From equation (9.98)

$$\ln K = \frac{nFE°}{RT}$$

$$= \frac{(1)(96485)(0.368)}{(8.314)(298.15)} = 14.3.$$

$$K = \frac{a_{Cu^{2+}}}{a^2_{Cu^+}} = 1.62 \times 10^6.$$

The large $K$ indicates that the point of equilibrium in the reaction is far toward products, and disproportionation will occur until $a_{Cu^+}$ becomes very small.

**Measurement of Equilibrium Constants:** Electrochemical cells can be used to measure equilibrium constants for chemical reactions. For example, consider the cell

$$\mathrm{Pt,H_2(g)\,|\,MOH}(m_1),\ \mathrm{MCl}(m_2)\,|\,\mathrm{AgCl(s),\ Ag(s)}.$$

This is a cell with the same electrodes shown in Figure 9.4, but with the HCl(aq) replaced by a mixture of MOH(aq), a metal hydroxide, and MCl(aq), a metal chloride electrolyte.

The half-cell reactions in the cell can be written as before

$$\begin{aligned} \tfrac{1}{2}\mathrm{H_2(g)} &= \mathrm{H^+(aq)} + \mathrm{e^-} \\ \mathrm{AgCl(s)} + \mathrm{e^-} &= \mathrm{Ag(s)} + \mathrm{Cl^-(aq)} \\ \hline \mathrm{AgCl(s)} + \tfrac{1}{2}\mathrm{H_2(g)} &= \mathrm{Ag(s)} + \mathrm{H^+(aq)} + \mathrm{Cl^-(aq)} \end{aligned}$$

With our choice of standard states, the emf is given by

$$E = E^\circ - \frac{RT}{F}\ln\frac{a_{\mathrm{H^+}}a_{\mathrm{Cl^-}}}{p_{\mathrm{H_2}}^{1/2}} \tag{9.109}$$

where $E^\circ$ is the same value determined earlier.

We can substitute for $a_{\mathrm{H^+}}$ in equation (9.109) from the relationship

$$a_{\mathrm{H^+}} = \frac{K_{\mathrm{w}}}{a_{\mathrm{OH^-}}},$$

where $K_{\mathrm{w}}$ is the equilibrium constant for the dissociation of water. The result is

$$E = E^\circ - \frac{RT}{F}\ln K_{\mathrm{w}} - \frac{RT}{F}\ln\frac{a_{\mathrm{Cl^-}}}{a_{\mathrm{OH^-}}} + \frac{RT}{2F}\ln p_{\mathrm{H_2}}. \tag{9.110}$$

Substituting $a_{\mathrm{Cl^-}} = \gamma_{\mathrm{Cl^-}}m_2$ and $a_{\mathrm{OH^-}} = \gamma_{\mathrm{OH^-}}m_1$ into equation (9.110) and rearranging gives

$$E - E^\circ + \frac{RT}{F}\ln\frac{m_2}{m_1} - \frac{RT}{2F}\ln p_{\mathrm{H_2}} = -\frac{RT}{F}\ln K_{\mathrm{w}} - \frac{RT}{F}\ln\frac{\gamma_{\mathrm{Cl^-}}}{\gamma_{\mathrm{OH^-}}} \tag{9.111}$$

or

$$\text{LHS} = -\frac{RT}{F}\ln K_\text{w} - \frac{RT}{F}\ln\frac{\gamma_{\text{Cl}^-}}{\gamma_{\text{OH}^-}}. \tag{9.112}$$

The extrapolation of a graph of LHS against $\text{I}_\text{m}^{1/2}$ (with the aid of the Debye–Hückel limiting law) to $\text{I}_\text{m}^{1/2} = 0$ gives an intercept with a value $-(RT/F)\ln K_\text{w}$.

As a final example, consider the cell

$$\text{Pt, H}_2\text{(g)} \mid \text{HA}(m_1),\ \text{NaA}(m_2),\ \text{NaCl}(m_3) \mid \text{AgCl(s), Ag(s)}$$

where HA is a weak acid with a dissociation constant given by

$$K_\text{a} = \frac{a_{\text{H}^+}a_{\text{A}^-}}{a_{\text{HA}}} \tag{9.113}$$

and NaA is the salt of this weak acid. The cell reaction is the same as in the previous example so that

$$E = E^\circ - \frac{RT}{F}\ln\frac{a_{\text{H}^+}a_{\text{Cl}^-}}{p_{\text{H}_2}^{1/2}}.$$

Substituting

$$a_{\text{HA}} = \gamma_{\text{HA}}m_1$$

$$a_{\text{A}^-} = \gamma_{\text{A}^-}m_2$$

$$a_{\text{Cl}^-} = \gamma_{\text{Cl}^-}m_3,$$

and $a_{\text{H}^+}$ from equation (9.113) gives (after rearranging),

$$E - E^\circ + \frac{RT}{F}\ln\frac{m_1m_3}{m_2} - \frac{RT}{2F}\ln p_{\text{H}_2} = -\frac{RT}{F}\ln K_\text{a} - \frac{RT}{F}\ln\left(\frac{\gamma_{\text{HA}}\gamma_{\text{Cl}^-}}{\gamma_{\text{A}^-}}\right) \tag{9.114}$$

or

$$\text{LHS} = -\frac{RT}{F}\ln K_\text{a} - \frac{RT}{F}\ln\left(\frac{\gamma_{\text{HA}}\gamma_{\text{Cl}^-}}{\gamma_{\text{A}^-}}\right). \tag{9.115}$$

Since a Henry's law standard state has been chosen for all three solutes, a graph of LHS against $I_m^{1/2}$ extrapolated to $I_m^{1/2} = 0$ gives $-(RT/F)\ln K_a$ as the intercept.

This last example provides a demonstration of the flexibility inherent in the choice of standard states. A strong electrolyte standard state is chosen for NaA(aq) and NaCl(aq) so that

$$\frac{a_{Na^+} a_{A^-}}{a_{NaA}} = 1$$

$$\frac{a_{Na^+} a_{Cl^-}}{a_{NaCl}} = 1$$

and $\Delta_r G^\circ = 0$ for the reactions

$$\mathrm{NaA(aq)} = \mathrm{Na^+(aq)} + \mathrm{A^-(aq)}$$

and

$$\mathrm{NaCl(aq)} = \mathrm{Na^+(aq)} + \mathrm{Cl^-(aq)}.$$

Tables giving values for $\Delta_f G^\circ$ for $Na^+$(aq) and $A^-$(aq) reflect the fact that $\Delta_f G^\circ(\mathrm{NaA, aq}) = \Delta_f G^\circ(\mathrm{A^-}) + \Delta_f G^\circ(\mathrm{Na^+})$.

However, for HA, we choose a weak electrolyte standard state so that

$$K_a = \frac{a_{H^+} a_{A^-}}{a_{HA}} \neq 1$$

and

$$\Delta_r G^\circ = -RT \ln K_a. \tag{9.116}$$

The determination of $K_a$ requires a measurement using a technique such as electrical conductivity, absorption of light, or, in our case, the emf of an electrochemical cell. When $K_a$ is determined, $\Delta_r G^\circ$ is obtained from equation (9.116), and tables giving $\Delta_f G^\circ$ must take into account this value. For example, $K_a = 1.75 \times 10^{-5}$ at 298.15 K and $\Delta_r G^\circ = 27.15\ \mathrm{kJ \cdot mol^{-1}}$ for the reaction[cc]

$$\mathrm{HOAc(aq)} = \mathrm{H^+(aq)} + \mathrm{OAc^-(aq)}.$$

[cc] We use HOAc to represent acetic acid and $OAc^-$ to represent the acetate ion.

Since $\Delta_f G^\circ(H^+, aq) = 0$ and $\Delta_f G^\circ(OAc^-, aq) = -369.31\ kJ{\cdot}mol^{-1}$, then $\Delta_f G^\circ(HOAc, aq) = -396.46\ kJ{\cdot}mol^{-1}$.

We should keep in mind that a strong electrolyte standard state could have been chosen for HOAc, in which case,

$$\begin{aligned}\Delta_f G^\circ(HOAc, aq) &= \Delta_f G^\circ(H^+, aq) + \Delta_f G^\circ(OAc^-, aq)\\ &= 0 - 369.31\ kJ{\cdot}mol^{-1}\\ &= -369.31\ kJ{\cdot}mol^{-1},\end{aligned}$$

and $a_{HOAc} = a_{H^+} a_{OAc^-} = \gamma_\pm^2 m^2$. With this (strong electrolyte) choice of standard state, we would find that $\gamma_\pm$ is one in the limit of infinite dilution (where even weak acids are completely dissociated), but $\gamma_\pm$ would start deviating from one at much lower concentrations than would be the case for an acid that we have designated as being strong, such as HCl(aq).

In assembling cells for making thermodynamic measurements, one should try not to combine half-cells in a manner that results in a **junction potential**. Figure 9.7 is a schematic representation of the Daniell cell, which is one with a junction potential. The half-cell reactions are

$$\begin{array}{ll}\text{Cathode:} & Cu^{2+}(aq) + 2e^- = Cu(s)\\ \text{Anode:} & \qquad Zn(s) = Zn^{2+}(aq) + 2e^-\\ \hline & Cu^{2+}(aq) + Zn(s) = Cu(s) + Zn^{2+}(aq).\end{array}$$

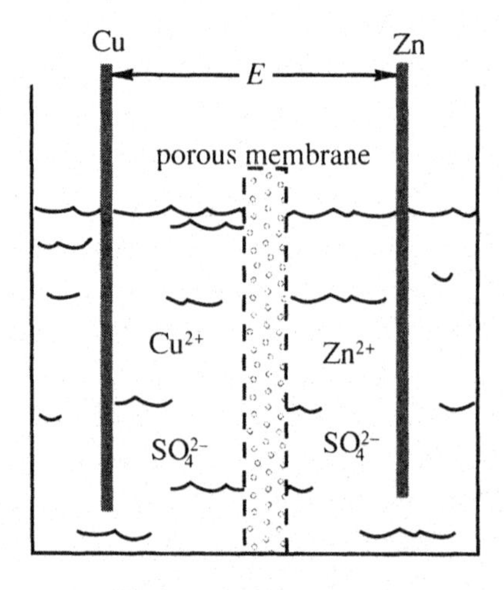


**Figure 9.7** The Daniell cell, an example of a cell with a junction potential.

As the cell is discharged, $Zn^{2+}$ ions are produced at the anode while $Cu^{2+}$ ions are used up at the cathode. To maintain electrical neutrality, $SO_4^{2-}$ ions must migrate through the porous membrane,[dd] which serves to keep the two solutions from mixing. The result of this migration is a potential difference across the membrane. This junction potential works in opposition to the cell voltage $E$ and affects the value obtained. Junction potentials are usually small, and in some cases, corrections can be made to $E$ if the transference numbers of the ions are known as a function of concentration.[ee] It is difficult to accurately make these corrections, and, if possible, cells with transference should be avoided when using cell measurements to obtain thermodynamic data.

# Exercises

E9.1 Given the following thermodynamic data at 298.15 K

| Substance | $\Delta_f G°/(kJ{\cdot}mol^{-1})$ | $\Delta_f H°/(kJ{\cdot}mol^{-1})$ |
|---|---|---|
| $Br_2(l)$ | 0 | 0 |
| $Br_2(g)$ | 3.110 | 30.907 |
| $Br(g)$ | 82.396 | 111.884 |

(a) Calculate the number of Br atoms in 1.00 $dm^3$ of gaseous $Br_2$ at 298.15 K and a total pressure of 100 kPa. The equilibrium reaction is

$$Br_2(g) = 2Br(g).$$

(b) Calculate the vapor pressure of $Br_2(l)$ at 298.15 K and a total pressure of 0.101 MPa.

E9.2 The enthalpies of combustion of quinone(s) and hydroquinone(s) at $p = 100$ kPa and $T = 298.15$ K are 2745.9 $kJ{\cdot}mol^{-1}$ and 2852.4 $kJ{\cdot}mol^{-1}$, respectively.[8] Entropies at $T = 298.15$ K computed from heat capacity data and the Third Law give $S_m^{\circ} = 161.3\ J{\cdot}K^{-1}{\cdot}mol^{-1}$ for quinone(s) and 137.1 $J{\cdot}K^{-1}{\cdot}mol^{-1}$ for hydroquinone(s).

[dd] The porous membrane is sometimes replaced with a salt bridge. The reader is referred to a text on electrochemical cells for more extended discussion.

[ee] For a discussion of the calculation of corrections for junction potential, see G. N. Lewis, M. Randall, K. S. Pitzer and L. Brewer, *Thermodynamics*, Second Edition, McGraw-Hill, New York, 1961, pp. 362–364.

Use enthalpy and entropy data in Table 9.1 to

(a) Compute $\Delta_f H^\circ_m$, the standard enthalpy of formation at $T = 298.15$ K, for quinone, and for hydroquinone.

(b) Compute $\Delta_r H^\circ$ for the reduction of quinone to hydroquinone at $T = 298.15$ K. The reaction is

$$C_6H_4O_2 \text{ (quinone)} + H_2(g) = C_6H_4(OH)_2 \text{ (hydroquinone)}$$

(c) Compute $\Delta_r S^\circ$ and $\Delta_r G^\circ$ for this reduction at $T = 298.15$ K.

E9.3 The deamination of aspartic acid is a reversible reaction catalyzed by the enzyme aspartase. The equilibrium constant for the reaction over a range of temperatures[9] can be expressed by the equation

$$\ln K = 18.854 - \frac{5331.6}{T} - 0.02360T$$

(a) Derive an expression for $\Delta_r H^\circ$ as a function of temperature.

(b) Calculate $\Delta_r G^\circ$, $\Delta_r H^\circ$, $\Delta_r S^\circ$, and $\Delta_r C^\circ_p$ at $T = 298.15$ K.

E9.4 (a) Use the data in Table 9.1 to calculate $\Delta_r H^\circ$, $\Delta_r S^\circ$, and $\Delta_r G^\circ$ at $T = 298.15$ K for the reaction

$$2Hg(g) + O_2(g) = 2HgO(s).$$

(b) The heat capacities in the temperature range from 298.15 to 600 K can be represented by

$$C^\circ_{p,m}(Hg, g)/(J{\cdot}K^{-1}{\cdot}mol^{-1}) = 20.79$$

$$C^\circ_{p,m}(O_2, g)/(J{\cdot}K^{-1}{\cdot}mol^{-1}) = 26.40 + 0.0096T$$

$$C^\circ_{p,m}(HgO, s)/J{\cdot}K^{-1}{\cdot}mol^{-1}) = 26.11 + 0.067T.$$

Assume that the gases are ideal and find the equilibrium constant for the reaction at 600 K.

(c) A closed vessel of fixed volume was kept at 600 K. It initially contained $O_2$ at a pressure of 100 kPa. A sample of HgO(s) was then introduced into the container without changing the volume available to the gases. At equilibrium, some HgO(s) was still present. Find the pressure of Hg(g) at equilibrium.

(d) The vapor pressure of Hg(l) at 600 K is 53.3 kPa. Find the equilibrium constant at 600 K for the reaction

$$2Hg(l) + O_2(g) = 2HgO(s).$$

E9.5 The variation with temperature of the emf of the cell

$$H_2(g, p = 100 \text{ kPa}) \mid HBr(a_2 = 1) \mid Hg_2Br_2(s), Hg(l)$$

is given by the expression[10]

$$E/(V) = 0.13970 - 8.1 \times 10^{-5}(T - 298.15)$$
$$- 3.6 \times 10^{-6}(T - 298.15)^2$$

where $T$ is the Kelvin temperature.

(a) Write the cell reaction and calculate $\Delta_r G^\circ$, $\Delta_r H^\circ$, and $\Delta_r S^\circ$ at 298.15 K.
(Remember that oxidation occurs at the left electrode while reduction occurs at the right electrode.)

(b) Given the following standard entropies at 298.15 K

| Substance | $S^\circ_m/(J{\cdot}K^{-1}{\cdot}mol^{-1})$ |
|---|---|
| $H_2(g)$ | 130.680 |
| $Hg(l)$ | 76.02 |
| $Hg_2Br_2(s)$ | 218.752 |

Calculate the entropy of $Br^-(aq)$.

## Problems

P9.1 Calculate the equilibrium pressure of $CO_2(g)$ resulting from the reaction

$$MgCO_3(s) = MgO(s) + CO_2(g)$$

when the system is maintained at 700 K and a total pressure of 100 MPa. The following information may be useful:

$$\Delta_r H^\circ(298 \text{ K}) = 116.92 \text{ kJ}$$

$$\Delta_r S^\circ(298 \text{ K}) = 174.85 \text{ J}{\cdot}\text{K}^{-1}$$

$$\Delta_r V^\circ_s(298 \text{ K}) = -16.77 \text{ cm}^3,$$

where $\Delta_r V^\circ_s(298 \text{ K}) = V^\circ_m(MgO) - V^\circ_m(MgCO_3)$.

The fugacity coefficient for $CO_2$(g) at this temperature and total pressure is 1.188 and the heat capacity change for the reaction is given by

$$\Delta_r C_p^\circ/(\text{J}\cdot\text{K}^{-1}) = 71.912 - 0.05616T - \frac{1386.1}{T^{1/2}} + \frac{2.0766 \times 10^6}{T^2}.$$

P9.2 The free energy change for the reaction between native and denatured chymotrypsinogen has been measured as a function of temperature and pressure. The reaction can be described as:

N (native) = D (denatured).

The following values[11] were found for the given thermodynamic functions at $T = 273.15$ K and $p = 0.101$ MPa.

$$\Delta_r G_m^\circ = 10{,}590\ \text{J}\cdot\text{mol}^{-1}$$

$$\Delta_r S_m^\circ = -950\ \text{J}\cdot\text{mol}^{-1}$$

$$\Delta_r V_m^\circ = -14.3\ \text{cm}^3\cdot\text{mol}^{-1}$$

$$\left(\frac{\partial \Delta_r V_m}{\partial T}\right)_p = 1.32\ \text{cm}^3\cdot\text{K}^{-1}\cdot\text{mol}^{-1}$$

$$\left(\frac{\partial \Delta_r V_m}{\partial p}\right)_T = -5.19\ \text{cm}^6\cdot\text{J}\cdot\text{mol}^{-1}$$

$$\Delta_r C_{p,\text{m}} = 15{,}900\ \text{J}\cdot\text{K}^{-1}\cdot\text{mol}^{-1}.$$

Assume that

$$\Delta_r C_{p,\text{m}},\ \left(\frac{\partial \Delta_r V_m}{\partial T}\right)_p,\ \text{and}\ \left(\frac{\partial \Delta_r V_m}{\partial p}\right)_T$$

are constant, and calculate $\Delta_r G$ for the reaction at $T = 308.15$ K and $p = 300$ MPa.

P9.3 The potentials (emfs) of lithium amalgam electrodes in the cell

$$\text{Li(amalgam, } x_2) \mid \text{LiCl in acetonitrile} \mid \text{Li(amalgam, } x_2' = 0.0003239)$$

at 298.15 K are as follows.

| $x_2$ | $E/(\text{V})$ | $x_2$ | $E/(\text{V})$ |
|---|---|---|---|
| 0.0003239 | 0.0000 | 0.008779 | 0.0894 |
| 0.001846 | 0.0458 | 0.01300 | 0.1006 |
| 0.002345 | 0.0517 | 0.02265 | 0.1189 |
| 0.004218 | 0.0684 | | |

(a) Plot an appropriate function of $E$ vs. $x_2$ and get the activity of Li in the 0.0003239 mole fraction amalgam.
(b) Calculate the activity coefficients of the other amalgams given in the table above and make a graph of $\gamma$ against $x_2$.

P9.4 The emf $E$ at $T = 298.15$ K of the cell

$$\text{Zn(s)} \mid \text{ZnCl}_2(m) \mid \text{AgCl(s), Ag(s)}$$

at various molalities $m$ of $ZnCl_2$ were found to be as follows[12]

| $m$ | $E/(\text{V})$ | $m/(\text{mol·kg}^{-1})$ | $E(\text{V})$ |
|---|---|---|---|
| 0.002941 | 1.1983 | 0.04242 | 1.10897 |
| 0.007814 | 1.16502 | 0.09048 | 1.08435 |
| 0.01236 | 1.14951 | 0.2211 | 1.05559 |
| 0.02144 | 1.13101 | 0.4499 | 1.03279 |

(a) Use this information and values given in this chapter to determine the standard potential $E^\circ$ of the $Zn \mid Zn^{2+}$ electrode. (Remember that the standard half-cell potential is the $E^\circ$ for the half-cell combined with the $H_2/H^+$ half cell.)
(b) Calculate $\ln \gamma_\pm$ at each of the molalities given in the table and make a graph of $\ln \gamma_\pm$ against $m^{1/2}$. Include on the graph a line indicating the prediction from the extended form of the Debye–Hückel equation, and decide the molality range over which this equation gives a reasonable approximation.

## References

1. F. D. Rossini, *Chemical Thermodynamics*, John Wiley and Sons, Inc., New York, 1950, p. 352.
2. M. Bastos, S. Nilsson, M. D. M. C. Ribeiro Da Silva, M. A. V. Ribeiro Da Silva, and I. Wadsö, "Thermodynamic Properties of Glycerol. Enthalpies of Combustion and Vaporization and the Heat Capacity at 298.15 K. Enthalpies of Solution in Water at 288.15, 298.15, and 308.15 K", *J. Chem. Thermodyn.*, **20**, 1353–1359 (1988).
3. W. T. Murray and P. A. G. O'Hare, "Thermochemistry of Inorganic Sulfur Compounds II. Standard Enthalpy of Formation of Germanium Disulfide", *J. Chem. Thermodyn.*, **16**, 335–341 (1988).
4. P. A. G. O'Hare, "Thermochemistry of Uranium Compounds XV. Calorimetric Measurements on $UCl_4$, $UO_2Cl_2$, and $UO_2F_2$, and the Standard Molar Enthalpy of Formation at 298.15 K of $UCl_4$", *J. Chem. Thermodyn.*, **17**, 611–622 (1985).
5. See for example, M. W. Chase Jr., C. A. Davies, J. R. Downey Jr., D. J. Frurip, R. A. McDonald, and A. N. Syverud, "JANAF Thermochemical Tables, Third Edition," *J. Phys. Chem. Ref. Data.*, **14**, (1995), Supplement No. 1. Also, see D. D. Wagman, W. H. Evans, V. B. Parker, R. H. Schumm, I. Halow, S. M. Bailey, K. L. Churney, and R. L. Nuttall, "The NBS Tables of Chemical Thermodynamic Properties. Selected Values for Inorganic and $C_1$ and $C_2$ Organic Substances in SI Units", *J. Phys. Chem. Ref. Data*, **11**, (1982), Supplement No. 2.
6. J. D. Hale, R. M. Izatt, and J. J. Christensen, "A Calorimetric Study of the Heat of Ionization of Water at 25°", *J. Phys. Chem.*, **67**, 2605–2608 (1963).
7. G. A. Linhart, "The Applicability of the Precipitated Silver–Silver Chloride Electrode to the Measurement of the Activity of Hydrochloric Acid in Extremely Dilute Solutions", *J. Am. Chem. Soc.*, **41**, 1175–1180 (1919).
8. G. Pilcher and L. E. Sutton, "The Heats of Combustion of Quinol and *p*-Benzoquinone and the Thermodynamic Quantities of Oxidation–Reduction Reaction", *J. Chem. Soc.*, **1956**, 2695–2700 (1956).
9. J. L. Bada and S. L. Miller, "Equilibrium Constant for the Reversible Deamination of Aspartic Acid", *Biochemistry* **7**, 3403–3408 (1968).
10. G. Scatchard and R. F. Tefft, "Electromotive Force Measurements on Cells Containing Zinc Chloride. The Activity Coefficients of the Chlorides of the Bivalent Metals", *J. Am. Chem. Soc.*, **52**, 2272–2281 (1930).
11. S. A. Hawley, "Reversible Pressure–Temperature Denaturation of Chymotrypsinogen", *Biochemistry*, **10**, 2436–2442 (1971).
12. G. Scatchard and R. F. Tefft, "Electromotive Force Measurements on Cells Containing Zinc Chloride. The Activity Coefficients of the Chlorides of the Bivalent Metals", *J. Am. Chem. Soc.*, **52**, 2272–2281 (1930).

# Chapter 10

# Statistical Thermodynamics

By now we should be convinced that thermodynamics is a science of immense power. But it also has serious limitations. Our fifty million equations predict what — but they tell us nothing about why or how. For example, we can predict for water, the change in melting temperature with pressure, and the change of vapor fugacity with temperature; or determine the point of equilibrium in a chemical reaction; but we cannot use thermodynamic arguments to understand why we end up at a particular equilibrium condition.

The reason is that classical thermodynamics tells us nothing about the atomic or molecular state of a system. We use thermodynamic results to infer molecular properties, but the evidence is circumstantial. For example, we can infer why a (hydrocarbon + alkanol) mixture shows large positive deviations from ideal solution behavior, in terms of the breaking of hydrogen bonds during mixing, but our description cannot be backed up by thermodynamic equations that involve molecular parameters.

Statistical thermodynamics provides the relationships that we need in order to bridge this gap between the macro and the micro. Our most important application will involve the calculation of the thermodynamic properties of the ideal gas, but we will also apply the techniques to solids. The procedure will involve calculating $U - U_0$, the internal energy above zero Kelvin, from the energy of the individual molecules. Enthalpy differences and heat capacities are then easily calculated from the internal energy. Boltzmann's equation

$$S = \mathrm{k} \ln W$$

provides the link for calculating entropy from the molecular properties. Combinations of $S$ and $U$ (or $H$) enable us to calculate $A$ (or $G$), and all of the thermodynamic properties are obtained.

## 10.1 Energy Levels Of An Ideal Gas Molecule

For the ideal gas, we know that attractions and repulsions between molecules are zero and hence the potential energy is zero. We express this as

$$\left(\frac{\partial U}{\partial V}\right)_T = \left(\frac{\partial U}{\partial p}\right)_T = 0.$$

When energy is added to the ideal gas as heat or work, the energy is added to the molecules of the gas as internal energy, and a change $\mathrm{d}U$ in internal energy results such that

$$\mathrm{d}U = \delta q + \delta w.$$

This energy increase can take different forms. It can be added as translational kinetic energy to speed up the movement "to and fro" of the molecules; it can be added to the rotations of the molecules to get them to spin faster; it can be added to increase the amplitude of the vibrational oscillations of the molecules; and it can be added to excite electrons to higher energy states in the atoms or molecules. Other forms of internal energy are also possible, but the above are the most common.

When energy is added to a gas, it is distributed among these different types of energy levels. The **Principle of Equipartition of Energy** comes into play in helping to understand how the energy is divided. This principle, based on classical mechanics, states that energy is distributed or partitioned equally among the different forms of internal energy. However, this principle is not followed, except at very high temperatures. (That quantum mechanics could explain this failure was one of its first successes.) Energy can only be added in amounts equal to differences between distinct energy levels in the atom or molecule, and energy differences between different kinds of energy levels vary greatly. Therefore, the addition of energy into the different types of energy levels depends upon the form of the energy used in the excitation. The addition of small units of energy cannot be used to excite energy modes where the differences between levels are large. At high temperatures, the amount of thermal energy available to excite the different modes is large, so that the energy can be partitioned among all modes, and the classical equipartition theorem is found to be a good approximation.

Quantum mechanics gives us a description of the kinds of energy levels found in atoms and molecules. We will spend some time describing the different types of energy levels so that we can calculate the distribution of energy among them. We note, now, that of the types of energy levels we will discuss, atoms possess only translational and electronic energies, while molecules have translational, rotational, vibrational and electronic energy modes.

**Translational Energy Levels:** To predict the nature of translational energy levels, one assumes that the molecules are confined to a rectangular box with sides

$L_x$, $L_y$, and $L_z$, so that the volume of the box is $L_xL_yL_z$. The potential energy of the "particle" inside the box is zero but goes to infinity at the walls. The quantum mechanical solution in the $x$ direction to this "particle in a box" problem gives

$$\epsilon_{\text{trans},\,x} = \frac{n_x^2 h^2}{8mL_x^2} \tag{10.1}$$

where $h$ is Planck's constant, $m$ is the mass of the particle, and $n_x$ is a quantum number with values

$$n_x = 1, 2, 3, \ldots$$

Similar expressions can be written for $\epsilon_{\text{trans},\,y}$ and $\epsilon_{\text{trans},\,z}$, the energies in the $y$ and $z$ directions. The total translational energy is the sum of the contributions in the three directions.

$$\epsilon_{\text{trans}} = \epsilon_{\text{trans},\,x} + \epsilon_{\text{trans},\,y} + \epsilon_{\text{trans},\,z}.$$

In a cubic box with dimensions $L_x = L_y = L_z = L$, the total translational energy becomes

$$\epsilon_{\text{trans}} = \frac{h^2}{8mL^2}\,(n_x^2 + n_y^2 + n_z^2)$$

or

$$\epsilon_{\text{trans}} = \frac{h^2}{8mV^{2/3}}\,(n_x^2 + n_y^2 + n_z^2) \tag{10.2}$$

where $V$ is the volume of the container. Calculations using equation (10.2) show that for a gas occupying a box with a volume on the order of cubic meters, the translational energy levels are very closely spaced. That is, the energy differences between successive levels are extremely small.

**Rotational Energy Levels:** The rotational energy of a molecule depends upon the molecular geometry. For a linear molecule that behaves as a rigid rotator,[a]

---

[a] A rigid rotator is one for which the bond length, and, hence, the moment of inertia, does not change with the rotational energy level. Later in the chapter, we will see how to correct rotational energy levels for deviations from the rigid rotator approximation.

quantum mechanics predicts that the energy levels would be given by

$$\epsilon_{\text{rot}} = \frac{h^2}{8\pi^2 I} J(J+1) \tag{10.3}$$

where $h$ is Planck's constant, $I$ is the moment of inertia along an axis perpendicular to the axis of the molecule and through the center of mass, and

$$J = 0, 1, 2, ... \tag{10.4}$$

is the quantum number that designates the energy level. The rotational energy levels have a degeneracy $g_J$ given by

$$g_J = 2J + 1. \tag{10.5}$$

Thus, only one level is present for $J = 0$, but three levels with the same energy occur when $J = 1$, five at $J = 2$, and so on. The degenerate levels are distinguished by a quantum number $K$ given by

$$K = 0, \pm 1, \pm 2, ... \pm J. \tag{10.6}$$

Thus, for $J = 1$, the three equivalent levels have $K = 0$, $+1$ and $-1$.[b]

For a nonlinear molecule the rotational energy levels are a function of three **principal moments of inertia** $I_A$, $I_B$ and $I_C$. These are moments of inertia around three mutually orthogonal axes that have their origin (or intersection) at the center of mass of the molecule. They are oriented so that the products of inertia are zero. The relationship between the three moments of inertia, and hence the energy levels, depends upon the geometry of the molecules.

In a **spherical top**, $I_A = I_B = I_C$. Examples are $CH_4$ and $SF_6$. For a spherical top, the energy levels are given by the same equation as for the linear molecule [equations (10.3) and (10.4)] and the degeneracy is given by equation (10.5).

In a **symmetrical top**, $I_A \neq I_B = I_C$. Examples are $NH_3$ and $CHCl_3$. The energy levels for a symmetrical top are given by

$$\epsilon_r = \frac{h^2}{8\pi^2 I_B} J(J+1) + \frac{h^2}{8\pi^2}\left(\frac{1}{I_A} - \frac{1}{I_B}\right) K^2 \tag{10.7}$$

[b] The energy of a rigid rotator depends only upon $J$, but other properties such as the angular momentum, depend also upon $K$. Thus, the degenerate energy levels are not identical.

where

$$J = 0, 1, 2, \ldots \tag{10.8}$$

$$K = 0, \pm 1, \pm 2, \ldots \pm J. \tag{10.9}$$

Symmetrical tops are of two types. A **prolate spheroid** (football shape) in which $I_A < I_B$, and an **oblate spheroid** (pancake shape) in which $I_A > I_B$. Again, there are $2J + 1$ sets of energy levels for each $J$, but they are no longer degenerate, as can be seen from equation (10.7), which includes both the $J$ and $K$ quantum number.

Finally, an **asymmetric top** is one in which all three principal moments of inertia are different. The energy levels are given by

$$\epsilon_r = J(J+1)\frac{(a+c)}{2} + \frac{(a-c)}{2}\epsilon'(\kappa) \tag{10.10}$$

where

$$a = \frac{h^2}{8\pi^2 I_A}$$

$$b = \frac{h^2}{8\pi^2 I_B}$$

$$c = \frac{h^2}{8\pi^2 I_C}$$

with $I_A < I_B < I_C$, and $\kappa$ is the parameter of asymmetry defined as

$$\kappa = \frac{2b - a - c}{a - c}.$$

The quantity $\epsilon'(\kappa)$ is a separate function of $\kappa$ for each energy level. Tabulations can be found in the literature.[c] Again, $2J + 1$ energy levels are present for each value of $J$.

---

[c] See for example, C. H. Townes and A. L. Schawlow, *Microwave Spectroscopy*, McGraw-Hill Book Company, Inc., New York, 1955, pp. 83–110, T. M. Sugden and C. N. Kenney, *Microwave Spectroscopy of Gases*, D. Van Nostrand Company Ltd., London, 1965, pp. 46–64.

Unless the moment of inertia is small, the energy difference between rotational energy levels is not large, although not as small as the difference between translational energy levels.

**Vibrational Energy Levels:** A diatomic molecule has a single set of vibrational energy levels resulting from the vibration of the two atoms around the center of mass of the molecule. A vibrating molecule is usually approximated by a harmonic oscillator[d] for which

$$\epsilon_{\mathrm{vib}} = h\nu(\mathrm{v} + 1/2) \qquad (10.11)$$

where $\nu$, the **fundamental vibrational frequency**, is related to the force constant, $k$, and the reduced mass, $\mu$, by

$$\nu = \frac{1}{2\pi}\sqrt{\frac{k}{\mu}}. \qquad (10.12)$$

The vibrational quantum number, v, has values

$$\mathrm{v} = 0, 1, 2, \ldots \qquad (10.13)$$

For a polyatomic molecule, the complex vibrational motion of the atoms can be resolved into a set of fundamental vibrations. Each fundamental vibration, called a normal mode, describes how the atoms move relative to each other. Every normal mode has its own set of energy levels that can be represented by equation (10.11). A linear molecule has $(3\eta - 5)$ such fundamental vibrations, where $\eta$ is the number of atoms in the molecule. For a nonlinear molecule, the number of fundamental vibrations is $(3\eta - 6)$.

The assignment of $(3\eta - 5)$ vibrational modes for a linear molecule and $(3\eta - 6)$ vibrational modes for a nonlinear molecule comes from a consideration of the number of degrees of freedom in the molecule. It requires $3\eta$ coordinates to completely specify the position of all $\eta$ atoms in the molecule, and each coordinate results in a degree of freedom. Three coordinates ($x$, $y$, and $z$) specify the movement of the center of mass of the molecule in space. They set the translational degrees of freedom, since translational motion is associated with movement of the molecule as a whole. Two internal coordinates (angles) are required to specify the orientation of the axis of a linear molecule during rotation, while three angles are required for a nonlinear

---

[d] Later in the chapter, we will see how to correct the vibrational energy levels of diatomic molecules for deviations from harmonic oscillator behavior.

molecule. These coordinates account for the rotational degrees of freedom. The remaining $(3\eta - 5)$ coordinates for a linear molecule or $(3\eta - 6)$ coordinates for a nonlinear molecule specify the bond lengths and bond angles during vibration and give the vibrational degrees of freedom. A fundamental vibration with its characteristic fundamental vibrational frequency (and set of energy levels) is present for each of these vibrational degrees of freedom.

As an example, $CO_2$ (a linear molecule) has four fundamental vibrations while $H_2O$ (a nonlinear molecule) has three fundamental vibrations (normal

**Table 10.1** Moments of inertia and rotational constants of some common molecules.

The values given are for a molecule that corresponds to the natural abundance of each isotope since these "average values" will give the thermodynamic properties of a mole of the naturally occurring substance.*

For diatomic molecules, $\tilde{B}_0$ is the rotational constant to use with equation (10.125), while $\tilde{B}_e$ applies to equation (10.124). They are related by $\tilde{B}_0 = \tilde{B}_e - \frac{1}{2}\tilde{\alpha}$. The moment of inertia $I_0(\text{kg·m}^2)$ is related to $B_0(\text{cm}^{-1})$ through the relationship $I_0 = h/(8 \times 10^{-2}\pi^2 B_0 c)$, with $h$ and $c$ expressed in SI units. For polyatomic molecules, $I_A$, $I_B$, and $I_C$ are the moments of inertia to use with Table 10.4 where the rigid rotator approximation is assumed. For diatomic molecules, $I_0$ is used with Table 10.4 to calculate values to which we add the anharmonicity and nonrigid rotator corrections.

**(a) Diatomic molecules**

| Molecule | $\tilde{B}_0/(\text{cm}^{-1})$ | $I_0/(10^{-47}\ \text{kg·m}^2)$ | $\tilde{B}_e/(\text{cm}^{-1})$ | $10^3\tilde{\alpha}/(10^{-3}\ \text{cm}^{-1})$ |
|---|---|---|---|---|
| $Br_2$ | 0.081948 | 341.59 | 0.082107 | 0.31873 |
| CO | 1.9215 | 14.568 | 1.9302 | 17.46 |
| $Cl_2$ | 0.24339 | 115.01 | 0.24415 | 1.5163 |
| $H_2$ | (59.304) | 0.47203 | (60.800) | (2993) |
| HBr | 8.34954 | 3.3526 | 8.46620 | 233.33 |
| HCl | 10.4326 | 2.6832 | 10.5844 | 303.7 |
| HF | 20.5577 | 1.3617 | 20.9555 | 795.8 |
| HI | 6.426 | 4.356 | 6.512 | 171.5 |
| ICl | 0.11272 | 248.34 | 0.11298 | 0.5275 |
| $I_2$ | 0.037333 | 749.81 | 0.037395 | 0.12435 |
| $N_2$ | 2.001 | 14.00 | (2.010) | (18.7) |
| NO | 1.6953 | 16.512 | 1.7042 | 17.8 |
| NaCl | 0.21611 | 129.53 | 0.21691 | 1.598 |
| OH | 18.514 | 1.5120 | 18.871 | 714 |
| $O_2$ | 1.43765 | 19.471 | 1.44562 | 15.933 |

*(continued)*

**Table 10.1** *Continued*

**(b) Polyatomic molecules**

*Linear molecules*

| Molecule | $I_A/(10^{-47}$ kg·m$^2)$ | Molecule | $I_A/(10^{-47}$ kg·m$^2)$ |
|---|---|---|---|
| HCN | 18.8585 | $CS_2$ | 256.42 |
| $N_2O$ | 66.4882 | $C_2H_2$ | 23.7864 |
| $CO_2$ | 71.4988 | | |

*Nonlinear molecules*

| Molecule | $I_AI_BI_C/(10^{-138}$kg$^3$·m$^6)$ | Molecule | $I_AI_BI_C/(10^{-138}$kg$^3$·m$^6)$ |
|---|---|---|---|
| $CH_4$ | 0.1499 | $NH_3$ | 0.0348 |
| $CCl_4$ | 115003 | $POCl_3$ | 10184.1 |
| $H_2$ | (0.0058577) | $CFCl_3$ | 57360.2 |
| $SO_2$ | 107.0 | $NO_2$ | 15.423 |

* Most values were taken from M. W. Chase, Jr., C. A. Davies, J. R. Downey, Jr., D. J. Frurip, R. A. McDonald, and A. N. Syverud, "JANAF Thermochemical Tables, Third Edition", *J. Phys. Chem. Ref. Data*, **14**, Supplement No. 1, 1985. A few (in parentheses) came from G. Hertzberg, *Molecular Spectra and Molecular Structure, I. Spectra of Diatomic Molecules*, and *II. Infrared and Raman Spectra of Polyatomic Molecules*, Van Nostrand Reinhold Co., New York, 1950 and 1945.

modes). The vibrations in $H_2O$ can be described as a combination of the three fundamental types:

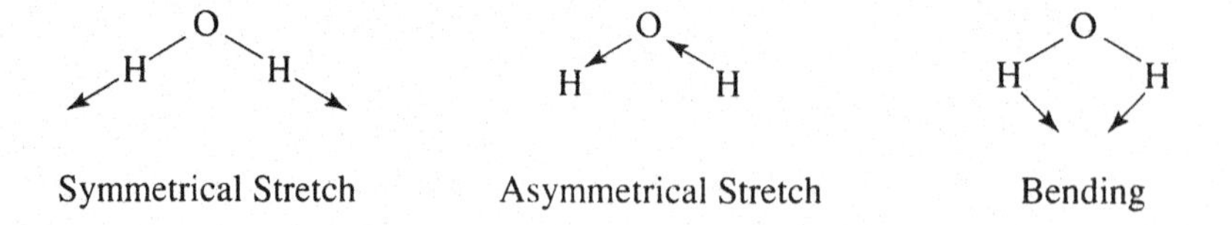


Each type of vibration (normal mode) has associated with it a fundamental vibrational frequency and a set of energy levels.

The vibrational energy levels associated with a single normal mode have a degeneracy of one. However, molecules with high symmetry may have several normal modes with the same frequency. For example, $CO_2$ has two bending modes, with the motion of one perpendicular to the motion of the other. Such modes are often referred to as **degenerate modes**, but there is a subtle difference between the use of degeneracy here than as it is used in the rotational or electronic energy levels.

**Electronic Energy Levels:** The energy differences between electronic energy levels are generally large unless unpaired electrons are present, while degeneracies higher than one can also occur. In the absence of unpaired electrons, we will find that, unless we are at very high temperatures, all of the atoms or molecules that we will consider, are in the ground (lowest energy) electronic state with a degeneracy of one, and we will not need to worry about increases in internal energy resulting from adding energy to electronic energy levels.

**Table 10.2** Fundamental vibrational frequencies of some common molecules.

The values given are for a molecule that corresponds to the natural abundance of each isotope since these "average values" will give the thermodynamic properties of a mole of the naturally occurring substance.*

For diatomic molecules, $\tilde{\omega}_0$ is the vibrational constant to use with equation (10.125) for calculating anharmonicity and nonrigid rotator corrections, while $\tilde{\omega}_e$ and $\tilde{\omega}_e x_e$ apply to equation (10.124). They are related by $\tilde{\omega}_0 = \tilde{\omega}_e - 2\tilde{\omega}_e x_e$. For polyatomic molecules, $\omega$ is the fundamental vibrational frequency, assuming the harmonic oscillator approximation. It is the value to use when making calculations with the equations in Table 10.4. For diatomic molecules $\tilde{\omega}_0$ is used with Table 10.4 to calculate the values to which we add the anharmonicity and nonrigid rotator corrections.

**(a) Diatomic molecules**

| Molecule | $\tilde{\omega}_0/(cm^{-1})$ | $\tilde{\omega}_e/(\text{cm}^{-1})$ | $\tilde{\omega}_e x_e/(\text{cm}^{-1})$ |
|---|---|---|---|
| $Br_2$ | 323.166 | 325.321 | 1.07742 |
| CO | 2142.61 | 2169.52 | 13.453 |
| $Cl_2$ | 554.362 | 559.751 | 2.6943 |
| $H_2$ | 4159.44 | (4395.24) | (117.90) |
| HBr | 2558.73 | 2649.182 | 45.225 |
| HCl | 2785.47 | 2889.59 | 52.06 |
| HF | 3958.63 | 4138.73 | 90.05 |
| HI | 2229.60 | 2309.06 | 39.73 |
| ICl | 379.28 | 382.18 | 1.450 |
| $I_2$ | 213.3156 | 214.5481 | 0.616259 |
| $N_2$ | (2330.70) | (2359.61) | (14.456) |
| NO | 1875.66 | 1903.60 | 13.97 |
| NaCl | 360.18 | 363.62 | 1.72 |
| OH | 3569.59 | 3735.21 | 82.81 |
| $O_2$ | 1556.231 | 1580.193 | 11.981 |

*(continued)*

**Table 10.2** *Continued*

**(b) Polyatomic molecules**
Numbers in parentheses designate the number of normal modes that have this frequency.

| Molecule | $\tilde{\omega}/(\text{cm}^{-1})$ | Molecule | $\tilde{\omega}/(\text{cm}^{-1})$ |
|---|---|---|---|
| *Linear molecules* | | | |
| HCN | 713.5(2), 2096.3, 3311.5 | $CS_2$ | 396.7(2), 656.6, 1523 |
| $N_2O$ | 589.2(2), 1276.5, 2223.7 | $C_2H_2$ | 611.6(2), 729.3(2), 1973.8, 3281.9, 3373.7 |
| $CO_2$ | 667.30(2), 1384.86, 2349.30 | | |
| *Nonlinear molecules* | | | |
| $CH_4$ | 1306(3), 1534.0(2), 2916.5, 3018.7(3) | $NH_3$ | 1022, 1691(2), 3506, 3577(2) |
| $CCl_4$ | 218(2), 314(3), 458, 776(3) | $NO_2$ | 756.8, 1357.8, 1665.5 |
| $H_2O$ | 1594.7, 3651.1, 3755.9 | $POCl_3$ | 193(2), 267(2), 337, 486 581(2), 1290 |
| $SO_2$ | 517.69, 1151.38, 1361.76 | $CFCl_3$ | 241(2), 349.5, 398(2), 535.3, 847(2), 1085 |

* Most values were taken from M. W. Chase, Jr., C. A. Davies, J. R. Downey, Jr., D. J. Frurip, R. A. McDonald, and A. N. Syverud, "JANAF Thermochemical Tables, Third Edition," *J. Phys. Chem. Ref. Data*, **14**, Supplement No. 1, 1985. For the diatomic molecules, a few (in parentheses) came from G. Hertzberg, *Molecular Spectra and Molecular Structure, I. Spectra of Diatomic Molecules*, and *II. Infrared and Raman Spectra of Polyatomic Molecules*, Van Nostrand Reinhold Co., New York, 1950 and 1945.

Tables 10.1, 10.2, and 10.3[e] summarize moments of inertia (rotational constants), fundamental vibrational frequencies (vibrational constants), and differences in energy between electronic energy levels for a number of common molecules or atoms.[f] The values given in these tables can be used to calculate the rotational, vibrational, and electronic energy levels. They will be useful as we calculate the thermodynamic properties of the ideal gas.

---

[e] Tables A4.2 through A4.4 summarize the same information.

[f] Atomic gases such as He have only translational levels, except at very high temperatures where electronic levels may become important. Atomic gases such as O or Cl have unpaired electrons and electronic levels do become important.

**Table 10.3** Electronic energy levels of some common molecules or atoms with unpaired electrons*

| Atom or molecule | $g_i$ | $\epsilon_i/(\text{cm}^{-1})$ | Atom or molecule | $g_i$ | $\epsilon_i/(\text{cm}^{-1})$ |
|---|---|---|---|---|---|
| $O_2$ | 3 | 0 | I | 4 | 0 |
| | 2 | 7882.39 | | 2 | 7603.15 |
| | 1 | 13120.91 | Cl | 4 | 0 |
| O | 5 | 0 | | 2 | 882.36 |
| | 3 | 158.265 | Na | 2 | 0 |
| | 1 | 226.977 | | 2 | 16956.172 |
| | 5 | 15867.862 | | 4 | 16973.368 |
| | 1 | 33792.583 | | 2 | 25739.991 |
| $NO_2$ | 2 | 0 | H | 2 | 0 |
| NO | 2 | 0 | | | |
| | 2 | 121.1 | | | |

* Values were taken from M. W. Chase, Jr., C. A. Davies, J. R. Downey, Jr., D. J. Frurip, R. A. McDonald, and A. N. Syverud, "JANAF Thermochemical Tables, Third Edition", *J. Phys. Chem. Ref. Data*, **14**, Supplement No. 1, 1985.

## 10.2 Distribution Of Energy Among Energy Levels

The relationship between the different types of energy levels in an ideal gas can be represented schematically as shown in Figure 10.1. Two electronic energy levels A and B are indicated. Associated with each electronic energy level are sets of vibrational energy levels, with one set for each normal mode of vibration. For the sake of simplicity, we have represented the vibrational levels for a single mode, where the levels are indicated by the quantum number $\text{v} = 0, 1, 2, 3, \ldots$ and $\text{v}' = 0, 1, 2, 3, \ldots$ . This diagram represents the situation found for a diatomic molecule where only one vibrational mode is present. Note that we have shown the vibrational levels as being equally spaced, which is the prediction of the harmonic oscillator approximation. In real molecules, the energy spacings get progressively smaller as v increases.

Within each vibrational energy level, the molecule has a series of rotational energy levels designated as $J = 0, 1, 2, \ldots$ in the A electronic state and $J' = 0, 1, 2, \ldots$ in the B electronic state. Note that energy spacing between the rotational levels increases with increasing $J$ as predicted by equation (10.3) for the rigid rotator. In turn, each rotational energy level would have a series of translational energy levels, but they are too close in energy to be separated in this figure.

As shown in Figure 10.1, the energy spacings between the vibrational levels in the two electronic levels are not the same. This is to be expected

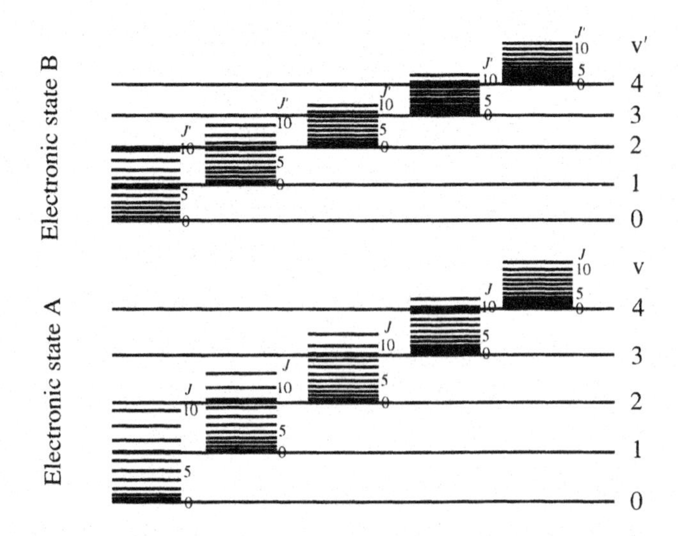


**Figure 10.1** Schematic energy-level diagram for a molecule. Two electronic levels A and B are present, with their vibrational levels (v) and rotational levels ($J$). The relative separation of electronic and vibrational levels is generally much greater than we have shown here.

since the spacing is given by $h\nu$, and $\nu$ depends upon the force constant as shown in equation (10.12). Changes in the character of the chemical bond result in changes in the force constant. In the terminology of simple molecular orbital theory, for example, excitation of electrons into nonbonding or antibonding orbitals leads to an electronic state that is less tightly bound, and therefore the force constant is reduced from that observed in the ground state.

The spacing between rotational levels also changes with the electronic and vibrational level. The bond lengths and hence, the moments of inertia, change with the electronic bonding, and this changes the rotational energy spacing. Bond lengths do not change with the vibration if it is harmonic, but anharmonicity effects stretch the bond as the energy level increases, leading to larger moments of inertia with a corresponding decrease in the separation between the rotational energy levels.

For the ideal gas, only the lowest translational, rotational, vibrational, and electronic energy level is occupied at zero Kelvin.[g] When energy is added to increase the temperature, it is distributed among the different types of energy levels. Thus, some of the energy increases the translational motion and is

[g] Note that at zero Kelvin, the molecule has a zero-point vibrational energy of ($\frac{1}{2}h\nu$) and translational energy equal to ($3h^2/8mL^2$). $U - U_0$ is the energy we add above these amounts to get to a temperature $T$.

absorbed into the translational energy levels. At the same time some is absorbed into the rotational levels and some into the vibrations. When unpaired electrons are present, electronic levels may be excited as well.

We want to calculate $U_T - U_0$, the increase in energy when the temperature of the gas is increased from 0 to $T$. An initial approach might be to calculate the amount of energy in each of the different kinds of energy levels for each molecule and then sum over the number of molecules. But this seems like an impossible task. With one mole of gas, we have $6 \times 10^{23}$ separate molecules, each with an entire series of energy levels. Solving the analytical equations of motion to sort out the energy in a three bodied problem is a difficult enough task, and we face a $6 \times 10^{23}$ bodied problem.

Actually, having such a very large number of molecules provides a solution to our problem. A given amount of energy can be absorbed into our system of molecules in many different ways. But some distributions are more probable than others. As we increase the number of energy states available, the difference in probability between different distributions increases, and one distribution begins to stand out as the one with highest probability. When we have the number of states corresponding to a mole of gas, one distribution becomes so much more probable than any other that we can be assured that it, or one very close to it, is the one that results. Our job is to use statistical methods to calculate this most probable distribution, and this calculation is not a difficult one. Let us consider some simple examples to see how this can be done.

Consider as an example a collection of three harmonic oscillators. We have seen that the energy levels for harmonic oscillators are equally spaced, with energy differences $h\nu$ between each successive level. Suppose we have $3h\nu$ units of energy shared among the three oscillators. Figure 10.2 shows the different ways that this energy can be distributed. In configuration (a) one of the oscillators takes all three units of energy while the other two oscillators get none. As can be seen, there are three different ways in which this can be accomplished if the three harmonic oscillators can be distinguished.

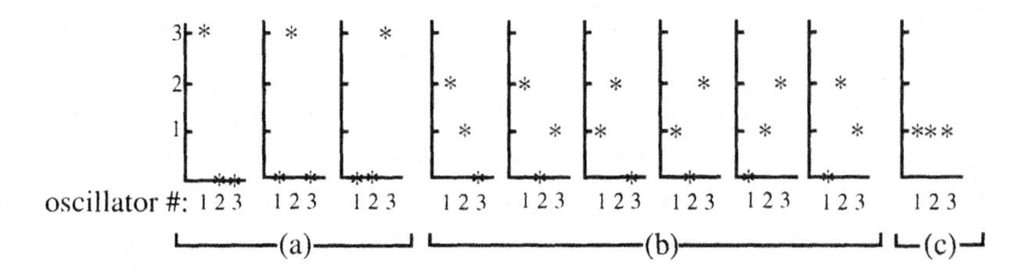


**Figure 10.2** Different ways that $3h\nu_0$ units of energy can be distributed among the energy levels of three equivalent harmonic oscillators (1, 2, and 3).

In configuration (b) one oscillator takes two units of energy, a second oscillator takes one unit and the third gets none. With distinguishable oscillators, there are six different ways that this can be accomplished. Finally, there is only one way [configuration (c)] that all three oscillators can each take one unit of energy. We conclude from Figure 10.2 that configuration (b) is a significantly more probable arrangement since it can be obtained in more ways. If we were going to predict the distribution of three units of energy among three harmonic oscillators, this is the one we would bet on. But we do not need to go through the exercise represented by Figure 10.2 to arrive at this equation since a mathematical relationship can be used to predict the number of different ways $W$ that a configuration can be obtained.[h] The equation is

$$W = \frac{N!}{\prod_i n_i!} \tag{10.14}$$

where $N$ is the total number of oscillators and $n_i$ is the number of oscillators in a particular energy level. The product is over the energy levels of the oscillators, $i = 1, 2, 3, \ldots$ Applying equation (10.14) for the conditions represented by Figure 10.2 gives the following[i]

$$\text{configuration (a): } W_\text{a} = \frac{3!}{2!0!0!1!} = 3$$

$$\text{configuration (b): } W_\text{b} = \frac{3!}{1!1!1!1!0!} = 6$$

$$\text{configuration (c): } W_\text{c} = \frac{3!}{0!3!0!0!} = 1,$$

which are the same results we obtained from Figure 10.2.

As a second example, consider the distribution of five units of energy among five distinguishable harmonic oscillators. The possible configurations are shown in Figure 10.3. We have not attempted to show in the figure the number of

[h] We will see later that $W$ is the thermodynamic probability related to entropy through the relationship $S = k \ln W$.

[i] $0! = 1$.

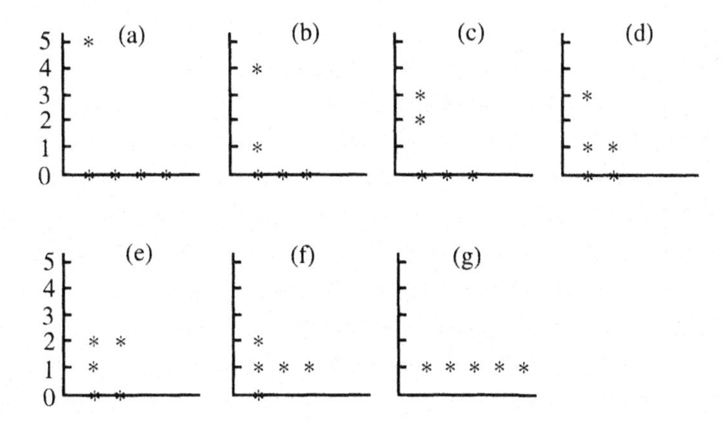


**Figure 10.3** Configurations leading to the distribution of five units of energy among five equivalent harmonic oscillators. The number of different ways that each configuration can be achieved are not shown. Equation (10.14) calculates this number. For example, we show that configuration (a) can be obtained in five different ways.

different ways each configuration can be realized. It can be calculated from equation (10.14) as shown below where we drop the 0!'s.

$$\text{configuration (a): } W_{\mathrm{a}} = \frac{5!}{1!4!} = 5$$

$$\text{configuration (b): } W_{\mathrm{b}} = \frac{5!}{3!1!1!} = 20$$

$$\text{configuration (c): } W_{\mathrm{c}} = \frac{5!}{3!1!1!} = 20$$

$$\text{configuration (d): } W_{\mathrm{d}} = \frac{5!}{2!2!1!} = 30$$

$$\text{configuration (e): } W_{\mathrm{e}} = \frac{5!}{2!2!1!} = 30$$

$$\text{configuration (f): } W_{\mathrm{f}} = \frac{5!}{1!3!1!} = 20$$

$$\text{configuration (g): } W_{\mathrm{g}} = \frac{5!}{5!} = 1.$$

As a final example, consider the distribution of five units of energy among ten harmonic oscillators. The possible configurations are given in Figure 10.4. The possible arrangements are as follows:

$$\text{configuration (a): } W_a = \frac{10!}{9!1!} = 10$$

$$\text{configuration (b): } W_b = \frac{10!}{8!1!1!} = 90$$

$$\text{configuration (c): } W_c = \frac{10!}{8!1!1!} = 90$$

$$\text{configuration (d): } W_d = \frac{10!}{7!2!1!} = 360$$

$$\text{configuration (e): } W_e = \frac{10!}{7!2!1!} = 360$$

$$\text{configuration (f): } W_f = \frac{10!}{6!3!1!} = 840$$

$$\text{configuration (g): } W_g = \frac{10!}{5!5!} = 252.$$

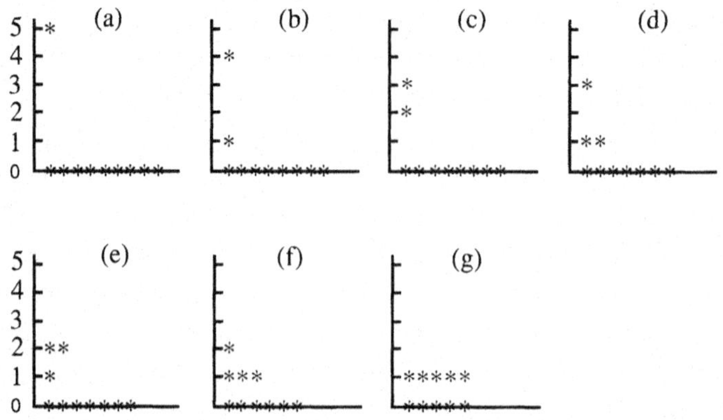

**Figure 10.4** Configurations leading to the distribution of five units of energy among ten equivalent harmonic oscillators.

In comparing the three examples, we can reach several conclusions. First, the number of total configurations (known as **microstates**) increases rapidly with the number of harmonic oscillators. In our three examples, it changed from 10 to 126 to 2002. One can calculate that an assembly of 1000 harmonic oscillators sharing 1000 units of energy possesses more than $10^{600}$ different microstates. This is an unimaginably large number, but small compared to the approximately $10^{10^{23}}$ microstates that would be present if a mole of harmonic oscillators shares a mole of energy units. This number is so large that it is essentially meaningless.

The second thing we note is the emergence of a dominant configuration. Figure 10.5 compares the number of microstates for the different configurations in our last example. Continuing calculations of this type with larger numbers of vibrations and more units of energy would verify that the bar graph sharpens as the number of harmonic oscillators increases, until one configuration dominates to the point that it is the only one that needs to be considered as a possible arrangement.

Finally, we have applied equation (10.14) to a collection of harmonic oscillators. But it can be applied to any collection of energy levels and units of energy with one modification. Equation (10.14) assumes that each level has an equal probability (as in a harmonic oscillator), and this is true only if $g_i$, the degeneracy, is one. The quantity $g_i$ is also known as the **statistical weight factor**. If it is greater than one, equation (10.14) must be multiplied by the $g_i$ for each

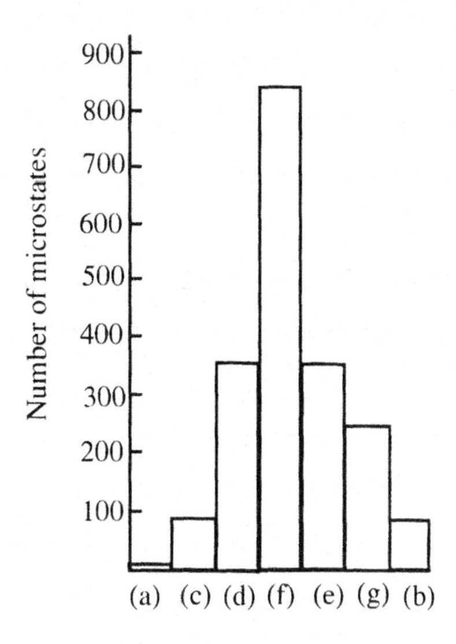


**Figure 10.5** Comparison of the number of microstates for the configurations shown in Figure 10.4.

energy level or $g_i^{n_i}$ for $n_i$ levels. The result is

$$W = N! \prod_i \frac{g_i^{n_i}}{n_i!}. \tag{10.15}$$

Equation (10.15) is now completely general and can be applied to our ideal gas with its variety of energy levels. What we want to do for distinguishable particles is calculate the most probable distribution of energy for $N$ gas molecules in which $n_1, n_2, n_3 \ldots n_i$ are the number of molecules occupying each of the energy levels in the collection of molecules, and $g_1, g_2, g_3, \ldots g_i$ are the degeneracies or statistical weight factors of these levels.

## 10.3 The Boltzmann Distribution Law

To calculate the most probable configuration, we start with equation (10.15) and take the logarithm of both sides to get

$$\ln W = \ln N! - \sum_i \ln n_i! + \sum_i n_i \ln g_i. \tag{10.16}$$

We next eliminate the factorial expressions by substituting Stirling's approximation[j] so that

$$\ln N! = N \ln N - N \tag{10.17}$$

and

$$\ln n_i! = n_i \ln n_i - n_i. \tag{10.18}$$

Stirling's approximation is a mathematical relationship that becomes more accurate as $N$ becomes larger, and becomes very precise for the large number of units we will consider as we work with moles of ideal gas. Substituting equations (10.17) and (10.18) into equation (10.16) gives

$$\ln W = N \ln N - N + \sum_i n_i \ln g_i + \sum_i n_i - \sum_i n_i \ln n_i. \tag{10.19}$$

[j] See Appendix 1 for a discussion of Stirling's approximation.

Since $\sum_i n_i = N$, equation (10.19) simplifies to

$$\ln W = N \ln N - \sum_i n_i \ln n_i + \sum_i n_i \ln g_i. \tag{10.20}$$

To find the most probable distribution (largest $W$) we make an infinitesimal displacement in $n_i$ (with $N$ held constant) and set the displacement equal to zero. That is

$$\delta W = 0.$$

The result is

$$\sum_i \ln g_i \, \delta n_i - \sum_i \ln n_i \, \delta n_i - \sum_i n_i \, \delta \ln n_i = 0. \tag{10.21}$$

Equation (10.21) can be simplified further by noting that since $N = \sum_i n_i$, then

$$\delta N = \sum_i \delta n_i = 0. \tag{10.22}$$

But

$$\sum_i \delta n_i = \sum_i \frac{n_i}{n_i} \, \delta n_i = \sum_i n_i \, \delta \ln n_i = 0.$$

Thus, the last term in equation (10.21) can be eliminated to give

$$\sum_i \ln g_i \, \delta n_i - \sum_i \ln n_i \, \delta n_i = 0. \tag{10.23}$$

Combining terms in equation (10.23) gives

$$\sum_i \ln \frac{n_i}{g_i} \, \delta n_i = 0. \tag{10.24}$$

We need one more equation before we can derive the Boltzmann distribution law. We note that the total energy $U - U_0$ is given by

$$U - U_0 = \sum_i n_i \epsilon_i$$

where $\epsilon_i$ is the energy of the $i$th level containing $n_i$ atoms or molecules. Since $U - U_0$ is constant

$$\delta(U - U_0) = \sum_i \epsilon_i \, \delta n_i = 0.$$

In summary, we have

$$\sum_i \ln \frac{n_i}{g_i} \, \delta n_i = 0 \tag{10.24}$$

$$\sum_i \delta n_i = 0 \tag{10.22}$$

$$\sum_i \epsilon_i \, \delta n_i = 0. \tag{10.25}$$

We can combine these three into one equation by using Lagrange's method of undetermined multipliers. To do so, we multiply equation (10.22) by $\alpha$ and equation (10.25) by $\beta$

$$\sum_i \alpha \, \delta n_i = 0 \tag{10.26}$$

$$\sum_i \beta\epsilon_i \, \delta n_i = 0 \tag{10.27}$$

where $\alpha$ and $\beta$ are universal constants (independent of $n_i$, $g_i$, and $\epsilon_i$). Equations (10.24), (10.26), and (10.27) are then added to give

$$\sum_i \left( \ln \frac{n_i}{g_i} + \alpha + \beta\epsilon_i \right) \delta n_i = 0. \tag{10.28}$$

Equation (10.28) is zero if $\delta n_i = 0$ for all $i$, or if

$$\ln \frac{n_i}{g_i} + \alpha + \beta\epsilon_i = 0 \tag{10.29}$$

for all $i$. The first is a trivial solution. The second is the one we want. We rewrite it in the form

$$n_i = g_i \exp(-\alpha)\exp(-\beta\epsilon_i). \tag{10.30}$$

We will be able to use this equation to find the most probable energy distribution after we obtain appropriate values for $\alpha$ and $\beta$.

### 10.3a Evaluation of $\alpha$

We set $\epsilon_0$ equal to zero in the ground state so that all the $\epsilon_i$ are expressed as energies relative to the ground state. Substitution of this condition into equation (10.30) gives

$$n_0 = g_0 \exp(-\alpha)$$

or

$$\exp(-\alpha) = \frac{n_0}{g_0}.$$

Substitution into equation (10.30) to eliminate $\alpha$ gives

$$\frac{n_i}{g_i} = \frac{n_0}{g_0} \exp(-\beta\epsilon_i). \tag{10.31}$$

Equation (10.31) can be put into another form. We start with the relationship

$$N = \sum n_i$$

and substitute for $n_i$ using equation (10.30) to get

$$N = \exp(-\alpha) \sum g_i \exp(-\beta\epsilon_i). \tag{10.32}$$

We define a quantity, $z$, called the **partition function** as

$$z = \sum g_i \exp(-\beta\epsilon_i) \tag{10.33}$$

and substitute into equation (10.32) to get

$$N = z\exp(-\alpha). \tag{10.34}$$

Dividing equation (10.30) by equation (10.34) eliminates $\exp(-\alpha)$ and gives

$$\frac{n_i}{N} = \frac{g_i\exp(-\beta\epsilon_i)}{z}. \tag{10.35}$$

Equation (10.35) tells us the fraction, $n_i/N$, of the molecules that are in energy state, $\epsilon_i$. The partition function is a measure of the extent to which energy is partitioned among the different states. We will return to a discussion of $z$ after we have determined a value for $\beta$.

## 10.3b Evaluation of $\beta$

The Lagrangian multiplier $\beta$ is a universal constant that is independent of the type of energy. Thus, if we can evaluate $\beta$ for one particular energy system, we will have a value of $\beta$ for all systems.

The energy system we choose to use in deriving an expression for $\beta$ is the translational energy of the ideal gas. From kinetic-molecular theory we know that $\overline{U}_{\text{trans}}$, the average translational energy is given by

$$\overline{U}_{\text{trans}} = \tfrac{3}{2}kT \tag{10.36}$$

where $k$ is the Boltzmann constant and $T$ is the absolute temperature. We can write an expression for $\overline{U}_{\text{trans}}$ by summing the energies of all the molecules and dividing by the total number of particles. The result is

$$\overline{U}_{\text{trans}} = \frac{\sum_i \epsilon_i n_i}{N}, \tag{10.37}$$

where $n_i$ is the number of molecules with translational energy $\epsilon_i$. Since $g_i = 1$ for translational energy levels, substitution of equation (10.35) into (10.37) gives

$$\overline{U}_{\text{trans}} = \frac{\sum_i \epsilon_i\exp(-\beta\epsilon_i)}{\sum_i \exp(-\beta\epsilon_i)}. \tag{10.38}$$

To evaluate the terms in equation (10.38) we relate $\epsilon_i$, the energy of a particular molecule to its momentum $\vec{p}_i$ (not to be confused with pressure $p$). Momentum is a vector quantity that is related to its components along the three coordinate axes by

$$p_i^2 = p_{x,i}^2 + p_{y,i}^2 = p_{z,i}^2. \tag{10.39}$$

These momentum components, in turn, are related to the contributions to $\epsilon_i$ in the three directions

$$\epsilon_{x,i} = \frac{p_{x,i}^2}{2m} \tag{10.40}$$

$$\epsilon_{y,i} = \frac{p_{y,i}^2}{2m} \tag{10.41}$$

$$\epsilon_{z,i} = \frac{p_{z,i}^2}{2m} \tag{10.42}$$

where $m$ is the mass of the molecule.

The average translational energy in the $x$ direction is given by

$$\overline{U}_{\text{trans},x} = \frac{\sum_i \epsilon_{x,i} n_i}{\sum_i n_i}. \tag{10.43}$$

Substitution of equations (10.35) and (10.40) into equation (10.43) gives

$$\overline{U}_{\text{trans},x} = \frac{\sum_i \frac{p_{x,i}^2}{2m} \exp\left(\frac{-\beta p_{x,i}^2}{2m}\right)}{\sum_i \exp\left(\frac{-\beta p_{x,i}^2}{2m}\right)}. \tag{10.44}$$

Since the energy difference between translational energy levels is very small and the sum is over a large number of particles, we can assume the energy distribution is continuous and replace the summations in equation (10.44) with integrals

$$\overline{U}_{\text{trans},x} = \frac{\dfrac{1}{2m}\displaystyle\int_{-\infty}^{\infty} p_x^2 \exp\left(-\frac{\beta p_x^2}{2m}\right) \mathrm{d}p_x}{\displaystyle\int_{-\infty}^{\infty} \exp\left(-\frac{\beta p_x^2}{2m}\right) \mathrm{d}p_x}. \tag{10.45}$$

The integrals in equation (10.45) are of a standard form given by

$$\int_{-\infty}^{\infty} x^2 \exp(-ax^2)\,\mathrm{d}x = \frac{1}{2}\sqrt{\frac{\pi}{a^3}} \tag{10.46}$$

$$\int_{-\infty}^{\infty} \exp(-ax^2)\,\mathrm{d}x = \sqrt{\frac{\pi}{a}}. \tag{10.47}$$

Substituting $a = \beta/2m$ into the standard forms of the integrals as given by equations (10.46) and (10.47) and evaluating $\overline{U}_{\text{trans},x}$ in equation (10.45) gives

$$\overline{U}_{\text{trans},x} = \frac{1}{2\beta}.$$

But

$$\overline{U}_{\text{trans}} = \overline{U}_{\text{trans},x} + \overline{U}_{\text{trans},y} + \overline{U}_{\text{trans},z}.$$

Because the three directions in space are equivalent

$$\overline{U}_{\text{trans},x} = \overline{U}_{\text{trans},y} = \overline{U}_{\text{trans},z}$$

and

$$\overline{U}_{\text{trans}} = 3\overline{U}_{\text{trans},x} = \frac{3}{2\beta}. \tag{10.48}$$

We can equate this result with $\overline{U}_{\text{trans}}$ in equation (10.36) to get

$$\frac{3}{2\beta} = \frac{3}{2}kT$$

so that

$$\beta = \frac{1}{kT}. \tag{10.49}$$

Substitution of equation (10.49) into equation (10.31) gives

$$\frac{n_i}{g_i} = \frac{n_0}{g_0}\exp\left(-\frac{\epsilon_i}{kT}\right). \tag{10.50}$$

Equation (10.50) is a form of the **Boltzmann Distribution Equation**. It relates the population of an excited energy state to the temperature (and the statistical weight factors). Its significance can be seen by referring to Figure 10.6 in which $n_i/n_0$ is plotted against $T$ for the vibrational and rotational energy levels in CO. Some of the conclusions that can be drawn from this figure are as follows:

(1) For both the vibrations and the rotations, the ratios $n_v/n_0$ and $n_J/n_0$ become zero at low temperatures. That is, only the ground state is populated.
(2) With increasing temperature, the population of the excited states increases. For vibrations, the limiting ratio at high temperatures is 1. That is, all levels become equally populated.[k] The temperature at which this occurs is very high. For rotation, the ratios approach a limiting value equal to the statistical weight factor of $(2J+1)$. Limiting values are approached at much lower temperatures for rotation than for vibration. This reflects the fact that the energy difference between rotational energy levels is much smaller than between vibrational energy levels. This relationship is shown in Figure 10.1.
(3) For vibration, the v = 0 energy level always has the largest population, although the relative population of the higher levels increases with

[k] The harmonic oscillator and rigid rotator approximations have been used in plotting Figures 10.6 and 10.7. These approximations are not valid at very high temperatures and the ratios shown in the figures do not accurately represent the real molecules under these temperature conditions.

increasing temperatures. This relationship is shown in Figure 10.7a. For rotation, the ground state generally does not have the largest population. The most highly populated level increases with increasing temperature. This effect is shown in Figure 10.7b. A maximum in the ratio $n_J/n_0$ results from two opposing effects: an increase in the statistical weight factor with increasing $J$ and a decrease in the exponential term with increasing energy.[1]

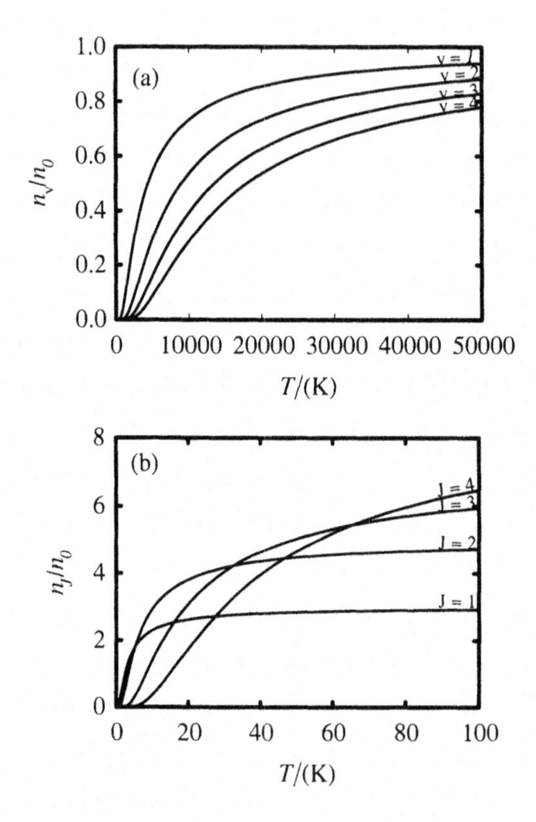


**Figure 10.6** Graph of the Boltzmann distribution function for the CO molecule in the ground electronic state for (a), the vibrational energy levels and (b), the rotational energy levels. Harmonic oscillator and rigid rotator approximations have been used in the calculations.

---

[1] We should remember that the statistical weight factor results from degenerate energy levels. Quantum mechanics distinguishes between the different degenerate levels with a quantum number $K$ that is a function of the angular momentum but not the energy. Thus, when $J = 1$, three levels are present that have the same energy, but different angular momentum. With $J = 2$, five levels are present, and so on. If one considers individual rotational energy levels, the population in each does decrease with increasing $T$.

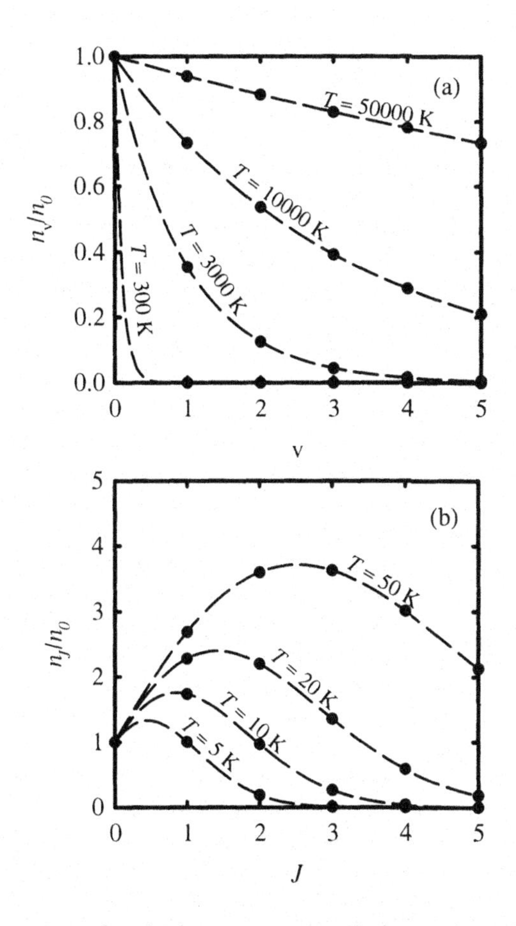


**Figure 10.7** The population of excited energy states relative to that of the ground state for the CO molecule as predicted by the Boltzmann distribution equation. Graph (a) gives the ratio for the vibrational levels while graph (b) gives the ratio for the rotational levels. Harmonic oscillator and rigid rotator approximations have been used in the calculations. The dots represent ratios at integral values of v and $J$, which are the only allowed values.

## 10.4 The Partition Function

In equation (10.33), we defined the partition function $z$ as

$$z = \sum g_i \exp(-\beta\epsilon_i). \qquad (10.33)$$

With

$$\beta = \frac{1}{kT}, \qquad (10.51)$$

the partition function becomes

$$z = \sum g_i \exp\left(-\frac{\epsilon_i}{kT}\right) \tag{10.52}$$

and equation (10.35) becomes

$$\frac{n_i}{N} = \frac{g_i}{z} \exp\left(-\frac{\epsilon_i}{kT}\right). \tag{10.53}$$

The partition function turns out to be a very useful quantity in our calculations, and it is important that we understand its properties. As we said earlier, the name comes from the realization that $z$ is a measure of the distribution of energy among excited states, as can be seen by writing $z$ as[m]

$$z = g_0 + g_1 \exp\left(-\frac{\epsilon_1}{kT}\right) + g_2 \exp\left(-\frac{\epsilon_2}{kT}\right) + \cdots \tag{10.54}$$

As we discussed earlier, at low temperatures, $kT \ll \epsilon_1, \epsilon_2, \ldots$ , each $\epsilon_i/kT$ is large, and all $\exp(-\epsilon_i/kT)$ terms are negligibly small, so that $z = g_0$. With increasing temperature, the exponential terms become larger and $z$ increases. In the limit of very high temperatures, $kT \gg \epsilon_1, \epsilon_2, \ldots, \exp(-\epsilon/kT) \cong \exp(-\epsilon_2/kT) \cong \cdots \cong 1$, and $z = g_0 + g_1 + g_2 + \cdots$ becomes a large number.

Another interesting characteristic of $z$ can be seen by noting that, although we have been considering an assembly of $N$ atoms or molecules, $z$ is independent of the number of units in the assembly. Thus, $z$ is defined in terms of the energy levels of a single unit and can be thought of as the partition function of a single unit. We can, of course, write a partition function for a collection of units, in which case the total partition function is the product of the partition functions of the individual units. This can be seen by considering, as an example, a

[m] In writing equation (10.54) we have assumed that $\varepsilon_0 = 0$ and $\varepsilon_1, \varepsilon_2, \ldots$ are energies above this ground state. This will be our standard procedure unless we state otherwise.

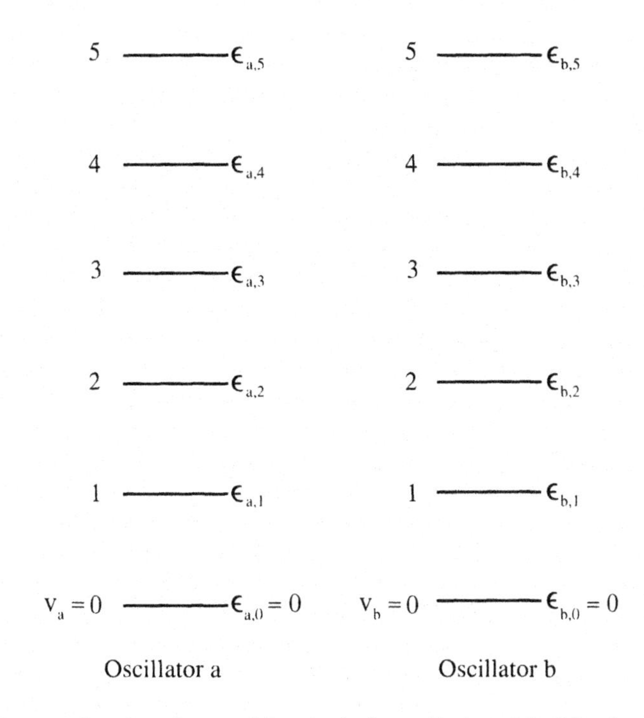


**Figure 10.8** Energy levels of two identical, but distinguishable, harmonic oscillators.

collection of two identical, but distinguishable, harmonic oscillators whose energy levels are shown in Figure 10.8. For these two units, the partition functions are given by

$$z_a = \exp\left(-\frac{\epsilon_{a,0}}{kT}\right) + \exp\left(-\frac{\epsilon_{a,1}}{kT}\right) + \exp\left(-\frac{\epsilon_{a,2}}{kT}\right) + \exp\left(-\frac{\epsilon_{a,3}}{kT}\right) + \cdots$$

$$z_b = \exp\left(-\frac{\epsilon_{b,0}}{kT}\right) + \exp\left(-\frac{\epsilon_{b,1}}{kT}\right) + \exp\left(-\frac{\epsilon_{b,2}}{kT}\right) + \exp\left(-\frac{\epsilon_{b,3}}{kT}\right) + \cdots$$

We can write an expression for $Z_{ab}$, the partition function for the combination of these two harmonic oscillators. It must take into account all of the possible combinations of energy levels. That is,

$$
\begin{aligned}
Z_{ab} = {} & \exp\left[-\frac{(\epsilon_{a,0}+\epsilon_{b,0})}{kT}\right] + \exp\left[-\frac{(\epsilon_{a,0}+\epsilon_{b,1})}{kT}\right] \\
& + \exp\left[-\frac{(\epsilon_{a,0}+\epsilon_{b,2})}{kT}\right] + \cdots \\
& + \exp\left[-\frac{(\epsilon_{a,1}+\epsilon_{b,0})}{kT}\right] + \exp\left[-\frac{(\epsilon_{a,1}+\epsilon_{b,1})}{kT}\right] \\
& + \exp\left[-\frac{(\epsilon_{a,1}+\epsilon_{b,2})}{kT}\right] + \cdots \\
& + \exp\left[-\frac{(\epsilon_{a,2}+\epsilon_{b,0})}{kT}\right] + \exp\left[-\frac{(\epsilon_{a,2}+\epsilon_{b,1})}{kT}\right] + \cdots \\
& + \exp\left[-\frac{(\epsilon_{a,3}+\epsilon_{b,0})}{kT}\right] + \cdots \\
& + \cdots
\end{aligned}
\tag{10.55}
$$

An examination of the terms will show that

$$Z_{ab} = z_a z_b.$$

Since a and b are identical, we can drop the subscripts and write that for two identical, but distinguishable, harmonic oscillators

$$Z = z^2.$$

We can generalize this expression to write that for $N$ identical units

$$Z = z^N. \tag{10.56}$$

Similarly, we can write that for a combination of translational, rotational, vibrational, and electronic energy levels for a single molecule[n]

$$z = z_{trans} \cdot z_{rot} \cdot z_{vib} \cdot z_{elect} \tag{10.57}$$

where

$$z_{trans} = z_{trans,x} \cdot z_{trans,y} \cdot z_{trans,z}. \tag{10.58}$$

We run into a complication when we attempt to write an expression similar to equation (10.56) for a combination of $N$ molecules that are not distinguishable.[o] When this happens, combinations such as $(\epsilon_{a,1} + \epsilon_{b,2})$ and $(\epsilon_{a,2} + \epsilon_{b,1})$ are the same and should not be counted twice. Thus, the total number of terms in the partition function should be decreased to eliminate such duplications. To determine how to correct for this duplication, consider three gas molecules a, b, and c with energy levels we will represent as $\epsilon_1$, $\epsilon_2$, and $\epsilon_3$. A total of $3! = 6$ different combinations of these energy levels can be written as follows

| Molecule a | Molecule b | Molecule c |
|---|---|---|
| $\epsilon_1$ | $\epsilon_2$ | $\epsilon_3$ |
| $\epsilon_1$ | $\epsilon_3$ | $\epsilon_2$ |
| $\epsilon_2$ | $\epsilon_1$ | $\epsilon_3$ |
| $\epsilon_2$ | $\epsilon_3$ | $\epsilon_1$ |
| $\epsilon_3$ | $\epsilon_1$ | $\epsilon_2$ |
| $\epsilon_3$ | $\epsilon_2$ | $\epsilon_1$ |

If the molecules are not distinguishable, these six combinations are the same and the combined partition function should be decreased by a factor of 6. In general, we can write that for $N$ indistinguishable units

$$Z = \frac{1}{N!} z^N = \frac{1}{N!} (z_{trans} \cdot z_{rot} \cdot z_{vib} \cdot z_{elect})^N$$

[n] Statistical weight factors are present in $z_{rot}$ and $z_{elect}$. The multiplication involving these partition functions still works, since the total statistical weight factor is the product of the statistical weight factors for the individual units. Thus $g_i = g_{i,trans} \cdot g_{i,rot} \cdot g_{i,vib} \cdot g_{i,elect} \cdots$

[o] Molecules in a gas are not distinguishable. Molecules in a solid, where we can give them coordinates in a crystal lattice, are distinguishable. We will return to this condition later when we use our statistical methods to describe vibrations in a solid.

where $Z$ is the total partition function for $N$ indistinguishable molecules whose individual partition functions are $z$.

We assign the $1/N!$ term to the translational partition function because all gases have translational motion, so that

$$Z_{\text{trans}} = \frac{1}{N!} z_{\text{trans}}^N.$$

But molecular gases also have rotation and vibration. We only make the correction for indistinguishability once. Thus, we do not divide by $N!$ to write the relationship between $Z_{\text{rot}}$, the rotational partition function of $N$ molecules, and $z_{\text{rot}}$, the rotational partition function for an individual molecule, if we have already assigned the $1/N!$ term to the translation. The same is true for the relationship between $Z_{\text{vib}}$ and $z_{\text{vib}}$. In general, we write for the total partition function $Z$ for $N$ units

$$Z = \frac{1}{N!} z^N \tag{10.59}$$

$$Z = Z_{\text{trans}} \cdot Z_{\text{rot}} \cdot Z_{\text{vib}} \cdot Z_{\text{elect}} \tag{10.60}$$

where[p]

$$Z_{\text{trans}} = \frac{1}{N!} z_{\text{trans}}^N \tag{10.61}$$

$$Z_{\text{rot}} = z_{\text{rot}}^N \tag{10.62}$$

$$Z_{\text{vib}} = z_{\text{vib}}^N \tag{10.63}$$

$$Z_{\text{elect}} = z_{\text{elect}}^N. \tag{10.64}$$

## 10.5 Relationship Between The Partition Function And The Thermodynamic Properties

We are now ready to relate the thermodynamic properties of the ideal gas to $Z$. We start with $U - U_0$, where $U_0$ is the energy when all the molecules are at zero

---

[p] For atoms, we do not include $Z_{\text{rot}}$ and $Z_{\text{vib}}$. For atoms or molecules without unpaired electrons, we will write $Z_{\text{elect}} = 1$ except at very high temperatures.

Kelvin and hence, in their lowest energy states. At a temperature $T$, the collection of $N$ molecules has an energy given by

$$U - U_0 = \sum n_i \epsilon_i. \tag{10.65}$$

Substituting for $n_i$ from equation (10.53) gives

$$U - U_0 = \frac{N}{z} \sum g_i \epsilon_i \exp\left(-\frac{\epsilon_i}{kT}\right). \tag{10.66}$$

Differentiation of equation (10.52) gives

$$\left(\frac{\partial z}{\partial T}\right)_V = \frac{\sum g_i \epsilon_i \exp\left(-\frac{\epsilon_i}{kT}\right)}{kT^2}$$

or

$$\sum g_i \epsilon_i \exp\left(-\frac{\epsilon_i}{kT}\right) = kT^2 \left(\frac{\partial z}{\partial T}\right)_V. \tag{10.67}$$

Substitution of equation (10.67) into equation (10.66) gives

$$U - U_0 = \frac{NkT^2}{z} \left(\frac{\partial z}{\partial T}\right)_V$$

or

$$U - U_0 = NkT^2 \left(\frac{\partial \ln z}{\partial T}\right)_V. \tag{10.68}$$

From equation (10.59)

$$\ln Z = N \ln z - \ln N!.$$

Differentiating gives

$$\left(\frac{\partial \ln Z}{\partial T}\right)_V = N\left(\frac{\partial \ln z}{\partial T}\right)_V. \tag{10.69}$$

Substituting equation (10.69) into equation (10.68) gives

$$U - U_0 = kT^2\left(\frac{\partial \ln Z}{\partial T}\right)_V. \tag{10.70}$$

We will use equation (10.70) to calculate the energy difference for one mole, where $N = N_A$, Avogadro's number and $Z = Z_m$, the partition function for a mole of molecules. Equation (10.70) then becomes

$$U_m - U_{0,m} = kT^2\left(\frac{\partial \ln Z_m}{\partial T}\right)_{V_m} \tag{10.71}$$

where

$$Z_m = \frac{1}{N_A\,!}\, z_m^{N_A}. \tag{10.72}$$

A similar relationship involving $H$ can be obtained from

$$H_m = U_m + pV_m$$

which, for the ideal gas becomes

$$H_m = U_m + RT \tag{10.73}$$

and

$$H_m - U_{0,m} = U_m - U_{0,m} + RT. \tag{10.74}$$

From equation (10.73) we see that $U_{0,m} = H_{0,m}$, and equation (10.74) becomes

$$H_m - H_{0,m} = U_m - U_{0,m} + RT. \tag{10.75}$$

Substitution of equation (10.71) into equation (10.75) gives

$$H_{\mathrm{m}} - H_{0,\mathrm{m}} = kT^2\left(\frac{\partial \ln Z_{\mathrm{m}}}{\partial T}\right)_{V_{\mathrm{m}}} + RT. \tag{10.76}$$

An expression relating heat capacity to $Z_{\mathrm{m}}$ can be obtained by differentiating equation (10.71)

$$\left[\frac{\partial(U_{\mathrm{m}} - U_{0,\mathrm{m}})}{\partial T}\right]_{V_{\mathrm{m}}} = \left(\frac{\partial U_{\mathrm{m}}}{\partial T}\right)_{V_{\mathrm{m}}} = C_{V,\mathrm{m}}$$

or

$$C_{V,\mathrm{m}} = 2kT\left(\frac{\partial \ln Z_{\mathrm{m}}}{\partial T}\right)_{V_{\mathrm{m}}} + kT^2\left(\frac{\partial^2 \ln Z_{\mathrm{m}}}{\partial T^2}\right)_{V_{\mathrm{m}}}. \tag{10.77}$$

Heat capacity at constant pressure is then easily obtained for the ideal gas from

$$C_{p,\mathrm{m}} = C_{V,\mathrm{m}} + R. \tag{10.78}$$

To obtain the entropy, we start with

$$S_{\mathrm{m}} = f(T, V_{\mathrm{m}})$$

so that

$$\mathrm{d}S_{\mathrm{m}} = \left(\frac{\partial S_{\mathrm{m}}}{\partial T}\right)_{V_{\mathrm{m}}} \mathrm{d}T + \left(\frac{\partial S_{\mathrm{m}}}{\partial V_{\mathrm{m}}}\right)_T \mathrm{d}V_{\mathrm{m}}.$$

At constant $V_{\mathrm{m}}$

$$\mathrm{d}S_{\mathrm{m}} = \left(\frac{\partial S_{\mathrm{m}}}{\partial T}\right)_{V_{\mathrm{m}}} \mathrm{d}T.$$

Substituting the relationship between heat capacity and entropy gives

$$dS_m = \frac{C_{V,m}}{T}\,dT. \tag{10.79}$$

Substitution of equation (10.77) into equation (10.79) gives

$$dS_m = 2k\left(\frac{\partial \ln Z_m}{\partial T}\right)_{V_m} dT + kT\left(\frac{\partial^2 \ln Z_m}{\partial T^2}\right)_{V_m} dT.$$

Integrating (at constant $V_m$) gives

$$\int_{S_{0,m}}^{S_m} dS_m = 2k\int_{\ln Z_{0,m}}^{\ln Z_m} d\ln Z_m + k\int_0^T T\left(\frac{\partial^2 \ln Z_m}{\partial T^2}\right)_{V_m} dT \tag{10.80}$$

where $S_{0,m}$ and $Z_{0,m}$ are the entropy and partition function of the ideal gas at zero Kelvin.

The last term in equation (10.80) can be integrated by parts

$$\int T\frac{\partial^2 \ln Z_m}{\partial T^2}\,dT = T\left(\frac{d\ln Z_m}{dT}\right)_{V_m} - \int d\ln Z_m.$$

Hence:

$$S_m - S_{0,m} = k\int_{\ln Z_{0,m}}^{\ln Z_m} d\ln Z_m + kT\left(\frac{\partial \ln Z_m}{\partial T}\right)_{V_m}$$

or

$$S_m - S_{0,m} = k\ln Z_m - k\ln Z_{0,m} + kT\left(\frac{\partial \ln Z_m}{\partial T}\right)_{V_m}. \tag{10.81}$$

At this point, we use the Boltzmann equation given in Chapter 1 to relate $S$ and $W$

$$S_{\mathrm{m}} = k \ln W_{\mathrm{m}}$$

which at zero Kelvin becomes

$$S_{0,\mathrm{m}} = k \ln W_{0,\mathrm{m}}.$$

But $W_{\mathrm{m}}$ is given by equation (10.15)

$$W_{\mathrm{m}} = N_{\mathrm{A}}! \prod_i \frac{g_i^{n_i}}{n_i!}$$

so that

$$S_{0,\mathrm{m}} = k \ln g_0^{N_{\mathrm{A}}}$$

since at zero Kelvin, all the molecules are in the ground state and $\prod n_i! = N_{\mathrm{A}}!$. But

$$g_0^{N_{\mathrm{A}}} = Z_{0,\mathrm{m}}$$

so that

$$S_{0,\mathrm{m}} = k \ln Z_{0,\mathrm{m}}.$$

Substitution into equation (10.81) causes terms to cancel to give[q]

$$S_{\mathrm{m}} = k \ln Z_{\mathrm{m}} + kT\left(\frac{\partial \ln Z_{\mathrm{m}}}{\partial T}\right)_{V_{\mathrm{m}}}. \tag{10.82}$$

We note from the derivation that $S_{\mathrm{m}}$ in equation (10.82) is the absolute entropy of the gas, since $S_{0,\mathrm{m}}$ has been cancelled by the constant of integration.

---

[q] Instead of starting with equation (10.79) to relate $Z$ to $S$, we could have started directly with $S = k \ln W$ and equation (10.15). The derivation we have used in which the Boltzmann relation is used only to evaluate the constant of integration is easier.

Equations involving the free energy are now easily obtained. We start with

$$A_m = U_m - TS_m$$

or

$$A_m - U_{0,m} = (U_m - U_{0,m}) - TS_m.$$

Substitution of equation (10.71) for $U_m - U_{0,m}$ gives

$$A_m - U_{0,m} = kT^2\left(\frac{\partial \ln Z_m}{\partial T}\right)_{V_m} - kT \ln Z_m - kT^2\left(\frac{\partial \ln Z_m}{\partial T}\right)_{V_m}$$

or

$$A_m - U_{0,m} = -kT \ln Z_m. \tag{10.83}$$

Since $H_{0,m} = U_{0,m}$, equation (10.83) can also be written as

$$A_m - H_{0,m} = -kT \ln Z_m. \tag{10.84}$$

An expression for the Gibbs free energy is obtained from

$$G_m = A_m + pV_m$$

$$= A_m + RT.$$

$$G_m - H_{0,m} = A_m - H_{0,m} + RT,$$

or

$$G_m - H_{0,m} = -kT \ln Z_m + RT. \tag{10.85}$$

## 10.6 Evaluation Of The Partition Function For The Ideal Gas

To review, we derived, in the previous section, the following relationships for one mole of ideal gas:

$$U_m - U_{0,m} = kT^2\left(\frac{\partial \ln Z_m}{\partial T}\right)_{V_m} \tag{10.71}$$

$$H_{\mathrm{m}} - H_{0,\mathrm{m}} = kT^2 \left( \frac{\partial \ln Z_{\mathrm{m}}}{\partial T} \right)_{V_{\mathrm{m}}} + RT \tag{10.76}$$

$$C_{V,\mathrm{m}} = 2kT \left( \frac{\partial \ln Z_{\mathrm{m}}}{\partial T} \right)_{V_{\mathrm{m}}} + kT^2 \left( \frac{\partial^2 \ln Z_{\mathrm{m}}}{\partial T^2} \right)_{V_{\mathrm{m}}} \tag{10.77}$$

$$C_{p,\mathrm{m}} = C_{V,\mathrm{m}} + R \tag{10.78}$$

$$S_{\mathrm{m}} = k \ln Z_{\mathrm{m}} + kT \left( \frac{\partial \ln Z_{\mathrm{m}}}{\partial T} \right)_{V_{\mathrm{m}}} \tag{10.82}$$

$$A_{\mathrm{m}} - U_{0,\mathrm{m}} = -kT \ln Z_{\mathrm{m}} \tag{10.84}$$

$$G_{\mathrm{m}} - H_{0,\mathrm{m}} = -kT \ln Z_{\mathrm{m}} + RT. \tag{10.85}$$

We will now derive expressions for $Z_{\mathrm{m}}$ that can then be substituted into the above equations to calculate the thermodynamic properties. In doing so, we note that, in all instances, these properties are related to the logarithm of $Z_{\mathrm{m}}$. Since the $Z$'s associated with different degrees of freedom are multiplied,

$$Z_{\mathrm{m}} = Z_{\mathrm{m,trans}} \cdot Z_{\mathrm{m,rot}} \cdot Z_{\mathrm{m,vib}} \cdot Z_{\mathrm{m,elect}},$$

then their logarithms are additive:

$$\ln Z_{\mathrm{m}} = \ln Z_{\mathrm{m,trans}} + \ln Z_{\mathrm{m,rot}} + \ln Z_{\mathrm{m,vib}} + \ln Z_{\mathrm{m,elec}}$$

and

$$\left( \frac{\partial \ln Z_{\mathrm{m}}}{\partial T} \right)_{V_{\mathrm{m}}} = \left( \frac{\partial \ln Z_{\mathrm{m,trans}}}{\partial T} \right)_{V_{\mathrm{m}}} + \left( \frac{\partial \ln Z_{\mathrm{m,rot}}}{\partial T} \right)_{V_{\mathrm{m}}} + \cdots$$

The implication is that the thermodynamic properties are also additive. Thus

$$U_{\mathrm{m}} - U_{0,\mathrm{m}} = (U_{\mathrm{m}} - U_{0,\mathrm{m}})_{\mathrm{trans}} + (U_{\mathrm{m}} - U_{0,\mathrm{m}})_{\mathrm{rot}} + \cdots$$

$$S_{\mathrm{m}} = S_{\mathrm{m,trans}} + S_{\mathrm{m,rot}} + \cdots$$

Our plan is to calculate $Z_{\mathrm{m,trans}}$, $Z_{\mathrm{m,rot}}$, $Z_{\mathrm{m,vib}}$, ... from which we can calculate $(U_{\mathrm{m}} - U_{0,\mathrm{m}})_{\mathrm{trans}}$, $(U_{\mathrm{m}} - U_{0,\mathrm{m}})_{\mathrm{rot}}$, $S_{\mathrm{m,trans}}$, $S_{\mathrm{m,rot}}$, ... Adding the contributions together then gives the total $(U_m - U_{0,m})$, $S_m$, ...[r] As we proceed with our calculations, it is important to keep in mind the approximations involved in our equations so that we can avoid conditions where serious errors can occur. For example, the rigid-rotator and harmonic oscillator approximations and the approximation that $Z_{elect} = 1$ for molecules without unpaired electrons are not valid at very high temperatures. Also, we will see that we have a problem using our procedure to calculate the rotational contributions to the properties at very low temperatures for a molecule with a small moment of inertia, such as $H_2$. It is satisfying to realize, however, that accurate calculations can be made if we avoid extreme conditions. In fact, this is one of the few instances in science where a macroscopic property can be calculated from molecular parameters with an accuracy that is often better than the accuracy that is possible from a direct experimental measurement.

## 10.6a Translational Partition Function

Suppose we confine a gas to a box of dimensions $L_x$, $L_y$, and $L_z$. In equation (10.1) we expressed the energy levels along the $x$ direction as

$$\epsilon_{trans,\,x} = \frac{n_x^2 h^2}{8mL_x^2}$$

where $m$ is the mass of the molecule and $n_x = 1, 2, 3, ...$ is the quantum number. For the particle in the box, the $g_i$ are 1. Hence the partition function is given by

$$z_{trans,\,x} = \sum_{n_x=1}^{\infty} \exp\left[-\frac{(n_x^2 - 1)h^2}{8mL_x^2 kT}\right].$$

Translational energy levels are very close together. We can neglect the 1 compared to $n_x^2$ and replace the sum by an integral to give

$$z_{trans,\,x} = \int_0^{\infty} \exp\left(-\frac{n_x^2 h^2}{8mL_x^2 kT}\right) dn_x. \tag{10.86}$$

[r] This procedure assumes that the translational, rotational, vibrational, and electronic energy levels are independent. This is not completely so. In the instance of diatomic molecules, we will see how to correct for the interaction. For more complicated molecules we will ignore the correction since it is usually a small effect.

The definite integral in equation (10.86) is of a standard form and is easily evaluated. The result is

$$z_{\text{trans},x} = \frac{(2\pi mkT)^{1/2}}{h} L_x . \tag{10.87}$$

Similar derivations in the $y$ and $z$ directions give

$$z_{\text{trans},y} = \frac{(2\pi mkT)^{1/2}}{h} L_y \tag{10.88}$$

$$z_{\text{trans},z} = \frac{(2\pi mkT)^{1/2}}{h} L_z . \tag{10.89}$$

But

$$z_{\text{trans}} = z_{\text{trans},x} \cdot z_{\text{trans},y} \cdot z_{\text{trans},z} .$$

Hence,

$$z_{\text{trans}} = \frac{(2\pi mkT)^{3/2}}{h^3} L_x L_y L_z$$

or

$$z_{\text{trans}} = \frac{(2\pi mkT)^{3/2} V}{h^3} \tag{10.90}$$

where $L_x L_y L_z = V$ is the volume of the container. Finally, substitution into equation (10.61) for one mole gives

$$Z_{m,\text{trans}} = \frac{1}{N_A !} \left[ \frac{(2\pi mkT)^{3/2} V_{\text{m}}}{h^3} \right]^{N_A} \tag{10.91}$$

where $V_{\text{m}}$ is the molar volume.

## 10.6b Rotational Partition Function

We will assume the rigid-rotator approximation to obtain $z_{\text{rot}}$. For the linear molecule we get

$$\epsilon_{\text{rot}} = J(J+1)\frac{h^2}{8\pi^2 I} \tag{10.92}$$

where $I$ is the moment of inertia and $J = 0, 1, 2, \ldots$ is the quantum number. For an unsymmetrical (AB) molecule, the statistical weight factor is given by

$$g_J = 2J + 1.$$

However, for a symmetrical (AA) molecule, we find that only half the energy levels are occupied. Thus, in molecules such as $H_2$ or $N_2$, $J = 0, 2, 4, 6, \ldots$ or $J = 1, 3, 5, 7, \ldots$ are allowed. To account for this difference, we correct $g_J$ by dividing by the symmetry number $\sigma$ so that

$$g_J = \frac{2J+1}{\sigma}. \tag{10.93}$$

The symmetry number is the number of indistinguishable rotated positions. For a homonuclear diatomic such as $H_2$, $\sigma = 2$ since H—H′ and H′—H are indistinguishable. For a heteronuclear diatomic, such as CO, $\sigma = 1$ since C—O and O—C are distinguishable. As another example, $\sigma = 2$ for O—C—O and $\sigma = 1$ for O—N—N, both of which are linear molecules.[s]

Substituting equations (10.92) and (10.93) into the equation for the partition function gives

$$z_{\text{rot}} = \frac{1}{\sigma}\sum_{J=0}^{\infty}(2J+1)\exp\left[-J(J+1)\frac{h^2}{8\pi^2 IkT}\right]. \tag{10.94}$$

---

[s] We will use the concept of a symmetry number again as we consider $g_J$ for more complicated molecules. In those cases, $\sigma_J$ can have values larger than 2.

Generally, the energy levels are close together, and we can replace the sum by an integral unless the temperature is low.[1] The result is

$$z_{\text{rot}} = \frac{1}{\sigma}\int_0^{\infty} (2J+1)\exp\left[-J(J+1)\frac{h^2}{8\pi^2 IkT}\right] \mathrm{d}J$$

$$= \frac{1}{\sigma}\int_0^{\infty} \exp\left[\frac{-(J^2+J)h^2}{8\pi^2 IkT}\right] \mathrm{d}(J^2+J). \tag{10.95}$$

The definite integral in equation (10.95) is of a standard form and is easily evaluated. The result is

$$z_{\text{rot}} = \frac{8\pi^2 IkT}{\sigma h^2}, \tag{10.96}$$

or for one mole,

$$Z_{\text{m, rot}} = \left[\frac{8\pi^2 IkT}{\sigma h^2}\right]^{N_A}. \tag{10.97}$$

A similar (although more involved) procedure can be used to obtain the partition function for a nonlinear molecule. The result is

$$z_{\text{rot}} = \left[\frac{8\pi^2(8\pi^3 I_A I_B I_C)^{1/2}(kT)^{3/2}}{\sigma h^3}\right] \tag{10.98}$$

or

$$Z_{\text{m, rot}} = \left[\frac{8\pi^2(8\pi^3 I_A I_B I_C)^{1/2}(kT)^{3/2}}{\sigma h^3}\right]^{N_A} \tag{10.99}$$

---

[1] Because of its small moment of inertia, the substitution of the integral for the sum results in errors in the calculations of $z_{\text{rot}}$ for $H_2$ at low temperatures. In that case, the laborious process of summing the energy levels must be used. $H_2$ and $D_2$ are the only molecular species for which this is a serious problem.

where $I_A$, $I_B$, and $I_C$ are the three principal moments of inertia as described earlier. The symmetry number $\sigma$ for a nonlinear molecule, which is again the number of indistinguishable rotations, can take on a range of values, depending upon the geometrical arrangement of the atoms in the molecule. It is easy to show that $\sigma$ has the following values for some common molecules.

| Molecule | $\sigma$ |
|---|---|
| $H_2O$ | 2 |
| $NH_3$ | 3 |
| $CHCl_3$ | 3 |
| $CH_4$ | 12 |

### 10.6c Vibrational Partition Function

We will use the harmonic oscillator approximation to derive an equation for the vibrational partition function. The quantum mechanical expression gives the vibrational energies as

$$\epsilon_{\text{vib}} = h\nu(\text{v} + 1/2)$$

where $\nu$ is the fundamental vibrational frequency and the quantum numbers, v, are 0, 1, 2, 3, ... However, in the derivation of the partition function, we have defined energies relative to the ground state, so that we must express the vibrational energy as a function of v with $\epsilon_{\text{vib, v}} - \epsilon_{\text{vib, 0}} = \text{v}h\nu$. The energy levels are still equally spaced, but now occur at integer multiples of $h\nu$, that is $h\nu$, $2h\nu$, $3h\nu$ ... Since the degeneracy of the vibrational levels are unity, the partition function can be written as

$$z_{\text{vib}} = 1 + \exp\left(-\frac{h\nu}{kT}\right) + \exp\left(-\frac{2h\nu}{kT}\right) + \exp\left(-\frac{3h\nu}{kT}\right) + \cdots$$

Rearranging and factoring gives

$$z_{\text{vib}} - 1 = \exp\left(-\frac{h\nu}{kT}\right)\left[1 + \exp\left(-\frac{h\nu}{kT}\right) + \cdots\right]$$

or

$$z_{\text{vib}} - 1 = \exp\left(-\frac{h\nu}{kT}\right) z_{\text{vib}}.$$

Solving for $z_{\text{vib}}$ gives

$$z_{\text{vib}} = \frac{1}{1 - \exp\left(-\dfrac{h\nu}{kT}\right)} \tag{10.100}$$

or, for a mole of harmonic oscillators

$$Z_{\text{m, vib}} = \left[\frac{1}{1 - \exp\left(-\dfrac{h\nu}{kT}\right)}\right]^{N_A}. \tag{10.101}$$

We have seen earlier that for a linear polyatomic molecule, the vibrational motions can be divided into $(3\eta - 5)$ fundamentals, where $\eta$ is the number of atoms. For a nonlinear molecule $(3\eta - 6)$ fundamentals are present. In either case, each fundamental vibration can be treated as a harmonic oscillator with a partition function given by equations (10.100) and (10.101). Thus,

$$z_{\text{vib}} = \prod_i \left(\frac{1}{1 - \exp(-h\nu_i/kT)}\right) \tag{10.102}$$

where the product has $(3\eta - 5)$ or $(3\eta - 6)$ terms, and

$$z_{\text{m. vib}} = z_{\text{vib}}^{N_A}. \tag{10.103}$$

### 10.6d Electronic Partition Function

For electronic systems without unpaired electrons, $g_0 = 1$ and $\epsilon_1, \epsilon_2, \epsilon_3$ ... are very large. Hence

$$z_{\text{elect}} = 1 + g_1 e^{-\epsilon_1/kT} + g_2 e^{-\epsilon_2/kT} + \cdots \approx 1.$$

When unpaired electrons are present, low-lying energy levels with multiple degeneracies may be present. Unlike the other three energy modes we have described, there is no general energy expression for electronic levels. The actual energy separations and degeneracies are different for each system, and a partition function must be written to reflect the energy levels of the particular atom or molecule. For example, the energy levels and degeneracies for the lowest levels of atomic oxygen are as follows:

| $g_i$ | $\epsilon_i/(\mathrm{cm}^{-1})$ |
|---|---|
| 5 | 0 |
| 3 | 157.4 |
| 1 | 226.1 |
| 5 | 15,807 |
| 1 | 33,662 |

With this set of energy levels, the electronic partition function is given by

$$z_{\text{elect}} = 5 + 3\exp\left(-\frac{157.4hc}{kT}\right) + \exp\left(-\frac{226.1hc}{kT}\right)$$

$$+ 5\exp\left(-\frac{15807hc}{kT}\right) + \exp\left(-\frac{33662hc}{kT}\right)$$

and

$$Z_{\text{m, elect}} = z_{\text{elect}}^{N_{\text{A}}}.$$

Note that the energy (written as $\tilde{\epsilon}$) is expressed in $\mathrm{cm}^{-1}$ (wave numbers), an energy unit often used by spectroscopists. The ratio $hc/k$ has a value of $1.43877 \times 10^{-2}$ m·K or 1.43876 cm·K. The latter value is the one to use when energy is expressed in $\mathrm{cm}^{-1}$.

It is instructive to consider the magnitude of each of the terms in the partition function, $z_{\text{elect}}$, at some temperature, for example, 1000 K. For atomic oxygen at this temperature,

$$z_{\text{elect}} = 5 + 3\exp\left(-\frac{157.4 \times 1.439}{1000}\right) + \exp\left(-\frac{226.1 \times 1.439}{1000}\right)$$

$$+ 5\exp\left(-\frac{15807 \times 1.439}{1000}\right) + \exp\left(-\frac{33662 \times 1.439}{1000}\right)$$

$$= 5 + 2.392 + 0.722 + 6.61 \times 10^{-10} + 9 \times 10^{-22} = 8.114.$$

Notice that each term in the partition function gets progressively smaller, and that the last two are negligibly small. These terms will only contribute to the partition function at much higher temperatures.

## 10.7 Calculation of the Thermodynamic Properties of the Ideal Gas

We now have equations for the partition functions for the ideal gas and equations for relating the partition functions to the thermodynamic properties. We are ready to derive the equations for calculating the thermodynamic properties from the molecular parameters. As an example, let us calculate $U_{\text{m}} - U_{0,\text{m}}$ for the translational motion of the ideal gas. We start with

$$Z_{\text{m, trans}} = \frac{1}{N_{\text{A}}!}\left[\frac{(2\pi mkT)^{3/2} V_{\text{m}}}{h^3}\right]^{N_{\text{A}}}.$$

Taking logarithms and putting the constants in a separate term gives

$$\ln Z_{\text{m, trans}} = \ln\left\{\frac{1}{N_{\text{A}}!}\left(\frac{2\pi mk}{h^3}\right)^{3/2} V_{\text{m}}\right\}^{N_{\text{A}}} + \frac{3}{2} N_{\text{A}} \ln T.$$

Differentiation gives

$$\left(\frac{\partial \ln Z_{\text{m, trans}}}{\partial T}\right)_{V_{\text{m}}} = \frac{3}{2}\frac{N_{\text{A}}}{T}.$$

From equation (10.71)

$$U_{\mathrm{m}} - U_{0,\mathrm{m}} = kT^2 \left( \frac{\partial \ln Z_{\mathrm{m,trans}}}{\partial T} \right)_{V_{\mathrm{m}}}$$

$$= (kT^2)\left( \frac{3}{2} \frac{N_{\mathrm{A}}}{T} \right)$$

$$= \frac{3}{2} RT$$

where $N_{\mathrm{A}}k = R$. This is, of course, the same value for the energy that we started with (obtained from kinetic-molecular theory) in the derivation of the Boltzmann distribution equation as we found a value for $\beta$. The answer is not surprising — but reassuring.

The thermodynamic functions of primary interest in chemistry are $H_{\mathrm{m}} - H_{0,\mathrm{m}}$, $C_{p,\mathrm{m}}$, $S_{\mathrm{m}}$, and $G_{\mathrm{m}} - H_{0,\mathrm{m}}$. The translational, rotational, and vibrational contributions are summarized in Table 10.4.[u] We will not attempt to derive all the equations in this table but will do enough to show how it is done.

## 10.7a Examples of the Derivation of the Contribution to the Thermodynamic Properties

**Translational Contribution to Entropy:** We start with equation (10.82)

$$S_{\mathrm{m}} = k \ln Z_{\mathrm{m}} + kT \left( \frac{\partial \ln Z_{\mathrm{m}}}{\partial T} \right)_{V_{\mathrm{m}}} \tag{10.82}$$

and substitute into it, equation (10.91)

$$Z_{\mathrm{m,trans}} = \frac{1}{N_{\mathrm{A}}!} \left[ \frac{(2\pi mkT)^{3/2} V_{\mathrm{m}}}{h^3} \right]^{N_{\mathrm{A}}}. \tag{10.91}$$

[u] Table 10.4 is also summarized in Appendix 4 as Table A4.1.

**Table 10.4** Thermodynamic functions of an ideal gas (in $J{\cdot}K^{-1}{\cdot}mol^{-1}$) (use $R = 8.314510\ J{\cdot}K^{-1}{\cdot}mol^{-1}$ and SI units for pressure, temperature, and all molecular data)

## Translation

$$S_{m,\,trans} = \frac{3}{2} R \ln M + \frac{5}{2} R \ln T - R \ln p + \underbrace{\left[\frac{5}{2} R + R \ln\left(\frac{(2\pi k)^{3/2}}{h^3 N_A^{5/2}}\right) + R \ln R\right]}_{+172.3005}$$

$$\left(\frac{G_m - H_{0,\,m}}{T}\right)_{trans} = -\frac{3}{2} R \ln M - \frac{5}{2} R \ln T + R \ln p + \underbrace{\left[-R \ln\left(\frac{(2\pi k)^{3/2}}{h^3 N_A^{5/2}}\right) - R \ln R\right]}_{-151.5142}$$

$$\left(\frac{U_m - U_{0,\,m}}{T}\right)_{trans} = \frac{3}{2} R = (C_{V,\,m})_{trans}$$

$$\left(\frac{H_m - H_{0,\,m}}{T}\right)_{trans} = \frac{5}{2} R = (C_{p,\,m})_{trans}$$

## Rotation (rigid molecule approximation)

*Linear polyatomic or diatomic molecules:*

$$S_{m,\,rot} = R \ln T + R \ln I - R \ln \sigma + \underbrace{\left[R \ln\left(\frac{8\pi^2 k}{h^2}\right) + R\right]}_{+877.3950}$$

$$\left(\frac{G_m - H_{0,\,m}}{T}\right)_{rot} = -R \ln T - R \ln I + R \ln \sigma + \underbrace{\left[-R \ln \frac{8\pi^2 k}{h^2}\right]}_{-869.0805}$$

$$\left(\frac{U_m - U_{0,\,m}}{T}\right)_{rot} = \left(\frac{H_m - H_{0,\,m}}{T}\right)_{rot} = (C_m)_{rot} = R$$

*(continued)*

**Table 10.4** *Continued*

*Nonlinear polyatomic molecules:*

$$S_{\mathrm{m,\,rot}} = \frac{3}{2} R \ln T + \frac{1}{2} R \ln I_A I_B I_C - R \ln \sigma + \underbrace{\left[ R \ln \frac{8\pi^2 (2\pi k)^{3/2}}{h^3} + \frac{3}{2} R \right]}_{1320.8515}$$

$$\left(\frac{U_{\mathrm{m}} - U_{0,\mathrm{m}}}{T}\right)_{\mathrm{rot}} = \left(\frac{H_{\mathrm{m}} - H_{0,\mathrm{m}}}{T}\right)_{\mathrm{rot}} = (C)_{\mathrm{rot}} = \frac{3}{2} R$$

$$\left(\frac{G_{\mathrm{m}} - H_{0,\mathrm{m}}}{T}\right)_{\mathrm{rot}} = -\frac{3}{2} R \ln T - \frac{1}{2} R \ln I_A I_B I_C + R \ln \sigma + \underbrace{\left[ -R \ln \frac{8\pi^2 (2\pi k)^{3/2}}{h^3} \right]}_{-1308.3797}$$

**Vibration (harmonic oscillator approximation)**

$$S_{\mathrm{m,\,vib}} = R \sum_{i=1}^{n} \left[ \frac{x_i}{\exp(x_i)} - \ln(1 - \exp(-x_i)) \right] \text{ where;}$$

$$x_i = \frac{hc\tilde{\omega}_i}{kT} = 1.43877 \frac{\tilde{\omega}_i}{T} \text{ (use } \tilde{\omega}_i \text{ in cm}^{-1}\text{)}$$

$$\left(\frac{G_{\mathrm{m}} - H_{0,\mathrm{m}}}{T}\right)_{\mathrm{vib}} = \sum_{i=1}^{n} R \ln(1 - \exp(-x_i))$$

$$\approx \sum_{i=1}^{n} \left( -R \ln \frac{kT}{hc\tilde{\omega}_i} \right) \quad \text{(High temp. approx.)}$$

$$\left(\frac{U_{\mathrm{m}} - U_{0,\mathrm{m}}}{T}\right)_{\mathrm{vib}} = \left(\frac{H_{\mathrm{m}} - H_{0,\mathrm{m}}}{T}\right)_{\mathrm{vib}} = R \sum_{i=1}^{n} \frac{x_i}{\exp(x_i) - 1}$$

$$(C_{\mathrm{m}})_{\mathrm{vib}} = R \sum_{i=1}^{n} \frac{x_i^2 \exp(x_i)}{[\exp(x_i) - 1]^2}$$

where $n = (3\eta - 6)$ or $(3\eta - 5)$, with $\eta$ equal to the number of atoms in the molecule.

Taking logarithms of equation (10.91) and separating terms gives

$$\ln Z_{\mathrm{m,\,trans}} = N_{\mathrm{A}} \ln \left[ \frac{(2\pi k)^{3/2}}{h^3} \right] + \frac{3}{2} N_{\mathrm{A}} \ln T + \frac{3}{2} N_{\mathrm{A}} \ln m + N_{\mathrm{A}} \ln V_{\mathrm{m}} - \ln N_{\mathrm{A}}!. \tag{10.104}$$

Taking a derivative gives

$$\left( \frac{\partial \ln Z_{\mathrm{m,\,trans}}}{\partial T} \right)_{V_{\mathrm{m}}} = \frac{3N_{\mathrm{A}}}{2T}. \tag{10.105}$$

Substituting equations (10.104) and (10.105) into equation (10.82), along with the following substitutions

$N_{\mathrm{A}}k = R$ (gas constant)

$N_{\mathrm{A}}m = M$ (molecular weight)

$\ln N_{\mathrm{A}}! = N_{\mathrm{A}} \ln N_{\mathrm{A}} - N_{\mathrm{A}}$ (Stirling's approximation)

and $V_{\mathrm{m}} = \dfrac{RT}{p}$ (ideal gas)

gives

$$S_{\mathrm{m,\,trans}} = \frac{3}{2} R \ln M + \frac{5}{2} R \ln T - R \ln p + \left\{ \frac{5}{2} R + R \ln \left[ \frac{(2\pi k)^{3/2}}{h^3 N_{\mathrm{A}}^{5/2}} \right] + R \ln R \right\}. \tag{10.106}$$

The constant term within the brackets has a value of 172.3005 in SI units ($\mathrm{J \cdot K^{-1} \cdot mol^{-1}}$).

**Translational and Rotational Contributions to Enthalpy for a Linear Molecule:** For translation

$$H_{\mathrm{m}} - H_{0,\mathrm{m}} = U_{\mathrm{m}} - U_{0,\mathrm{m}} + RT$$

$$= \frac{3}{2} RT + RT = \frac{5}{2} RT. \tag{10.107}$$

For rotation

$$H_{\mathrm{m}} - H_{0,\mathrm{m}} = U_{\mathrm{m}} - U_{0,\mathrm{m}}. \tag{10.108}$$

From equation (10.97)

$$Z_{\mathrm{m,\,rot}} = \left[\frac{8\pi^2 IkT}{\sigma h^2}\right]^{N_{\mathrm{A}}}.$$

Taking logarithms gives

$$\ln Z_{\mathrm{m,\,rot}} = N_{\mathrm{A}} \ln T + N_{\mathrm{A}} \ln \left[\frac{8\pi^2 Ik}{\sigma h^2}\right].$$

Differentiating gives

$$\left(\frac{\partial \ln Z_{\mathrm{m,\,rot}}}{\partial T}\right)_{V_{\mathrm{m}}} = \frac{N_{\mathrm{A}}}{T}.$$

Thus,

$$H_{\mathrm{m}} - H_{0,\mathrm{m}} = kT^2 \left(\frac{\partial \ln Z_{\mathrm{m,\,rot}}}{\partial T}\right)_{V_{\mathrm{m}}}$$

$$= \frac{N_{\mathrm{A}} kT^2}{T} = RT. \tag{10.109}$$

Note that we do not add $RT$ to $U_m - U_{0,m}$ to get $H_m - H_{0,m}$ for rotation. We only do this once in calculating the total $H_m - H_{0,m}$, and we already added $RT$ in calculating the translational contribution.

**Vibrational Contribution to the Gibbs Free Energy:** For a Linear Diatomic Molecule: From equations (10.84) and (10.101)

$$A_m - U_{0,m} = G_m - H_{0,m} = -kT \ln Z_{m,vib}$$

$$Z_{m,vib} = \left[\frac{1}{1 - \exp\left(-\dfrac{h\nu}{kT}\right)}\right]^{N_A}.$$

Substitution of $Z_{m,vib}$ from the second equation into the first gives

$$G_m - H_{0,m} = -kT \ln\left[\frac{1}{1 - \exp\left(-\dfrac{h\nu}{kT}\right)}\right]^{N_A}$$

$$= N_A kT \ln\left[1 - \exp\left(-\frac{h\nu}{kT}\right)\right]$$

$$= RT \ln\left[1 - \exp\left(-\frac{h\nu}{kT}\right)\right]. \qquad (10.110)$$

Again, we note that the $RT$ term is not added in changing from $(A_m - U_{0,m})$ to $(G_m - H_{0,m})$. This term is added only in the translational contribution.

**Calculation of Thermodynamic Properties:** We note that the translational contributions to the thermodynamic properties depend on the mass or molecular weight of the molecule, the rotational contributions on the moments of inertia, the vibrational contributions on the fundamental vibrational frequencies, and the electronic contributions on the energies and statistical weight factors for the electronic states. With the aid of this information, as summarized in Tables 10.1 to 10.3 for a number of molecules, and the thermodynamic relationships summarized in Table 10.4, we can calculate a thermodynamic property by adding together the translational, rotational, vibrational, and electronic contributions.

**Example 10.1** Calculate $S^\circ_m$ at $T = 298.15$ K for Ar behaving as an ideal gas.

**Solution:** Ar is a monatomic gas with $Z_{\text{m, elect}} = 1$. The translational contribution is the only one we need to consider. For Ar, $M = 0.039948$ kg·mol$^{-1}$. We want the standard state entropy when $p = 1.000 \times 10^5$ Pa. Substituting into the equation in Table 10.4 gives

$$S^\circ_{\text{m}} = \left(\frac{3}{2}\right)(8.3145)\ln 0.039948 + \left(\frac{5}{2}\right)(8.3145)\ln 298.15$$

$$- R \ln 1.000 \times 10^5 + 172.3005$$

$$= 154.85 \text{ J·K}^{-1}\text{·mol}^{-1}.$$

**Example 10.2** Calculate the entropy of (ideal) NO(g) at $T = 500$ K and $p = 10.00$ kPa.

**Solution:**

*Translation:*

$$M = 0.03001 \text{ kg·mol}^{-1},\ p = 10.00 \text{ kPa} \times 10^3 \text{ Pa·kPa}^{-1} = 1.000 \times 10^4 \text{ Pa}$$

$$S_{\text{m, trans}} = \frac{3}{2} R \ln 0.03001 + \frac{5}{2} R \ln 500 - R \ln 1.000 \times 10^4 + 172.30$$

$$= 181.17 \text{ J·K}^{-1}\text{·mol}^{-1}.$$

*Rotation:*

$$\sigma = 1,\ I = 16.51 \times 10^{-47} \text{ kg·m}^2 \text{ (from Table 10.1)}$$

$$S_{\text{m, rot}} = R \ln 500 + R \ln 16.51 \times 10^{-47} - R \ln 1 + 877.395$$

$$= 52.57 \text{ J·K}^{-1}\text{·mol}^{-1}.$$

*Vibration:*

$$\tilde{\omega} = 1876\ \text{cm}^{-1} \quad \text{(from Table 10.2)}^{\text{v}}$$

$$x = \frac{hc\tilde{\omega}}{kT} = \frac{(1.439)(1876)}{(500)} = 5.399$$

$$S_{\text{m, vib}} = R\left\{\frac{x}{[\exp(x) - 1]} - \ln[1 - \exp(-x)]\right\}$$

$$= R\left\{\frac{5.399}{[\exp(5.399) - 1]} - \ln[1 - \exp(-5.399)]\right\}$$

$$= 0.22\ \text{J·K}^{-1}\text{·mol}^{-1}.$$

*Electronic:*

From Table 10.3, $\tilde{\epsilon}_0 = 0$, $g_0 = 2$, $\tilde{\epsilon}_1 = 121.1\ \text{cm}^{-1}$, $g_1 = 2$

$$S_{\text{m. elect}} = k \ln Z_{\text{m. elect}} + kT\left(\frac{\partial \ln Z_{\text{m. elect}}}{\partial T}\right)_{V_{\text{m}}}$$

$$= k \ln z_{\text{elect}}^{N_{\text{A}}} + kT\left(\frac{\partial \ln z_{\text{elect}}^{N_{\text{A}}}}{\partial T}\right)_{V_{\text{m}}}$$

$$= N_{\text{A}}k \ln z_{\text{elect}} + N_{\text{A}}kT\left(\frac{\partial \ln z_{\text{elect}}}{\partial T}\right)_{V_{\text{m}}}$$

$$z_{\text{elect}} = 2 + 2 \exp\left(-121.1\frac{hc}{kT}\right) = 2 + 2 \exp\left(-\frac{174.3}{T}\right)$$

$$z_{\text{elect}} = 2 + 2 \exp\left(-\frac{174.3}{500}\right) = 3.411$$

---

[v] $\tilde{\omega}$ is the fundamental vibrational frequency in $\text{cm}^{-1}$. It is related to $\nu$ by $\tilde{\omega} = \nu/c$, where $c$ is the speed of light in $\text{cm·sec}^{-1}$.

$$\left(\frac{\partial \ln z_{\text{elect}}}{\partial T}\right)_{V_{\text{m}}} = \frac{1}{z_{\text{elect}}}\left(\frac{\partial z_{\text{elect}}}{\partial T}\right)_{V_{\text{m}}}$$

$$\left(\frac{\partial z_{\text{elect}}}{\partial T}\right)_{V_{\text{m}}} = \left[2\exp\left(-\frac{174.3}{T}\right)\right]\left(\frac{174.3}{T^2}\right)$$

$$= \left[2\exp\left(-\frac{174.3}{500}\right)\right]\left[\frac{174.3}{(500)^2}\right]$$

$$= 9.84 \times 10^{-4}$$

$$\left(\frac{\partial \ln z_{\text{elect}}}{\partial T}\right)_{V_{\text{m}}} = \frac{9.84 \times 10^{-4}}{3.411} = 2.88 \times 10^{-4}$$

$$S_{\text{m, elect}} = N_{\text{A}}k \ln z_{\text{elect}} + N_{\text{A}}kT\left(\frac{\partial \ln z_{\text{elect}}}{\partial T}\right)_{V_{\text{m}}}$$

$$= R \ln 3.411 + RT(2.88 \times 10^{-4})$$

$$= 11.40 \text{ J·K}^{-1}\text{·mol}^{-1}$$

$$S_{\text{m}} = S_{\text{m, trans}} + S_{\text{m, rot}} + S_{\text{m, vib}} + S_{\text{m, elect}}$$

$$= 181.17 + 52.57 + 0.22 + 11.40$$

$$= 245.36 \text{ J·K}^{-1}\text{·mol}^{-1}.$$

**Example 10.3** Calculate $C^{\circ}_{p,\text{m}}$ for $CO_2$ at 1000 K.

**Solution:**
Translation: $C_{p,\text{m, trans}} = \frac{5}{2}R$
Rotation: $C_{p,\text{m, rot}} = R$
Vibration: The vibrational contribution is calculated from the fundamental vibrational frequencies and the relationship in Table 10.4. $CO_2$ is a linear molecule with $(3\eta - 5) = 4$ fundamentals. The values of $\tilde{\omega}$ are obtained from

Table 10.2 and the contributions to the heat capacity are calculated and tabulated as follows

| $\tilde{\omega}/(\text{cm}^{-1})$ | $x = hc\tilde{\omega}/(\text{kT})$ | $C_{p,\text{vib}}/R$ |
|---|---|---|
| 667.3 | 0.960 | 0.927 |
| 667.3 | 0.960 | 0.927 |
| 1385 | 1.993 | 0.726 |
| 2349 | 3.380 | 0.417 |
| | | Total = 3.00 |

Thus

$$C_{p,\text{m}} = \tfrac{5}{2}R + R + 3.00R = 6.50R$$
$$= 54.02\ \text{J}\cdot\text{K}^{-1}\cdot\text{mol}^{-1}.$$

**Example 10.4** Use $Z_\text{m}$ to derive the ideal gas equation:

**Solution:** From Table 3.1

$$\left[\frac{\partial(A_\text{m} - A_{0,\text{m}})}{\partial V_\text{m}}\right]_T = \left(\frac{\partial A_\text{m}}{\partial V_\text{m}}\right)_T = -p.$$

But, from Table 10.4

$$A_\text{m} - A_{0,\text{m}} = -kT \ln Z_\text{m}\,.$$

Differentiating gives

$$\left[\frac{\partial(A_\text{m} - A_{0,\text{m}})}{\partial V_\text{m}}\right]_T = -kT\left(\frac{\partial \ln Z_\text{m}}{\partial V_\text{m}}\right)_T$$

so that

$$p = kT\left(\frac{\partial \ln Z_\text{m}}{\partial V_\text{m}}\right)_T. \qquad (10.111)$$

To get

$$\left(\frac{\partial \ln Z_{\mathrm{m}}}{\partial V_{\mathrm{m}}}\right)_T,$$

we write

$$Z_{\mathrm{m}} = Z_{\mathrm{m,\,trans}} \cdot Z_{\mathrm{m,\,rot}} \cdot Z_{\mathrm{m,\,vib}} \cdot Z_{\mathrm{m,\,elect}}$$

$$\ln Z_{\mathrm{m}} = \ln Z_{\mathrm{m,\,trans}} + \ln Z_{\mathrm{m,\,rot}} + \ln Z_{\mathrm{m,\,vib}} + \ln Z_{\mathrm{m,\,elect}}$$

$$\left(\frac{\partial \ln Z_{\mathrm{m}}}{\partial V_{\mathrm{m}}}\right)_T = \left(\frac{\partial \ln Z_{\mathrm{m,\,trans}}}{\partial V_{\mathrm{m}}}\right)_T + \left(\frac{\partial \ln Z_{\mathrm{m,\,rot}}}{\partial V_{\mathrm{m}}}\right)_T + \left(\frac{\partial \ln Z_{\mathrm{m,\,vib}}}{\partial V_{\mathrm{m}}}\right)_T + \left(\frac{\partial \ln Z_{\mathrm{m,\,elect}}}{\partial V_{\mathrm{m}}}\right)_T.$$

$$Z_{\mathrm{m,\,trans}} = \frac{1}{N_{\mathrm{A}}!}\left\{\frac{(2\pi mkT)^{3/2} V_{\mathrm{m}}}{h^3}\right\}^{N_{\mathrm{A}}}$$

$$\ln Z_{\mathrm{m,\,trans}} = \ln\frac{1}{N_{\mathrm{A}}!} + N_{\mathrm{A}} \ln \frac{(2\pi mkT)^{3/2}}{h^3} + N_{\mathrm{A}} \ln V_{\mathrm{m}}$$

$$\left(\frac{\partial \ln Z_{\mathrm{m,\,trans}}}{\partial V_{\mathrm{m}}}\right)_T = 0 + 0 + N_{\mathrm{A}}\left(\frac{\partial \ln V_{\mathrm{m}}}{\partial V_{\mathrm{m}}}\right)_T = \frac{N_{\mathrm{A}}}{V_{\mathrm{m}}}.$$

For the rotation, vibration, and electronic contributions

$$\left(\frac{\partial \ln Z_{\mathrm{m,\,rot}}}{\partial V_{\mathrm{m}}}\right)_T = \left(\frac{\partial \ln Z_{\mathrm{m,\,vib}}}{\partial V_{\mathrm{m}}}\right)_T = \left(\frac{\partial \ln Z_{\mathrm{m,\,elect}}}{\partial V_{\mathrm{m}}}\right)_T = 0$$

since only the translational partition function depends upon $V$.

Hence,

$$\left(\frac{\partial \ln Z_{\mathrm{m}}}{\partial V_{\mathrm{m}}}\right)_T = \frac{N_{\mathrm{A}}}{V_{\mathrm{m}}}. \tag{10.112}$$

Substituting equation (10.112) into equation (10.111) gives

$$p = kT\frac{N_{\mathrm{A}}}{V_{\mathrm{m}}}$$

or

$$p = \frac{RT}{V_{\mathrm{m}}}$$

since

$$N_{\mathrm{A}}k = R.$$

## 10.7b Corrections to Table 10.4 for Diatomic Molecules

Under most circumstances the equations given in Table 10.4 accurately calculate the thermodynamic properties of the ideal gas. The most serious approximations involve the replacement of the summation with an integral [equations (10.94) and (10.95)] in calculating the partition function for the rigid rotator, and the approximation that the rotational and vibrational partition functions for a gas can be represented by those for a rigid rotator and harmonic oscillator. In general, the errors introduced by these approximations are most serious for the diatomic molecule.[w] Fortunately, it is for the diatomic molecule that corrections are most easily calculated. It is also for these molecules that spectroscopic information is often available to make the corrections for anharmonicity and nonrigid rotator effects. We will summarize the relationships

[w] It is difficult to generalize the validity of the equations in Table 10.4 for polyatomic molecules because of the great variety of sizes and shapes that are possible. It appears that no significant errors arise in using Table 10.4 to calculate the thermodynamic properties of room temperature. Errors increase with decreasing temperature, but calculations appear to remain valid at temperatures, down to 100 K or lower.

that can be used to calculate these corrections, but will not attempt a detailed derivation leading to the equations.[x]

**Rotational Partition Function Corrections:** The rigid-rotator partition function given in equation (10.94) can be written as

$$z_{\mathrm{rot}} = \sum_{J=0} \frac{(2J+1)}{\sigma} \exp[-J(J+1)y] \tag{10.113}$$

where

$$y = \frac{h^2}{8\pi^2 IkT}. \tag{10.114}$$

In our earlier derivation we approximated equation (10.113) by the integral

$$z_{\mathrm{rot}} = \frac{1}{\sigma} \int_0^{\infty} \{\exp[-J(J+1)y]\}(2J+1)\,\mathrm{d}J \tag{10.115}$$

which has a value

$$z_{\mathrm{rot}} = \frac{1}{\sigma y}. \tag{10.116}$$

Equation (10.116) is valid for $y \leqslant 0.01$. At $T = 300$ K, $y = 0.0012$ for $Cl_2$, 0.0069 for $O_2$, 0.0095 for $N_2$, 0.050 for HCl, and 0.283 for $H_2$, demonstrating that $H_2$ creates a problem in this calculation, even at high temperatures, and molecules such as HCl present problems at room temperature and below.

A more elaborate treatment, which does not involve replacing the summation in equation (10.113) with an integral, gives

$$z_{\mathrm{rot}} = \frac{1}{\sigma y}\left(1 + \frac{y}{3} + \frac{y^2}{15} + \cdots\right). \tag{10.117}$$

[x] For a more detailed description of the derivation, see K. S. Pitzer, *Thermodynamics*, McGraw-Hill, Inc., New York, 1995, pp. 366–371.

Substitution into the equations relating $Z_{m,rot}(=z_{rot}^{N_A})$ to the thermodynamic properties gives

$$(H_m - H_{m,0})_{rot} = RT\left[1 - \frac{y}{3} - \frac{y^2}{45} - \cdots\right] \tag{10.118}$$

$$(S_m)_{rot} = R\left[1 - \ln y - \ln\sigma - \frac{y^2}{90} - \cdots\right] \tag{10.119}$$

$$(C_{p,m})_{rot} = R\left[1 + \frac{y^2}{45} + \cdots\right] \tag{10.120}$$

$$(G_m - H_{0,m}) = -RT\left[-\ln y - \ln\sigma + \frac{y}{3} + \frac{y^2}{90} + \cdots\right] \tag{10.121}$$

**Example 10.5** Calculate the rotational corrections to $(H_m - H_{0,m})_{rot}$ for HF at 300 K.

**Solution:** From Table 10.1, $I = 1.36 \times 10^{-47}$ kg·m$^2$. Since HF is not a symmetrical molecule, $\sigma = 1$. From equation (10.114)

$$y = \frac{(6.63 \times 10^{-34})^2}{(8\pi^2)(1.36 \times 10^{-47})(1.38 \times 10^{-23})(300)}$$

$$= 0.0989$$

$$(H_m - H_{m,0})_{rot} = RT\left[1 - \frac{0.0989}{3} - \frac{(0.0989)^2}{45} - \cdots\right]$$

$$= RT[1 - 0.033 - 0.0002 - \cdots].$$

Using Table 10.4, which neglects the correction terms ($-0.033$ and $-0.0002$) to calculate $(H_m - H_{m,0})$, would introduce an error of 3.3% in the calculation.

**Anharmonicity and Nonrigid Rotator Corrections:** With the rigid rotator and harmonic oscillator approximations, the combined energy for rotation and vibration can be written as

$$\frac{\epsilon}{hc} = \tilde{\omega}(\mathrm{v} + 1/2) + \tilde{B}J(J+1) \tag{10.122}$$

where $\tilde{\omega}$ is the fundamental vibrational frequency and $\tilde{B}$ is the rotational constant. They are given by

$$\tilde{\omega} = \frac{\nu}{c}$$

and

$$\tilde{B} = \frac{B}{c} = \frac{h}{8\pi^2 Ic}. \tag{10.123}$$

With $c$ given in cm·s$^{-1}$, both $\tilde{\omega}$ and $\tilde{B}$ have the units of cm$^{-1}$, which is the unit most commonly used by spectroscopists.

But a rotating molecule stretches as it is excited to higher rotational energy levels. This leads to errors in the calculation that assumes a rigid rotator. Also, the vibrational levels are not equally spaced in energy; the spacing actually decreases with increasing level. Furthermore, the molecule stretches as the vibrations increase. Both of these effects lead to deviations from the values obtained assuming a harmonic oscillator.

To correct for these effects, the combined energy due to rotation and vibration is often written as

$$\frac{\epsilon}{hc} = \tilde{\omega}_e(\mathrm{v} + 1/2) - x_e\tilde{\omega}_e(\mathrm{v} + 1/2)^2 + \tilde{B}_e J(J+1)$$
$$- \tilde{D}J^2(J+1)^2 - \tilde{\alpha}(\mathrm{v} + 1/2)J(J+1) + \cdots \tag{10.124}$$

The vibrational and rotational constants are now written as $\tilde{\omega}_e$ and $\tilde{B}_e$. They may be thought of as the values that correspond to the equilibrium interatomic distance of minimum potential energy.[y] The first and third terms are expressions

[y] It can be shown from theoretical considerations that relate the stretching to the centrifugal force, and to the force holding the atoms together, that $\tilde{D}$ and $\tilde{B}_e$ are related by

$$\tilde{D} = 4\frac{\tilde{B}_e^3}{\omega_e^2}.$$

that take the place of the harmonic oscillator and rigid rotator terms. The second term corrects for the fact that the spacing between vibrational levels decreases with increasing v. Thus, it is an anharmonicity correction.[z] The fourth term corrects for the fact that the molecule stretches with increased rotation. It is a nonrigid rotator correction. Finally, the fifth term corrects for the expansion of the molecule due to anharmonicity. It is a combination term that represents the effect on the rotational levels of stretching the molecules in higher vibrational levels.

Equation (10.124) gives the total rotational and vibrational energy of the diatomic molecule above the bottom of the potential energy well. In Section 10.6c we saw that for the derivation of the vibrational partition function for the harmonic oscillator, we needed to use energy levels relative to the ground state. That is, we do not want to include $\epsilon_0$, the zero point vibrational energy. An expression equivalent to equation (10.124) that gives the energy above the ground state (v = 0, $J = 0$) can be written as follows

$$\frac{\epsilon_{\mathrm{v},J} - \epsilon_0}{hc} = \tilde{\omega}_0 \mathrm{v} - x\tilde{\omega}_0 \mathrm{v}(\mathrm{v} - 1) + \tilde{B}_0 J(J+1) - \tilde{D}J^2(J+1)^2 - \tilde{\alpha}(\mathrm{v} + 1/2)J(J+1) + \cdots \tag{10.125}$$

where the coefficients are related to those in equation (10.124) by

$$\tilde{\omega}_0 = \tilde{\omega}_\mathrm{e} - 2\tilde{\omega}_\mathrm{e} x_\mathrm{e} \tag{10.126}$$

$$\tilde{B}_0 = \tilde{B}_\mathrm{e} - \frac{1}{2}\tilde{\alpha} \tag{10.127}$$

and

$$x = \frac{x_\mathrm{e}}{1 - 2x_\mathrm{e}} \cong x_\mathrm{e}. \tag{10.128}$$

---

[z] There are also higher-order correction terms that usually make a negligible contribution. We will not include them in our analysis.

Values for the constants for a few diatomic molecules are given in Tables 10.1 and 10.2.

The partition function corresponding to equation (10.125) can be written as

$$z = \sum_{v=0} \sum_{J=0} \frac{(2J+1)}{\sigma} \exp\left(-\frac{\epsilon_{v,J} - \epsilon_0}{kT}\right). \tag{10.129}$$

By starting with this partition function and going through considerable mathematical manipulation, one arrives at the following equations for calculating the corrections to the rigid rotator and harmonic oscillator values calculated from Table 10.4.[aa]

$$\left(\frac{H_m - H_{0,m}}{T}\right)_{corr} = \left(\frac{R}{u}\right)\left\{2x\left[\frac{u^2\{2u\exp(u) - \exp(u) + 1\}}{\{\exp(u) - 1\}^3}\right] + \delta\left[\frac{u^2\exp(u)}{\{\exp(u) - 1\}^2}\right] + 8\gamma\right\} \tag{10.130}$$

$$-\left(\frac{G_m - H_{0,m}}{T}\right)_{corr} = \left(\frac{R}{u}\right)\left\{2x\left[\frac{u^2}{\{\exp(u) - 1\}^2}\right] + \delta\left[\frac{u}{\exp(u) - 1}\right] + 8\gamma\right\} \tag{10.131}$$

$$(S_m)_{corr} = \frac{(H_m - H_{0,m})_{corr}}{T} - \frac{(G_m - H_{0,m})_{corr}}{T} \tag{10.132}$$

[aa] Equations for obtaining anharmonicity and nonrigid rotator corrections are also summarized in Table A4.5 of Appendix 4.

and

$$(C_{\text{m}})_{\text{corr}} = \left(\frac{R}{u}\right)\left\{4x\left[\frac{\{u^3 \exp(u)\}\{2u \exp(u) - 2 \exp(u) + u + 2\}}{\{\exp(\text{u}) - 1\}^4}\right]\right\}$$

$$+ \left(\frac{R}{u}\right)\left\{\delta\left[\frac{\{u^3 \exp(u)\}\{\exp(u) + 1\}}{\{\exp(u) - 1\}^3}\right] + 16\gamma\right\}, \qquad (10.133)$$

where[bb]

$$u = \frac{hc\tilde{\omega}_0}{kT} \qquad (10.134)$$

$$y = \frac{hc\tilde{B}_0}{kT} \qquad (10.135)$$

$$x = \frac{x_e}{1 - 2x_e} \approx \frac{x_e\tilde{\omega}_e}{\tilde{\omega}_e} \qquad (10.128)$$

$$\delta = \frac{\tilde{\alpha}}{\tilde{B}_0} \qquad (10.136)$$

and

$$\gamma = \frac{\tilde{B}_e}{\tilde{\omega}_e}. \qquad (10.137)$$

Equations (10.130) to (10.133) can be simplified for calculation of the corrections by first defining and calculating the functions $\nu$ and $w$ given by

$$\nu = \exp(u) - 1 \quad \text{and} \quad w = \exp(u) = \nu + 1.$$

---

[bb] Note that in equation (10.134), and the relationships that use this equation, we are using the symbol $u$ to represent a quantity to which we gave the symbol $x$ when we worked with the harmonic oscillator. We have done so because $x$ represents a different quantity in equations (10.124), (10.125), and (10.135). We will return to $x$ as a representation for a similar quantity when we discuss the Einstein and Debye theories of solids.

Substitution of these quantities into equations (10.130), (10.131), and (10.133) gives

$$-\left(\frac{G_{\mathrm{m}} - G_{0,\mathrm{m}}}{T}\right)_{\mathrm{corr}} = \left(\frac{R}{u}\right)\left\{\left[\frac{2u^2}{\nu^2}\right]x + \left[\frac{u}{\nu}\right]\delta + 8\gamma\right\}$$

$$\left(\frac{H_{\mathrm{m}} - H_{0,\mathrm{m}}}{T}\right)_{\mathrm{corr}} = \left(\frac{R}{u}\right)\left\{\left[\left(\frac{2u^2}{\nu^3}\right)(2uw - \nu)\right]x + \left[\frac{u^2 w}{\nu^2}\right]\delta + 8\gamma\right\}$$

$$(C_{\mathrm{m}})_{\mathrm{corr}} = \left(\frac{R}{u}\right)\left\{\left[\left(\frac{4u^3}{\nu^4}\right)(2uw^2 + uw - 2\nu w)\right]x + \left[\left(\frac{u^3}{\nu^3}\right)(w^2 + w)\right]\delta + 16\gamma\right\}.$$

With modern computers, the calculations are not difficult using either set of equations.

**Example 10.6:** Calculate $(G^\circ_{\mathrm{m}} - H^\circ_{0,\mathrm{m}})$ and $C^\circ_{p,\mathrm{m}}$ for $N_2$ (ideal gas) at 2000 K. Include corrections for anharmonicity and nonrigid rotator effects.

**Solution:** From Tables 10.1 and 10.2

$$\tilde{\omega}_0 = 2330.70\ \mathrm{cm}^{-1},\quad \tilde{\omega}_{\mathrm{e}} = 2359.61\ \mathrm{cm}^{-1},\quad \tilde{\omega}_{\mathrm{e}}x_{\mathrm{e}} = 14.456\ \mathrm{cm}^{-1}$$

$$\tilde{B}_0 = 2.000\ \mathrm{cm}^{-1},\quad \tilde{B}_{\mathrm{e}} = 2.010\ \mathrm{cm}^{-1}$$

$$\tilde{\alpha} = 0.0187\ \mathrm{cm}^{-1}.$$

From equations (10.134) to (10.137)

$$u = \frac{hc\tilde{\omega}_0}{kT} = \frac{(1.4387)(2330.70)}{(2000)} = 1.6766$$

$$y = \frac{hc\tilde{B}_0}{kT} = \frac{(1.4387)(2.000)}{(2000)} = 0.001439$$

$$x = \frac{x_{\mathrm{e}}\tilde{\omega}_{\mathrm{e}}}{\tilde{\omega}_{\mathrm{e}}} = \frac{14.46}{2359.6} = 0.0061$$

$$\delta = \frac{\tilde{\alpha}}{\tilde{B}_0} = \frac{0.0187}{2.000} = 0.00935$$

$$\gamma = \frac{\tilde{B}_e}{\tilde{\omega}_e} = \frac{2.010}{2359.6} = 0.00852.$$

Calculations that involve Table 10.4 and the corrections equations with

$$\nu = \exp(u) - 1 = \exp(1.6766) - 1 = 4.3473$$

and

$$w = \exp(u) = 5.3473$$

give the following results (broken down by type of correction) for $N_2(g)$ at $T = 2000$ K.

| | $-\frac{(G^\circ_m - H^\circ_{m,0})}{T} \Big/ (J{\cdot}K^{-1}{\cdot}mol^{-1})$ | $C^\circ_{p,m}/(J{\cdot}K^{-1}{\cdot}mol^{-1})$ |
|---|---|---|
| Translation ($M = 0.028016$ kg·mol$^{-1}$) | 169.09 | 20.786 |
| Rigid rotator ($y = 0.001439$) | 48.65 | 8.314 |
| Harmonic oscillator ($u = 1.6766$) | 1.72 | 6.611 |
| Corrections | | |
| Vibrational anharmonicity ($x = 0.0061$) | 0.01 | 0.096 |
| Rotational stretching ($\gamma = 0.000852$) | 0.03 | 0.067 |
| Vibration–rotation ($\delta = 0.00935$) | 0.02 | 0.088 |
| | 219.52 | 35.962 |

The corrections are small for $N_2$ since it has a large force constant. Other molecules would show larger effects.

Equations (10.130) to (10.133) give reliable corrections under most circumstances.[cc] As we alluded to earlier, an alternative is to sum the contributions from the actual energy levels to obtain the partition function.

[cc] The above corrections assume all of the molecules are in the ground electronic state so that the molecular parameters are the same for all molecules. The correction equations can give results in error when low-lying excited electronic states are present and significantly populated. An example is NO, which has an electronic state only 121.1 cm$^{-1}$ above the ground state. (See Table 10.3 for this and other examples.)

With the advent of modern high-speed computers, this is not difficult to do for diatomic molecules, and it is the procedure followed when energy level information is available to perform the summation. Similar procedures have been followed for some nonlinear molecules, although as we have noted earlier, Table 10.4 gives reliable values for these molecules under most circumstances. References can be found in the literature to formulas and tables for calculating corrections for selected nonlinear molecules.[1,2]

## 10.7c Contributions of Internal Rotation to the Thermodynamic Properties

Some molecules undergo an internal motion in which one part of the molecule rotates about a bond connecting it with the rest of the molecule. Some examples are

$H_3C$ — $CH_3$          $H_3C$ — O — H

$H_3C$ — $CCl_3$          $H_3C$ — $NO_2$

In ethane, for example, the two halves of the molecule rotate around the bond so that the molecule passes through eclipsed and staggered conformations and all the intermediate orientations. In methanol, the hydroxyl group swings around the methyl group.

As the parts of the molecule pass by one another, the potential energy associated with their interaction changes. Repulsive interactions that occur if parts of the molecule get too close will cause the potential energy to increase, while attractive interactions will reduce the potential energy.

The potential energy is often described in terms of an oscillating function like the one shown in Figure 10.9(a) where the minima correspond to the relative orientations in which the interactions are most favorable, and the maxima correspond to unfavorable orientations. In ethane, the minima would occur at the staggered conformation and the maxima at the eclipsed conformation. In symmetrical molecules like ethane, the potential function reflects the symmetry and has a number of equivalent maxima and minima. In less symmetric molecules, the function may be more complex and show a number of minima of various depths and maxima of various heights. For our purposes, we will consider only molecules with symmetric potential functions and designate the number of minima in a complete rotation as $n_r$. For molecules like ethane and $H_3C{-}CCl_3$, $n_r = 3$.

We are interested in understanding how this relative motion of two parts of the molecule contributes to its thermodynamic properties. It turns out that the thermodynamic treatment of this internal rotation depends upon the relative

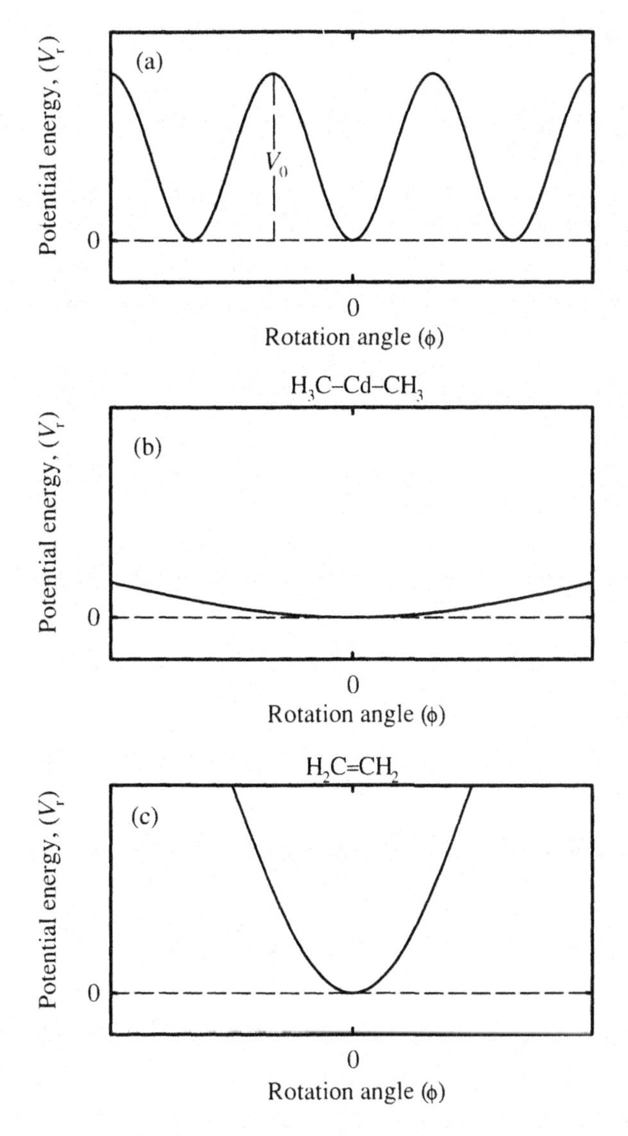


**Figure 10.9** Potential energy for internal rotation: (a), as a function of angle; (b), for a molecule such as dimethylcadmium with a small potential barrier; and (c), for a molecule such as ethene with a large potential barrier.

magnitude of $kT$ and the barrier height $V_0$. When the two parts of the molecule interact little — neither attractively nor repulsively — the well is flat and $V_0$ is small. When the interaction is strong and destabilizing, the potential rises steeply and the well is narrow. As an example of the first, the two methyl groups in dimethylcadmium are separated by the cadmium atom so that they may rotate almost independently of each other and the potential well is essentially flat as shown in Figure 10.9(b). In ethene, however, twisting the molecule about the C=C bond breaks the overlap of the orbitals leading to the $\pi$ bond and destabilizes the molecule; the potential barrier to rotation around this bond is steep with a narrow well as shown in Figure 10.9(c).

In calculating the thermodynamic properties of a molecule with internal rotation, a vibration is replaced by the internal rotation, leaving $(3\eta - 7)$ fundamental vibrations. The extent to which internal rotations contribute to the thermodynamic functions of a molecule is determined by the magnitude of $kT$, the energy available to thermally excite the molecule, relative to $V_0$. When the potential barrier is much greater than $kT$ at all except the highest of temperatures, the motion is treated as a (torsional) vibration, using the formulas we have developed for the harmonic oscillator. This would be the case for ethene. At the other extreme, when the potential well is flat and shallow, as in the case for dimethylcadmium, the thermal energy is sufficiently large to overcome the small barrier to rotation and the rotation can be treated as a free rotation. In the intermediate case, $kT$ and the barrier height are of comparable magnitude. This case is often referred to as "hindered" rotation. We will discuss briefly the quantum mechanical and statistical thermodynamic results for free and hindered rotation and see how the energy associated with these rotations affects the calculation of the thermodynamic properties.

**Free Rotation ($kT \gg V_0$)** The energy levels, $\epsilon_f$, for free internal rotation can be derived from quantum mechanics. They are

$$\epsilon_f = \frac{h^2}{8\pi^2 I_r} K^2 \qquad (10.138)$$

where $K = 0, \pm 1, \pm 2, \pm 3, \ldots$, and $I_r$ is a reduced moment of inertia for the internal rotation. When the molecular rotation involves a pair of symmetrical tops (for example, motion in ethane, 1,1,1-trichloroethane, and dimethylcadmium), the reduced moment of inertia is given by

$$I_r = \frac{I_A I_B}{I_A + I_B} \qquad (10.139)$$

where $I_A$ and $I_B$ are the moments of inertia of the individual tops about the axis of rotation. Internal rotations involving more complicated (usually less symmetric) motions can be found, but they are correspondingly more difficult to represent.

With equations (10.138) and (10.139) the partition function for free rotation can be written. However, when the internal rotation can be described by a potential energy with minima corresponding to equivalent orientations of the two parts of the molecule, the partition function must be modified by a factor analogous to the symmetry number in the rotation partition functions of linear and polyatomic molecules. If $n_f$ is the number of equivalent orientations for the free rotation, then the partition function $z_f$ is given by

$$z_f = \frac{1}{n_f} \sum_{K=-\infty}^{\infty} \exp\left( -\frac{h^2K^2}{8\pi^2 I_r kT} \right). \tag{10.140}$$

Because the quantity $(h^2/8\pi I_r k)$ is small at most $T$, the summation can be replaced by an integral over K in a procedure similar to that used to evaluate the rotational and translational partition functions earlier. The result is

$$z_f = \frac{1}{n_f} \left( \frac{8\pi^2 I_r kT}{h^2} \right)^{1/2} \tag{10.141}$$

or

$$Z_{m,f} = \left[ \frac{1}{n_f} \left( \frac{8\pi^2 I_r kT}{h^2} \right)^{1/2} \right]^{N_A}. \tag{10.142}$$

The thermodynamic properties can be calculated from $Z_{m,f}$ using the equations derived earlier. For example, the contribution to the heat capacity can be shown to be $\frac{1}{2}R$.

As an example of the application of this procedure, Pitzer[3] has calculated the thermodynamic properties for dimethylcadmium, including free rotation. At 298.15 K, he obtained the following result for the entropy:

| | $S^\circ_m/$ (J·K$^{-1}$·mol$^{-1}$) |
|---|---|
| Translation and overall rotation | 253.6 |
| Vibration | 36.7 |
| Free internal rotation | 12.2 |
| Total | 302.5 |

which agrees very well with the Third Law value[4] of $302.9 \pm 0.8$ J·K$^{-1}$·mol$^{-1}$.

**Hindered Rotation ($kT \approx V_0$)** With hindered rotation, the potential energy of the internal rotation is restricted by a potential barrier, $V_0$, whose magnitude varies as the two parts of the molecules rotate past each other in a cyclic fashion. For example, in the molecule $H_3C–CCl_3$, the potential varies as the hydrogen atoms on one carbon move past the chlorine atoms on the other.

In order to obtain the partition function for systems of this type (where the thermal energy and potential barrier are of the same magnitude), it is necessary to have the quantum mechanical energy levels associated with the barrier. Pitzer[5] has used a potential of the form

$$V_r = \frac{1}{2} V_0(1 - \cos n_f\phi) \tag{10.143}$$

to derive the mathematical function for the needed energy levels. In equation (10.143), $\phi$ is the rotational angle, $V_0$ is the height of the potential barrier, and $n_f$ is again the number of equivalent orientations. The expressions for the energy levels are not simple, however, and their use in the partition function and the subsequent calculation of thermodynamic contributions is beyond the scope of our work here. Pitzer[6] has tabulated results for the various thermodynamic quantities as a function of two variables: $V_0/RT$ and $1/z_f$, where $z_f$ is the value the partition function would have as a free rotator. This table is reproduced in Appendix 4 as Table A4.6. The reader is referred to Pitzer's original tabulation for further details.

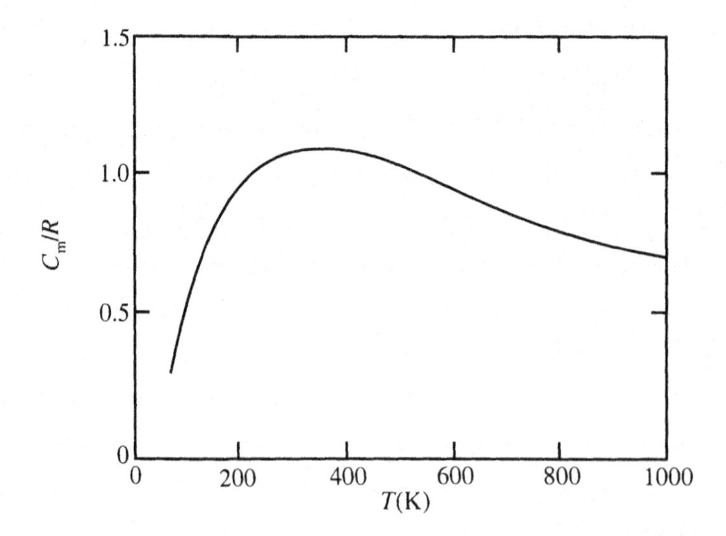


**Figure 10.10** Internal rotation contribution to the heat capacity of $CH_3–CCl_3$ as a function of temperature. Reprinted from K. S. Pitzer, *Thermodynamics*, McGraw-Hill, Inc., New York, 1995, p. 374. Reproduced with permission of the McGraw-Hill Companies.

One of the more interesting results of these calculations is the contribution to the heat capacity. Figure 10.10 shows the temperature dependence of this contribution to the heat capacity for $CH_3$–$CCl_3$ as calculated from Pitzer's tabulation with $I_r = 5.25 \times 10^{-47}$ kg·m$^2$ and $V_0/R = 1493$ K. The heat capacity increases initially, reaches a maximum near the value expected for an anharmonic oscillator, but then decreases asymptotically to the value of $\frac{1}{2}R$ expected for a free rotator as $kT$ increases above $V_0$. The total entropy calculated for this molecule at 286.53 K is 318.86 J·K$^{-1}$·mol$^{-1}$, which compares very favorably with the value of $318.94 \pm 0.6$ J·K$^{-1}$·mol$^{-1}$ calculated from Third Law measurements.[7]

## 10.8 Calculation of the Thermodynamic Properties of Solids

So far, we have used the statistical approach to calculate the thermodynamic properties of an ideal gas. Translational, rotational, vibrational, and electronic contributions were included, along with internal rotations where applicable.

We will now move to the opposite condition — from the very disordered state of the gas to the highly ordered state of the solid. In doing so, we will initially limit our discussion to atomic solids. Examples are solid Pb, C(diamond) and Ar.

### 10.8a The Einstein Heat Capacity Equation

One of the first attempts to calculate the thermodynamic properties of an atomic solid assumed that the solid consists of an array of spheres occupying the lattice points in the crystal. Each atom is "rattling" around in a hole at the lattice site. Adding energy (usually as heat) increases the motion of the atom, giving it more kinetic energy. The heat capacity, which we know is a measure of the ability of the solid to absorb this heat, varies with temperature and with the substance.[8] Figure 10.11, for example, shows how the heat capacity $C_{V,m}$ for the atomic solids Ag and C(diamond) vary with temperature.[dd, ee] The heat capacity starts at a value of zero at zero Kelvin, then increases rapidly with temperature, and levels out at a value of $3R$ (24.94 J·K$^{-1}$·mol$^{-1}$). The

---

[dd] For a solid, $C_{V,m}$ and $C_{p,m}$ do not differ significantly, especially at low temperatures. In equation (3.68) we showed that

$$C_{p,m} - C_{V,m} = \frac{\alpha^2 V_m T}{\kappa}.$$

[ee] In Chapter 4, we saw examples of the variation of $C_{p,m}$ with $T$ for molecular systems.

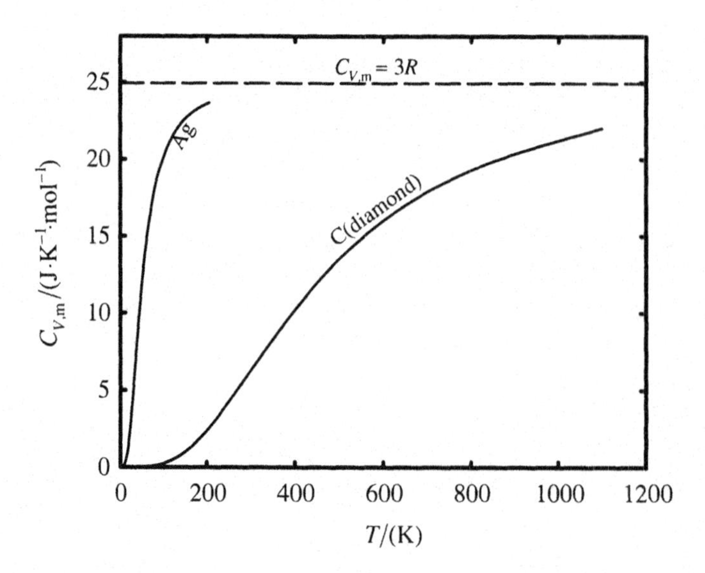


**Figure 10.11** Heat capacity $C_{V,m}$ for Ag and C(diamond) as a function of temperature $T$.

shape of the $C_{V,m}$ against $T$ curve is the same for different atomic solids, but in general we find that the harder and more rigid the crystal, the more slowly the $C_{V,m}$ curve increases with temperature. Thus, $C_{V,m}$ for diamond will eventually obtain a value of $3R$, but it does so at a much higher temperature than does silver.

Einstein[9] was the first to propose a theory for describing the heat capacity curve. He assumed that the atoms in the crystal were three-dimensional harmonic oscillators. That is, the motion of the atom at the lattice site could be resolved into harmonic oscillations, with the atom vibrating with a frequency in each of the three perpendicular directions. If this is so, then the energy in each direction is given by the harmonic oscillator term in Table 10.4

$$U_m - U_{0,m} = \frac{RTx}{\exp(x) - 1} \tag{10.144}$$

where

$$x = \frac{h\nu}{kT} = \frac{hc\tilde{\omega}}{kT}. \tag{10.145}$$

The quantity $(hc\tilde{\omega}/k)$ has the units of temperature. It is often written as $\theta_E$, the Einstein temperature, in which case

$$x = \frac{\theta_E}{T}. \tag{10.146}$$

In three dimensions, equation (10.144) becomes

$$U_m - U_{0,m} = RT \sum_{i=1}^{3} \frac{x_i}{\exp(x_i) - 1}.$$

Since $\nu$ is the same in each direction we get

$$U_m - U_{0,m} = 3RT\left[\frac{x}{\exp(x) - 1}\right]. \tag{10.147}$$

Differentiating equation (10.147) gives the heat capacity

$$C_{V,m} = \left[\frac{\partial(U_m - U_{0,m})}{\partial T}\right]_{V_m}.$$

The result is

$$C_{V,m} = 3R\left[\frac{x^2 \exp(x)}{\{\exp(x) - 1\}^2}\right]. \tag{10.148}$$

Equation (10.148) correctly predicts the qualitative dependence of $C_{V,m}$ on $T$. It is easy to show the following limiting values

$$\text{As } T \rightarrow 0, \quad x \rightarrow \infty \quad \text{and} \quad \frac{x^2 \exp(x)}{\{\exp(x) - 1\}^2} \rightarrow 0$$

$$\text{so that} \quad C_{V,m} \rightarrow 0$$

$$\text{As } T \rightarrow \infty, \quad x \rightarrow 0 \quad \text{and} \quad \frac{x^2 \exp(x)}{\{\exp(x) - 1\}^2} \rightarrow 1$$

$$\text{so that} \quad C_{V,m} \rightarrow 3R.$$

But at low temperatures, equation (10.148) does not quantitatively predict the shape of the $C_{V,\mathrm{m}}$ against $T$ curve. For example, Figure 10.12 compares the experimental value of $C_{V,\mathrm{m}}$ for diamond with that predicted from equation (10.148). It is apparent that the Einstein equation predicts that $C_{V,\mathrm{m}}$ for diamond will decrease too rapidly at low temperatures. Similar results would be obtained for Ag and other atomic solids.

## 10.8b The Debye Heat Capacity Equation

Equation (10.148) does not correctly predict $C_{V,\mathrm{m}}$ at low temperatures because it assumes all the atoms are vibrating with the same frequency. In the ideal gas, this is a good assumption, and in the previous section we used an equation similar to (10.148) to calculate the vibrational contribution to the heat capacity of an ideal gas.

But in a solid, the motions of the atoms are coupled together. The vibrational motion involves large numbers of atoms, and the frequency for these vibrations is less than the value used in equation (10.148). Thus, vibrations occur over a range of frequencies from $\nu = 0$ to a maximum $\nu = \nu_{\mathrm{m}}$. When this occurs the energy can be calculated by summing over all the frequencies. The

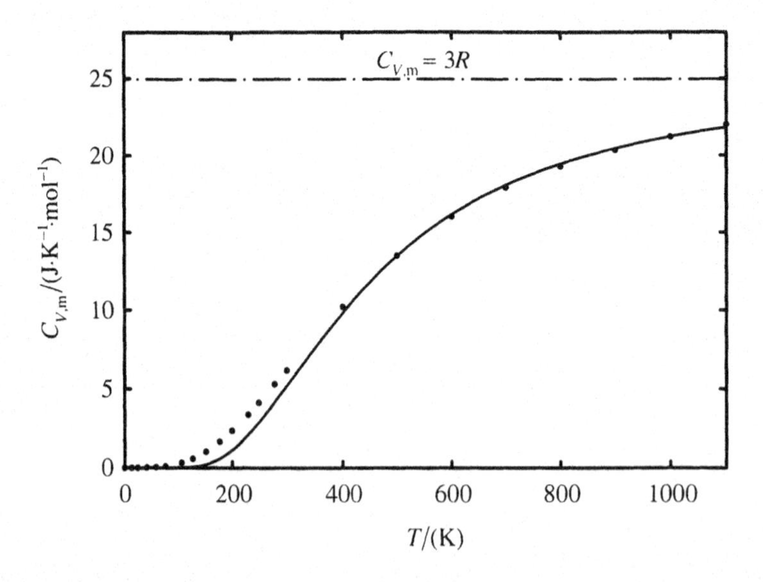


**Figure 10.12** Comparison for diamond of the experimental $C_{V,\mathrm{m}}$ (circles) and the prediction of the Einstein heat capacity equation with $\theta_{\mathrm{E}} = 1400$ K (solid line). The experimental results below $T = 300$ K are closely spaced in temperature, and not all are shown in the figure.

resulting relationship expressed in integral form is

$$U_{\mathrm{m}} - U_{0,\mathrm{m}} = \int_0^{\nu_{\mathrm{m}}} \frac{h\nu}{\exp\left(\dfrac{h\nu}{kT}\right) - 1} \, \mathrm{d}n \tag{10.149}$$

where $\mathrm{d}n$ is the number of vibrational modes with frequencies between $\nu$ and $\nu + \mathrm{d}\nu$.

To integrate equation (10.149) we must know how $\mathrm{d}n$ is related to $\nu$. Debye assumed that a crystal is a continuous medium that supports standing (stationary) waves with frequencies varying continuously from $\nu = 0$ to $\nu = \nu_{\mathrm{m}}$. The situation is similar to that for a black-body radiator, for which it can be shown that

$$\mathrm{d}n = \frac{12\pi V}{c^3} \nu^2 \, \mathrm{d}\nu \tag{10.150}$$

where $V$ is the volume that confines vibrations and $c$ is the velocity of elastic waves in the crystal. Substitution of equation (10.150) into equation (10.149) gives

$$U_{\mathrm{m}} - U_{0,\mathrm{m}} = \int_0^{\nu_{\mathrm{m}}} \frac{12\pi V}{c^3} \frac{h\nu^3}{\exp\left(\dfrac{h\nu}{kT}\right) - 1} \, \mathrm{d}\nu. \tag{10.151}$$

We can eliminate $c$ from equation (10.151) by integrating equation (10.150)

$$\int_0^{n_{\mathrm{max}}} \mathrm{d}n = \int_0^{\nu_{\mathrm{m}}} \frac{12\pi V}{c^3} \nu^2 \, \mathrm{d}\nu, \tag{10.152}$$

making use of the normalization condition that for $N_{\mathrm{A}}$ (one mole) of atoms, the number of vibrations in each direction is $N_{\mathrm{A}}$ so that the total is $3N_{\mathrm{A}}$. Hence

$$\int_0^{n_{\mathrm{max}}} \mathrm{d}n = 3N_{\mathrm{A}}.$$

Integrating the right side of equation (10.152) and setting it equal to $3N_A$ gives

$$c^3 = \frac{4\pi}{3N_A} V\nu_m^3. \quad (10.153)$$

Subsitution of equation (10.153) into equation (10.151) gives

$$U_m - U_{0,m} = \frac{9N_A h}{\nu_m^3} \int_0^{\nu_m} \frac{\nu^3 \, d\nu}{\exp\left(\frac{h\nu}{kT}\right) - 1}. \quad (10.154)$$

Differentiation of equation (10.154) with respect to $T$ gives an equation for $C_{V,m}$. The result is

$$C_{V,m} = \frac{9N_A h^2}{kT^2\nu_m^3} \int_0^{\nu_m} \frac{\nu^4 \exp\left(\frac{h\nu}{kT}\right)}{\left[\exp\left(\frac{h\nu}{kT}\right) - 1\right]^2} d\nu. \quad (10.155)$$

To simplify equation (10.155) we write

$$x = \frac{h\nu}{kT} \quad (10.156)$$

and

$$x_m = \frac{h\nu_m}{kT} = \frac{\theta_D}{T} \quad (10.157)$$

where $\theta_D$ is known as the Debye temperature. Equation (10.155) then becomes (with $N_A k = R$)

$$\frac{C_{V,m}}{3R} = \frac{3}{(\theta_D/T)^3} \int_0^{\theta_D/T} \frac{x^4 \exp(x)}{\{\exp(x) - 1\}^2} dx. \quad (10.158)$$

The evaluation of the integral in equation (10.158) can only be done numerically except in limiting cases. As $T \rightarrow \infty$, the upper limit becomes small, and

$$\frac{x^4 \exp(x)}{\{\exp(x) - 1\}^2} \rightarrow x^2,$$

in which case the integration leads to $\frac{1}{3}(\theta_D/T)^3$. Therefore

$$\frac{C_{V,m}}{3R} \rightarrow 1 \quad \text{as} \quad T \rightarrow \infty.$$

This is the same high temperature limit predicted by the Einstein equation and is the limit approached by experimental results for monatomic solids.[ff]

To find the limiting value of $C_{V,m}/3R$ as $T \rightarrow 0$, we integrate equation (10.158) by parts

$$\frac{3}{(\theta_D/T)^3} \int_0^{\theta_D/T} \frac{x^4 \exp(x)}{\{\exp(x) - 1\}^2} \mathrm{d}x = \frac{3}{(\theta_D/T)^3} \int_0^{\theta_D/T} x^4 \, \mathrm{d}\left(\frac{-1}{\{\exp(x) - 1\}}\right)$$

$$= \frac{3}{(\theta_D/T)^3} \left[\int_0^{\theta_D/T} \frac{4x^3}{\{\exp(x) - 1\}} \mathrm{d}x - \frac{x^4}{\{\exp(x) - 1\}}\right].$$

Therefore

$$\frac{C_{V,m}}{3R} = \frac{12}{(\theta_D/T)^3} \int_0^{\theta_D/T} \frac{x^3}{\{\exp(x) - 1\}} \mathrm{d}x - \frac{3x}{\{\exp(x) - 1\}}.$$

As $T \rightarrow 0$

$$\frac{3x}{\{\exp(x) - 1\}} = \frac{3(\theta_D/T)}{\left[\exp\left(\dfrac{\theta_D}{T}\right) - 1\right]} \rightarrow 0$$

[ff] The law of Dulong and Petit was a formulation of this experimentally observed limit.

and

$$\int_0^{\theta_D/T} \frac{x^3}{\{\exp(x) - 1\}} \, dx \rightarrow \int_0^{\infty} \frac{x^3}{\{\exp(x) - 1\}} \, dx = 3!\xi(4)$$

where $\xi(4)$ is a Reimann zeta function equal to

$$\xi(4) = 1 + \frac{1}{2^4} + \frac{1}{3^4} + \cdots = \frac{\pi^4}{90}.$$

Combining equations gives the limiting relationship at low temperatures

$$\frac{C_{V,\mathrm{m}}}{3R} = \frac{12}{(\theta_\mathrm{D}/T)^3} \cdot \frac{3!\pi^4}{90} \tag{10.159}$$

or

$$\frac{C_{V,\mathrm{m}}}{3R} = \frac{4\pi^4}{5}\left(\frac{T}{\theta_\mathrm{D}}\right)^3,$$

in which case

$$C_{V,\mathrm{m}} = \frac{12\pi^4 R}{5}\left(\frac{T}{\theta_\mathrm{D}}\right)^3. \tag{10.160}$$

Intermediate values for $C_{V,\mathrm{m}}$ can be obtained from a numerical integration of equation (10.158). When all are put together the complete heat capacity curve with the correct limiting values is obtained. As an example, Figure 10.13 compares the experimental $C_{V,\mathrm{m}}$ for diamond with the Debye prediction. Also shown is the prediction from the Einstein equation (shown in Figure 10.12), demonstrating the improved fit of the Debye equation, especially at low temperatures.

Equation (10.160) predicts that a graph of $C_{V,\mathrm{m}}/T$ against $T^2$ should be a straight line with intercept at $T = 0$ since

$$\frac{C_{V,\mathrm{m}}}{T} = \frac{12\pi^4 R}{5\theta_\mathrm{D}^3} T^2. \tag{10.161}$$

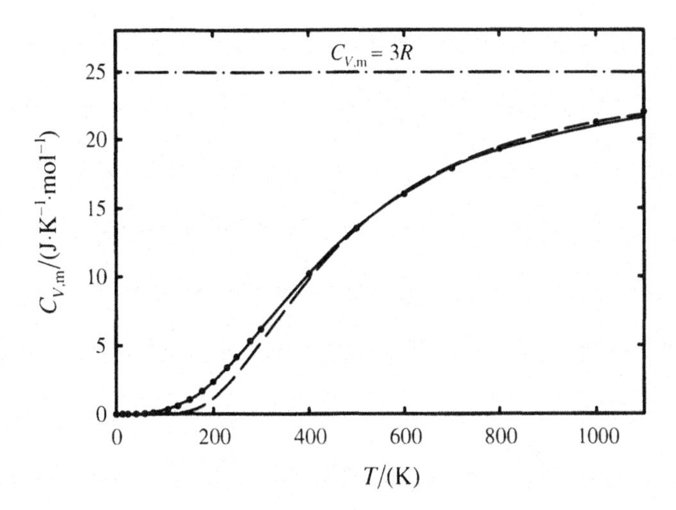


**Figure 10.13** Comparison of the experimental $C_{V,m}$ for diamond (circles), with the Einstein prediction with $\theta_E = 1400$ K (dashed line) and the Debye prediction with $\theta_D = 1890$ K (solid line). The experimental results below $T = 300$ K are closely spaced in temperature, and not all are shown in the figure.

Figure 10.14a shows such a plot for Kr.[10] The straight line below $T^2 \cong 4\ \mathrm{K}^2$ ($T \cong 2$ K) demonstrates the validity of equation (10.160). A graph similar to the one shown in Figure 10.14a was used in Chapter 4 to extrapolate $C_p$ results[gg] to zero Kelvin when we used the Third Law to obtain absolute entropies.

The Debye temperature $\theta_D$ can be calculated from the slope of the line. The value obtained for Kr is 72 K. This small $\theta_D$ results from the weak van der Waals forces that hold the Kr atoms together in the solid.

It is apparent from Figure 10.14a that significant deviations from equation (10.160) occur at higher temperatures. Barron and Morrison[11] report from an analysis of experimental results and model systems that the $T^3$ relationship should be accurately followed for $T \leqslant \theta_D/100$, with significant deviations occurring for $T \leqslant \theta_D/50$. This latter limit is approximately the temperature where we have shown in Figure 10.14a that deviations become important for Kr.

Figure 10.14b shows a similar graph for Cu.[12] The straight line shown in this figure demonstrates that deviations begin for Cu at $T^2 > 35\ \mathrm{K}^2$ ($T > 6$ K). The $\theta_D = 345$ K obtained for Cu from this line reflects the stronger bonding present in this metallic solid when compared to Kr. With the larger $\theta_D$, deviations become important at significantly higher temperatures for Cu than

[gg] At these very low temperatures, $C_{p,m}$ and $C_{V,m}$ are equal.

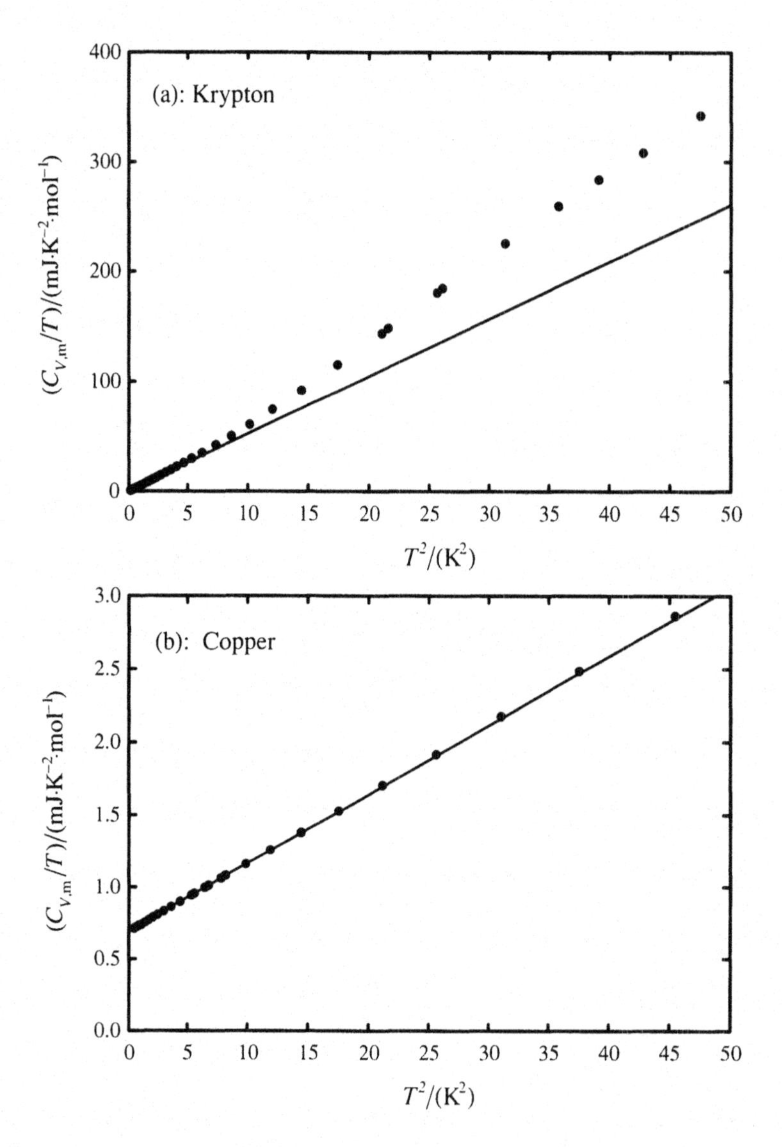


**Figure 10.14** Graph showing the limiting behavior at low temperatures of the heat capacity of (a), krypton, a nonconductor, and (b), copper, a conductor. The straight line in (a) follows the prediction of the Debye low-temperature heat capacity equation. In (b), the heat capacity of the conduction electrons displaces the Debye straight line so that it does not go to zero at 0 K.

for Kr.[hh] Because of the inverse (cubic) relationship between $C_{V,\text{m}}$ and $\theta_\text{D}$, at a given temperature, the heat capacity is much larger for Kr than for Cu.

---

[hh] As with Kr, the deviations for Cu begin to become significant for $T \leqslant \theta_\text{D}/50$.

A complication shown in Figure 10.14b is that the graph of $C_{V,m}/T$ against $T^2$ does not extrapolate to zero at $T = 0$. In Chapter 4, we indicated that at very low temperatures, a heat capacity term with $C_{V,m}$ (or $C_{p,m}$) proportional to $T$ becomes important for metals with free electrons (such as Cu). In that case, we wrote

$$C_{V,m} = aT^3 + \gamma T$$

or

$$\frac{C_{V,m}}{T} = aT^2 + \gamma$$

and a graph of $C_{V,m}/T$ against $T^2$ should extrapolate to $\gamma$ at $T = 0$. For Cu, the value obtained from Figure 10.14b is $\gamma = 6.93 \times 10^{-4}\ \text{J}\cdot\text{K}^{-2}\cdot\text{mol}^{-1}$.

The Debye temperature, $\theta_D$, can be calculated from the elastic properties of the solid. Required are the molecular weight, molar volume, compressibility, and Poisson's ratio.[ii] More commonly, $\theta_D$ is obtained from a fit of experimental heat capacity results to the Debye equation as shown above. Representative values for $\theta_D$ are as follows:

| Substance | $\theta_D/(\text{K})$ |
|---|---|
| Kr | 72 |
| Pb | 90.3 |
| Au | 169 |
| Ag | 213 |
| Cu | 345 |
| Al | 389 |
| Fe | 417 |
| Diamond | 1890 |

[ii] For a calculation of $\theta_D$, see R. H. Fowler, *Statistical Thermodynamics*, Second Edition, Cambridge University Press, 1936, p. 127. In Section 1.5a of Chapter 1 we defined the compressibility and cautioned that this compressibility can be applied rigorously only for gases, liquids, and isotropic solids. For anisotropic solids where the effect of pressure on the volume would not be the same in the three perpendicular directions, more sophisticated relationships are required. Poisson's ratio is the ratio of the strain of the transverse contraction to the strain of the parallel elongation when a rod is stretched by forces applied at the end of the rod in parallel with its axis.

From an examination of the values, it is easy to see the correlation of $\theta_D$ with the hardness of the solid. Figure 10.15 compares the fit of the Debye equation for several elements using the $\theta_D$ values given in the table. Reasonably good fits are obtained, although not perfect. One must keep in mind that the assumptions used in the derivation of equation (10.158) are approximations and simplifications of the "real" behavior of the solid. It is pleasing to see how well this equation works under the circumstances. Of even more interest is the observation that the low temperature $T^3$ relationship approximates the behavior of more complicated molecular and ionic solids at low temperatures. This is the basis for using this relationship for extrapolating heat capacity values to zero Kelvin in Third Law entropy calculations.

Once equation (10.158) has been obtained for relating $C_{V,m}$ to $T$ for a Debye solid, equations relating $(U_m - U_{0,m})$, $(H_m - U_{0,m})$, and $S_m$ to $T$ can be derived. Tables of values, expressed in terms of $\theta_D/T$, can be found in Table A4.7, Appendix 4, with more extensive tables found in the literature[jj] to calculate these thermodynamic properties.

### 10.8c Contribution to the Heat Capacity of Solids from Low-lying Electronic Levels: The Schottky Effect

We have concentrated our discussion of solids on non-metallic systems where there are only paired electrons. In such systems, the excited electronic states are very much higher in energy than the ground state, typically in excess of

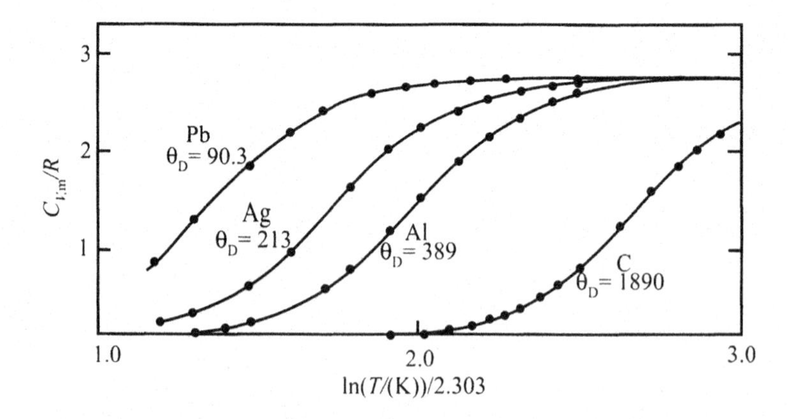


**Figure 10.15** Comparison of the fit of the Debye heat capacity equation for several elements. Reproduced from K. S. Pitzer, *Thermodynamics*, McGraw-Hill, Inc., New York, 1995, p. 78. Reproduced with permission of the McGraw-Hill Companies.

[jj] K. S. Pitzer, *Thermodynamics*, McGraw-Hill, Inc., New York, 1995, pp. 449–504.

10,000 $cm^{-1}$. As previous examples for gases[kk] have shown, such states do not begin to contribute to the thermodynamic functions until temperatures exceed several thousand Kelvin. However, in solid systems with unpaired electrons, as in transition and lanthanide metal compounds, excited electronic states exist at much lower energies. Thermal excitation into these low-lying states contributes to the thermodynamic functions. We will illustrate this contribution for the heat capacity where it is sometimes referred to as a Schottky heat capacity term.[ll]

We have seen that for the electronic partition function there is no closed form expression (as there is for translation, rotation, and vibration) and one must know the energy and degeneracy of each state. That is,

$$z_{\text{elect}} = \sum_i g_i \exp\left(-\frac{\epsilon_i}{kT}\right) \tag{10.162}$$

$$Z_{\text{m, elect}} = z_{\text{elect}}^{N_A} \tag{10.163}$$

where $z_{\text{elect}}$ is the atomic partition function, with the summation taken over all the electronic states, and $Z_{\text{m, elect}}$ is the molar electronic partition function. At low temperatures, or if the energies are very high, most of the terms in the partition function are negligible and can be ignored. Furthermore, if unpaired electrons are not present, $g_0 = 1$ so that $z_{\text{elect}} = Z_{\text{m, elect}} = 1$.

Using equation (10.77), the electronic contribution to the heat capacity can be obtained by appropriate differentiation of the partition function:

$$C_{\text{elect}} = 2kT\left(\frac{\partial \ln Z_{\text{m, elect}}}{\partial T}\right)_{V_m} + kT^2\left(\frac{\partial^2 \ln Z_{\text{m, elect}}}{\partial T^2}\right)_{V_m}. \tag{10.77}$$

Since $Z_{\text{m, elect}} = (z_{\text{elect}})^{N_A}$, $\ln Z_{\text{m, elect}} = N_A \ln z_{\text{elect}}$ and

$$C_{\text{elect}} = 2RT\left(\frac{\partial \ln z_{\text{elect}}}{\partial T}\right)_{V_m} + RT^2\left(\frac{\partial^2 \ln z_{\text{elect}}}{\partial T^2}\right)_{V_m}. \tag{10.164}$$

---

[kk] In our discussion of gases, we considered contributions from electronic states. Our discussion here, although applied to solids, is also applicable to gases.

[ll] Other types of low-lying energy levels can also lead to Schottky effects. For example, rotational energy levels can give rise to a Schottky effect. We will limit our discussion here to electronic levels.

Given that

$$\left(\frac{\partial \ln z_{\text{elect}}}{\partial T}\right)_{V_m} = \frac{1}{z_{\text{elect}}}\left(\frac{\partial z_{\text{elect}}}{\partial T}\right)_{V_m}$$

the electronic contribution to the molar heat capacity can be shown to be

$$C_{\text{elect}}/R = \left[\frac{\sum_i g_i\left(\frac{\epsilon_i}{kT}\right)^2 \exp\left(-\frac{\epsilon_i}{kT}\right)}{z_{\text{elect}}}\right] - \left[\frac{\sum_i g_i\left(\frac{\epsilon_i}{kT}\right) \exp\left(-\frac{\epsilon_i}{kT}\right)}{z_{\text{elect}}}\right]^2. \tag{10.165}$$

As we saw earlier for gaseous atomic oxygen,[mm] each exponential term begins to contribute to the electronic partition function only when $T$ is on the same order of magnitude as $\epsilon_i/k$ (or 1.439 $\tilde{\epsilon}_i$ for energies in $\text{cm}^{-1}$). Thus it is reasonable to expect that the electronic contribution to the heat capacity will be significant only in this region as well. Figure 10.16 shows the Schottky or electronic heat capacity contribution for a simple two-level system with one excited state, where both the ground and excited state degeneracies are unity. Curves are shown for excited states 30, 300 and 1000 $\text{cm}^{-1}$ above the ground state. Each curve starts at $C_{\text{elect}} = 0$ and increases to a maximum. Note that, since the degeneracies of the electronic states are all unity, the maximum value of the heat capacity is the same for each curve.

In Figure 10.17 we see the effect on the Schottky heat capacity for a simple two level system, but with different degeneracies of ground and excited states. We note that the temperature at which the maximum heat capacity occurs is independent of degeneracy. But, increasing the degeneracy of the ground state decreases the heat capacity maximum, while increasing the degeneracy of the excited state increases the maximum value of the heat capacity. This effect can be understood by consideration of the summations in equation (10.162). Going from a nondegenerate to a degenerate ground state increases $z_{\text{elect}}$ but, because

[mm] See Section 10.6d.

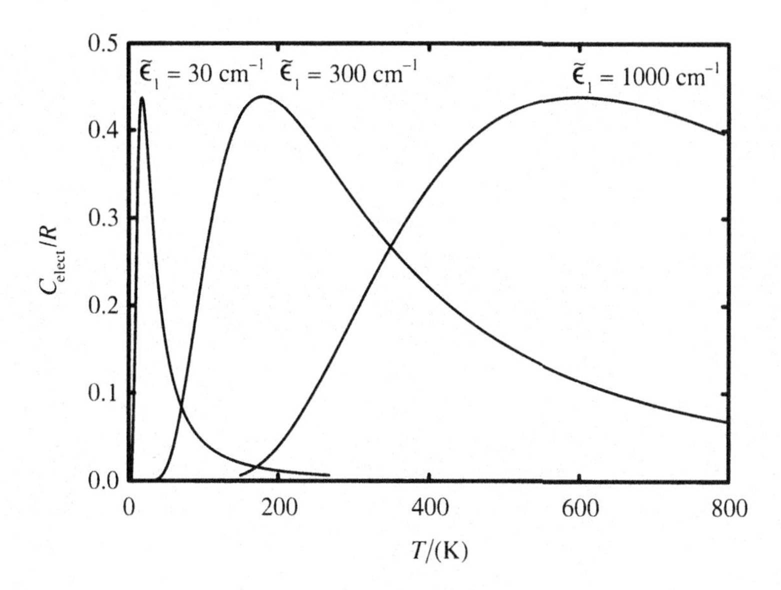


**Figure 10.16** The Schottky heat capacity curve for a single excited state with energies of 30, 300 and 1000 cm$^{-1}$ and a degeneracy of one.

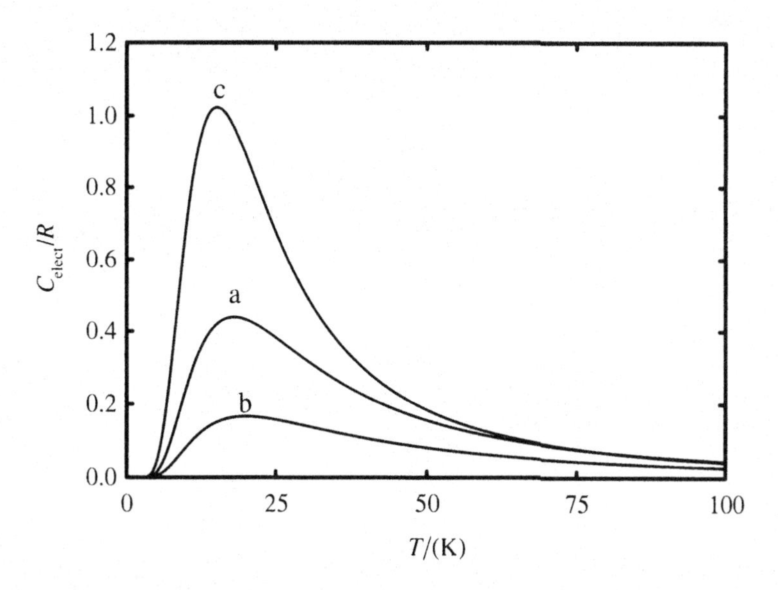


**Figure 10.17** The effect of degeneracy on a Schottky heat capacity from a single excited state 30 cm$^{-1}$ above the ground state. (a), $g_0 = g_1 = 1$; (b), $g_0 = 3$, $g_1 = 1$; and (c), $g_0 = 1$, $g_1 = 3$.

$\epsilon_0 = 0$, it does not affect either summation in equation (10.165). Since $z_{elec}$ appears in the denominator of both terms in equation (10.165), the net effect is to reduce the value of $C_{elect}$. However, when the excited states are degenerate, both summation terms are increased because the degeneracy factor is multiplied by the non-zero energy factor. The effect on $z_{elec}$ is not as large and the net result is an increase in $C_{elect}$.

As examples of the behavior observed in more complicated (real) systems, we show the Schottky heat capacities arising from thermal excitation into excited electronic states of the $Eu^{3+}$ and $Pr^{3+}$ ions in $Eu(OH)_3$ and $Pr(OH)_3$. Table 10.5 shows the energy levels and degeneracies of the lowest excited states of these ions, and Figure 10.18 illustrates the heat capacity contribution arising from these states. The differences in the two curves reflect the different energies and degeneracies associated with the two sets of states. The $Pr^{3+}$ ion has a very low-lying electronic level that gives rise to the sharp peak with a maximum at 7.1 K. Thermal occupation of the next states, separated by more than 100 $cm^{-1}$ from the first excited state, leads to a second maximum near 105 K. The ground state of $Pr^{3+}$ in $Pr(OH)_3$ is doubly degenerate, while that of $Eu^{3+}$ in $Eu(OH)_3$ is non-degenerate, and the $C_{elect}$ maximum is smaller for $Pr^{3+}$ then for $Eu^{3+}$. Because the lowest excited state in $Eu(OH)_3$ is much higher than that of the $Pr(OH)_3$, the maximum is at a higher temperature and the peak is broader. The curve for $Eu(OH)_3$ drops more slowly from the maximum because the electronic contributions from the two states near 1000 $cm^{-1}$ are beginning to be significant at the same temperature at which that from the 300 and 400 $cm^{-1}$ states begin to decrease.

**Table 10.5** Energies and degeneracies of electronic states arising from the lanthanide ions in $Eu(OH)_3$ and $Pr(OH)_3$.*

| $Eu(OH)_3$ | | $Pr(OH)_3$ | |
|---|---|---|---|
| $\varepsilon_i/(cm^{-1})$ | $g_i$ | $\varepsilon_i/(cm^{-1})$ | $g_i$ |
| 0 | 1 | 0 | 2 |
| 335 | 2 | 11 | 1 |
| 436 | 1 | 120 | 2 |
| 1012 | 2 | 175 | 1 |
| 1028 | 2 | 270 | 2 |
| 1151 | 1 | 380 | 1 |
| | | 6865 | 1 |

* Data taken from R. D. Chirico, E. F. Westrum, Jr., J. B. Gruber, and J. Warmkessel, *J. Chem. Thermodyn.*, **11**, 835–850 (1979); and R. D. Chirico and E. F. Westrum, Jr. *J. Chem. Thermodyn.*, **12**, 79–85 (1980).

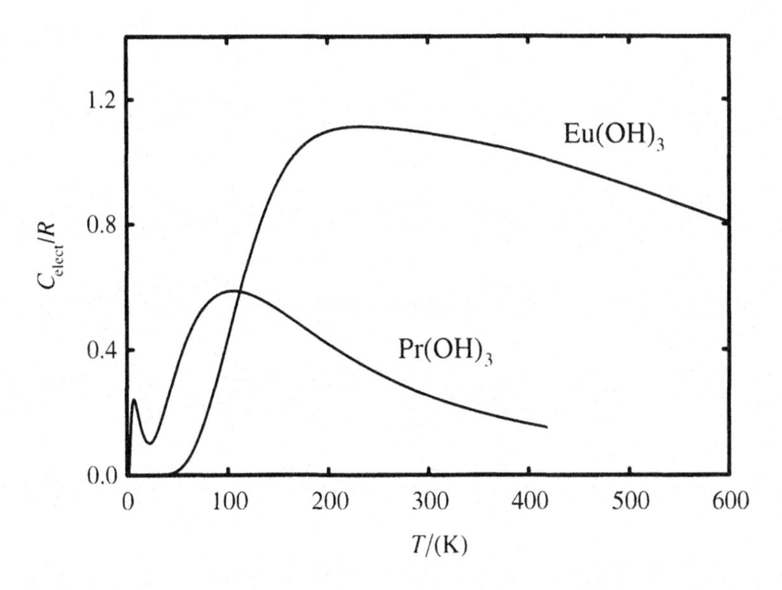


**Figure 10.18** The electronic heat capacity in $Eu(OH)_3$ and $Pr(OH)_3$.

The electronic contribution is generally only a relatively small part of the total heat capacity in solids. In a few compounds like $Pr(OH)_3$ with excited electronic states just a few wavenumbers above the ground state, the Schottky anomaly occurs at such a low temperature that other contributions to the total heat capacity are still small, and hence, the Schottky anomaly shows up. Even in compounds like $Eu(OH)_3$ where the excited electronic states are only several hundred wavenumbers above the ground state, the Schottky maximum occurs at temperatures where the total heat capacity curve is dominated by the vibrational modes of the solid, and a peak is not apparent in the measured heat capacity. In compounds where the electronic and lattice heat capacity contributions can be separated, calorimetric measurements of the heat capacity can provide a useful check on the accuracy of spectroscopic measurements of electronic energy levels.

## Exercises

Data summarized in Tables 10.1 to 10.3 can be used to solve the exercises and problems given in this chapter. Unless specifically stated otherwise, the rigid rotator and harmonic oscillator approximations (and hence, Table 10.4) and the assumption of ideal gas can be used.

E10.1 Calculate the most highly populated rotational energy level in $I_2(g)$ at $T = 500$ K.

E10.2 Calculate the heat capacity and entropy of $Cl_2(g)$ at $T = 500$ K and $p = 0.101$ MPa. Include the anharmonicity and nonrigid rotator corrections.

E10.3 Calculate the heat capacity and entropy of NO(g) at $T = 500$ K and $p = 0.101$ MPa. You may neglect the anharmonicity and nonrigid rotator corrections.

E10.4 Calculate the heat capacity and entropy of $CO_2$(g) at $T = 500$ K and $p = 0.101$ MPa.

E10.5 Calculate the heat capacity and entropy of $NO_2$(g) at $T = 500$ K and $p = 0.101$ MPa.

E10.6 For the diatomic molecule $Na_2$, $S^\circ_m = 230.476\ J{\cdot}K^{-1}{\cdot}mol^{-1}$ at $T = 300$ K, and $256.876\ J{\cdot}K^{-1}{\cdot}mol^{-1}$ at $T = 600$ K. Assume the rigid rotator and harmonic oscillator approximations and calculate $\tilde{\omega}$, the fundamental vibrational frequency and $r$, the interatomic separation between the atoms in the molecule. For a diatomic molecule, the moment of inertia is given by $I = \mu r^2$, where $\mu$ is the reduced mass given by

$$\frac{1}{\mu} = \frac{1}{m_1} + \frac{1}{m_2}$$

with $m_1$ and $m_2$ the masses of the two atoms.

E10.7 Carbonyl sulfide (OCS) is a linear molecule with a moment of inertia of $137 \times 10^{-40}\ g{\cdot}cm^2$. The three fundamental vibrational frequencies are 521.50, 859.2, and 2050.5 $cm^{-1}$, but one is degenerate and needs to be counted twice in calculating the entropy. A Third Law measurement of the entropy of OCS (ideal gas) at the normal boiling point of $T = 222.87$ K and $p = 0.101325$ MPa gives a value of $219.9\ J{\cdot}K^{-1}{\cdot}mol^{-1}$. Use this result to decide which vibrational frequency should be given double weight.

E10.8 Calculate the temperature at which the $J = 4$ and $J = 5$ rotational levels of HCN would be equally populated.

E10.9 A gas contains molecules with two nondegenerate energy levels that are separated by $72.3\ kJ{\cdot}mol^{-1}$. Calculate the temperature when there are 10 times as many molecules in the lower state as in the upper.

E10.10 (a) For $O_2$(g) at $p = 1.00$ bar and $T = 500$ K, calculate and compare $z_{trans}$, $z_{rot}$, $z_{vib}$, $z_{elect}$, and $z_{total}$.

(b) At $T = 500$ K, calculate the fraction of the molecules in the $v = 2$ vibrational state. Also, calculate the fraction in the first excited electronic state.

E10.11 The electronic energies and degeneracies for the first two states of atomic fluorine are given below. All other states are at such high energies that their contribution is negligible.

| $\tilde{\epsilon}_i/(\text{cm}^{-1})$ | $g_i$ |
|---|---|
| 0 | 4 |
| 404 | 2 |

(a) Write the partition function and calculate the temperature at which one tenth of the molecules would be in the first excited electronic state.
(b) Calculate the enthalpy difference $(H^\circ - H_0^\circ)$ for atomic fluorine gas at 500 K.

E10.12 Use the information in Table 10.3 to
(a) Calculate the electronic contribution to the entropy of Cl at $T = 2000$ K.
(b) Calculate $C_{p,\text{m}}$, the total heat capacity, of Cl at $T = 2000$ K.

## Problems

P10.1 At what temperature would $N_2$(g) and $I_2$(g) have the same standard state entropy?

P10.2 Given the following molecular data for perchloryl fluoride. {Perchloryl fluoride ($ClO_3F$) is tetrahedral with three oxygens and a fluorine bonded to a central chlorine.}
Molecular mass

$$M = 0.102449 \text{ kg·mol}^{-1}$$

Moments of inertia

$$I_A I_B I_C = 4.29248 \times 10^{-135} \text{ kg}^3\text{·m}^6$$

Vibrational frequencies and degeneracies (in parentheses):

| $\tilde{\omega}/(\text{cm}^{-1})$ | $\tilde{\omega}/(\text{cm}^{-1})$ |
|---|---|
| 1061 (1) | 1315 (2) |
| 715 (1) | 589 (2) |
| 549 (1) | 405 (2) |

Using these molecular parameters, calculate the standard state entropy $S_\text{m}^\circ$ for $ClO_3F$ (g) at the normal boiling temperature of 226.48 K and compare with the value determined from the Third Law in problem P4.3. Explain any discrepancy between the results.

P10.3 Given the following properties of mercury

Normal boiling point = 629.88 K
$\Delta_{vap}H_m$ at normal boiling point = 59,107 J·mol$^{-1}$
Normal melting point = 234.29 K
$\Delta_{fus}H_m$ = 2295 J·mol$^{-1}$ at the normal melting point

Heat capacities of liquid Hg as a function of temperature

| $T$/(K) | $C_{p,m}$ of liquid Hg/ (J·K$^{-1}$·mol$^{-1}$) | $T$/(K) | $C_{p,m}$ of liquid Hg/ (J·K$^{-1}$·mol$^{-1}$) |
|---|---|---|---|
| 250 | 28.35 | 400 | 27.36 |
| 260 | 28.28 | 450 | 27.20 |
| 270 | 28.20 | 500 | 27.13 |
| 280 | 28.12 | 550 | 27.12 |
| 290 | 28.04 | 600 | 27.15 |
| 298.15 | 27.99 | 650 | 27.22 |
| 300 | 27.96 | 700 | 27.34 |
| 350 | 27.61 | 750 | 27.51 |

(a) Hg(g) is monatomic with no unpaired electrons. Calculate $S^\circ_m$ for Hg(g) at the normal boiling temperature. (Assume the gas behaves ideally.)

(b) Calculate $S^\circ_m$ for Hg(l) at $T$ = 298.15 K.

P10.4 The calorimetric measurement of the entropy of ideal ethyl chloride at 0.101 MPa and the normal boiling point of 285.37 K is 273.3 J·K$^{-1}$·mol$^{-1}$. The fundamental vibrational frequencies (degeneracies in parentheses) are 377, 655, 790, 970, 1050, 1070, 1290, 1319, 1385, 1400, 1450(2), and 3000(5) cm$^{-1}$. The three principal moments of inertia are $I_1 = 27.63 \times 10^{-49}$ kg·m$^2$, $I_2 = 150.3 \times 10^{-49}$ kg·m$^2$, and $I_3 = 167.1 \times 10^{-49}$ kg·m$^3$. The reduced moment of inertia is $5.800 \times 10^{-49}$ kg·m$^3$. Determine the potential barrier to internal rotation in ethyl chloride (in J·mol$^{-1}$) by using Table A4.6, Appendix 4, or tables from the literature.[13]

P10.5 The thermodynamic functions for solid, liquid, and gaseous carbonyl chloride ($COCl_2$) obtained from Third Law and statistical calculations are given in the following tables. The melting point of carbonyl chloride is 145.5 K. $\Delta H_0^\circ$ for the vaporization of liquid $COCl_2$ at 0 Kelvin is 38,187 J·mol$^{-1}$. The density of liquid $COCl_2$ as a function of the Kelvin temperature $T$ is given by

$$\rho/(\text{g·cm}^{-3}) = 1.42014 - 0.0023120\,(T - 273.15) - 2.872 \times 10^{-6}\,(T - 273.15)^2$$

(a) What is the temperature of the normal boiling point of $COCl_2$?
(b) What is the enthalpy of vaporization of $COCl_2$ at the normal boiling point?
(c) What is the vapor pressure of liquid $COCl_2$ at 200 K? (Assume the gas behaves ideally.)
(d) What is the entropy of liquid $COCl_2$ at $T = 200$ K and $p = 10.00$ MPa? You may assume that the density of liquid $COCl_2$ is independent of pressure.

**Table 1** Thermodynamic functions of gaseous $COCl_2$ (in J·K$^{-1}$·mol$^{-1}$)

| $T/(\text{K})$ | $C_{p,\text{m}}^\circ$ | $S_\text{m}^\circ$ | $-(G_\text{m}^\circ - H_{\text{m},0}^\circ)/T$ | $(H_\text{m}^\circ - H_{\text{m},0}^\circ)/T$ |
|---|---|---|---|---|
| 15 | 33.259 | 169.728 | 136.465 | 33.259 |
| 25 | 33.259 | 186.715 | 153.457 | 33.259 |
| 40 | 33.292 | 202.351 | 169.088 | 33.263 |
| 60 | 33.681 | 215.903 | 182.581 | 33.321 |
| 80 | 34.819 | 225.739 | 192.196 | 33.543 |
| 100 | 36.539 | 233.685 | 199.723 | 33.962 |
| 120 | 38.681 | 240.530 | 205.966 | 34.564 |
| 140 | 41.070 | 246.668 | 211.346 | 35.321 |
| 145.4 | 41.727 | 247.852 | 212.681 | 35.547 |
| 160 | 43.547 | 252.316 | 216.120 | 36.196 |
| 180 | 46.007 | 257.588 | 220.438 | 37.150 |
| 200 | 48.363 | 262.554 | 224.400 | 38.154 |
| 220 | 50.585 | 267.270 | 228.087 | 39.183 |
| 240 | 52.643 | 271.763 | 231.543 | 40.221 |
| 260 | 54.587 | 276.052 | 234.802 | 41.250 |
| 280.7 | 56.317 | 280.299 | 238.003 | 42.296 |
| 290 | 57.066 | 282.148 | 239.388 | 42.760 |
| 298.15 | 57.702 | 283.734 | 240.576 | 43.158 |
| 400 | 63.948 | 301.645 | 253.923 | 47.723 |

**Table 2** Thermodynamic functions for solid $COCl_2$ (in $J{\cdot}K^{-1}{\cdot}mol^{-1}$)

| $T/(K)$ | $C^{\circ}_{p,m}$ | $S^{\circ}_{m}$ | $-(G^{\circ}_{m} - H^{\circ}_{m,0})/T$ |
|---|---|---|---|
| 15 | 8.243 | 0.343 | 2.343 |
| 25 | 20.564 | 3.105 | 7.192 |
| 40 | 34.280 | 8.213 | 15.004 |
| 60 | 43.928 | 15.941 | 23.217 |
| 80 | 50.120 | 23.481 | 29.217 |
| 100 | 54.907 | 30.522 | 33.869 |
| 120 | 59.973 | 37.049 | 37.802 |
| 140 | 64.957 | 43.150 | 41.325 |
| 145.4 | 66.295 | 44.723 | 42.221 |

**Table 3** Thermodynamic functions of liquid $COCl_2$ (in $J{\cdot}K^{-1}{\cdot}mol^{-1}$)

| $T/(K)$ | $C^{\circ}_{p,m}$ | $S^{\circ}_{m}$ | $-(G^{\circ}_{m} - H^{\circ}_{m,0})/T$ |
|---|---|---|---|
| 145.4 | 104.834 | 44.723 | 81.693 |
| 160 | 102.801 | 52.656 | 83.718 |
| 180 | 101.039 | 62.639 | 85.726 |
| 200 | 99.621 | 71.751 | 87.182 |
| 220 | 99.466 | 80.115 | 88.303 |
| 240 | 99.880 | 87.839 | 89.245 |
| 260 | 100.441 | 95.014 | 90.086 |
| 280.7 | 100.675 | 101.952 | 90.851 |

P10.6 Given that at $T = 298.15$ K, $\Delta_f H^{\circ} = 17.506$ kJ·mol$^{-1}$ for ICl(g) and 62.421 kJ·mol$^{-1}$ for $I_2$(g)
For the reaction

$$I_2(g) + Cl_2(g) = 2ICl(g)$$

(a) Calculate $\Delta_r H^{\circ}$, $\Delta_r G^{\circ}$, and $\Delta_r S^{\circ}$ at $T = 500$ K. Neglect anharmonicity and nonrigid rotator corrections.
(b) Calculate $K_p$ for the above reaction at 500 K and calculate $p_{I_2}$, $p_{Cl_2}$ and $p_{ICl}$ at equilibrium at a total pressure of 1.00 bar and a temperature of 500 K when equal numbers of moles of $I_2$(g) and $Cl_2$(g) are mixed.
(c) At $T = 298.15$ K, $\Delta_f G^{\circ}$ of I(g) and $I_2$(g) are 70.174 kJ·mol$^{-1}$ and 19.325 kJ·mol$^{-1}$, respectively. Is the dissociation of $I_2$(g) into I(g) in

the reaction mixture at $T = 500$ K a complication that should be considered in the equilibrium calculation in (b).

P10.7 (a) Use the Debye heat capacity values in Table A4.7, Appendix 4, or tables in the literature[14] to calculate $C_{V,\mathrm{m}}$ for Kr(s) from 0 to 116 K and make a graph of $C_V$ against $T$. For Kr(s) $\theta_\mathrm{D} = 60$ K.

(b) Use the $C_{V,\mathrm{m}}$ results obtained in (a) to calculate by graphical integration the entropy of solid krypton at the normal melting point of 116.5 K and compare with values obtained directly from Table A4.7, or from the literature.

P10.8 Use the parameters (including the anharmonicity and nonrigid rotator corrections) in Tables 10.1 to 10.3 to calculate $S^\circ_\mathrm{m}$ and $C^\circ_{p,\mathrm{m}}$ for $O_2$(g) at 1000 K.

P10.9 We have seen that the heat capacity of silver at low temperatures can be represented by the expression

$$C_p = aT^3 + \gamma T$$

where $aT^3$ is the lattice contribution and $\gamma T$ is the electronic contribution. From the Debye heat capacity equation, we find that the coefficient $a$ is given by

$$a = \frac{12\pi^4 R}{5\theta_\mathrm{D}^3}$$

where $\theta_\mathrm{D}$ is the Debye temperature. Given the following heat capacities for Ag metal at low temperatures

| $T$/(K) | $C_{V,\mathrm{m}}$/(J·K$^{-1}$·mol$^{-1}$) | $T$/(K) | $C_{V,\mathrm{m}}$/(J·K$^{-1}$·mol$^{-1}$) |
|---|---|---|---|
| 1.35 | 0.00106 | 8.00 | 0.0987 |
| 2.00 | 0.00261 | 10.00 | 0.1987 |
| 3.00 | 0.00657 | 12.00 | 0.3473 |
| 4.00 | 0.0127 | 14.00 | 0.5590 |
| 5.00 | 0.0213 | 16.00 | 0.8452 |
| 6.00 | 0.0373 | 20.00 | 1.672 |
| 7.00 | 0.0632 | | |

plot the data in a way that will allow you to obtain values for $\gamma$ and $\theta_\mathrm{D}$.

## References


1. R. E. Pennington and K. A. Kobe, "Contributions of Vibrational Anharmonicity and Rotation–Vibration Interaction to Thermodynamic Functions", *J. Chem. Phys.*, **22**, 1442–1447 (1954).
2. H. W. Woolley, "Calculation of Thermodynamic Functions for Polyatomic Molecules", *J. Res. Natl. Bur. Stand.*, **56**, 105–110 (1956).
3. K. S. Pitzer, *Thermodynamics*, McGraw-Hill, Inc., New York, 1995, p. 373.
4. J. C. M. Li, "The Thermodynamic Properties of Cadmium Dimethyl: Heat Capacities from 14 to 291 °K, Heats of Transition, Fusion, and Vaporization, Vapor Pressure up to 296 °K and the Entropy of Ideal Gas", *J. Am. Chem. Soc.*, **78**, 1081–1083 (1956).
5. K. S. Pitzer, "Potential Energies for Rotation about Single Bonds", *Discuss. Faraday Soc.*, **10**, 66–73 (1951).
6. K. S. Pitzer, *Thermodynamics*, McGraw-Hill, New York, 1995, pp. 371–374 and Appendix 12.
7. T. R. Rubin, B. H. Levendahl, and D. M. Yost, "The Heat Capacity, Heat of Transition, Vaporization, Vapor Pressure and Entropy of 1,1,1-Trichloroethane", *J. Am. Chem. Soc.*, **66**, 279–282 (1944).
8. The heat capacity of silver was taken from C. Kittel, *Solid State Physics*, Wiley, New York, 1956. The heat capacities of diamond were taken from J. E. Desnoyers, and J. A. Morrison, "The Heat Capacity of Diamond between 12.8 and 222 °K", *Phil. Mag.*, **3**, 42–48 (1958); and A. C. Victor, "Heat Capacity of Diamond at High Temperatures", *J. Chem. Phys.*, **36**, 1903–1911 (1962).
9. A. Einstein, "Planck's Theory of Radiation and the Theory of Specific Heats", *Ann. Physik.*, **22**, 180–190 (1907).
10. Data obtained from L. Finegold and N. E. Phillips, "Low-Temperature Heat Capacities of Solid Argon and Krypton", *Phys. Rev.*, **177**, 1383–1391 (1969).
11. T. H. K. Barron and J. A. Morrison, "On the Specific Heat of Solids at Low Temperatures", *Canad. J. Phys.*, **35**, 799–810 (1957).
12. Unpublished data obtained from B. F. Woodfield and N. E. Phillips, University of California, Berkeley, California.
13. See G. N. Lewis, M. Randall, K. S. Pitzer, and L. Brewer, *Thermodynamics*, McGraw-Hill Book Company, New York, 1961, table 27.8 on page 441.
14. See G. N. Lewis, M. Randall, K. S. Pitzer, and L. Brewer, *Thermodynamics*, McGraw-Hill Book Company, New York, 1961, Appendix 5, p. 660.

## Appendix 1

# Mathematics for Thermodynamics

Thermodynamic derivations and applications are closely associated with changes in properties of systems. It should not be too surprising, then, that the mathematics of differential and integral calculus are essential tools in the study of this subject. The following topics summarize the important concepts and mathematical operations that we will use.

## A1.1 Operations with Derivatives and Integrals

The following derivatives and integrals are commonly encountered, where $x$, $y$, $u$, and $v$ are variables, $a$ is a constant, and $n$ an integer.

### Derivatives

$$\frac{d(au)}{dx} = a\frac{du}{dx}$$

$$\frac{d(u+v)}{dx} = \frac{du}{dx} + \frac{dv}{dx}$$

$$\frac{d(uv)}{dx} = u\frac{dv}{dx} + v\frac{du}{dx}$$

$$\frac{d\frac{u}{v}}{dx} = \frac{v\frac{du}{dx} - u\frac{dv}{dx}}{v^2}$$

$$\frac{du^n}{dx} = nu^{n-1}\frac{du}{dx}$$

$$\frac{d\frac{1}{u}}{dx} = -\frac{1}{u^2}\frac{du}{dx}$$

$$\frac{d\ln u}{dx} = \frac{1}{u}\frac{du}{dx}$$

$$\frac{de^u}{dx} = e^u\frac{du}{dx}$$

## Indefinite integrals

$$\int a \, \mathrm{d}x = a \int \mathrm{d}x = ax + \text{const} \qquad \int (u + v) \, \mathrm{d}x = \int u \, \mathrm{d}x + \int v \, \mathrm{d}x$$

$$\int x \, \mathrm{d}x = \frac{1}{2} x^2 + \text{const} \qquad \int \frac{\mathrm{d}x}{x^2} = -\frac{1}{x} + \text{const}$$

$$\int x^n \, \mathrm{d}x = \frac{1}{n+1} x^{n+1} + \text{const} \qquad \int \frac{\mathrm{d}x}{x} = \ln x + \text{const}$$

## Integration by parts

$$\int u \, \mathrm{d}v = uv - \int v \, \mathrm{d}u$$

## A1.2 Total Differentials and Relationships Between Partial Derivatives

Let us consider a function $Z$ that depends on two variables, $X$ and $Y$, and signify this with the notation $Z = f(X, Y)$. In addition to designating $Z$ as a function, we may also refer to $Z$ as the **dependent** variable, and $X$ and $Y$ as the **independent** variables. We can write a differential expression $\mathrm{d}Z$ that tells us the change in the dependent variable $Z$ arising from small changes in the independent variables, $\mathrm{d}X$ and $\mathrm{d}Y$. The result is

$$\mathrm{d}Z = \left(\frac{\partial Z}{\partial X}\right)_Y \mathrm{d}X + \left(\frac{\partial Z}{\partial Y}\right)_X \mathrm{d}Y. \tag{A1.1}$$

The quantities $\mathrm{d}X$ and $\mathrm{d}Y$ are called **differentials**, the coefficients in front of $\mathrm{d}X$ and $\mathrm{d}Y$ are called **partial derivatives**,[a] and $\mathrm{d}Z$ is referred to as a **total differential** because it gives the total change in $Z$ arising from changes in both $X$ and $Y$. If $Z$ were to depend upon additional variables, additional terms would be included in equation (A1.1) to represent the changes in $Z$ arising from changes in those variables. For much of our discussion, two variables describe the processes of interest, and therefore, we will limit our discussion to two independent variables, with the exception of the description of Pfaffian differentials in

[a] We refer to partial derivatives as the partial of $Z$ with respect to $X$ at constant $Y$, partial of $Z$ with respect to $Y$ at constant $X$, and so on.

Section A1.5 where it will be useful to consider a more general case involving three independent variables.

The total differential given by equation (A1.1) is a useful starting point for many thermodynamic derivations. If we consider a process in which $X$ and $Y$ are changed such that $Z$ remains constant, then $\mathrm{d}Z = 0$, and equation (A1.1) can be rearranged to yield

$$\frac{\mathrm{d}X_Z}{\mathrm{d}Y_Z} = -\frac{\left(\dfrac{\partial Z}{\partial Y}\right)_X}{\left(\dfrac{\partial Z}{\partial X}\right)_Y} \tag{A1.2}$$

where the subscript $Z$ on the differentials is to remind us that these changes were made for a special process, one in which $Z$ was held constant. We can show that

$$\frac{\mathrm{d}X_Z}{\mathrm{d}Y_Z} = \left(\frac{\partial X}{\partial Y}\right)_Z$$

by considering $X = X(Y, Z)$, for which the total differential

$$\mathrm{d}X = \left(\frac{\partial X}{\partial Y}\right)_Z \mathrm{d}Y + \left(\frac{\partial X}{\partial Z}\right)_Y \mathrm{d}Z$$

can be written. Specifying constant $Z$ ($\mathrm{d}Z = 0$) and dividing by $\mathrm{d}Y$ gives

$$\frac{\mathrm{d}X_Z}{\mathrm{d}Y_Z} = \left(\frac{\partial X}{\partial Y}\right)_Z. \tag{A1.3}$$

In our thermodynamic derivations, we will routinely make use of equations of the type represented by equation (A1.3) to replace $\mathrm{d}X_Z/\mathrm{d}Y_Z$ with $(\partial X/\partial Y)_Z$. We may also represent this ratio of differentials as $(\mathrm{d}X/\mathrm{d}Y)_Z$ in which the direct equality with the partial derivative $(\partial X/\partial Y)_Z$ is more immediately evident. That is

$$\left(\frac{\mathrm{d}X}{\mathrm{d}Y}\right)_Z = \left(\frac{\partial X}{\partial Y}\right)_Z.$$

If we now return to equation (A1.2) and substitute (A1.3), we get an important relationship

$$\left(\frac{\partial X}{\partial Y}\right)_Z = -\frac{\left(\frac{\partial Z}{\partial Y}\right)_X}{\left(\frac{\partial Z}{\partial X}\right)_Y}. \qquad (A1.4)$$

This equation allows us to calculate the partial derivative $(\partial X/\partial Y)_Z$ when it may be difficult or impossible to obtain an explicit relationship $X = X(Y, Z)$, but $Z = Z(X, Y)$ can be written explicitly and its derivatives can be evaluated. We will find this to be useful when we consider equations of state like the van der Waals equation that cannot be solved for $V$ as an explicit function of $T$ and $p$, and we are interested in $(\partial V/\partial T)_p$ or $(\partial V/\partial p)_T$.

As long as the subscripted variable (the variable held constant in the partial derivative) remains the same, partial derivatives can be treated in many ways like fractions. For example

$$\left(\frac{\partial Z}{\partial X}\right)_Y = \frac{1}{\left(\frac{\partial X}{\partial Z}\right)_Y} \qquad (A1.5)$$

and

$$\left(\frac{\partial Z}{\partial X}\right)_Y = \left(\frac{\partial Z}{\partial Q}\right)_Y \left(\frac{\partial Q}{\partial X}\right)_Y \qquad (A1.6)$$

where $Q = Q(X, Y)$.

With the help of equation (A1.5), equation (A1.4) can be rearranged to give

$$\left(\frac{\partial Z}{\partial X}\right)_Y \left(\frac{\partial Y}{\partial Z}\right)_X \left(\frac{\partial X}{\partial Y}\right)_z = -1. \qquad (A1.7)$$

Some students find equation (A1.7) easier to remember than equation (A1.4) because it can be generated from the application of a series of permutations.

One can start by writing any partial derivative involving $X$, $Y$, and $Z$ and produce the next one by making the substitution: $Z$ goes into the derivative where $X$ was, $X$ goes where $Y$ was, and $Y$ goes where $Z$ was. The third derivative is assembled by taking the second derivative and following the same rule. The permutation process can be pictured in the cyclic form:

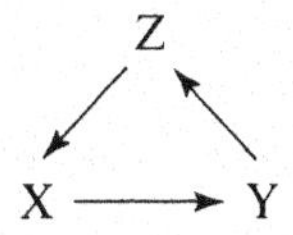


For well-behaved[b] functions, the order of differentiation does not matter, so that

$$\left(\frac{\partial}{\partial X}\left(\frac{\partial Z}{\partial Y}\right)_X\right)_Y = \left(\frac{\partial}{\partial Y}\left(\frac{\partial Z}{\partial X}\right)_Y\right)_X. \qquad \text{(A1.8)}$$

We will find this equation to be useful later in our discussion of state functions and exact differentials.

Occasionally, we will find it necessary to consider derivatives of the form $(\partial Z/\partial X)_Q$ where $Q = Q(X, Y)$. This derivative can be obtained by starting with the total differential for d$Z$ given by equation (A1.1), dividing by d$X$, and specifying constant $Q$ to get

$$\left(\frac{\mathrm{d}Z}{\mathrm{d}X}\right)_Q = \left(\frac{\partial Z}{\partial X}\right)_Y\left(\frac{\mathrm{d}X}{\mathrm{d}X}\right)_Q + \left(\frac{\partial Z}{\partial Y}\right)_X\left(\frac{\mathrm{d}Y}{\mathrm{d}X}\right)_Q,$$

which can be simplified with the help of equation (A1.3) to give

$$\left(\frac{\partial Z}{\partial X}\right)_Q = \left(\frac{\partial Z}{\partial X}\right)_Y + \left(\frac{\partial Z}{\partial Y}\right)_X\left(\frac{\partial Y}{\partial X}\right)_Q, \qquad \text{(A1.9)}$$

---

[b] Functions that are single-valued, continuous and differentiable are said to be well-behaved. Thermodynamic functions generally are well-behaved, except in the vicinity of first-order and continuous phase transactions. At the temperature and pressure of a first-order phase transition, the thermodynamic functions $V$, $U$, $H$, and $S$ are double-valued; that is, each phase has a different value of these functions.

since

$$\left(\frac{dX}{dX}\right)_Q = 1.$$

## A1.3 Intensive and Extensive Variables

In our study of thermodynamics we frequently work with the thermodynamic quantities: pressure $p$, temperature $T$, volume $V$, internal energy $U$, entropy $S$, enthalpy $H$, Helmholtz free energy $A$, and Gibbs free energy $G$.[c] The first two differ from the rest in that they are intensive variables. That is, they do not depend upon the quantity of matter present in the system. The others are extensive and do depend upon the amount. If the system is a pure substance, these quantities can be converted to intensive variables by dividing by the number of moles, $n$, to give the molar quantity. Thus, the molar volume $V_m = V/n$ is intensive, as is the molar entropy $S_m = S/n$. In general, the intensive molar quantity is given by $Z_m = Z/n$, where $Z$ is any of the extensive thermodynamic properties. For mixtures, the number of moles of the various components is a variable, and the effect of composition on the thermodynamic functions must be considered. There, the intensive variable is $(\partial Z/\partial n_i)_{T,p,n_{j\neq i}}$, and it is known as $\bar{Z}_i$, the partial molar property. We describe this quantity in detail in Chapter 5.

Another distinction that we make among the thermodynamic functions is to describe $p$, $V$, $T$, $U$, and $S$ as the fundamental properties of thermodynamics. The other quantities, $H$, $A$, and $G$ are derived properties, in that they are defined in terms of the fundamental properties, with

$$H = U + pV,$$

$$A = U - TS,$$

and

$$G = U + pV - TS.$$

[c] Other functions can also be included, if we consider the effect of surface area, gravitational field, etc. Derivatives of the thermodynamic functions such as the heat capacity, compressibility, and the coefficient of thermal expansion are also involved in the thermodynamic relationships that we derive and use extensively.

## A1.4 State Functions and Exact Differentials; Inexact Differentials and Line Integrals

### A1.4a State Functions

Generally, for a pure substance in which the composition is constant, only two of the thermodynamic quantities listed above need be specified as independent variables to uniquely define the system. In the presence of significant gravitational, electric, or magnetic fields, or where the surface area or chemical composition of the system is variable, additional quantities may be needed to fix the state of the system. We will limit our discussion to situations where these additional variables are held constant, and hence, do not need to be considered.

Most often, we will choose the independent variables to be those quantities we control in the laboratory. The usual thermodynamic choices are ($p$ and $T$) or ($V$ and $T$). Then, we measure changes in the thermodynamic properties of the system as these variables are altered. Thus, for a pure substance, writing

$$Z = f(p, T) \quad \text{or} \quad Z = f(V, T)$$

indicates that the value of $Z$ for a given substance will be uniquely determined once we have specified values for ($p$ and $T$) or ($V$ and $T$).[d] Relationships such as equations (A1.4) and (A1.7) enable us to change our assignment of dependent and independent variables, and we will have occasions to consider conditions of constant enthalpy or constant entropy, for example, where it is more convenient to take $H$ or $S$ as independent variables.

Quantities like $V$, $U$, $S$, $H$, $A$, and $G$ are properties of the system. That is, once the state of a system is defined, their values are fixed. Such quantities are called **state functions**. If we let $Z$ represent any of these functions, then it does not matter how we arrive at a given state of the system, $Z$ has the same value. If we designate $Z_1$ to be the value of $Z$ at some state 1, and $Z_2$ to be the value of $Z$ at another state 2, the difference $\Delta Z = Z_2 - Z_1$ in going from state 1 to state 2 is the same, no matter what process we take to get from one state to the other. Thus, if we go from state 1 through a series of intermediate steps, for which the changes in $Z$ are given by $\Delta Z_1, \Delta Z_2, \ldots, \Delta Z_i$, and eventually end up in state 2,

[d] We do not have to choose $p$, $V$, and $T$ as the independent variables. In general, choosing any two variables fixes the values of all the rest. Thus we can write $G = G(p, S)$ or $H = H(G, U)$. This is usually not done, since we have no easy way of varying $S$, $G$, and $U$ in the laboratory. We do sometimes choose $p$ or $V$ as dependent variables, but $T$ is almost always used as an independent variable. Thus, we may write $p = p(V, T)$ or $V = V(p, T)$, but we almost never write $T = T(p, V)$.

then $\Delta Z$ for the entire process is given by

$$\Delta Z = \sum_i \Delta Z_i. \tag{A1.10}$$

For a chemical process, equation (A1.10) is known as Hess's law, which states that the overall change $\Delta Z$ for a chemical process is the sum of the changes in $Z$ for the individual steps whose sum yields the desired reaction. This property is often used to calculate $\Delta_r H$ or $\Delta_r G$ for a chemical reaction by summing the enthalpy or Gibbs free energy changes for a series of reactions that add together to give the overall reaction.

When we go through a cyclic process and end up where we started, $\Delta Z = 0$, since we have returned to the same state, and therefore, the sum of changes in $Z$ for the steps in the closed cyclic path must also equal 0. We can write this mathematically as

$$\sum_{\text{path}}^{\text{closed}} \Delta Z_i = 0. \tag{A1.11}$$

This is true for finite changes, $\Delta Z$, or for infinitesimal changes, $\mathrm{d}Z$, where we replace the summation in equation (A1.11) with an integral sign and a circle to represent an integration over a closed path. That is

$$\oint \mathrm{d}Z = 0. \tag{A1.12}$$

Let us illustrate and extend this concept with a simple example. Consider the three paths that connect two states as shown in Figure A1.1. The sum of path

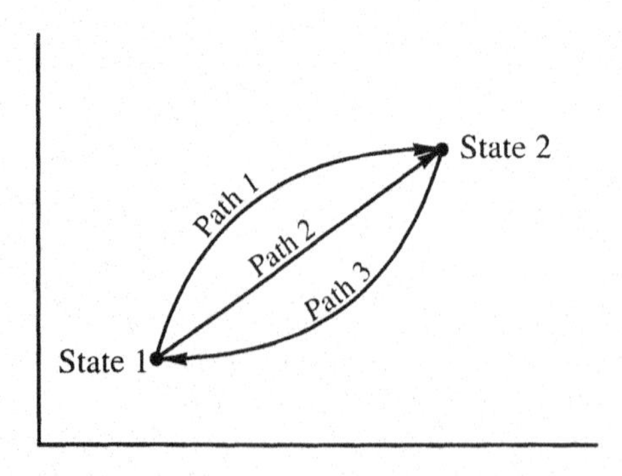


**Figure A1.1** Comparison of three different paths connecting states 1 and 2.

1 + path 3 constitutes a cycle and so the sum

$$\int_{\text{path 1}} \mathrm{d}Z + \int_{\text{path 3}} \mathrm{d}Z = 0, \tag{A1.13}$$

according to equation (A1.12), if $Z$ is a state function. Likewise, the sum of path 2 + path 3 constitutes a cycle and so

$$\int_{\text{path 2}} \mathrm{d}Z + \int_{\text{path 3}} \mathrm{d}Z = 0. \tag{A1.14}$$

Subtraction of equation (A1.14) from equation (A1.13) gives

$$\int_{\text{path 1}} \mathrm{d}Z - \int_{\text{path 2}} \mathrm{d}Z = 0 \tag{A.15}$$

or

$$\int_{\text{path 1}} \mathrm{d}Z = \int_{\text{path 2}} \mathrm{d}Z = \Delta Z. \tag{A1.16}$$

Equation (A1.16) shows that the integral evaluated over any two paths that connect states 1 and 2 must be equal. The value of the integral, $\Delta Z$, cannot depend upon the path but must be associated with the choice of states so that $\Delta Z = Z_2 - Z_1$. This is consistent with our earlier definition of a state function.

As a specific application of equation (A1.16), consider the following example.

**Example A1.1:** Calculate and compare $\Delta V$ for the following three expansions of one mole of ideal gas.

(a) A constant temperature (isothermal) expansion from $p = 0.2$ MPa and $V_{\text{m},1} = 25\ \text{dm}^3{\cdot}\text{mol}^{-1}$ to $p = 0.1$ MPa and $V_{\text{m},2} = 50\ \text{dm}^3{\cdot}\text{mol}^{-1}$.

(b) (i) A constant volume cooling at $V_{\text{m},1} = 25\ \text{dm}^3{\cdot}\text{mol}^{-1}$ from $p = 0.2$ MPa to $p = 0.1$ MPa, followed by (ii) a constant pressure expansion at $p = 0.1$ MPa from $V_{\text{m},1} = 25\ \text{dm}^3{\cdot}\text{mol}^{-1}$ to $V_{\text{m},2} = 50\ \text{dm}^3{\cdot}\text{mol}^{-1}$.

(c) (i) A constant pressure expansion at $p = 0.2$ MPa from $V_{\text{m},1} = 25\ \text{dm}^3{\cdot}\text{mol}^{-1}$ to $V_{\text{m},2} = 50\ \text{dm}^3{\cdot}\text{mol}^{-1}$, followed by (ii) a constant volume cooling at $V_{\text{m},2} = 50\ \text{dm}^3{\cdot}\text{mol}^{-1}$ from $p = 0.2$ MPa to $p = 0.1$ MPa.

**Solution** The steps followed in the three processes are shown in Figure A1.2. For process (a)

$$\Delta V = V_{m,2} - V_{m,1} = 50\ \text{dm}^3\cdot\ \text{mol}^{-1} - 25\ \text{dm}^3\cdot\ \text{mol}^{-1}$$

$$= 25\ \text{dm}^3\cdot\text{mol}^{-1}.$$

For process (b)

(i) $\Delta V_i = 0$
(ii) $\Delta V_{ii} = V_{m,2} - V_{m,1} = 50\ \text{dm}^3\cdot\text{mol}^{-1} - 25\ \text{dm}^3\cdot\text{mol}^{-1}$
$= 25\ \text{dm}^3\cdot\text{mol}^{-1}$
$\Delta V_{total} = \Delta V_i + \Delta V_{ii} = 25\ \text{dm}^3\cdot\text{mol}^{-1}.$

For process (c)

(i) $\Delta V_i = V_{m,2} - V_{m,1} = 50\ \text{dm}^3\cdot\text{mol}^{-1} - 25\ \text{dm}^3\cdot\text{mol}^{-1}$
$= 25\ \text{dm}^3\cdot\text{mol}^{-1}$
(ii) $\Delta V_{ii} = 0$
$\Delta V_{total} = \Delta V_i + \Delta V_{ii} = 25\ \text{dm}^3\cdot\text{mol}^{-1}.$

In this simple example, we note that $\Delta V$ is the same for all three processes. This is necessary, since it is evident from Figure A1.2 that all three processes start and end at the same state. Since $V$ is a state function, any other path would give the same result.

When the differential expression is for a change in a state function, $dZ$, equation (A1.12) follows almost trivially. It gains importance, however, in that

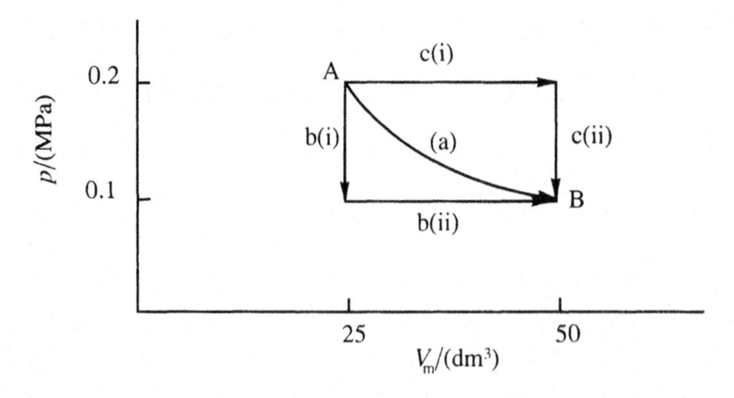


**Figure A1.2** Graph of $p$ against $V_m$ for the ideal gas. All three processes start at the same point A and end up at the same point B.

it provides us a way to determine whether some differential expression not derived directly from a state function, but obtained from some other means, is associated with a state function.

In order to develop this concept, it is convenient to distinguish between a differential expression, d$Z$, associated with a state function, $Z$, and a general differential expression whose possible connection with a state function is yet to be established. Let us introduce the notation $\delta Q$ to represent this latter differential expression.

Suppose then, we encounter a general differential expression and want to know whether it is associated with a state function. The behavior of this differential expression integrated over a closed path provides a means to answer this question. Two possibilities need be considered.

$$\oint \delta Q = 0 \tag{A1.17}$$

and

$$\oint \delta Q \neq 0. \tag{A1.18}$$

We have already established that an integral of a differential expression associated with a state function is zero over a closed path. Now, we must consider whether the converse of that statement is true. That is, if the integral of a general differential expression over a closed path is found to be zero, this expression is the differential expression of some state function. To answer the question, let us reconsider the example described in Figure (A1.1) and equations (A1.13) and (A1.14) and assume that equation (A1.17) is true for all closed (cyclic) paths. Then, for a path 1 and a path 3 that connect the same two states, 1 and 2,

$$\int_{\text{path 1}} \delta \mathrm{Q} + \int_{\text{path 3}} \delta \mathrm{Q} = 0, \tag{A1.19}$$

and for any path 2 that connects the same two states,

$$\int_{\text{path 2}} \delta \mathrm{Q} + \int_{\text{path 3}} \delta \mathrm{Q} = 0. \tag{A1.20}$$

Subtraction of (A1.20) from (A1.19) and rearrangement yields the same result obtained earlier:

$$\int_{\text{path 1}} \mathrm{d}Q = \int_{\text{path 2}} \mathrm{d}Q = \Delta Q. \tag{A1.21}$$

Thus, the assumption that an integration of a differential expression over a closed path is zero leads to a conclusion that an integration between two different, but fixed, states is independent of path. But, this property coincides with those we have ascribed to state functions. Thus, we have shown that a differential for which equation (A1.17) is true must correspond to the differential of some state function.

### A1.4b Exact and Inexact Differentials

By similar reasoning, one can show that differential expressions for which equation (A1.18) is true must yield integrals between two fixed states whose values depend upon the path. Such differential expressions cannot be associated with state functions because of the dependence upon path. Therefore, equations (A1.17) and (A1.18) distinguish between differentials that can ultimately be associated with state functions and that cannot. Expressions for which equation (A1.17) is true are called **exact** differentials while those for which equation (A1.18) is true are called **inexact** differentials.

It is not convenient to test for exactness by showing that equation (A1.17) is true for all closed paths, but an easier test can be developed. Consider a general differential expression, $\delta Q$, for a quantity $Q$ that is associated with the variables $X$ and $Y$:

$$\delta Q = M\,\mathrm{d}X + N\,\mathrm{d}Y \tag{A1.22}$$

where $M$ and $N$ are functions of $X$ and $Y$. If $\delta Q$ is an exact differential, then $\delta Q$ is equivalent to the total differential $\mathrm{d}Z$ for some function $Z$ that also depends on $X$ and $Y$. Thus

$$\delta Q = \mathrm{d}Z.$$

Using this relationship to combine equations (A1.1) and (A1.22) gives

$$M\,\mathrm{d}X + N\,\mathrm{d}Y = \left(\frac{\partial Z}{\partial X}\right)_Y \mathrm{d}X + \left(\frac{\partial Z}{\partial Y}\right)_X \mathrm{d}Y.$$

Comparing coefficients we see that $M = (\partial Z/\partial X)_Y$ and $N = (\partial Z/\partial Y)_X$. Differentiating $M$ with respect to $Y$ at constant $X$, and $N$ with respect to $X$ at constant $Y$ gives

$$\left(\frac{\partial M}{\partial Y}\right)_X = \left(\frac{\partial}{\partial Y}\left(\frac{\partial Z}{\partial X}\right)_Y\right)_X \tag{A1.23}$$

and

$$\left(\frac{\partial N}{\partial X}\right)_Y = \left(\frac{\partial}{\partial X}\left(\frac{\partial Z}{\partial Y}\right)_X\right)_Y. \tag{A1.24}$$

If $Z$ is well-behaved, then the order of differentiation does not matter, as we have seen in equation (A1.8), and equations (A1.23) and (A1.24) can be combined to give

$$\left(\frac{\partial M}{\partial Y}\right)_X = \left(\frac{\partial N}{\partial X}\right)_Y. \tag{A1.25}$$

Equation (A1.25) is known as the **Maxwell relation**. If this relationship is found to hold for $M$ and $N$ in a differential expression of the form of equation (A1.22), then $\delta Q = \mathrm{d}Q$ is exact, and some state function exists for which $\mathrm{d}Q$ is the total differential. We will consider a more general form of the Maxwell relationship for differentials in three dimensions later.

### A1.4c Line Integrals

As we have seen above, a path must be specified to integrate an inexact differential expression between two states, because different paths give different integration results. Since the value of the integral

$$Q = \int_{\text{state 1}}^{\text{state 2}} \delta Q$$

depends upon the details of how we get from state 1 to state 2, $Q$ must be associated with a process rather than ascribed to individual states. In mathematics, an integral of this type is known as a **line integral**, since its value depends upon the path (or line) followed in the integration. Heat and work are examples of thermodynamic quantities that depend upon the path and are calculated from line integrals.

In this text, we will not need to consider line integrals of the form

$$L = \int M \, dX + \int N \, dY$$

where the integration is taken over both variables. Rather, either by definition, or by choice of path, one integration will be eliminated. Therefore, we will write these line integrals in the general form

$$L = \int_1^2 f(x, y, \ldots) \, dx. \tag{A1.26}$$

When $f(x, y, \ldots)$ is plotted against $x$, the area under the curve is the integral. The line integral $L$ depends upon the path since $f(x, y, \ldots)$ depends upon the path. Figure A.1.3 is a graph of three different functions $f(x, y, \ldots)$ plotted against $x$. The area under each curve between A and B gives the value of the line integrals $L_a$, $L_b$, and $L_c$. Since the areas are different, the integrals are different, and the value of the integral depends upon the path. A specific example will demonstrate this effect.

**Example A.2** What is the value of the line integral

$$L = \int_A^B \frac{1}{x + y} \, dx$$

along each of the three paths in Figure A1.4?

**Solution**

(a) Following the curve $y = x^2$, we get

$$L = \int_A^B \frac{1}{x + y} \, dx = \int_1^2 \frac{dx}{x + x^2} = \int_1^2 \left( \frac{1}{x} - \frac{1}{1 + x} \right) dx$$

$$L = \ln x - \ln(1 + x) \big|_1^2 = \ln \frac{4}{3}$$

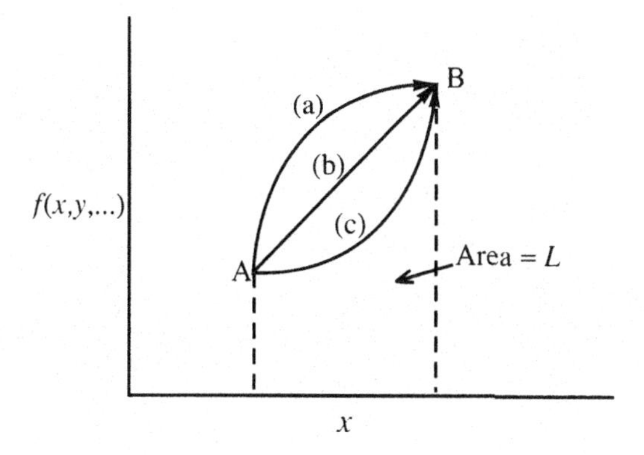


**Figure A1.3** The area between A and B under the curves (a), (b), and (c) gives the value for the line integrals. Since the areas are different, the values of the line integrals are different and depend upon the path.

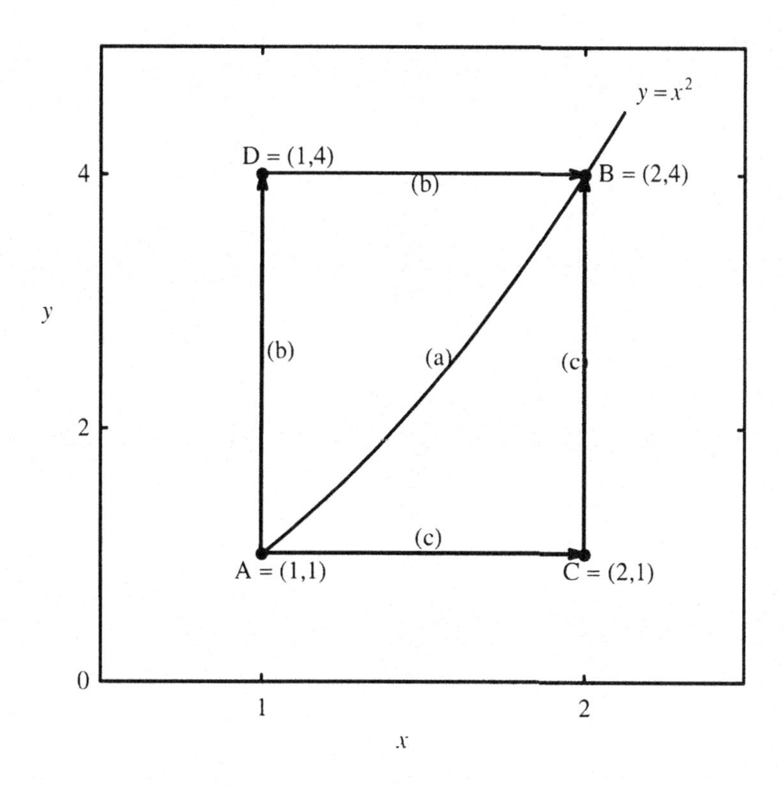


**Figure A1.4** Comparison of the line integrals obtained in (a), going from point A = (1, 1) to B = (2, 4) along the line $y = x^2$; (b) going from A = (1, 1) to D = (1, 4) and then from D = (1, 4) to B = (2, 4); and (c), going from A = (1, 1) to C = (2, 1) and then from C = (2, 1) to B = (2, 4).

(b) Following the paths AD + DB:

Along path AD, $\mathrm{d}x = 0$ so that $L_1 = 0$

Along path DB, $y = 4$ so

$$L_2 = \int_A^B \frac{1}{x+y}\,\mathrm{d}x = \int_1^2 \frac{\mathrm{d}x}{x+4}$$

$$= \ln(x+4)\,|_1^2 = \ln\frac{6}{5}$$

$$L = L_1 + L_2 = 0 + \ln\frac{6}{5} = \ln\frac{6}{5}$$

(c) Along the paths AC + CB:
Along path AC, $y = 1$ so that

$$L_1 = \int_A^B \frac{1}{x+y}\,\mathrm{d}x = \int_1^2 \frac{\mathrm{d}x}{x+1}$$

$$= \ln(x+1)\,|_1^2 = \ln\frac{3}{2}$$

And along path CB, $\mathrm{d}x = 0$ so that $L_2 = 0$

$$L = L_1 + L_2 = \ln\frac{3}{2} + 0 = \ln\frac{3}{2}$$

## A1.5 Pfaffian Differentials

We started our discussion of differentials in Section A1.4 and return to it now to develop some additional concepts. We start with differential expressions that contain three variables, because the results are more general than in the simpler two-dimensional case.[e]

[e] Extension to four or more variables does not add anything new. The conditions described here for three variables apply equally as well to any number of variables of four or more.

## A1.5a Pfaffian Differential Expressions in Three Dimensions

A differential expression of the form

$$M(x, y, z)\,\mathrm{d}x + N(x, y, z)\,\mathrm{d}y + P(x, y, z)\,\mathrm{d}z = \delta Q(x, y, z) \tag{A1.27}$$

is said to be a Pfaffian or linear differential expression in three dimensions. There are three possible behaviors for such an expression:

(a) $\delta Q$ is exact. That is, the differential expression represents the total differential of a function $F(x, y, z)$ such that $\delta Q = \mathrm{dF}$, with

$$M(x, y, z) = \left(\frac{\partial F}{\partial x}\right)_{y,z}, \qquad N(x, y, z) = \left(\frac{\partial F}{\partial y}\right)_{x,z}, \qquad \text{and}$$

$$P(x, y, z) = \left(\frac{\partial F}{\partial z}\right)_{x,y}. \tag{A1.28}$$

(b) $\delta Q$ is inexact, but integrable. That is, an integrating factor $\lambda(x, y, z)$ exists that can convert the inexact expression of $\delta Q$ into an exact one: $\lambda(x, y, z) \cdot \delta Q = \mathrm{d}F$.

(c) $dQ$ is inexact and no integrating factor exists.

## A1.5b Maxwell Relations in Three Dimensions

Reciprocity relations that guarantee exactness of three-dimensional Pfaffians are

$$\left(\frac{\partial M}{\partial y}\right)_{x,z} = \left(\frac{\partial N}{\partial x}\right)_{y,z}, \quad \left(\frac{\partial N}{\partial z}\right)_{x,y} = \left(\frac{\partial P}{\partial y}\right)_{x,z},$$

$$\left(\frac{\partial M}{\partial z}\right)_{x,y} = \left(\frac{\partial P}{\partial x}\right)_{y,z}. \tag{A1.29}$$

That is, if these three relationships are satisfied simultaneously for a given Pfaffian, the Pfaffian is exact, and some function $F(x, y, z)$ exists such that the total differential $\mathrm{d}F = \delta Q$.

The necessary (and sufficient) condition for a three-dimensional Pfaffian to be inexact but integrable is

$$M\left[\left(\frac{\partial P}{\partial y}\right)_{x,z}-\left(\frac{\partial N}{\partial z}\right)_{x,y}\right]+N\left[\left(\frac{\partial M}{\partial z}\right)_{x,y}-\left(\frac{\partial P}{\partial x}\right)_{y,z}\right]$$
$$+P\left[\left(\frac{\partial N}{\partial x}\right)_{y,z}-\left(\frac{\partial M}{\partial y}\right)_{x,z}\right]=0. \quad \text{(A1.30)}$$

If neither of the two sets of relations holds, then the Pfaffian expression can be considered to be inexact and non-integrable.

### A1.5c Differential Equations, Solution Curves, and Solution Surfaces

Three-dimensional Pfaffian expressions of the form

$$\delta Q = 0$$

are called Pfaffian differential equations. If the Pfaffian $\delta Q$ is exact, and, thus equal to the total differential $\mathrm{d}F$ of some function $F$, then the solutions to the differential equation are a family of surfaces $F(x, y, z) = \text{constant}$. Each surface is said to be a **solution surface** of the Pfaffian. To illustrate this concept and its implications, consider the Pfaffian differential equation

$$\mathrm{d}Q = (x-a)\,\mathrm{d}x + (y-b)\,\mathrm{d}y + (z-c)\,\mathrm{d}z = 0.$$

Application of the Maxwell relations {equation (A1.28)} will show that this differential is exact. Integration leads to a family of surfaces

$$F(x, y, z) = r^2 = (x-a)^2 + (y-b)^2 + (z-c)^2,$$

which is an equation for a sphere centered on $(a, b, c)$ with a radius $r$, where $r^2$ is the constant of integration.

One solution surface is a sphere with a radius of 5.00 centered at the origin ($a = b = c = 0$). The equation of this sphere is

$$x^2 + y^2 + z^2 = 25.00.$$

Other surfaces centered at the same point include those with radii of 5.01 and 4.99. Note that all the surfaces corresponding to different radii are parallel and non-intersecting. That is, no point can be on two surfaces simultaneously.

Within each solution surface are numerous subsets of points that also satisfy the differential equation $\delta Q = \mathrm{d}F = 0$. These subsets are referred to as **solution curves** of the Pfaffian. The curve $z = 0$, $x^2 + y^2 = 25.00$ is one of the solution curves for our particular solution surface with radius = 5.00. Others would include $x = 0$, $y^2 + z^2 = 25.00$, and $y = 0$, $x^2 + z^2 = 25.00$. Solution curves on the same solution surface can intersect. For example, our first two solution curves intersect at two points (5, 0, 0) and (−5, 0, 0). However, solution curves on one surface cannot be solution curves for another surface since the surfaces do not intersect. That two solution surfaces to an exact Pfaffian differential equation cannot intersect and that solution curves for one surface cannot be solution curves for another have important consequences as we see in our discussion of the Carathéodory formulation of the Second Law of Thermodynamics.

When the Pfaffian expression is inexact but integrable, then an integrating factor $\lambda$ exists such that $\lambda \cdot \delta Q = \mathrm{d}S$, where $\mathrm{d}S$ is an exact differential and the solution surfaces are $S$ = constant. While solution surfaces do not exist for the inexact differential $\delta Q$, solution curves do exist. The solution curves to $\mathrm{d}S = 0$ will also be solution curves to $\delta Q = 0$. Since solution curves for $\mathrm{d}S$ on one surface do not intersect those on another surface, a solution curve for $\delta Q = 0$ that lies on one surface cannot intersect another solution curve for $\delta Q = 0$ that lies on a different surface.

Thus, exact or integrable Pfaffians lead to non-intersecting solution surfaces, which requires that solution curves that lie on different solution surfaces cannot intersect. For a given point **p**, there will be numerous other points in very close proximity to **p** that cannot be connected to **p** by a solution curve to the Pfaffian differential equation. No such condition exists for non-integrable Pfaffians, and, in general, one can construct a solution curve from one point to any other point in space. (However, the process might not be a trivial exercise.)

## A1.5d Pfaffian Differential Expressions in Two Dimensions

In two dimensions, the Pfaffian differential expression reduces to the form we saw earlier

$$M(x, y)\,\mathrm{d}x + N(x, y)\,\mathrm{d}y = \delta Q(x, y). \qquad \text{(A1.31)}$$

For the two-dimensional Pfaffian, condition (c) above is no longer applicable. That is, the Pfaffian differential expression is either (a), exact, or (b), integrable with an integrating factor.

## A1.6 Euler's Theorem

Euler's theorem states that if a function $f(x, y, z, \ldots)$ is homogeneous of degree $n$, then

$$x\left(\frac{\partial f}{\partial x}\right)_{y,z} + y\left(\frac{\partial f}{\partial y}\right)_{x,z} + z\left(\frac{\partial f}{\partial z}\right)_{x,y} + \cdots = nf(x, y, z \ldots). \qquad \text{(A1.32)}$$

A homogeneous function of degree $n$ is one for which

$$f(kx, ky, kz) = k^n f(x, y, z). \qquad \text{(A1.33)}$$

As a simple example, consider the function

$$f(x, y) = ax^2 + bxy + cy^2.$$

Then

$$\begin{aligned} f(kx, ky) &= a(kx)^2 + b(kx)(ky) + c(ky)^2 \\ &= k^2ax^2 + k^2bxy + k^2cy^2 \\ &= k^2(ax^2 + bxy + cy^2) \\ &= k^2 f(x, y). \end{aligned}$$

Thus, our function is homogeneous of degree two.

As an example of a function that is not homogeneous, consider

$$f(x, y) = ax^2 + bx + cy^2.$$

Then

$$f(kx, ky) = a(kx)^2 + b(kx) + c(ky)^2.$$

The constant $k$ cannot be factored out of this function, and so the function is not homogeneous.

The extensive thermodynamic variables are homogeneous functions of degree one in the number of moles, and Euler's theorem can be used to relate the composition derivatives of these variables.

## A1.7 Graphical Integrations

We will be interested in many occasions upon obtaining the result of an integration represented by

$$\int_A^B y \, dx.$$

Two simple numerical methods are often used to determine the area under the curve that equals the desired integral. They involve the use of the **trapezoidal rule** and **Simpson's rule**.

### A1.7a The Trapezoidal Rule

To approximate the area under the curve shown in Figure A1.5(a), we can divide the total area into $n$ areas with equally-spaced widths $w$ along the $x$ axis. The heights of the vertical segments, $y_0, y_1, y_2$, etc. mark the endpoints of the intervals and their intersection with the curve whose area is to be calculated. Lines connect adjacent endpoints and each enclosed area is a trapezoid with rectangular bases and triangular tops. The area of each of these trapezoids is given by the width of its base $w$ multiplied by the average height of a vertical side. Thus, the first area in figure A1.5(a) would be $(y_0 + y_1)/2w$, the second area would be $(y_1 + y_2)/2w$, and so forth. Adding all of the areas gives

$$A = \left(\frac{y_0}{2} + y_1 + y_2 + \cdots + y_{n-2} + y_{n-1} + \frac{y_n}{2}\right) w,$$

an approximation to the area under the curve and therefore the value of the integral. This result can be rewritten as

$$\int_A^B y \, dx \cong \frac{w}{2}(y_0 + 2y_1 + 2y_2 + \cdots 2y_{n-1} + y_n). \tag{A1.34}$$

Equation (A1.34) is a generalized statement of the trapezoidal rule and it can be used to find an approximate value for the integral. Close inspection of Figure A1.5a will show that there are small areas under the curve that are not included in the trapezoids, and these will lead to errors in the area calculated by applying equation (A1.34). The error arises because linear segments are used to connect the successive $y$ values as they intersect the curve but the curve itself is not linear. We can minimize this error by dividing the area into a large enough number of sections that the curve is very nearly linear over the width of a section.

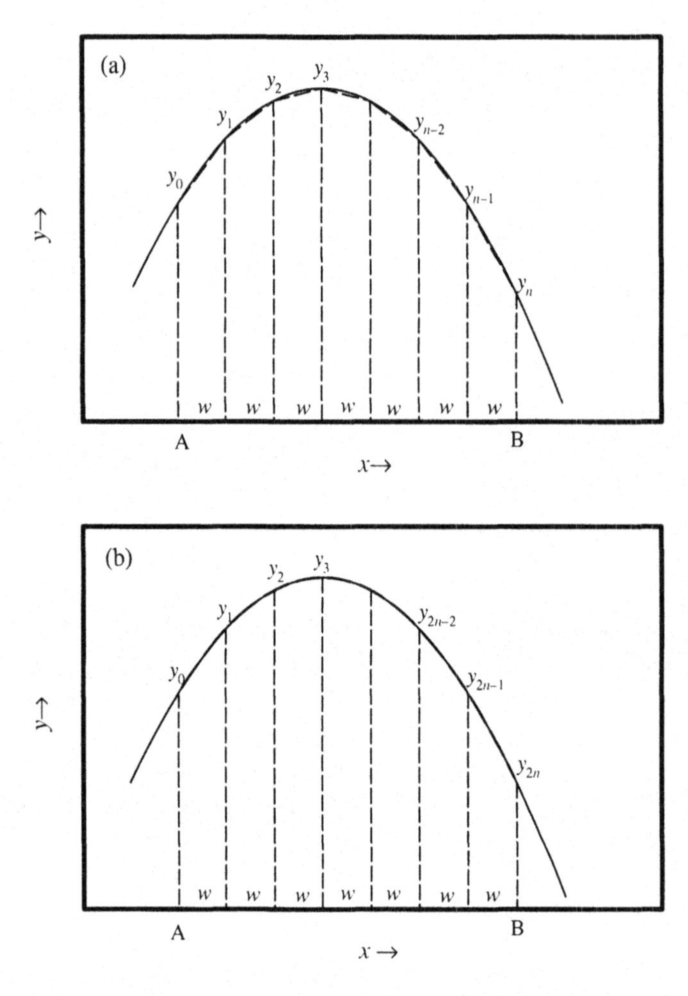


**Figure A1.5** Graphical integrations using (a) the trapezoidal rule and (b) Simpson's rule.

## A1.7b Simpson's Rule

Another approach is to use Simpson's rule instead of the trapezoidal rule for the integration. With Simpson's rule, three successive points, e.g, $(y_0, y_1, y_2)$, are assumed to be connected by a parabola instead of a straight line. The area between A and B is divided into $2n$ equally spaced subintervals of width $w$ as shown in Figure A1.5(b). The area under a parabola with three equidistant $y$ values $y_0, y_1, y_2$, separated by a width $w$ is given by

$$A = \frac{w}{3}(y_0 + 4y_1 + y_2).$$

With the exception of the first and last segments, $y_0$ and $y_{2n}$, that appear in only one segment, each endpoint of the three-point intervals appears twice, first at the end of one interval and then at the beginning of the next. Thus, the sum of the areas from the $2n$ intervals is given by

$$\int_A^B y\,dx \cong \frac{w}{3}(y_0 + 4y_1 + 2y_2 + 4y_3 + \cdots + 2y_{2n-2} + 4y_{2n-1} + y_{2n}). \tag{A1.35}$$

The advantage of equation (A1.35) over equation (A1.34) is that it approximates the line segment with a curve (parabola) instead of a straight line, which generally allows the line segment to better follow the actual line, and increases the accuracy with which the area can be determined.

## A1.8 Stirling's Approximation

The quantity $n!$ is given by

$$n! = \prod_{i=1}^{n} i. \tag{A1.36}$$

When $n$ becomes large, it is difficult to evaluate $n!$ and **Stirling's approximation** is often employed. To find the value for $n!$, we take the logarithms of both sides of equation (A1.36) to obtain

$$\ln n! = \sum_{i=1}^{n} \ln i. \tag{A1.37}$$

If $n$ is a large number, the sum in equation (A1.37) can be replaced by an integral

$$\ln n! = \int_1^n \ln i\,di. \tag{A1.38}$$

The integration of equation (A1.38) gives

$$\ln n! = (i \ln i - i)\,\big|_1^n$$

or

$$\ln n! = n \ln n - n + 1. \tag{A1.39}$$

For large $n$, 1 is small and can be neglected to give

$$\ln n! = n \ln n - n. \tag{A1.40}$$

This is the form in which we will use Stirling's approximation.[f]

[f]Alternate derivations that do not go directly from equations (A1.37) to (A1.39) give the expression

$$\ln n! = n \ln n - n + \tfrac{1}{2} \ln n + \tfrac{1}{2} \ln 2\pi. \tag{A1.41}$$

For the large values of $n$ that we will be using, equation (A1.41) also reduces to equation (A1.40).

## Appendix 2

# The International Temperature Scale of 1990[1]

The ability to measure temperature and temperature differences accurately and reproducibly is essential to the experimental study of thermodynamics. A thermometer constructed with an ideal gas as its working fluid yields temperatures that correspond to the fundamental thermodynamic temperature scale. However, such thermometers are extremely difficult to use, are not amenable to miniaturization, and are very expensive. Therefore, other means to measure temperatures that reproduce the ideal gas or thermodynamic temperature scale (Kelvin) have had to be developed. The international temperature scale represents a method to determine temperatures over a wide range with measuring devices that are easier to use than the ideal gas thermometer. The goal is to make temperature measurements that correspond to the thermodynamic temperature as accurately as possible.

The international temperature scale is based upon the assignment of temperatures to a relatively small number of "fixed points", conditions where three phases, or two phases at a specified pressure, are in equilibrium, and thus are required by the Gibbs phase rule to be at constant temperature. Different types of thermometers (for example, He vapor pressure thermometers, platinum resistance thermometers, platinum/rhodium thermocouples, blackbody radiators) and interpolation equations have been developed to reproduce temperatures between the fixed points and to generate temperature scales that are continuous through the intersections at the fixed points.

Approximately every twenty years, the international temperature scale is updated to incorporate the most recent measurements of the equilibrium thermodynamic temperature of the fixed points, to revise the interpolation equations, or to change the specifications of the interpolating measuring devices. The latest of these scales is the international temperature scale of 1990 (ITS-90). It supersedes the earlier international practical temperature scale of 1968 (IPTS-68), along with an interim scale (EPT-76). These temperature scales replaced earlier versions (ITS-48 and ITS-27).

The ITS-90 scale is designed to give temperatures $T_{90}$ that do not differ from the Kelvin Thermodynamic Scale by more than the uncertainties associated with the measurement of the fixed points on the date of adoption of ITS-90 (January 1, 1990), to extend the low-temperature range previously covered by EPT-76, and to replace the high-temperature thermocouple measurements of IPTS-68 with platinum resistance thermometry. The result is a scale that has better agreement with thermodynamic temperatures, and much better continuity, reproducibility, and accuracy than all previous international scales.

Temperatures on ITS-90, as on earlier scales, are defined in terms of fixed points, interpolating instruments, and equations that relate the measured property of the instrument to temperature. The report on ITS-90 of the Consultative Committee on Thermometry is published in *Metrologia* and in the *Journal of Research of the National Institute of Standards and Technology*.[2] The description that follows is extracted from those publications.[3] Two additional documents by CCT further describe ITS-90: *Supplementary Information for the ITS-90*; and *Techniques for Approximating the ITS-90*.[4]

**Table A2.1** Fixed points for the International Temperature Scale, ITS-90

| Equilibrium state | $T_{90}^{*}$/(K) |
|---|---|
| Triple point of equilibrium** $H_2$ | 13.8033 |
| Triple point of Ne | 24.5561 |
| Triple point of $O_2$ | 54.3584 |
| Triple point of Ar | 83.8058 |
| Triple point of Hg | 234.3156 |
| Triple point of $H_2O$ | 273.16*** |
| Melting point of Ga | 302.9146 |
| Freezing point of In | 429.7485 |
| Freezing point of Sn | 505.078 |
| Freezing point of Zn | 692.677 |
| Freezing point of Al | 933.473 |
| Freezing point of Ag | 1234.93 |
| Freezing point of Au | 1337.33 |
| Freezing point of Cu | 1357.77 |

* $T_{90}$ is the ITS-90 temperature in Kelvins.
** Equilibrium distribution of ortho and para states.
*** Defined value.

## A2.1 Fixed Points

The fixed reference points for ITS-90 (temperatures at specified equilibrium states) are given in Table A2.1. These reference points are selected to calibrate thermometers over different temperature ranges as we describe later.

## A2.2 Choice of Thermometer

The type of thermometer used to interpolate between the reference points depends on the temperature interval.

### A2.2a Temperature Interval 0.65 to 5.0 K

Temperature $T_{90}$ (the ITS-90 temperature), in the range from 0.65 to 5.0 K is defined by vapor pressure versus temperature relations of He. A $^3$He vapor pressure thermometer is used to obtain $T_{90}$ in the region from 0.65 to 3.2 K, while a $^4$He vapor pressure thermometer gives $T_{90}$ from 1.25 to 2.1768 K ($\lambda$ point), and 2.1768 to 5.0 K.

The form of the vapor pressure–temperature relation is

$$T_{90} = A_0 + \sum_{i=1}^{9} A_i\{[\ln p - B]/C\}^i \qquad \text{(A2.1)}$$

where $T_{90}$ is in Kelvins and $p$ is the vapor pressure in pascals. The values of the coefficients $A_1$ to $A_9$ and constants $A_0$, $B$, and $C$ are given in Table A2.2.

**Table A2.2** Values of the coefficients and constants for equation (A2.1) used to define the ITS-90 Temperature Scale

| Coefficient or constant | $^3$He 0.65–3.2 K | $^4$He 1.25–2.1768 K | $^4$He 2.1768–5.0 K |
|---|---|---|---|
| $A_0$ | 1.053447 | 1.392408 | 3.146631 |
| $A_1$ | 0.980106 | 0.527153 | 1.357655 |
| $A_2$ | 0.676380 | 0.166756 | 0.413923 |
| $A_3$ | 0.372692 | 0.050988 | 0.091159 |
| $A_4$ | 0.151656 | 0.026514 | 0.016349 |
| $A_5$ | −0.002263 | 0.001975 | 0.001826 |
| $A_6$ | 0.006596 | −0.017976 | −0.004325 |
| $A_7$ | 0.088966 | 0.005409 | −0.004973 |
| $A_8$ | −0.004770 | 0.013259 | 0 |
| $A_9$ | −0.054943 | 0 | 0 |
| $B$ | 7.3 | 5.6 | 10.3 |
| $C$ | 4.3 | 2.9 | 1.9 |

## A2.2b Temperature Interval 3.0 to 24.5561 K

Temperature $T_{90}$ in the range between 3.0 and 24.5561 K is defined in terms of $^3He$ or $^4He$ constant volume gas thermometers (CVGT), calibrated at the triple points of Ne and $H_2$, and at a temperature between 3.0 and 5.0 K that has been obtained from vapor pressure versus temperature relations for He.

In the temperature range between 4.2 and 24.5561 K (triple point of neon), $T_{90}$ is defined by the equation

$$T_{90} = a + bp + cp^2 \tag{A2.2}$$

where $p$ is the pressure in the CVGT, and $a$, $b$, and $c$ are coefficients determined by calibration at the three specified temperatures.

When the $^4He$ CVGT is used between 3.0 and 4.2 K (or the $^3He$ CVGT between 3.0 and 24.5561 K), gas imperfection must be taken into account and equation (A2.2) becomes

$$T_{90} = \frac{a + bp + cp^2}{1 + B_x(T_{90})N/V} \tag{A2.3}$$

where $p$, $a$, $b$, and $c$ are the same as in equation (A2.2), $B_x(T_{90})$ is the second virial coefficient for $^3He[B_3(T_{90})]$ or $^4He[B_4(T_{90})]$, and $N/V$ is the gas density (moles per cubic meter) in the CVGT bulb. The values of $B_x(T_{90})$ at any given temperature are calculated from equations specified in the official ITS-90 document referenced earlier and also in the *National Institute of Standards and Technology's* (NIST) Technical Note 1265.

## A2.2c Temperature Interval 13.8033 to 1234.93 K

The thermometer used in the large temperature interval between 13.8033 and 1234.93 K is a platinum resistance thermometer calibrated at fixed reference points. Temperatures are expressed in terms of $W(T_{90})$, which is the ratio of the resistance $R(T_{90})$ of the thermometer at temperature $T_{90}$ and the resistance at the triple point of water $R(273.16)$, as seen in equation (A2.4):

$$W(T_{90}) = R(T_{90})/R(273.16\ \text{K}). \tag{A2.4}$$

The strain-free coil of pure platinum used in the resistance thermometer must meet one of the following specifications: $W(302.9146\ K) \geqslant 1.11807$ or $W(234.3156\ K) \leqslant 0.844235$. If the resistance thermometer is to be used over the entire range 13.8033 to 1234.93 K, it must also meet the requirement that $W(1234.93\ K) \geqslant 4.2844$.

Temperatures $T_{90}$ are calculated from the equation

$$W(T_{90}) = W_r(T_{90}) + \Delta W(T_{90}) \tag{A2.5}$$

in which $W(T_{90})$ is the resistance ratio observed and $W_r(T_{90})$ is the value calculated from a reference function. The difference between the value obtained at $T_{90}$ with a given platinum resistance thermometer and the reference function value at the same temperature is $\Delta W(T_{90})$, and is called the **deviation function**.

There are two reference functions, one for the range 13.8033 to 273.16 K and the other for the range 273.15 to 1234.93 K. The reference function $W_r(T_{90})$

**Table A2.3** Values of the coefficients $A_i$, $B_i$, $C_i$, and $D_i$ and of the constants $A_0$, $B_0$, $C_0$, and $D_0$ in the reference functions in equations (A2.6) and (A2.7) and in the inverse functions given by equations (A2.8) and (A2.9). These functions, coefficients, and constants are part of the definition of ITS-90.

| Constant or coefficient | Value | Constant or coefficient | Value |
|---|---|---|---|
| $A_0$ | −2.13534729 | $B_{12}$ | −0.029201193 |
| $A_1$ | 3.18324720 | $B_{13}$ | −0.091173542 |
| $A_2$ | −1.80143597 | $B_{14}$ | 0.001317696 |
| $A_3$ | 0.71727204 | $B_{15}$ | 0.026025526 |
| $A_4$ | 0.50344027 | | |
| $A_5$ | −0.61899395 | | |
| $A_6$ | −0.05332322 | $C_0$ | 2.78157254 |
| $A_7$ | 0.28021362 | $C_1$ | 1.64650916 |
| $A_8$ | 0.10715224 | $C_2$ | −0.13714390 |
| $A_9$ | −0.29302865 | $C_3$ | −0.00649767 |
| $A_{10}$ | 0.04459872 | $C_4$ | −0.00234444 |
| $A_{11}$ | 0.11868632 | $C_5$ | 0.00511868 |
| $A_{12}$ | −0.05248134 | $C_6$ | 0.00187982 |
| | | $C_7$ | −0.00204472 |
| | | $C_8$ | −0.00046122 |
| $B_0$ | 0.183324722 | $C_9$ | 0.00045724 |
| $B_1$ | 0.240975303 | | |
| $B_2$ | 0.209108771 | | |
| $B_3$ | 0.190439972 | $D_0$ | 439.932854 |
| $B_4$ | 0.142648498 | $D_1$ | 472.418020 |
| $B_5$ | 0.077993465 | $D_2$ | 37.684494 |
| $B_6$ | 0.012475611 | $D_3$ | 7.472018 |
| $B_7$ | −0.032267127 | $D_4$ | 2.920828 |
| $B_8$ | −0.075291522 | $D_5$ | 0.005184 |
| $B_9$ | −0.056470670 | $D_6$ | −0.963864 |
| $B_{10}$ | 0.076201285 | $D_7$ | −0.188732 |
| $B_{11}$ | 0.123893204 | $D_8$ | 0.191203 |
| | | $D_9$ | 0.049025 |

for temperatures in the range 13.8033 to 273.16 K is

$$\ln[W_r(T_{90})] = A_0 + \sum_{i=1}^{12} A_i\{[\ln(T_{90}/273.16) + 1.5]/1.5\}^i \tag{A2.6}$$

and for the temperature range 273.15 to 1234.93 K, it is

$$W_r(T_{90}) = C_0 + \sum_{i=1}^{9} C_i\left(\frac{T_{90} - 754.15}{481}\right)^i. \tag{A2.7}$$

Inverses of equations (A2.6) and (A2.7) that are explicit in temperature can also be used. The specified inverse of equation (A2.6), equivalent to within $\pm 0.0001$ K, is

$$\frac{T_{90}}{273.16} = B_0 + \sum_{i=1}^{15} B_i\left(\frac{[W_r(T_{90})]^{1/6} - 0.65}{0.35}\right)^i. \tag{A2.8}$$

The inverse of equation (A2.7), equivalent to within $\pm 0.00013$ K, is

$$T_{90} - 273.15 = D_0 + \sum_{i=1}^{9} D_i\left(\frac{W_r(T_{90}) - 2.64}{1.64}\right)^i. \tag{A2.9}$$

The values for the constants $A_0$, $B_0$, $C_0$, and $D_0$ and the coefficients $A_i$, $B_i$, $C_i$, and $D_i$ are given in Table A2.3.

## A2.3 The Deviation Function

The deviation function $\Delta W(T_{90})$ is obtained as a function of $T_{90}$ for various temperature intervals by calibration of the platinum resistance thermometer, using specified fixed points from Table A2.1. The form of the $\Delta W(T_{90})$ function is dependent on the temperature range in which the thermometer is being calibrated. For example, in the temperature subrange from 234.3156 to 302.9146 K, the form of the deviation function is

$$\Delta W_5(T_{90}) = a_5[W(T_{90}) - 1] + b_5[W(T_{90}) - 1]^2. \tag{A2.10}$$

The coefficients $a_5$ and $b_5$ are obtained by calibrating the thermometer at the triple points of mercury (234.3156 K) and water (273.16 K) and the melting point of gallium (302.9146 K).

**Table A2.4** Temperature subranges, deviation functions, and calibration points over the temperature range covered by platinum resistance thermometry for ITS-90.

| Temperature subrange/(K) | Deviation function $\Delta W(T_{90}) = W(T_{90}) - W_r(T_{90}) =$ | Calibration (fixed) points* used to determine coefficients in the deviation function |
|---|---|---|
| 13.8033–273.16 | $a_1[W(T_{90}) - 1] - b_1[W(T_{90}) - 1]^2 + \sum_{i=1}^{5} c_i[\ln W(T_{90})]^{i+2}$ | Triple point (tp) of $H_2$, Ne, $O_2$, Ar, Hg, and 2 more** |
| 24.5561–273.16 | $a_2[W(T_{90}) - 1] + b_2[W(T_{90}) - 1]^2 + \sum_{i=1}^{3} c_i[\ln W(T_{90})]^i$ | tp of $H_2$, Ne, $O_2$, Ar, Hg |
| 54.3584–273.16 | $a_3[W(T_{90}) - 1] + b_3[W(T_{90}) - 1]^2 + c_1[\ln W(T_{90})]^2$ | tp of $O_2$, Ar, Hg |
| 83.8058–273.16 | $a_4[W(T_{90}) - 1] + b_4[W(T_{90}) - 1]\ln W(T_{90})$ | tp of Ar, Hg |
| 273.15–1234.93 | $a_6[W(T_{90}) - 1] + b_6[W(T_{90}) - 1]^2 + c_6[W(T_{90}) - 1]^3 + d[W(T_{90}) - W(933.473\ \text{K})]^2$ | Freezing point (fp) of Sn, Zn, Al, Ag |
| 273.15–933.473 | $a_7[W(T_{90}) - 1] + b_7[W(T_{90}) - 1]^2 + c_7[W(T_{90}) - 1]^3$ | fp of Sn, Zn, Al |
| 273.15–692.677 | $a_8[W(T_{90}) - 1] + b_8[W(T_{90}) - 1]^2$ | fp of Sn, Zn |
| 273.15 – 505.078 | $a_9[W(T_{90}) - 1] + b_9[W(T_{90}) - 1]^2$ | fp of In, Sn |
| 273.15–429.7485 | $a_{10}[W(T_{90}) - 1]$ | fp of In |
| 273.15–302.9146 | $a_{11}[W(T_{90}) - 1]$ | Melting point (mp) of Ga |
| 234.3156–302.9146 | $a_5[W(T_{90}) - 1] + b_5[W(T_{90}) - 1]^2$ | tp of Hg; mp of Ga |

* In addition to the fixed points listed, calibration at the triple point (tp) of $H_2O$ is required.

** Two additional points near 17.0 and 20.3 K are required. These may be determined by using either the constant volume gas thermometer or by vapor pressure measurements of $H_2$.

Table A2.4 is a tabulation of subranges in the temperature region 13.8033 to 1234.93 K, together with the form of the deviation equation that applies to each, and the calibration points from which the coefficients in the deviation equation are to be obtained.

In summary, to obtain $T_{90}$ from a platinum resistance thermometer, one selects the range of interest, calibrates the thermometer at the fixed points specified for those ranges, and uses the appropriate function to calculate $\Delta W(T_{90})$ to be used in equation (A2.5). Companies are available that perform these calibrations and provide tables of $W(T_{90})$ versus $T_{90}$ that can be interpolated to give $T_{90}$ for a measured $W(T_{90})$.

## A2.4 Measurement of Temperatures Above 1234.93 K

At temperatures above the melting point of silver (1234.93 K), radiation thermometry is used. The equation that applies is

$$\frac{L_\lambda(T_{90})}{L_\lambda[T_{90}(X)]} = \frac{\exp[c_2/\lambda T_{90}(X)] - 1}{\exp(c_2/\lambda T_{90}) - 1} \tag{A2.11}$$

in which $L_\lambda(T_{90})$ is the spectral concentration of the radiance of a blackbody at wavelength $\lambda$ at $T_{90}$ and $L_\lambda[T_{90}(X)]$ is the same quantity at $T_{90}(X)$, where $X$ is the freezing point of silver, copper, or gold on ITS-90. The constant $c_2$ has the value 0.014388 m·K (the same as for IPTS-68).

## A2.5 Correction of Existing Data to ITS-90

The descriptions of ITS-90 in the references given at the end of this appendix include a table of differences[a] between $T_{90}$ and $T_{76}$ at 1 K intervals between 5 and 26 K; $T_{90}$ and $T_{68}$ at 1 K intervals from 4 to 100 K, and at 10 K intervals between 100 and 200 K; $t_{90}$ and $t_{68}$ at 10 °C intervals from −190 to 1000 °C, and $t_{90}$ and $t_{68}$ in 100 °C intervals from 1000 to 3900 °C. These data can be used to update IPTS-68 temperatures to ITS-90.

Figure A2.1 is a graphical representation of the differences $(T_{90} - T_{68})$ (in Kelvins) or $(t_{90} - t_{68})$ (in degrees Celsius) as a function of $t_{90}$ (in degrees Celsius). The corrections in the region from about 500 to 1000 °C (773 to 1273 K) are seen to be substantial, varying all the way from about +0.35 to −0.20 °C.

---

[a] $T_{76}$ is the temperature on the interim EPT-76 scale, while $T_{68}$ and $T_{48}$ are the IPTS-68 and ITS-48 temperatures (all in Kelvins). The $t_{90}$, $t_{68}$ and $t_{48}$ are the corresponding temperatures in °C.

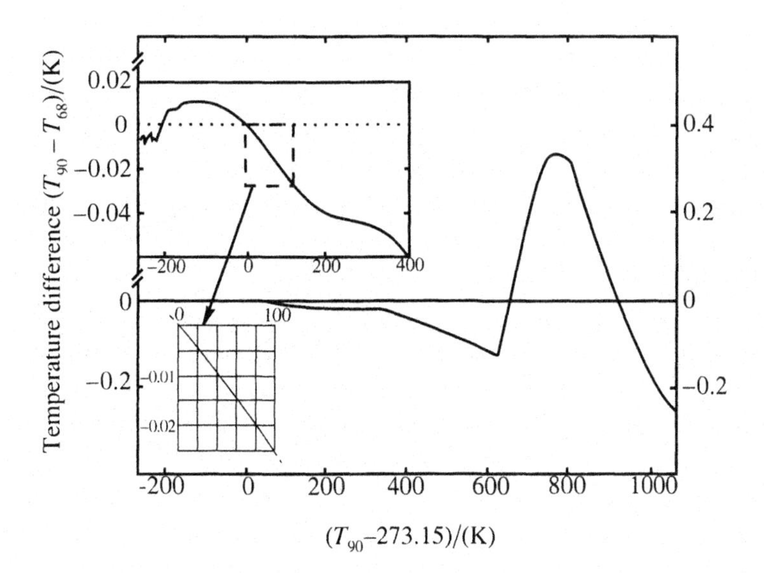


**Figure A2.1** Differences between $T_{90}$ and $T_{68}$ as a function of $T_{90}/(K)$. Reprinted with permission from H. Preston-Thomas, *Metrologia*, **27**, 4 (1990).

From Figure A2.1 we can see that differences between $t_{90}$ and $t_{68}$ in the temperature range from −100 to 200 °C vary almost linearly with temperature. Similar relationships occur with $t_{48}$. One of the authors[5] has reported that correction of $T_{48}$ and $T_{68}$ measurements to $T_{90}$ in the temperature range from 140 to 400 K can be made with an accuracy better than ±0.002 K by using the equation

$$(T_{90} - T_{mn})/(K) = \sum_{j=1}^{6} a_j[10^{-3}(T_{mn} - 273.150)]^j$$

where $T_{mn}$ (in Kelvins) is the ITS temperature, $T_{48}$ for $mn = 48$ or $T_{68}$ for $mn = 68$. The coefficients are

$mn = 68$: $a_1 = 0.224$, $a_2 = -0.84$, $a_3 = 2.8$, $a_4 = 19$, $a_5 = 0$, $a_6 = 0$

$mn = 48$: $a_1 = -0.732$, $a_2 = 4.07$, $a_3 = 50.3$, $a_4 = -508$, $a_5 = -780$, $a_6 = 14600$.

## References

1. This discussion follows closely a description of ITS-90 provided by one of the authors as part of J. B. Ott and J. R. Goates, "'Temperature Measurements with Application to Phase

Equilibria Studies,' Chapter 6, *Physical Methods of Chemistry*," B. W. Rossiter and R. C. Baitzold, ed., Wiley-Interscience, New York, 1992, pp. 463–471.
2. The original reports on ITS-90 are found in B. W. Mangum, "Special Report on the International Temperature Scale of 1990. Report on the 17th Session of the Consultative committee on Thermometry," *J. Res. Natl. Inst. Stand. Technol.*, **95**, 69 (1990); H. Preston-Thomas, "The International Scale of 1990 (ITS-90)," *Metrologia*, **27**, 3 (1990).
3. An excellent summary of ITS-90 can be found in M. L. McGlashan, "The International Temperature Scale of 1990 (ITS-90)," *J. Chem. Thermodyn.*, **22**, 653 (1990).
4. The supplementary documents are *Supplementary Information for the ITS-90*. International Bureau of Weights and Measures: Pavillon de Breteuil, F-92312 Sèvres, France, 1990; *Techniques for Approximating the ITS-90*. International Bureau of Weights and Measures: Pavillon de Breteuil, F-92312 Sèvres, France, 1990.
5. J. B. Ott and J. R. Goates, "Summary of Melting and Transition Temperatures of Pure Substances and Congruent and Incongruent Melting Temperatures of Molecular Addition Compounds," *J. Chem. Eng. Data*, **41**, 669–677 (1996).

# Appendix 3

# Equations of State For Gases

## A3.1 The Ideal Gas

The simplest equation of state for gases is the ideal gas equation

$$pV_{\mathrm{m}} = RT. \tag{A3.1}$$

An alternate expression of equation (A3.1) is

$$z = \frac{pV_{\mathrm{m}}}{RT} = 1 \tag{A3.2}$$

where $z$ is the compressibility factor.[a] Experimental measurements show that $z$ depends upon $p$ and $T$, and other equations of state have been developed to express this dependence.

## A3.2 The Virial Equation

A statistical mechanical treatment shows that the equation of state of a real gas can be expressed as a power series in $1/V_{\mathrm{m}}$ as given by equation (A3.3)

$$pV_{\mathrm{m}} = RT\left[1 + \frac{\mathrm{B}(T)}{V_{\mathrm{m}}} + \frac{\mathrm{C}(T)}{V_{\mathrm{m}}^2} + \frac{\mathrm{D}(T)}{V_{\mathrm{m}}^3} + \cdots\right]. \tag{A3.3}$$

---

[a] An equivalent definition of the ideal gas is given by the equation

$$\left(\frac{\partial U_{\mathrm{m}}}{\partial V_{\mathrm{m}}}\right)_T = 0.$$

From a molecular point of view, this equation implies that the internal energy of the gas does not depend upon the separation of the gaseous molecules, potential energy due to attractions and repulsions between the molecules is not present, and the internal energy is a function only of the temperature.

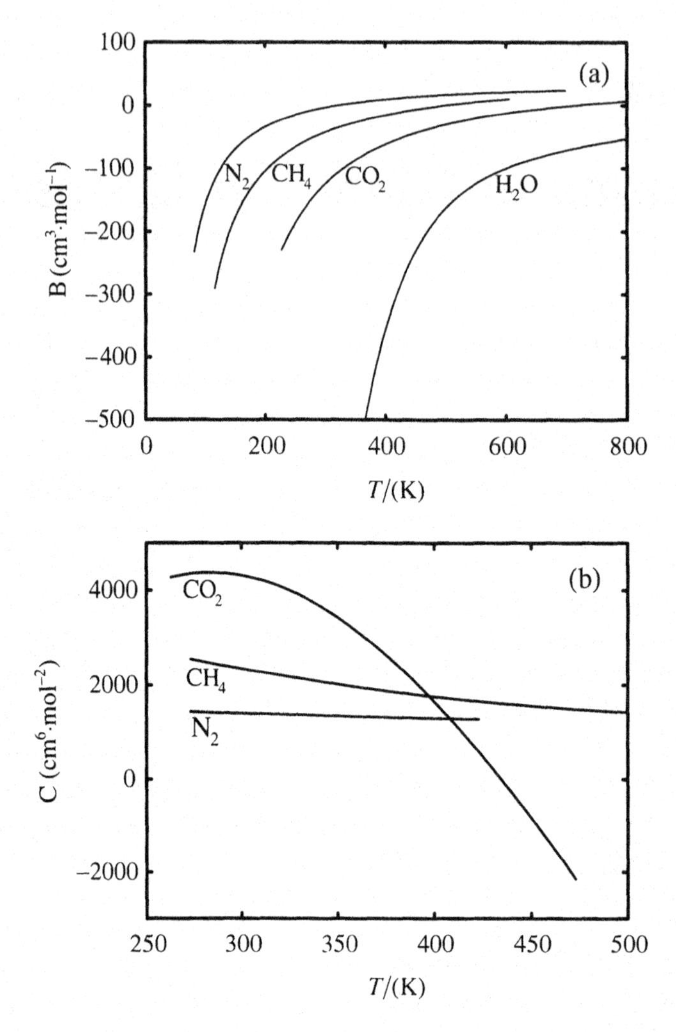


**Figure A3.1** Examples of (a) the second virial coefficient and (b) the third virial coefficient [from equation (A3.3)] as a function of temperature for several gases.

An equivalent expression is

$$z = 1 + \frac{\mathrm{B}(T)}{V_{\mathrm{m}}} + \frac{\mathrm{C}(T)}{V_{\mathrm{m}}^2} + \frac{\mathrm{D}(T)}{V_{\mathrm{m}}^3} + \cdots \tag{A3.4}$$

where $z = pV_{\mathrm{m}}/RT$ is again the compressibility factor. Equation (A3.3) is known as the **virial equation of state**. The coefficients, B, C, D ... are known as the second, third, fourth ... virial coefficients. They are functions only of temperature and can be calculated with limited success by assuming an interaction potential between the molecules. The second virial coefficient can be calculated from pair-wise interaction between the molecules. The third virial coefficient results from interactions between three molecules at a time and is more difficult to calculate. Higher order virial coefficients are still more difficult to calculate and are not usually considered. In fact, calculations of B and C do not give reliable values except, perhaps, for the simplest of systems, and the virial coefficients are usually determined experimentally. Extensive tabulations can be found in the literature.[1] Temperature does have a large effect on B and C. Figure A3.1 shows examples of B and C as a function of temperature for several gases. Note that B shows a similar temperature dependence for all the gases compared, but C shows more variability between different gases.

## A3.3 The Virial Equation Explicit in Pressure

An alternate form of the virial equation uses a power series in $p$. The result is

$$pV_m = RT + B(T)p + C(T)p^2 + D(T)p^3 + \cdots \tag{A3.5}$$

This form of the virial equation usually does not represent the properties of the gas as well as does equation (A3.3), but it is often more useful since it can be solved explicitly for $V_m$. The coefficients in equations (A3.3) and (A3.5) are related through

$$B = \mathrm{B} \tag{A3.6}$$

$$C = \frac{(\mathrm{C} - \mathrm{B}^2)}{RT}. \tag{A3.7}$$

If $p$ is not high, terms beyond the second and third virial coefficients in equation (A3.3) and (A3.5) are usually small and can be neglected. This is fortunate, since experimental data are usually not accurate enough to give reliable values for the higher order terms. At low pressures, equation (A3.5) is often used and truncated after the second virial coefficient so that

$$V_m = \frac{RT}{p} + B. \tag{A3.8}$$

## A3.4 Other Equations of State

A number of different ($p$, $V_m$, $T$) equations of states have been proposed and are described in the literature. Walas[2] summarizes 56, in addition to the ones that we will now describe in more detail.

The **modified Berthelot equation** is one that is often employed. It can be written in the form of a virial equation with

$$pV_{\mathrm{m}} = RT\left[1 + \frac{9pT_{\mathrm{c}}}{128p_{\mathrm{c}}T}\left(1 - \frac{6T_{\mathrm{c}}^2}{T^2}\right)\right] \tag{A3.9}$$

where $T_{\mathrm{c}}$ and $p_{\mathrm{c}}$ are the critical temperature and pressure of the gas. It takes the form of equation (A3.8) with

$$B = \frac{9RT_{\mathrm{c}}}{128P_{\mathrm{c}}}\left(1 - \frac{6T_{\mathrm{c}}^2}{T^2}\right). \tag{A3.10}$$

As we shall see, its usefulness is limited to low pressures.

The **Dieterici equation** is one that effectively represents the $(p, V_{\mathrm{m}}, T)$ properties of gases under certain conditions. It has the form

$$p = \frac{RT\exp(-a/V_{\mathrm{m}}RT)}{V_{\mathrm{m}} - b} \tag{A3.11}$$

where $a$ and $b$ are constants. The Dieterici equation has not been as popular as some other equations, perhaps due to the difficulties of handling the exponential term. This problem has largely disappeared with the advent of computers. The constants $a$ and $b$ in equation (A3.11) can be eliminated by expressing the equation in a reduced form. The result is

$$p_{\mathrm{r}} = \frac{T_{\mathrm{r}}\exp(2 - 2/T_{\mathrm{r}}V_{\mathrm{r}})}{2V_{\mathrm{r}} - 1} \tag{A3.12}$$

where the reduced variables are related to the critical pressure, temperature, and volume by

$$p_{\mathrm{r}} = p/p_{\mathrm{c}} \tag{A3.13}$$

$$T_{\mathrm{r}} = T/T_{\mathrm{c}} \tag{A3.14}$$

$$V_{\mathrm{r}} = V/V_{\mathrm{c}}. \tag{A3.15}$$

## A3.5 Cubic Equations of State

Equations of state that are cubic in volume are often employed, since they, at least qualitatively, reproduce the dependence of the compressibility factor $z$ on $p$ and $T$. Four commonly used cubic equations of state are the van der Waals, Redlich–Kwong, Soave, and Peng–Robinson. All four can be expressed in a reduced form that eliminates the constants $a$ and $b$. However, the reduced equations for the last two still include the acentric factor $\omega$ that is specific for the substance. In writing the reduced equations, coefficients can be combined to simplify the expression. For example, the reduced form of the Redlich–Kwong equation is

$$p_r = \frac{3T_r}{V_r - 3\Omega_b} - \frac{9\Omega_a}{T_r^{0.5}V_r(V_r + 3\Omega_b)}.$$

But

$$\Omega_a = 1/[9(2^{1/3} - 1)] = 0.427480$$

and

$$\Omega_b = (2^{1/3} - 1)/3 = 0.086640$$

so that

$$p_r = \frac{3T_r}{V_r - 0.2599} - \frac{3.8473}{T_r^{0.5}V_r(V_r + 0.2599)}.$$

The four cubic equations of state are summarized in Table A3.1.

### A3.5a Comparison of Cubic Equations of State

We are interested in comparing the effectiveness of the various equations of state in predicting the $(p, V, T)$ properties. We will limit our comparisons to $T_r \geqslant 1$ since for $T_r < 1$ condensations to the liquid phase occur. Prediction of (vapor + liquid) equilibrium would be of interest, but these predictions present serious problems, since in some instances the equations of state do not converge for $T_r < 1$.

The principle of **corresponding states** will be used as the basis for comparing the different equations of state. This principle states that, to a reasonable approximation, all gases show the same $(p, V_m, T)$ behavior when compared in terms of the reduced variables. The extent to which this principle is followed is

**Table A3.1** Summary of cubic equations of state*

| Equation | Equation of state | Reduced equation of state |
| --- | --- | --- |
| van der Waals: | $p = \frac{RT}{V_m - b} - \frac{a}{V_m^2}$ | $p_r = \frac{8T_r}{3V_r - 1} - \frac{3}{V_r^2}$ |
| Redlich–Kwong: | $p = \frac{RT}{V_m - b} - \frac{a}{T^{1/2}V_m(V_m + b)}$ | $p_r = \frac{3T_r}{V_r - 0.2599} - \frac{3.8473}{T_r^{0.5}V_r(V_r + 0.2599)}$ |
| Soave: | $p = \frac{RT}{V_m - b} - \frac{a\alpha}{V_m(V_m + b)}$ | $p_r = \frac{3T_r}{V_r - 0.2599} - \frac{3.8473\alpha}{V_r(V_r + 0.2599)}$ |
| | with $\alpha = [1 + (0.48508 + 1.55171\omega - 0.15613\omega^2)(1 - T_r^{0.5})]^2$ | |
| Peng–Robinson: | $p = \frac{RT}{V_m - b} - \frac{a\alpha}{V^2 + 2bV - b^2}$ | $p_r = \frac{3.2573T_r}{V_r - 0.2534} - \frac{4.8514\alpha}{V_r^2 + 0.5068V_r - 0.0642}$ |
| | with $\alpha = [1 + (0.37464 + 1.54226\omega - 0.26992\omega^2)(1 - T_r^{0.5})]^2$ | |

* The constants $a$ and $b$ are, of course, not the same for the different equations of state.

shown in Figure A3.2, where the compressibility factors $z$ for a number of gases are graphed against the reduced pressure at a series of reduced temperatures. A line representing the best average value for the compressibility factor is also shown. This figure shows that all the gases, to a reasonable approximation, fall on the same line for a given reduced temperature. Thus, we can use reduced temperature isotherms given by these lines to represent the experimental $(p, V_m, T)$ properties of the gas, against which we can compare the predictions of the various equations of state.

Figure A3.3 compares the experimental (corresponding states) results with the predictions from the van der Waals, modified Berthelot, Dieterici, and Redlich–Kwong equations of state.[b] The comparison is not so direct for the Soave and Peng–Robinson equations of state, since the reduced equation still includes $\omega$, the acentric factor. Figure A3.4 compares the corresponding states line, with the prediction from the Soave equation, using four different values of $\omega$. The acentric factors chosen are those for $H_2$ ($\omega = -0.218$), $CH_4$ ($\omega = 0.011$),

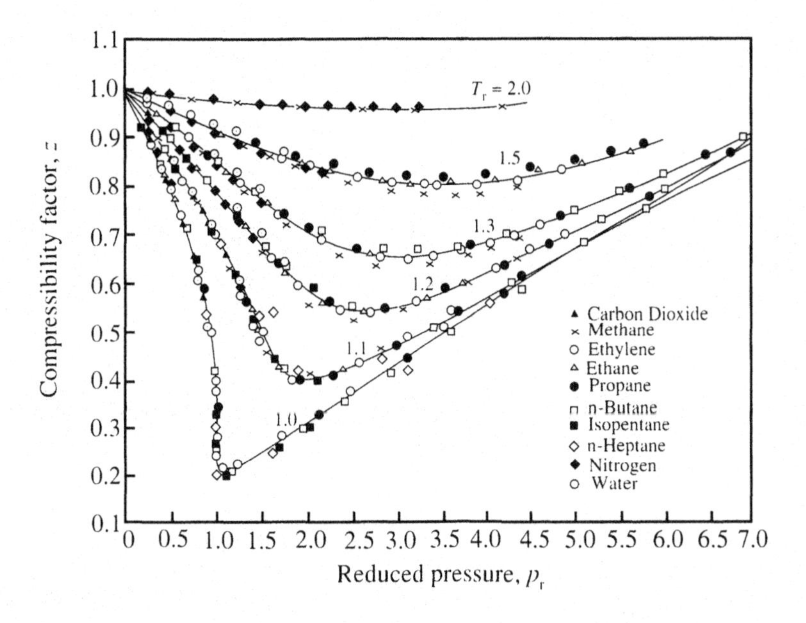


**Figure A3.2** Graph of the compressibility factor $z$ for a number of gases versus their reduced pressure at several reduced temperatures. Reprinted with permission, taken from Goug-Jen Su, *Ind. Eng. Chem.*, **38**, 803 (1946), the data illustrate the validity of the principle of corresponding states. The line is Goug-Jen Su's estimate of the average value for $z$.

[b] The assistance of Mr. Jason Moore in preparing the comparisons shown in Figures A3.3 to A3.5 is acknowledged and greatly appreciated.

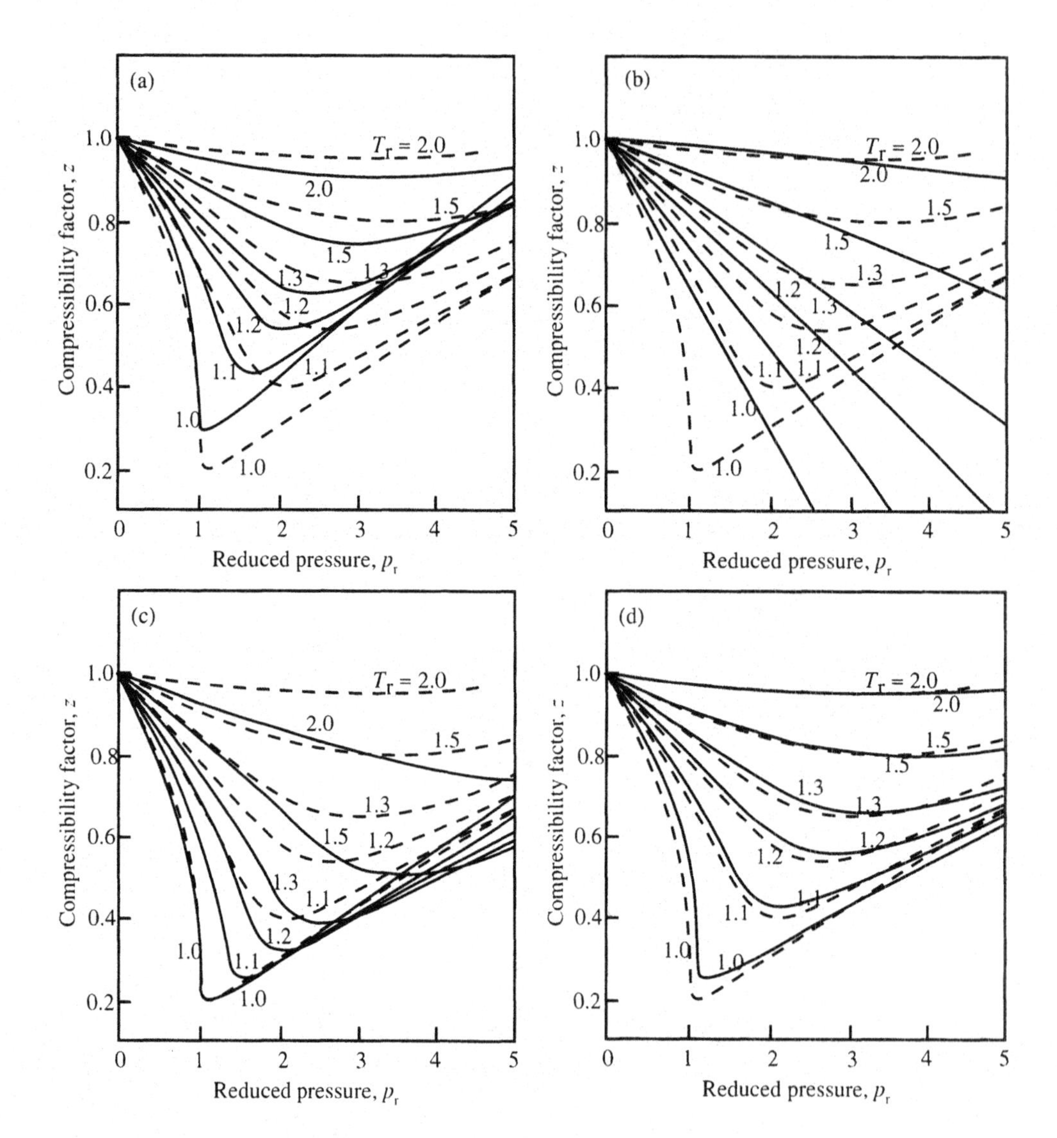


**Figure A3.3** Comparison of the experimental $z$ (dashed lines) with the $z$ values calculated from the (a) van der Waals, (b) modified Berthelot, (c) Dieterici, and (d) Redlich–Kwong equations of state expressed in reduced form.

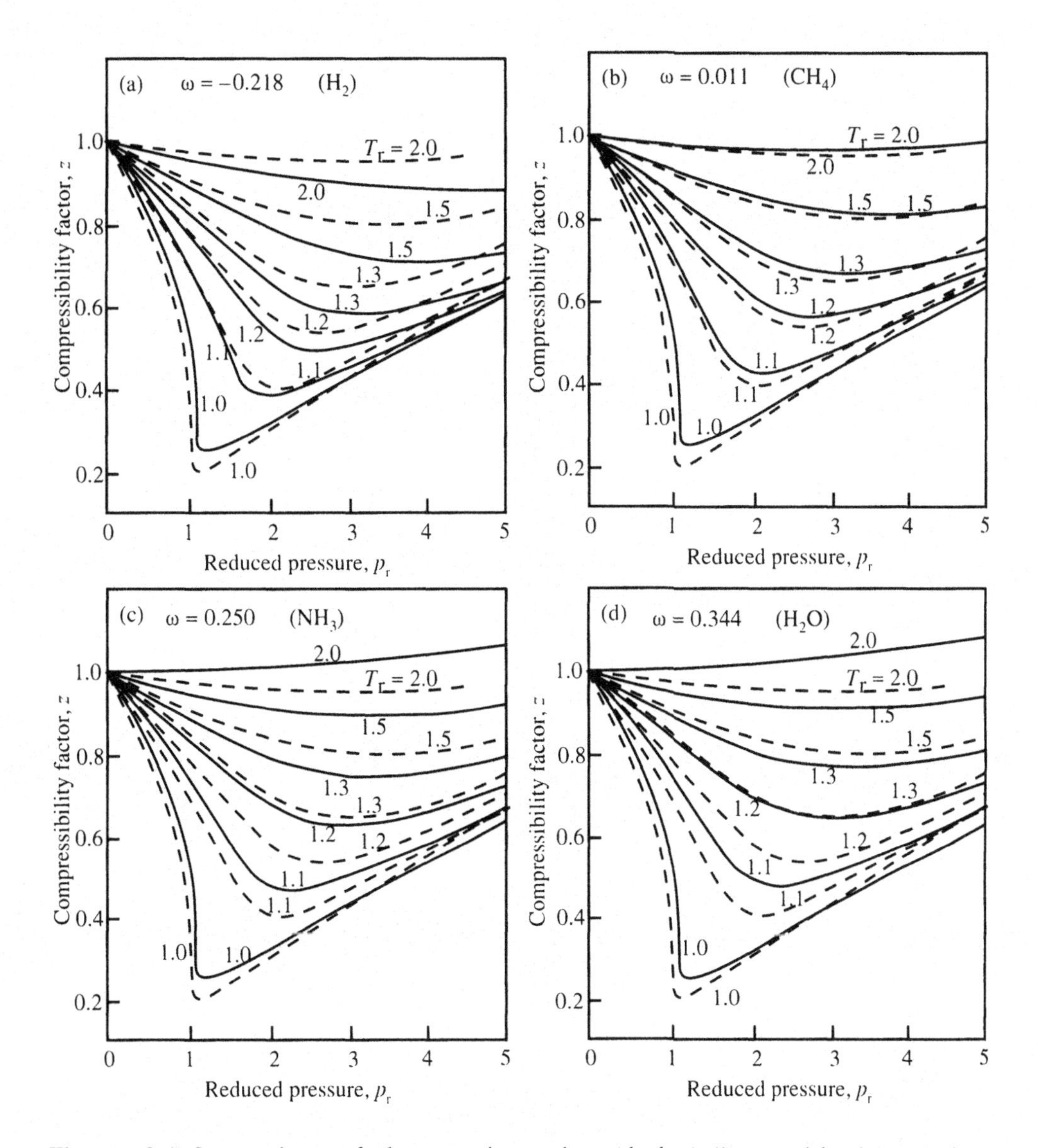


**Figure A3.4** Comparison of the experimental $z$ (dashed lines) with the $z$ values calculated from the Soave equation of state (solid lines). Values for the acentric factor are (a) $\omega = -0.218$ (the value for $H_2$), (b) $\omega = 0.011$ (the value for $CH_4$), (c) $\omega = 0.250$ (the value for $NH_3$), and (d) $\omega = 0.344$ (the value for $H_2O$).

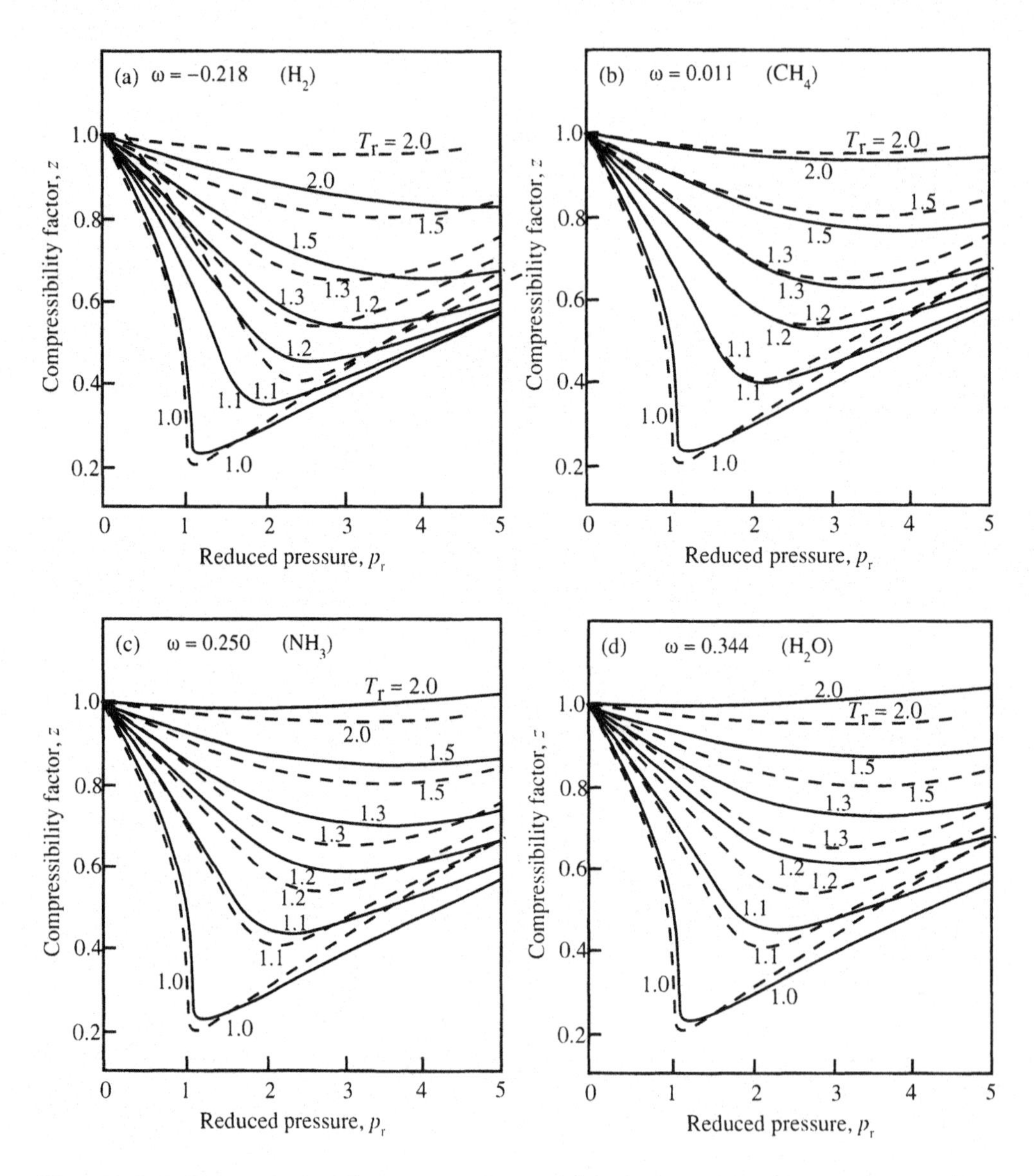


**Figure A3.5** Comparison of the experimental $z$ (dashed lines) with the $z$ values calculated from the Peng–Robinson equation of state (solid lines). Values for the acentric factor are (a) $\omega = -0.218$ (the value for $H_2$), (b) $\omega = 0.011$ (the value for $CH_4$), (c) $\omega = 0.250$ (the value for $NH_3$), and (d) $\omega = 0.344$ (the value for $H_2O$).

$NH_3$ ($\omega = 0.250$), and $H_2O$ ($\omega = 0.344$). Thus, results for a wide range of acentric factors are compared. In Figure A3.5, we make the same comparisons with the Peng–Robinson equation.

An examination of Figures A3.3 to A3.5 shows the strengths and weaknesses of each of the equations. The following are some conclusions that can be drawn.

1. The van der Waals equation gives a correct qualitative description, but it does not do well in quantitatively predicting $z$, except at low $p_r$.
2. As expected, the modified Berthelot equation incorporates a linear relationship of $z$ to $p_r$ and thus works only at low $p$. Even there the prediction is not exceptionally good.[c]
3. The Dieterici equation works well near the critical region, but it deviates significantly at higher reduced temperatures.
4. The Redlich–Kwong equation predicts the correct $z$ reasonably well over almost the entire $T_r$ and $p_r$ range shown in Figure A3.3. It appears that at $p_r > 5$, significant deviations may occur, but overall, this equation seems to give the best fit of the equations compared.
5. For both the Soave and Peng–Robinson equations, the fit is best for $\omega = 0$. The Soave equation, which essentially reduces to the Redlich–Kwong equation when $\omega = 0$, does a better job of predicting than does the Peng–Robinson equation. The acentric factors become important when phase changes occur, and it is likely that the Soave and Peng–Robinson equations would prove to be more useful when $T_r < 1$.

## References


1. An exhaustive compilation of experimental $B(T)$, along with some $C(T)$, is given by J. H. Dymond and E. B. Smith, *The Virial Coefficients of Pure Gases and Mixtures, A Critical Compilation*, Clarendon Press, Oxford, Oxford (1980). Other compilations are by V. B. Kogan, *Heterogeneous Equilibria* (in Russian), Izdatelstvo "Khimiya" Leningrad 1918, and by Landolt-Börnstein, *Numerical Data and Functional Relationships in Science and Technology*, Vol II, 1, 1970.
2. S. M. Walas, *Phase Equilibria in Chemical Engineering*, Butterworth Publishers, Boston, 1985, pp. 6–9.


[c] A number of years ago, one of the authors published a paper [*J. Chem. Ed.*, **48**, 515–517 (1971)] in which it was shown that the modified Berthelot equation could be improved significantly if the coefficients were changed so that it became

$$pV_m = RT\left[1 + \frac{pT_c}{17p_cT}\left(1 - \frac{15T_c^2}{2T^2}\right)\right].$$

## Appendix 4

# Calculations From Statistical Thermodynamics

Table A4.1 summarizes the equations needed to calculate the contributions to the thermodynamic functions of an ideal gas arising from the various degrees of freedom, including translation, rotation, and vibration (see Section 10.7). For most monatomic gases, only the translational contribution is used. For molecules, the contributions from rotations and vibrations must be included. If unpaired electrons are present in either the atomic or molecular species, so that degenerate electronic energy levels occur, electronic contributions may also be significant; see Example 10.2. In molecules where internal rotation is present, such as those containing a methyl group, the internal rotation contribution replaces a vibrational contribution. The internal rotation contributions to the thermodynamic properties are summarized in Table A4.6.

Tables A4.2 and A4.3 summarize moments of inertia and fundamental vibrational frequencies of some common molecules, while Table A4.4 gives electronic energy levels for some common molecules or atoms with unpaired electrons.

Table A4.5 summarizes the equations for calculating anharmonicity and nonrigid rotator corrections for diatomic molecules. These corrections are to be added to the thermodynamic properties calculated from the equations given in Table A4.1 (which assume harmonic oscillator and rigid rotator approximations).

Table A4.6 gives the internal rotation contributions to the heat capacity, enthalpy and Gibbs free energy as a function of the rotational barrier $V$. It is convenient to tabulate the contributions in terms of $V/RT$ against $1/z_{\mathrm{f}}$, where $z_{\mathrm{f}}$ is the partition function for free rotation [see equation (10.141)]. For details of the calculation, see Section 10.7c.

Table A4.7 summarizes the thermodynamics properties of monatomic solids as calculated by the Debye model. The values are expressed in terms of $\Phi_{\mathrm{D}}/T$, where $\Phi_{\mathrm{D}}$ is the Debye temperature. See Section 10.8 for details of the calculations. Tables A4.5 to A4.7 are adapted from K. S. Pitzer, *Thermodynamics*, McGraw-Hill, New York, 1995.

**Table A4.1** Thermodynamic functions of an ideal gas. (Use $R = 8.314510$ J·K$^{-1}$·mol$^{-1}$ and SI units for pressure, temperature, and all molecular data.)

*Translation*

$$S_{\mathrm{m,\,trans}} = \frac{3}{2}R\ln M + \frac{5}{2}R\ln T - R\ln p + \underbrace{\left[\frac{5}{2}R + R\ln\left(\frac{(2\pi k)^{3/2}}{h^3N^{5/2}}\right) + R\ln R\right]}_{+172.3005}$$

$$\left(\frac{G_{\mathrm{m}} - H_{0,\mathrm{m}}}{T}\right)_{\mathrm{trans}} = -\frac{3}{2}R\ln M - \frac{5}{2}R\ln T + R\ln p + \underbrace{\left[-R\ln\left(\frac{(2\pi k)^{3/2}}{h^3N^{5/2}}\right) - R\ln R\right]}_{-151.5142}$$

$$\left(\frac{U_{\mathrm{m}} - U_{0,\mathrm{m}}}{T}\right)_{\mathrm{trans}} = \frac{3}{2}R = (C_{V,\mathrm{m}})_{\mathrm{trans}}$$

$$\left(\frac{H_{\mathrm{m}} - H_{0,\mathrm{m}}}{T}\right)_{\mathrm{trans}} = \frac{5}{2}R = (C_{p,\mathrm{m}})_{\mathrm{trans}}$$

*Rotation (Rigid molecule approximation)*

Linear polyatomic or diatomic molecules

$$S_{\mathrm{m,\,rot}} = R\ln T + R\ln I - R\ln\sigma + \underbrace{\left[R\ln\left(\frac{8\pi^2k}{h^2}\right) + R\right]}_{+877.3950}$$

$$\left(\frac{G_{\mathrm{m}} - H_{0,\mathrm{m}}}{T}\right)_{\mathrm{rot}} = -R\ln T - R\ln I + R\ln\sigma + \underbrace{\left[-R\ln\frac{8\pi^2k}{h^2}\right]}_{-869.0805}$$

$$\left(\frac{U_{\mathrm{m}} - U_{0,\mathrm{m}}}{T}\right)_{\mathrm{rot}} = \left(\frac{H_{\mathrm{m}} - H_{0,\mathrm{m}}}{T}\right)_{\mathrm{rot}} = (C_{\mathrm{m}})_{\mathrm{rot}} = R$$

**Table A4.1** *Continued*

Nonlinear polyatomic molecules

$$S_{\mathrm{m,\,rot}} = \frac{3}{2} R \ln T + \frac{1}{2} R \ln I_A I_B I_C - R \ln \sigma + \underbrace{\left[ R \ln \frac{8\pi^2 (2\pi k)^{3/2}}{h^3} + \frac{3}{2} R \right]}_{+1320.8515}$$

$$\left( \frac{U_{\mathrm{m}} - U_{0,\mathrm{m}}}{T} \right)_{\mathrm{rot}} = \left( \frac{H_{\mathrm{m}} - H_{0,\mathrm{m}}}{T} \right)_{\mathrm{rot}} = (C)_{\mathrm{rot}} = \frac{3}{2} R$$

$$\left( \frac{G_{\mathrm{m}} - H_{0,\mathrm{m}}}{T} \right)_{\mathrm{rot}} = -\frac{3}{2} R \ln T - \frac{1}{2} R \ln I_A I_B I_C + R \ln \sigma + \underbrace{\left[ -R \ln \frac{8\pi^2 (2\pi k)^{3/2}}{h^3} \right]}_{-1308.3797}$$

*Vibration (harmonic oscillator approximation)*

$$S_{\mathrm{m,\,vib}} = R \sum_{i=1}^{n} \left[ \frac{x_i}{\exp(x_i) - 1} - \ln(1 - \exp(-x_i)) \right]; \quad x_i = \frac{hc\tilde{\omega}_i}{kT} = 1.43877 \frac{\tilde{\omega}_i}{T} \text{ (use } \tilde{\omega}_i \text{ in cm}^{-1}\text{)}$$

$$\left( \frac{G_{\mathrm{m}} - H_{0,\mathrm{m}}}{T} \right)_{\mathrm{vib}} = \sum_{i=1}^{n} R \ln(1 - \exp(-x_i)) \approx \sum_{i=1}^{n} - R \ln \frac{kT}{hc\omega_i} \quad \text{(High temp. approx.)}$$

$$\left( \frac{U_{\mathrm{m}} - U_{0,\mathrm{m}}}{T} \right)_{\mathrm{vib}} = \left( \frac{H_{\mathrm{m}} - H_{0,\mathrm{m}}}{T} \right)_{\mathrm{vib}} = R \sum_{i=1}^{n} \frac{x_i}{\exp(x_i) - 1}$$

$$(C_{\mathrm{m}})_{\mathrm{vib}} = R \sum_{i=1}^{n} \frac{x_i^2 \exp(x_i)}{(\exp(x_i) - 1)^2}$$

where $n = (3\eta - 6)$ or $(3\eta - 5)$, with $\eta$ equal to the number of atoms in the molecule

**Table A4.2** Moments of inertia and rotational constants of some common molecules.

The values given are for a molecule that corresponds to the natural abundance of each isotope since these "average values" will give the thermodynamic properties of a mole of the naturally occurring substance.*

For diatomic molecules, $\tilde{B}_0$ is the rotational constant to use with equation (10.125), while $\tilde{B}_e$ applies to equation (10.124). They are related by $\tilde{B}_0 = \tilde{B}_e - \frac{1}{2}\tilde{\alpha}$. The moment of inertia $I_0/(\text{kg·m}^2)$ is related to $B_0/(\text{cm}^{-1})$ through the relationship $I_0 = h/(8 \times 10^{-2}\pi^2 B_0 c)$ with $h$ and $c$ expressed in SI units. For polyatomic molecules, $I_A$, $I_B$, and $I_C$ are the moments of inertia to use with Table A4.1 where the rigid rotator approximation is assumed. For diatomic molecules, $I_0$ is used with Table A4.1 to calculate the thermodynamic properties assuming the rigid rotator approximation. The anharmonicity and nonrigid rotator corrections are added to this value.

**Table A4.2a** Diatomic molecules

| Molecule | $\tilde{B}_0/(\text{cm}^{-1})$ | $I_0/(10^{-47}\ \text{kg·m}^2)$ | $\tilde{B}_e/(\text{cm}^{-1})$ | $10^3\tilde{\alpha}/(10^{-3}\ \text{cm}^{-1})$ |
|---|---|---|---|---|
| $Br_2$ | 0.081948 | 341.59 | 0.082107 | 0.31873 |
| CO | 1.9215 | 14.568 | 1.9302 | 17.46 |
| $Cl_2$ | 0.24339 | 115.01 | 0.24415 | 1.5163 |
| $H_2$ | (59.304) | 0.47203 | (60.800) | (2993) |
| HBr | 8.34954 | 3.3526 | 8.46620 | 233.33 |
| HCl | 10.4326 | 2.6832 | 10.5844 | 303.7 |
| HF | 20.5577 | 1.3617 | 20.9555 | 795.8 |
| HI | 6.426 | 4.356 | 6.512 | 171.5 |
| Cl | 0.11272 | 248.34 | 0.11298 | 0.5275 |
| $I_2$ | 0.037333 | 749.81 | 0.037395 | 0.12435 |
| $N_2$ | 2.001 | 14.00 | (2.010) | (18.7) |
| NO | 1.6953 | 16.512 | 1.7042 | 17.8 |
| NaCl | 0.21611 | 129.53 | 0.21691 | 1.598 |
| OH | 18.514 | 1.5120 | 18.871 | 714 |
| $O_2$ | 1.43765 | 19.471 | 1.44562 | 15.933 |

**Table A4.2b** Polyatomic molecules
Linear molecules

| Molecule | $I_A/(10^{-47}\text{kg}\cdot\text{m}^2)$ | Molecule | $I_A/(10^{-47}\text{kg}\cdot\text{m}^2)$ |
|---|---|---|---|
| HCN | 18.8585 | $CS_2$ | 256.42 |
| $N_2O$ | 66.4882 | $C_2H_2$ | 23.7864 |
| $CO_2$ | 71.4988 | | |

Nonlinear molecules

| Molecule | $I_AI_BI_C/(10^{-138}\text{kg}^3\cdot\text{m}^6)$ | Molecule | $I_AI_BI_C/(10^{-138}\text{kg}^3\cdot\text{m}^6)$ |
|---|---|---|---|
| $CH_4$ | 0.1499 | $NH_3$ | 0.0348 |
| $CCl_4$ | 115003 | $POCl_3$ | 10184.1 |
| $H_2O$ | (0.0058577) | $CFCl_3$ | 57360.2 |
| $SO_2$ | 107.0 | $NO_2$ | 15.423 |

* Most values were taken from M. W. Chase, Jr., C. A. Davies, J. R. Downey, Jr., D. J. Frurip, R. A. McDonald, and A. N. Syverud, "JANAF Thermochemical Tables, Third Edition", *J. Phys. Chem. Ref. Data*, **14**, Supplement No. 1, 1985. For the diatomic molecules, a few values (in parentheses) came from G. Hertzberg, *Molecular Spectra and Molecular Structure, I. Spectra of Diatomic Molecules*, and *II. Infrared and Raman Spectra of Polyatomic Molecules*, Van Nostrand Reinhold Co., New York, 1950 and 1945.

**Table A4.3** Fundamental vibrational frequencies of some common molecules.

The values given are for a molecule that corresponds to the natural abundance of each isotope since these "average values" will give the thermodynamic properties of a mole of the naturally occurring substance.*

For diatomic molecules, $\tilde{\omega}_0$ is the vibrational constant to use with equation (10.125) for calculating anharmonicity and nonrigid rotator corrections, while $\tilde{\omega}_e$ and $\tilde{\omega}_e x_e$ apply to equation (10.124). They are related by $\tilde{\omega}_0 = \tilde{\omega}_e - 2\tilde{\omega}_e x_e$. For polyatomic molecules, $\tilde{\omega}$ is the fundamental vibrational frequency, assuming the harmonic oscillator approximation. It is the value to use when making calculations with the equations in Table A4.1. For diatomic molecules, $\tilde{\omega}_0$ is used with Table A4.1 to calculate the thermodynamic values assuming the rigid rotator and harmonic oscillator approximations are valid. Corrections for anharmonicity and the nonrigid rotator are added to these values.

**Table A4.3a** Diatomic molecules

| Molecule | $\tilde{\omega}_0/(\text{cm}^{-1})$ | $\tilde{\omega}_e/(\text{cm}^{-1})$ | $\tilde{\omega}_e x_e/(\text{cm}^{-1})$ |
|---|---|---|---|
| $Br_2$ | 323.166 | 325.321 | 1.07742 |
| CO | 2142.61 | 2169.52 | 13.453 |
| $Cl_2$ | 554.362 | 559.751 | 2.6943 |
| $H_2$ | 4159.44 | (4395.24) | (117.90) |
| HBr | 2558.73 | 2649.182 | 45.225 |
| HCl | 2785.47 | 2889.59 | 52.06 |
| HF | 3958.63 | 4138.73 | 90.05 |
| HI | 2229.60 | 2309.06 | 39.73 |
| ICl | 379.28 | 382.18 | 1.450 |
| $I_2$ | 213.316 | 214.5481 | 0.616259 |
| $N_2$ | (2330.70) | (2359.61) | (14.456) |
| NO | 1875.66 | 1903.60 | 13.97 |
| NaCl | 360.18 | 363.62 | 1.72 |
| OH | 3569.59 | 3735.21 | 82.81 |
| $O_2$ | 1556.231 | 1580.193 | 11.981 |

**Table A4.3b** Polyatomic molecules
The numbers in parentheses designate the number of normal modes that have this frequency.

Linear molecules

| Molecule | $\tilde{\omega}/(\text{cm}^{-1})$ | Molecule | $\tilde{\omega}/(\text{cm}^{-1})$ |
|---|---|---|---|
| HCN | 713.5(2), 2096.3, 3311.5 | $CS_2$ | 396.7(2), 656.6, 1523 |
| $N_2O$ | 589.2(2), 1276.5, 2223.7 | $C_2H_2$ | 611.6(2), 729.3(2), 1973.8, 3281.9, 3373.7 |
| $CO_2$ | 667.30(2), 1384.86, 2349.30 | | |

Nonlinear molecules

| Molecule | $\tilde{\omega}/(\text{cm}^{-1})$ | Molecule | $\tilde{\omega}/(\text{cm}^{-1})$ |
|---|---|---|---|
| $CH_4$ | 1306(3), 1534.0(2), 2916.5, 3018.7(3) | $NH_3$ | 1022, 1691(2), 3506, 3577(2) |
| $CCl_4$ | 218(2), 314(3), 458, 776(3) | $NO_2$ | 756.8, 1357.8, 1665.5 |
| $H_2O$ | 1594.7, 3651.1, 3755.9 | $POCl_3$ | 193(2), 267(2), 337, 486, 581(2), 1290 |
| $SO_2$ | 517.69, 1151.38, 1361.76 | $CFCl_3$ | 241(2), 349.5, 398(2), 535.3, 847(2), 1085 |

* Most values were taken from M. W. Chase, Jr., C. A. Davies, J. R. Downey, Jr., D. J. Frurip, R. A. McDonald, and A. N. Syverud, "JANAF Thermochemical Tables, Third Edition", *J. Phys. Chem. Ref. Data*, **14**, Supplement No. 1, 1985. For the diatomic molecules a few values (in parentheses) came from G. Hertzberg, *Molecular Spectra and Molecular Structure, I. Spectra of Diatomic Molecules*, and *II. Infrared and Raman Spectra of Polyatomic Molecules*, Van Nostrand Reinhold Co., New York, 1950 and 1945.

**Table A4.4** Electronic energy levels of some common molecules or atoms with unpaired electrons*.

| Atom or molecule | $g_i$ | $\tilde{\epsilon}_i/(\mathrm{cm}^{-1})$ | Atom or molecule | $g_i$ | $\tilde{\epsilon}_i/(\mathrm{cm}^{-1})$ |
|---|---|---|---|---|---|
| $O_2$ | 3 | 0 | I | 4 | 0 |
| | 2 | 7882.39 | | 2 | 7603.15 |
| | 1 | 13120.91 | Cl | 4 | 0 |
| O | 5 | 0 | | 2 | 882.36 |
| | 3 | 158.265 | Na | 2 | 0 |
| | 1 | 226.977 | | 2 | 16956.172 |
| | 5 | 15867.862 | | 4 | 16973.368 |
| | 1 | 33792.583 | | 2 | 25739.991 |
| $NO_2$ | 2 | 0 | H | 2 | 0 |
| NO | 2 | 0 | | | |
| | 2 | 121.1 | | | |

* Values were taken from M. W. Chase, Jr., C. A. Davies, J. R. Downey, Jr., D. J. Frurip, R. A. McDonald, and A. N. Syverud, "JANAF Thermochemical Tables, Third Edition", *J. Phys. Chem. Ref. Data*, **14**, Supplement No. 1, 1985.

**Table A4.5** Anharmonic oscillator and nonrigid rotator corrections

The following equations are used to calculate the anharmonicity and nonrigid rotator corrections to the thermodynamic properties of diatomic molecules.

$$\left(\frac{H_{\mathrm{m}} - H_{0,\mathrm{m}}}{T}\right)_{\mathrm{corr}} = \left(\frac{R}{u}\right)\left\{2x\left[\frac{u^2\{2u\exp(u) - \exp(u) + 1\}}{\{\exp(u) - 1\}^3}\right] + \delta\left[\frac{u^2\exp(u)}{\{\exp(u) - 1\}^2}\right] + 8\gamma\right\}$$

$$-\left(\frac{G_{\mathrm{m}} - H_{0,\mathrm{m}}}{T}\right)_{\mathrm{corr}} = \left(\frac{R}{u}\right)\left\{2x\left[\frac{u^2}{\{\exp(u) - 1\}^2}\right] + \delta\left[\frac{u}{\exp(u) - 1}\right] + 8\gamma\right\}$$

$$(S_{\mathrm{m}})_{\mathrm{corr}} = \frac{(H_{\mathrm{m}} - H_{0,\mathrm{m}})_{\mathrm{corr}}}{T} - \frac{(G_{\mathrm{m}} - H_{0,\mathrm{m}})_{\mathrm{corr}}}{T}$$

**Table A4.5** *Continued*

and

$$(C_{\rm m})_{\rm corr} = \left(\frac{R}{u}\right)\left\{4x\left[\frac{\{u^3\exp(u)\}\{2u\exp(u) - 2\exp(u) + u + 2\}}{\{\exp(u) - 1\}^4}\right]\right\}$$

$$+ \left(\frac{R}{u}\right)\left\{\delta\left[\frac{\{u^3\exp(u)\}\{\exp(u) + 1\}}{\{\exp(u) - 1\}^3}\right] + 16\gamma\right\},$$

where

$$u = \frac{hc\tilde{\omega}_0}{kT}, \quad y = \frac{hc\tilde{B}_0}{kT}, \quad x = \frac{x_{\rm e}}{1 - 2x_{\rm e}} \approx \frac{x_{\rm e}\tilde{\omega}_{\rm e}}{\tilde{\omega}_{\rm e}}, \quad \delta = \frac{\tilde{\alpha}}{\tilde{B}_0}, \quad \text{and} \quad \gamma = \frac{\tilde{B}_{\rm e}}{\tilde{\omega}_{\rm e}}.$$

The above equations can be simplified for calculation of the corrections by first defining and calculating the functions $\nu$ and $w$ given by

$$\nu = \exp(u) - 1 \quad \text{and} \quad w = \exp(u) = \nu + 1.$$

Substitution of these quantities into the above equations gives

$$-\left(\frac{G_{\rm m} - G_{0,\rm m}}{T}\right)_{\rm corr} = \left(\frac{R}{u}\right)\left\{\left[\frac{2u^2}{\nu^2}\right]x + \left[\frac{u}{\nu}\right]\delta + 8\gamma\right\}$$

$$\left(\frac{H_{\rm m} - H_{0,\rm m}}{T}\right)_{\rm corr} = \left(\frac{R}{u}\right)\left\{\left[\left(\frac{2u^2}{\nu^3}\right)(2uw - \nu)\right]x + \left[\frac{u^2w}{\nu^2}\right]\delta + 8\gamma\right\}$$

$$(S_{\rm m})_{\rm corr} = \frac{(H_{\rm m} - H_{0,\rm m})_{\rm corr}}{T} - \frac{(G_{\rm m} - H_{0,\rm m})_{\rm corr}}{T}$$

$$(C_{\rm m})_{\rm corr} = \left(\frac{R}{u}\right)\left\{\left[\left(\frac{4u^3}{\nu^4}\right)(2uw^2 + uw - 2\nu w)\right]x\right.$$

$$\left. + \left[\left(\frac{u^3}{\nu^3}\right)(w^2 + w)\right]\delta + 16\gamma\right\}$$

* With modern high-speed computers, the corrections can easily be calculated from either set of equations.

**Table A4.6** Contributions to the thermodynamic properties due to internal rotation as a function of $V$, the rotational barrier, and $z_f$, the partition function for free rotation.

Heat capacity $(C_m)/(J \cdot K^{-1} \cdot mol^{-1})$

$1/z_f$

| $V/RT$ | 0 | 0.05 | 0.1 | 0.15 | 0.2 | 0.25 | 0.3 | 0.35 | 0.4 | 0.45 | 0.5 | 0.55 | 0.6 | 0.65 | 0.7 | 0.75 | 0.8 | 0.85 | 0.9 | 0.95 |
|---|---|---|---|---|---|---|---|---|---|---|---|---|---|---|---|---|---|---|---|---|
| 0.0 | 4.159 | 4.159 | 4.159 | 4.159 | 4.159 | 4.159 | 4.159 | 4.159 | 4.159 | 4.159 | 4.159 | 4.159 | 4.159 | 4.159 | 4.159 | 4.159 | 4.159 | 4.159 | 4.159 | 4.159 |
| 0.2 | 4.1986 | 4.197 | 4.197 | 4.192 | 4.188 | 4.184 | 4.180 | 4.176 | 4.176 | 4.176 | 4.184 | 4.184 | 4.184 | 4.184 | 4.184 | 4.184 | 4.184 | 4.180 | 4.180 | 4.180 |
| 0.4 | 4.3212 | 4.322 | 4.318 | 4.310 | 4.301 | 4.289 | 4.284 | 4.272 | 4.263 | 4.255 | 4.259 | 4.255 | 4.247 | 4.238 | 4.234 | 4.226 | 4.217 | 4.213 | 4.205 | 4.201 |
| 0.6 | 4.5191 | 4.519 | 4.515 | 4.502 | 4.489 | 4.469 | 4.456 | 4.435 | 4.418 | 4.397 | 4.389 | 4.376 | 4.356 | 4.335 | 4.314 | 4.293 | 4.272 | 4.255 | 4.243 | 4.230 |
| 0.8 | 4.7844 | 4.782 | 4.774 | 4.761 | 4.740 | 4.720 | 4.690 | 4.661 | 4.628 | 4.598 | 4.569 | 4.535 | 4.498 | 4.464 | 4.427 | 4.389 | 4.351 | 4.314 | 4.289 | 4.268 |
| 1.0 | 5.1057 | 5.100 | 5.092 | 5.071 | 5.046 | 5.017 | 4.979 | 4.937 | 4.891 | 4.841 | 4.786 | 4.732 | 4.678 | 4.623 | 4.565 | 4.510 | 4.456 | 4.402 | 4.351 | 4.314 |
| 1.5 | 6.0701 | 6.063 | 6.042 | 6.004 | 5.954 | 5.891 | 5.820 | 5.732 | 5.640 | 5.540 | 5.435 | 5.326 | 5.217 | 5.096 | 4.987 | 4.874 | 4.774 | 4.665 | 4.561 | 4.477 |
| 2.0 | 7.0199 | 7.092 | 7.058 | 7.000 | 6.925 | 6.828 | 6.720 | 6.586 | 6.448 | 6.297 | 6.130 | 5.958 | 5.782 | 5.611 | 5.439 | 5.263 | 5.096 | 4.937 | 4.795 | 4.657 |
| 2.5 | 8.0387 | 8.021 | 7.983 | 7.899 | 7.807 | 7.699 | 7.535 | 7.347 | 7.184 | 6.987 | 6.774 | 6.535 | 6.293 | 6.058 | 5.828 | 5.611 | 5.393 | 5.180 | 4.979 | 4.795 |
| 3.0 | 8.7818 | 8.765 | 8.711 | 8.627 | 8.506 | 8.351 | 8.167 | 7.950 | 7.724 | 7.506 | 7.247 | 6.958 | 6.682 | 6.410 | 6.134 | 5.862 | 5.594 | 5.339 | 5.092 | 4.870 |
| 3.5 | 9.2994 | 9.280 | 9.222 | 9.121 | 8.979 | 8.812 | 8.594 | 8.347 | 8.092 | 7.820 | 7.544 | 7.226 | 6.920 | 6.611 | 6.301 | 5.991 | 5.694 | 5.410 | 5.130 | 4.874 |
| 4.0 | 9.6186 | 9.598 | 9.523 | 9.410 | 9.259 | 9.071 | 8.828 | 8.569 | 8.284 | 7.979 | 7.673 | 7.339 | 7.004 | 6.665 | 6.330 | 6.004 | 5.686 | 5.381 | 5.084 | 4.803 |
| 4.5 | 9.7730 | 9.749 | 9.673 | 9.540 | 9.364 | 9.163 | 8.908 | 8.627 | 8.326 | 7.996 | 7.665 | 7.318 | 6.962 | 6.602 | 6.259 | 5.912 | 5.577 | 5.268 | 4.958 | 4.665 |
| 5.0 | 9.8102 | 9.782 | 9.699 | 9.560 | 9.376 | 9.146 | 8.870 | 8.602 | 8.251 | 7.908 | 7.565 | 7.188 | 6.824 | 6.456 | 6.096 | 5.745 | 5.406 | 5.079 | 4.770 | 4.469 |
| 6.0 | 9.6893 | 9.652 | 9.552 | 9.393 | 9.171 | 8.912 | 8.615 | 8.280 | 7.920 | 7.544 | 7.159 | 6.753 | 6.360 | 5.979 | 5.615 | 5.251 | 4.908 | 4.586 | 4.276 | 3.992 |
| 7.0 | 9.4768 | 9.439 | 9.322 | 9.142 | 8.895 | 8.598 | 8.255 | 7.878 | 7.477 | 7.063 | 6.644 | 6.222 | 5.816 | 5.422 | 5.050 | 4.686 | 4.351 | 4.025 | 3.724 | 3.456 |
| 8.0 | 9.2717 | 9.226 | 9.096 | 8.891 | 8.611 | 8.280 | 7.899 | 7.481 | 7.046 | 6.594 | 6.142 | 5.715 | 5.280 | 4.870 | 4.494 | 4.134 | 3.799 | 3.489 | 3.201 | 2.946 |
| 9.0 | 9.1052 | 9.054 | 8.912 | 8.678 | 8.364 | 7.987 | 7.565 | 7.109 | 6.640 | 6.167 | 5.699 | 5.230 | 4.786 | 4.385 | 4.000 | 3.636 | 3.301 | 3.000 | 2.728 | 2.481 |
| 10.0 | 8.9776 | 8.924 | 8.761 | 8.506 | 8.163 | 7.757 | 7.301 | 6.820 | 6.305 | 5.782 | 5.280 | 4.816 | 4.372 | 3.946 | 3.556 | 3.201 | 2.879 | 2.586 | 2.326 | 2.088 |
| 12.0 | 8.8086 | 8.740 | 8.548 | 8.251 | 7.853 | 7.376 | 6.845 | 6.284 | 5.711 | 5.159 | 4.632 | 4.138 | 3.669 | 3.238 | 2.853 | 2.510 | 2.209 | 1.937 | 1.703 | 1.498 |
| 14.0 | 8.7082 | 8.632 | 8.406 | 8.046 | 7.590 | 7.054 | 6.468 | 5.858 | 5.247 | 4.653 | 4.092 | 3.577 | 3.113 | 2.694 | 2.318 | 2.004 | 1.720 | 1.473 | 1.268 | 1.096 |
| 16.0 | 8.6429 | 8.552 | 8.297 | 7.895 | 7.381 | 6.786 | 6.142 | 5.485 | 4.837 | 4.222 | 3.653 | 3.134 | 2.674 | 2.268 | 1.912 | 1.619 | 1.356 | 1.138 | 0.958 | 0.812 |
| 18.0 | 8.5969 | 8.498 | 8.205 | 7.753 | 7.184 | 6.535 | 5.845 | 5.155 | 4.477 | 3.845 | 3.264 | 2.749 | 2.297 | 1.908 | 1.582 | 1.305 | 1.084 | 0.900 | 0.732 | 0.602 |
| 20.0 | 8.5626 | 8.452 | 8.134 | 7.644 | 7.021 | 6.318 | 5.577 | 4.845 | 4.146 | 3.502 | 2.933 | 2.427 | 1.996 | 1.628 | 1.322 | 1.071 | 0.870 | 0.703 | 0.565 | 0.456 |

Enthalpy $(H_m/T)/(J \cdot K^{-1} \cdot mol^{-1})$
$1/z_f$

| $V/RT$ | 0 | 0.05 | 0.1 | 0.15 | 0.2 | 0.25 | 0.3 | 0.35 | 0.4 | 0.45 | 0.5 | 0.55 | 0.6 | 0.65 | 0.7 | 0.75 | 0.8 | 0.85 | 0.9 | 0.95 |
|---|---|---|---|---|---|---|---|---|---|---|---|---|---|---|---|---|---|---|---|---|
| 0.0 | 4.159 | 4.159 | 4.159 | 4.159 | 4.159 | 4.159 | 4.159 | 4.159 | 4.159 | 4.159 | 4.159 | 4.159 | 4.159 | 4.159 | 4.159 | 4.159 | 4.159 | 4.159 | 4.159 | 4.159 |
| 0.2 | 4.9472 | 4.778 | 4.628 | 4.494 | 4.393 | 4.318 | 4.276 | 4.247 | 4.217 | 4.201 | 4.184 | 4.167 | 4.159 | 4.159 | 4.159 | 4.151 | 4.151 | 4.146 | 4.142 | 4.138 |
| 0.4 | 5.6547 | 5.439 | 5.226 | 5.021 | 4.816 | 4.628 | 4.489 | 4.397 | 4.335 | 4.289 | 4.247 | 4.209 | 4.180 | 4.159 | 4.151 | 4.142 | 4.134 | 4.134 | 4.125 | 4.121 |
| 0.6 | 6.2814 | 6.012 | 5.749 | 5.485 | 5.234 | 4.979 | 4.761 | 4.598 | 4.485 | 4.389 | 4.310 | 4.243 | 4.201 | 4.163 | 4.142 | 4.130 | 4.117 | 4.109 | 4.100 | 4.096 |
| 0.8 | 6.8308 | 6.510 | 6.201 | 5.904 | 5.607 | 5.322 | 5.067 | 4.841 | 4.661 | 4.506 | 4.385 | 4.293 | 4.222 | 4.167 | 4.117 | 4.100 | 4.084 | 4.075 | 4.067 | 4.063 |
| 1.0 | 7.3065 | 6.945 | 6.594 | 6.255 | 5.933 | 5.623 | 5.335 | 5.067 | 4.833 | 4.628 | 4.456 | 4.343 | 4.243 | 4.167 | 4.109 | 4.067 | 4.038 | 4.025 | 4.017 | 4.012 |
| 1.5 | 8.2048 | 7.766 | 7.335 | 6.920 | 6.531 | 6.159 | 5.795 | 5.464 | 5.146 | 4.870 | 4.615 | 4.431 | 4.263 | 4.130 | 4.025 | 3.954 | 3.899 | 3.858 | 3.833 | 3.828 |
| 2.0 | 8.7600 | 8.247 | 7.757 | 7.289 | 6.845 | 6.427 | 6.025 | 5.648 | 5.293 | 4.979 | 4.686 | 4.422 | 4.205 | 4.025 | 3.883 | 3.782 | 3.707 | 3.653 | 3.615 | 3.598 |
| 2.5 | 9.0625 | 8.498 | 7.950 | 7.443 | 6.954 | 6.485 | 6.058 | 5.653 | 5.272 | 4.933 | 4.619 | 4.318 | 4.067 | 3.858 | 3.690 | 3.556 | 3.460 | 3.393 | 3.351 | 3.330 |
| 3.0 | 9.1939 | 8.573 | 7.987 | 7.435 | 6.908 | 6.422 | 5.966 | 5.527 | 5.121 | 4.770 | 4.435 | 4.134 | 3.866 | 3.640 | 3.464 | 3.310 | 3.192 | 3.113 | 3.063 | 3.046 |
| 3.5 | 9.2186 | 8.548 | 7.920 | 7.335 | 6.782 | 6.263 | 5.782 | 5.335 | 4.920 | 4.552 | 4.209 | 3.904 | 3.632 | 3.393 | 3.201 | 3.042 | 2.916 | 2.828 | 2.774 | 2.757 |
| 4.0 | 9.1826 | 8.468 | 7.799 | 7.176 | 6.598 | 6.058 | 5.561 | 5.109 | 4.690 | 4.310 | 3.962 | 3.648 | 3.372 | 3.134 | 2.933 | 2.766 | 2.636 | 2.548 | 2.489 | 2.469 |
| 4.5 | 9.1174 | 8.360 | 7.653 | 7.000 | 6.397 | 5.832 | 5.326 | 4.862 | 4.439 | 4.050 | 3.699 | 3.389 | 3.113 | 2.874 | 2.669 | 2.506 | 2.372 | 2.280 | 2.222 | 2.201 |
| 5.0 | 9.0416 | 8.247 | 7.506 | 6.824 | 6.197 | 5.623 | 5.096 | 4.619 | 4.192 | 3.803 | 3.448 | 3.138 | 2.866 | 2.628 | 2.427 | 2.259 | 2.125 | 2.029 | 1.966 | 1.946 |
| 6.0 | 8.8969 | 8.025 | 7.226 | 6.494 | 5.824 | 5.217 | 4.665 | 4.180 | 3.736 | 3.343 | 2.987 | 2.694 | 2.427 | 2.188 | 1.992 | 1.828 | 1.699 | 1.602 | 1.540 | 1.510 |
| 7.0 | 8.7810 | 7.845 | 6.987 | 6.209 | 5.502 | 4.870 | 4.305 | 3.799 | 3.356 | 2.962 | 2.611 | 2.318 | 2.054 | 1.828 | 1.640 | 1.481 | 1.356 | 1.264 | 1.197 | 1.167 |
| 8.0 | 8.6960 | 7.699 | 6.791 | 5.971 | 5.234 | 4.581 | 3.996 | 3.485 | 3.033 | 2.640 | 2.297 | 2.008 | 1.757 | 1.540 | 1.364 | 1.213 | 1.092 | 1.000 | 0.933 | 0.900 |
| 9.0 | 8.6345 | 7.577 | 6.623 | 5.770 | 5.004 | 4.330 | 3.732 | 3.213 | 2.766 | 2.381 | 2.042 | 1.761 | 1.519 | 1.305 | 1.142 | 1.004 | 0.883 | 0.799 | 0.736 | 0.703 |
| 10.0 | 8.5893 | 7.477 | 6.477 | 5.586 | 4.799 | 4.109 | 3.506 | 2.992 | 2.544 | 2.155 | 1.828 | 1.548 | 1.314 | 1.125 | 0.967 | 0.837 | 0.728 | 0.644 | 0.586 | 0.552 |
| 12.0 | 8.5291 | 7.318 | 6.243 | 5.289 | 4.464 | 3.749 | 3.117 | 2.611 | 2.171 | 1.803 | 1.490 | 1.238 | 1.021 | 0.845 | 0.711 | 0.598 | 0.506 | 0.435 | 0.381 | 0.351 |
| 14.0 | 8.4914 | 7.184 | 6.029 | 5.029 | 4.171 | 3.443 | 2.812 | 2.305 | 1.883 | 1.527 | 1.243 | 1.004 | 0.816 | 0.661 | 0.531 | 0.431 | 0.351 | 0.301 | 0.259 | 0.234 |
| 16.0 | 8.4651 | 7.071 | 5.862 | 4.812 | 3.920 | 3.180 | 2.565 | 2.063 | 1.648 | 1.314 | 1.042 | 0.828 | 0.657 | 0.531 | 0.410 | 0.318 | 0.255 | 0.213 | 0.184 | 0.159 |
| 18.0 | 8.4454 | 6.971 | 5.703 | 4.611 | 3.707 | 2.958 | 2.347 | 1.854 | 1.452 | 1.134 | 0.883 | 0.686 | 0.536 | 0.414 | 0.322 | 0.251 | 0.197 | 0.151 | 0.121 | 0.109 |
| 20.0 | 8.4308 | 6.887 | 5.561 | 4.439 | 3.519 | 2.761 | 2.155 | 1.669 | 1.284 | 0.987 | 0.757 | 0.577 | 0.439 | 0.335 | 0.255 | 0.197 | 0.151 | 0.117 | 0.092 | 0.075 |

(continued)

**Table A4.6** *Continued*

Gibbs free energy $(-G_m/T)/(J \cdot K^{-1} \cdot mol^{-1})$

$1/z_f$

| $V/RT$ | 0.25 | 0.3 | 0.35 | 0.4 | 0.45 | 0.5 | 0.55 | 0.6 | 0.65 | 0.7 | 0.75 | 0.8 | 0.85 | 0.9 | 0.95 |
|---|---|---|---|---|---|---|---|---|---|---|---|---|---|---|---|
| 0.0 | 11.523 | 10.008 | 8.728 | 7.619 | 6.640 | 5.761 | 4.979 | 4.243 | 3.582 | 2.971 | 2.406 | 1.854 | 1.351 | 0.870 | 0.427 |
| 0.2 | 11.339 | 9.870 | 8.623 | 7.544 | 6.586 | 5.724 | 4.945 | 4.222 | 3.565 | 2.958 | 2.385 | 1.845 | 1.343 | 0.866 | 0.423 |
| 0.4 | 10.975 | 9.606 | 8.427 | 7.385 | 6.456 | 5.615 | 4.870 | 4.171 | 3.523 | 2.925 | 2.364 | 1.833 | 1.331 | 0.862 | 0.414 |
| 0.6 | 10.535 | 9.238 | 8.134 | 7.146 | 6.268 | 5.477 | 4.753 | 4.075 | 3.456 | 2.874 | 2.322 | 1.803 | 1.318 | 0.854 | 0.406 |
| 0.8 | 10.067 | 8.812 | 7.766 | 6.845 | 6.033 | 5.297 | 4.598 | 3.962 | 3.364 | 2.803 | 2.272 | 1.774 | 1.297 | 0.837 | 0.402 |
| 1.0 | 9.606 | 8.385 | 7.381 | 6.523 | 5.770 | 5.079 | 4.418 | 3.816 | 3.251 | 2.707 | 2.201 | 1.720 | 1.264 | 0.816 | 0.393 |
| 1.5 | 8.535 | 7.406 | 6.477 | 5.732 | 5.063 | 4.473 | 3.920 | 3.410 | 2.929 | 2.460 | 2.013 | 1.586 | 1.159 | 0.745 | 0.351 |
| 2.0 | 7.611 | 6.540 | 5.690 | 4.992 | 4.402 | 3.879 | 3.418 | 2.983 | 2.573 | 2.180 | 1.791 | 1.414 | 1.042 | 0.669 | 0.310 |
| 2.5 | 6.820 | 5.812 | 5.008 | 4.364 | 3.816 | 3.356 | 2.950 | 2.577 | 2.234 | 1.900 | 1.569 | 1.247 | 0.916 | 0.590 | 0.264 |
| 3.0 | 6.163 | 5.188 | 4.431 | 3.824 | 3.318 | 2.908 | 2.544 | 2.218 | 1.916 | 1.632 | 1.356 | 1.079 | 0.799 | 0.510 | 0.222 |
| 3.5 | 5.607 | 4.674 | 3.946 | 3.356 | 2.904 | 2.523 | 2.197 | 1.912 | 1.653 | 1.406 | 1.163 | 0.929 | 0.690 | 0.439 | 0.176 |
| 4.0 | 5.125 | 4.238 | 3.544 | 2.983 | 2.565 | 2.205 | 1.904 | 1.644 | 1.418 | 1.205 | 1.000 | 0.795 | 0.586 | 0.368 | 0.142 |
| 4.5 | 4.740 | 3.870 | 3.197 | 2.665 | 2.272 | 1.937 | 1.665 | 1.423 | 1.213 | 1.033 | 0.858 | 0.678 | 0.490 | 0.310 | 0.113 |
| 5.0 | 4.406 | 3.552 | 2.912 | 2.414 | 2.021 | 1.707 | 1.452 | 1.243 | 1.059 | 0.895 | 0.741 | 0.582 | 0.427 | 0.264 | 0.084 |
| 6.0 | 3.845 | 3.046 | 2.452 | 1.996 | 1.644 | 1.360 | 1.142 | 0.962 | 0.808 | 0.674 | 0.548 | 0.431 | 0.310 | 0.188 | 0.050 |
| 7.0 | 3.427 | 2.661 | 2.105 | 1.682 | 1.360 | 1.117 | 0.912 | 0.757 | 0.623 | 0.515 | 0.418 | 0.326 | 0.234 | 0.134 | 0.033 |
| 8.0 | 3.075 | 2.360 | 1.841 | 1.448 | 1.151 | 0.925 | 0.749 | 0.607 | 0.494 | 0.402 | 0.326 | 0.251 | 0.176 | 0.100 | 0.021 |
| 9.0 | 2.791 | 2.109 | 1.623 | 1.255 | 0.983 | 0.778 | 0.623 | 0.502 | 0.397 | 0.326 | 0.259 | 0.197 | 0.134 | 0.079 | 0.017 |
| 10.0 | 2.552 | 1.908 | 1.443 | 1.105 | 0.849 | 0.665 | 0.519 | 0.418 | 0.331 | 0.264 | 0.205 | 0.155 | 0.109 | 0.063 | 0.008 |
| 12.0 | 2.180 | 1.590 | 1.172 | 0.874 | 0.657 | 0.502 | 0.385 | 0.297 | 0.226 | 0.176 | 0.138 | 0.105 | 0.075 | 0.042 | 0.004 |
| 14.0 | 1.891 | 1.343 | 0.971 | 0.707 | 0.519 | 0.385 | 0.289 | 0.218 | 0.159 | 0.126 | 0.096 | 0.067 | 0.050 | 0.029 | 0.000 |
| 16.0 | 1.657 | 1.155 | 0.816 | 0.582 | 0.418 | 0.301 | 0.222 | 0.163 | 0.117 | 0.088 | 0.067 | 0.050 | 0.033 | 0.017 | 0.000 |
| 18.0 | 1.469 | 1.004 | 0.695 | 0.490 | 0.343 | 0.243 | 0.176 | 0.126 | 0.092 | 0.067 | 0.050 | 0.033 | 0.025 | 0.013 | 0.000 |
| 20.0 | 1.318 | 0.883 | 0.602 | 0.410 | 0.285 | 0.197 | 0.138 | 0.096 | 0.071 | 0.050 | 0.038 | 0.025 | 0.017 | 0.008 | 0.000 |

**Table A4.7** The Debye thermodynamic functions expressed in terms of $\theta_D/T$

Debye heat capacity $C_{V,\mathrm{m}}/3R$ as a function of $\theta_\mathrm{D}/T$

when $\dfrac{\Phi_\mathrm{D}}{T} \geqslant 16, \dfrac{C_{V,\mathrm{m}}}{3R} = 77.927\left(\dfrac{T}{\Phi_\mathrm{D}}\right)^3$

| $\Phi_\mathrm{D}/T$ | 0.0 | 0.1 | 0.2 | 0.3 | 0.4 | 0.5 | 0.6 | 0.7 | 0.8 | 0.9 | 1.0 |
|---|---|---|---|---|---|---|---|---|---|---|---|
| 0.0 | 1.0000 | 0.9995 | 0.9980 | 0.9955 | 0.9920 | 0.9876 | 0.9822 | 0.9759 | 0.9687 | 0.9606 | 0.9517 |
| 1.0 | 0.9517 | 0.9420 | 0.9315 | 0.9203 | 0.9085 | 0.8960 | 0.8828 | 0.8692 | 0.8550 | 0.8404 | 0.8254 |
| 2.0 | 0.8254 | 0.8100 | 0.7943 | 0.7784 | 0.7622 | 0.7459 | 0.7294 | 0.7128 | 0.6961 | 0.6794 | 0.6628 |
| 3.0 | 0.6628 | 0.6461 | 0.6296 | 0.6132 | 0.5968 | 0.5807 | 0.5647 | 0.5490 | 0.5334 | 0.5181 | 0.5031 |
| 4.0 | 0.5031 | 0.4883 | 0.4738 | 0.4595 | 0.4456 | 0.4320 | 0.4187 | 0.4057 | 0.3930 | 0.3807 | 0.3686 |
| 5.0 | 0.3686 | 0.3569 | 0.3455 | 0.3345 | 0.3237 | 0.3133 | 0.3031 | 0.2933 | 0.2838 | 0.2745 | 0.2656 |
| 6.0 | 0.2656 | 0.2569 | 0.2486 | 0.2405 | 0.2326 | 0.2251 | 0.2177 | 0.2107 | 0.2038 | 0.1972 | 0.1909 |
| 7.0 | 0.1909 | 0.1847 | 0.1788 | 0.1730 | 0.1675 | 0.1622 | 0.1570 | 0.1521 | 0.1473 | 0.1426 | 0.1382 |
| 8.0 | 0.1382 | 0.1339 | 0.1297 | 0.1257 | 0.1219 | 0.1182 | 0.1146 | 0.1111 | 0.1078 | 0.1046 | 0.1015 |
| 9.0 | 0.1015 | 0.09847 | 0.09558 | 0.09280 | 0.09011 | 0.08751 | 0.08500 | 0.08259 | 0.08025 | 0.07800 | 0.07582 |
| 10.0 | 0.07582 | 0.07372 | 0.07169 | 0.06973 | 0.06783 | 0.06600 | 0.06424 | 0.06253 | 0.06087 | 0.05928 | 0.05773 |
| 11.0 | 0.05773 | 0.05624 | 0.05479 | 0.05339 | 0.05204 | 0.05073 | 0.04946 | 0.04823 | 0.04705 | 0.04590 | 0.04478 |
| 12.0 | 0.04478 | 0.04370 | 0.04265 | 0.04164 | 0.04066 | 0.03970 | 0.03878 | 0.03788 | 0.03701 | 0.03617 | 0.03535 |
| 13.0 | 0.03535 | 0.03455 | 0.03378 | 0.03303 | 0.03230 | 0.03160 | 0.03091 | 0.03024 | 0.02959 | 0.02896 | 0.02835 |
| 14.0 | 0.02835 | 0.02776 | 0.02718 | 0.02661 | 0.02607 | 0.02553 | 0.02501 | 0.02451 | 0.02402 | 0.02354 | 0.02307 |
| 15.0 | 0.02307 | 0.02262 | 0.02218 | 0.02174 | 0.02132 | 0.02092 | 0.02052 | 0.02013 | 0.01975 | 0.01938 | 0.01902 |

(*continued*)

**Table A4.7** *Continued*

Debye energy content $(U_m - U_{0,m})/3RT$ as a function of $\theta_D/T$

$$\text{when } \frac{\Phi_D}{T} \geqslant 16, \ \frac{U_m - U_{0,m}}{3RT} = 19.482\left(\frac{T}{\Phi_D}\right)^3$$

| $\Phi_D/T$ | 0.0 | 0.1 | 0.2 | 0.3 | 0.4 | 0.5 | 0.6 | 0.7 | 0.8 | 0.9 | 1.0 |
|---|---|---|---|---|---|---|---|---|---|---|---|
| 0.0 | 1.0000 | 0.9630 | 0.9270 | 0.8920 | 0.8580 | 0.8250 | 0.7929 | 0.7619 | 0.7318 | 0.7026 | 0.6744 |
| 1.0 | 0.6744 | 0.6471 | 0.6208 | 0.5954 | 0.5708 | 0.5471 | 0.5243 | 0.5023 | 0.4811 | 0.4607 | 0.4411 |
| 2.0 | 0.4411 | 0.4223 | 0.4042 | 0.3868 | 0.3701 | 0.3541 | 0.3388 | 0.3241 | 0.3100 | 0.2965 | 0.2836 |
| 3.0 | 0.2836 | 0.2712 | 0.2594 | 0.2481 | 0.2373 | 0.2269 | 0.2170 | 0.2076 | 0.1986 | 0.1900 | 0.1817 |
| 4.0 | 0.1817 | 0.1739 | 0.1664 | 0.1592 | 0.1524 | 0.1459 | 0.1397 | 0.1338 | 0.1281 | 0.1227 | 0.1176 |
| 5.0 | 0.1176 | 0.1127 | 0.1080 | 0.1036 | 0.09930 | 0.09524 | 0.09137 | 0.08768 | 0.08415 | 0.08079 | 0.07758 |
| 6.0 | 0.07758 | 0.07452 | 0.07160 | 0.06881 | 0.06615 | 0.06360 | 0.06118 | 0.05886 | 0.05664 | 0.05453 | 0.05251 |
| 7.0 | 0.05251 | 0.05057 | 0.04873 | 0.04696 | 0.04527 | 0.04366 | 0.04211 | 0.04063 | 0.03921 | 0.03786 | 0.03656 |
| 8.0 | 0.03656 | 0.03532 | 0.03413 | 0.03298 | 0.03189 | 0.03084 | 0.02983 | 0.02887 | 0.02794 | 0.02705 | 0.02620 |
| 9.0 | 0.02620 | 0.02538 | 0.02459 | 0.02384 | 0.02311 | 0.02241 | 0.02174 | 0.02109 | 0.02047 | 0.01987 | 0.01930 |
| 10.0 | 0.01930 | 0.01874 | 0.01821 | 0.01769 | 0.01720 | 0.01672 | 0.01626 | 0.01581 | 0.01538 | 0.01497 | 0.01457 |
| 11.0 | 0.01457 | 0.01418 | 0.01381 | 0.01345 | 0.01311 | 0.01277 | 0.01245 | 0.01213 | 0.01183 | 0.01153 | 0.01125 |
| 12.0 | 0.01125 | 0.01098 | 0.01071 | 0.01045 | 0.01020 | 0.00996 | 0.00973 | 0.00950 | 0.00928 | 0.00907 | 0.00886 |
| 13.0 | 0.00886 | 0.00866 | 0.00846 | 0.00827 | 0.00809 | 0.00791 | 0.00774 | 0.00757 | 0.00741 | 0.00725 | 0.00710 |
| 14.0 | 0.00710 | 0.00695 | 0.00680 | 0.00666 | 0.00652 | 0.00639 | 0.00626 | 0.00613 | 0.00601 | 0.00589 | 0.00577 |
| 15.0 | 0.00577 | 0.00566 | 0.00555 | 0.00544 | 0.00533 | 0.00523 | 0.00513 | 0.00503 | 0.00494 | 0.00485 | 0.00476 |

Debye Helmholtz free energy $(A_m - U_{0,m})/3RT$ as a function of $\theta_D/T$

when $\frac{\Phi_D}{T} \geqslant 16, -\frac{A_m - U_{0,m}}{3RT} = 6.494\left(\frac{T}{\Phi_D}\right)^3$

| $\Phi_D/T$ | 0.0 | 0.1 | 0.2 | 0.3 | 0.4 | 0.5 | 0.6 | 0.7 | 0.8 | 0.9 | 1.0 |
|---|---|---|---|---|---|---|---|---|---|---|---|
| 0.0 | ∞ | 2.6732 | 2.0168 | 1.6476 | 1.3956 | 1.2077 | 1.0602 | 0.9403 | 0.8405 | 0.7560 | 0.6835 |
| 1.0 | 0.6835 | 0.6205 | 0.5653 | 0.5166 | 0.4734 | 0.4349 | 0.4003 | 0.3692 | 0.3410 | 0.3156 | 0.2925 |
| 2.0 | 0.2925 | 0.2714 | 0.2522 | 0.2346 | 0.2185 | 0.2037 | 0.1901 | 0.1776 | 0.1661 | 0.1554 | 0.1456 |
| 3.0 | 0.1456 | 0.1365 | 0.1281 | 0.1203 | 0.1130 | 0.1063 | 0.1000 | 0.09423 | 0.08882 | 0.08377 | 0.07906 |
| 4.0 | 0.07906 | 0.07467 | 0.07057 | 0.06674 | 0.06316 | 0.05981 | 0.05667 | 0.05373 | 0.05097 | 0.04839 | 0.04596 |
| 5.0 | 0.04596 | 0.04368 | 0.04154 | 0.03952 | 0.03763 | 0.03584 | 0.03416 | 0.03258 | 0.03108 | 0.02967 | 0.02834 |
| 6.0 | 0.02834 | 0.02709 | 0.02590 | 0.02477 | 0.02371 | 0.02271 | 0.02175 | 0.02085 | 0.02000 | 0.01918 | 0.01841 |
| 7.0 | 0.01841 | 0.01768 | 0.01699 | 0.01633 | 0.01570 | 0.01510 | 0.01454 | 0.01400 | 0.01348 | 0.01299 | 0.01252 |
| 8.0 | 0.01252 | 0.01208 | 0.01165 | 0.01124 | 0.01085 | 0.01048 | 0.01013 | 0.00979 | 0.00946 | 0.00915 | 0.00886 |
| 9.0 | 0.00886 | 0.00857 | 0.00830 | 0.00804 | 0.00779 | 0.00755 | 0.00731 | 0.00709 | 0.00688 | 0.00667 | 0.00648 |
| 10.0 | 0.00648 | 0.00629 | 0.00611 | 0.00593 | 0.00576 | 0.00560 | 0.00544 | 0.00529 | 0.00515 | 0.00501 | 0.00487 |
| 11.0 | 0.00487 | 0.00474 | 0.00462 | 0.00450 | 0.00438 | 0.00427 | 0.00416 | 0.00405 | 0.00395 | 0.00385 | 0.00376 |
| 12.0 | 0.00376 | 0.00366 | 0.00357 | 0.00349 | 0.00340 | 0.00332 | 0.00325 | 0.00317 | 0.00310 | 0.00302 | 0.00296 |
| 13.0 | 0.00296 | 0.00289 | 0.00282 | 0.00276 | 0.00270 | 0.00264 | 0.00258 | 0.00253 | 0.00247 | 0.00242 | 0.00237 |
| 14.0 | 0.00237 | 0.00232 | 0.00227 | 0.00222 | 0.00217 | 0.00213 | 0.00209 | 0.00204 | 0.00200 | 0.00196 | 0.00192 |
| 15.0 | 0.00192 | 0.00189 | 0.00185 | 0.00181 | 0.00178 | 0.00174 | 0.00171 | 0.00168 | 0.00165 | 0.00162 | 0.00159 |

*(continued)*

**Table A4.7** *Continued*

Debye entropy $S_m/3R$ as a function of $\theta_D/T$

when $\dfrac{\Phi_D}{T} \geqslant 16$, $\dfrac{S_m}{3R} = 25.976\left(\dfrac{T}{\Phi_D}\right)^3$

| $\Phi_D/T$ | 0.0 | 0.1 | 0.2 | 0.3 | 0.4 | 0.5 | 0.6 | 0.7 | 0.8 | 0.9 | 1.0 |
|---|---|---|---|---|---|---|---|---|---|---|---|
| 0.0 | $\infty$ | 3.6362 | 2.9438 | 2.5396 | 2.2536 | 2.0327 | 1.8531 | 1.7022 | 1.5723 | 1.4587 | 1.3579 |
| 1.0 | 1.3579 | 1.2676 | 1.1861 | 1.1120 | 1.0442 | 0.9820 | 0.9246 | 0.8714 | 0.8222 | 0.7763 | 0.7336 |
| 2.0 | 0.7336 | 0.6937 | 0.6564 | 0.6214 | 0.5886 | 0.5578 | 0.5289 | 0.5017 | 0.4761 | 0.4519 | 0.4292 |
| 3.0 | 0.4292 | 0.4077 | 0.3875 | 0.3683 | 0.3503 | 0.3332 | 0.3171 | 0.3018 | 0.2874 | 0.2737 | 0.2608 |
| 4.0 | 0.2608 | 0.2486 | 0.2370 | 0.2260 | 0.2156 | 0.2057 | 0.1964 | 0.1875 | 0.1791 | 0.1711 | 0.1636 |
| 5.0 | 0.1636 | 0.1564 | 0.1496 | 0.1431 | 0.1369 | 0.1311 | 0.1255 | 0.1203 | 0.1152 | 0.1105 | 0.1059 |
| 6.0 | 0.1059 | 0.1016 | 0.09750 | 0.09358 | 0.08986 | 0.08631 | 0.08293 | 0.07971 | 0.07664 | 0.07371 | 0.07092 |
| 7.0 | 0.07092 | 0.06826 | 0.06572 | 0.06329 | 0.06097 | 0.05876 | 0.05665 | 0.05463 | 0.05270 | 0.05085 | 0.04908 |
| 8.0 | 0.04908 | 0.04739 | 0.04578 | 0.04423 | 0.04274 | 0.04132 | 0.03996 | 0.03866 | 0.03741 | 0.03621 | 0.03506 |
| 9.0 | 0.03506 | 0.03395 | 0.03289 | 0.03187 | 0.03090 | 0.02996 | 0.02905 | 0.02818 | 0.02735 | 0.02655 | 0.02577 |
| 10.0 | 0.02577 | 0.02503 | 0.02431 | 0.02362 | 0.02296 | 0.02232 | 0.02170 | 0.02111 | 0.02053 | 0.01998 | 0.01944 |
| 11.0 | 0.01944 | 0.01893 | 0.01843 | 0.01795 | 0.01749 | 0.01704 | 0.01660 | 0.01618 | 0.01578 | 0.01539 | 0.01501 |
| 12.0 | 0.01501 | 0.01464 | 0.01428 | 0.01394 | 0.01361 | 0.01328 | 0.01297 | 0.01267 | 0.01237 | 0.01209 | 0.01181 |
| 13.0 | 0.01181 | 0.01155 | 0.01129 | 0.01103 | 0.01079 | 0.01055 | 0.01032 | 0.01010 | 0.00988 | 0.00967 | 0.00946 |
| 14.0 | 0.00946 | 0.00926 | 0.00907 | 0.00888 | 0.00870 | 0.00852 | 0.00834 | 0.00818 | 0.00801 | 0.00785 | 0.00770 |
| 15.0 | 0.00770 | 0.00754 | 0.00740 | 0.00725 | 0.00711 | 0.00697 | 0.00684 | 0.00671 | 0.00659 | 0.00646 | 0.00634 |

# Index

CPSIA information can be obtained
at www.ICGtesting.com
Printed in the USA
BVOW04*0812220817
492271BV00017B/29/P